ISBN 0-8169-0340-9

# Heat Transfer—Denver 1985
## *Nayeem M. Farukhi, editor*

R.L. Adams
S. Adibhatla
P.K. Agarwal
Nasser Alavizadeh
M.J. Antal
A.A. Arafa
Rene Arrazola
J. Ayer
P. Basu
L.C. Beavis
Y.C. Belentepe
M.J. Beliveau
R.C. Birkebak
M. Boggs
R.T. Bosworth
M. Bottoni
M. Burger
G.W. Caille
C. Carachalios
S.H. Chan
Y. Chang
S.K. Chaturvedi
Falin F. Chen
H.N. Chi
T.H. Chien
R.H. Clarke

L.D. Clements
F.W. Cox
Roger Crane
M.E. Crawford
C. Crowe
R.L. Curran
W.R. Curtis
H.C. daSilva
C. DeBellis
R. Dieh
K.R. Diller
H.M. Domanus
M.K. Drost
Floyd Dunn
A.M. Eaton
G.G. Elia
G. Evans
P.K. Falcone
I.H. Farag
S.T. Free
A. Ganguli
W.E. Genetti
C.C. Gentry
T. Ginsberg
Ali Goshayeshi

R. Greif
F.S. Gunnerson
O.J. Hahn
L.J. Hayes
E.R. Hosler
W. Houf
J.M. Hruby
G.C. Huang
P. Hull
A.J. Hunt
J.L. Jechura
D.D. Lanning
Y.Y. Lee
R.W. Lyczkowski
Bahram Mahbod
J.J. Marano
V.C. Mei
C.C. Miao
E.E. Michaelides
F. Miller
M.Moss
K.F. Neusen
J.E. O'Brien
J.W. Palen
A.J. Pearlstein

W.P. Prueter
A.T. Raissi
J.D. Reichert
M.K. Richardson
J.M. Robertson
R. Russo
A.L. Schor
W.T. Sha
S. Shakerin
W.M. Small
G.S. Srikantiah
Taraneh Tabesh
J. Taborek
A.M. Terpolilli
N.E. Todreas
S.S. Tung
A. Ungan
H. Unger
R. Viskanta
M.L. Wade
J.R. Wall
S.S. Wang
K.L. Watson
J.R. Welty
W. Yuen

AIChE Symposium Series
1985
Published by
American Institute of Chemical Engineers

Copyright 1985

American Institute of Chemical Engineers
345 East 47 Street, New York, N.Y. 10017

*AIChE shall not be responsible for statements or opinions advanced in papers or printed in its publications.*

**Library of Congress Cataloging in Publication Data**
Main entry under title:

Heat transfer—Denver, 1985.

(AIChE symposium series ; no. 245, v. 81)
Papers accepted for presentation at the 23rd National Heat Transfer Conference in Denver, Aug. 4-7, 1985.
1. Heat—Transmission—Congresses. I. Farukhi, N. M. II. National Heat Transfer Conference (23rd : 1985 : Denver, Colo.) III. Series.
QC319.8.H425 1985              621.402'2              85-13374
ISBN 0-8169-0340-9

Authorization to photocopy items for internal or personal use, or the internal or personal use of specific clients, is granted by AIChE for libraries and other users registered with the Copyright Clearance Center (CCC) Transactional Reporting Service, provided that the $2.00 fee per copy is paid directly to CCC, 21 Congress St., Salem, MA 01970. This consent does not extend to copying for general distribution, for advertising or promotional purposes, for inclusion in a publication, or for resale.

Articles published before 1978 are subject to the same copyright conditions and the fee is $2.00 for each article. AIChE Symposium Series fee code: 0065-8812/83 $2.00

Printed in the United States of America by
Twin Production & Design

# FOREWORD

This Symposium Series volume contains AIChE-sponsored session papers accepted for presentation at the 23rd National Heat Transfer Conference. The papers are grouped by conference sessions. AIChE papers that were presented in sessions jointly sponsored with ASME are available in bound volumes printed by ASME.

Due to the short time frame that we work with in getting this volume ready for distribution at the conference, the task of editing the papers published in this volume is actually shared with the session organizers as well as the reviewers. Without their input and efforts this volume could not be published by the due date. I would like to thank them for serving as co-editors of this volume.

I would also like to thank the AIChE publication staff, especially Maura Mullen, for providing invaluable assistance during my past three years as editor of this volume. I will miss working with her in the future since my term as editor will end later this year. I am sure, however, that she will provide the same dedicated effort to the next editor(s) of this volume.

Nayeem M. Farukhi, *editor*
Westinghouse Electric Corp.
Nuclear Technology Division
Pittsburgh, Pennsylvania

# REVIEWERS

| | | |
|---|---|---|
| H. Akbari | E.D. Hughes | A.R. Peters |
| R.W. Alperi | J.H. Kim | R.L. Pons |
| J. Althof | J.G. Knudson | D. Poulikakous |
| R. Anderson | F. Kreith | J. Reeves |
| M.A. Bergougnou | P.G. Kroeger | V. Roberts |
| W.R. Bohl | C. Kutscher | S.C. Saxena |
| J. Bowyer | W.H. Lee | A.L. Schor |
| L.L. Briggs | B.W. LeTourneau | G.S. Srikantiah |
| R.A. Cassanova | F.W. Lincoln | H.B. Stewart |
| Y. Cengel | N. Lior | J.-P. Sursock |
| J.C. Chen | G.O.G. Lof | Y.S. Tang |
| J.L.S. Chen | R.W. Lyczkowski | L.N. Tao |
| N.S. Crewal | V.K. Mathur | W.J. Thomson |
| D.L. Crum | A.E. McGarity | R. Trusty |
| J. Duffie | R.D. Mitchell | R. Viskanta |
| I.H. Farag | R.A. Newby | L.A. Waldman |
| C. Flood | R.G. Nix | C.S. Wang |
| T.P. Galloway | S. Oh | K.Y. Wang |
| W. Genetti | C.B. Panchal | J.R. Welty |
| J. Grace | B. Parsons | O. Weres |
| M.A. Grolmes | T. Penny | S.B. Woodruff |
| B. Gupta | J. Perona | J. Wright |
| G.R. Hadley | A. Pesaran | F. Zangrando |
| J.B. Howard | | E.N. Ziegler |

# CONTENTS

## HEAT TRANSFER FOR ALTERNATE ENERGY TECHNOLOGIES
*Y.S. Tang, Chairman, V. Roberts, Co-chairman*

THERMALLY CONDUCTIVE SILICONE BASED MATERIALS FOR ATTACHING CONCENTRATOR SOLAR CELLS TO HEAT SINKS ................................................. L.C. Beavis, M. Moss   1

THEORETICAL HEAT PUMP GROUND COIL ANALYSIS WITH VARIABLE GROUND FARFIELD BOUNDARY CONDITIONS ........................................................ V.C. Mei   7

FUEL PROCESSING ANALYSES FOR FUEL CELL POWER SYSTEMS .................. G.G. Elia   13

SYNTHETIC BRINE FOULING OF A GEOTHERMAL HEAT EXCHANGER ...... S.H. Chan, K.F. Neusen, C. DeBellis   19

## HEAT AND MASS TRANSFER DURING FLUIDIZED BED COMBUSTION
*W.E. Genetti, Chairman, Y.Y. Lee, Co-chairman*

AN ANALYTICAL MODEL FOR THE PREDICTION OF RADIATIVE HEAT TRANSFER IN LARGE-PARTICLE, GAS FLUIDIZED BEDS WITH AN EMBEDDED HORIZONTAL TUBE (Abstract Only) ............................. Bahram Mahbod, Taraneh Tabesh, Ali Goshayeshi   26

A SINGLE CHAR PARTICLE COMBUSTION MODEL OF $SO_2$ RETENTION BY LIGNITE ASH DURING FLUIDIZED BED COMBUSTION ................................ F.W. Cox, W.E. Genetti, Y.Y. Lee   27

LOCAL HEAT TRANSFER COEFFICIENTS FOR HORIZONTAL TUBE ARRAYS IN HIGH TEMPERATURE LARGE PARTICLE FLUIDIZED BEDS
—AN EXPERIMENTAL STUDY .............. Ali Goshayeshi, J.R. Welty, R.L. Adams, Nasser Alavizadeh   34

EVOLUTION AND COMBUSTION OF VOLATILES FROM A SINGLE COAL PARTICLE ............. P.K. Agarwal   41

THE CONTRIBUTION OF GAS CONVECTION TO TOTAL HEAT TRANSFER FOR A HORIZONTAL CYLINDER SUBMERGED IN A FLUIDIZED BED ........................... J.E. O'Brien, M.L. Wade, A.M. Terpolilli   48

DENSE BED, SPLASH ZONE, AND FREEBOARD HEAT TRANSFER IN A FLUIDIZED BED COMBUSTOR ................................................. S. Adibhatla, M. Boggs   55

HEAT TRANSFER IN TURBULENT FLUIDIZED BEDS (Abstract Only) ................. P. Basu, R. Dieh   62

## HEAT TRANSFER IN GLASS
*I.H. Farag, Chairman, R.L. Curran, Co-chairman*

HEAT TRANSFER IN GLASS ............................................... R. Viskanta   63

EFFECT OF AIR BUBBLING ON CIRCULATION AND HEAT TRANSFER IN A GLASS MELTING TANK .................................................. A. Ungan, R. Viskanta   70

HEAT TRANSFER SIMULATION DURING FORMING OF GLASS CONTAINERS ........................................ I.H. Farag, M.J. Beliveau, R.L. Curran   77

HEAT TRANSFER ANALYSIS OF A BINARY COMPOUND (CHROME ORE-MAGNESIA) IN SOLIDIFICATION THROUGH A TEMPERATURE RANGE ........................................... Y.C. Belentepe   84

## SINGLE- AND TWO-PHASE PROCESS HEAT TRANSFER
*R.S. Kistler, Chairman, A.C. Pauls, Co-chairman*

| | |
|---|---|
| AN IMPROVED HEAT TRANSFER CORRELATION FOR LAMINAR FLOW OF HIGH PRANTDL NUMBER LIQUIDS IN HORIZONTAL TUBES ............ J.W. Palen, J. Taborek | 90 |
| DESIGN AND OPTIMIZATION OF HEAT EXCHANGERS FOR BATCH HEATING BY THE NTU-EFFECTIVENESS METHOD ............ Roger Crane, Rene Arrazola | 97 |
| RODBAFFLE EXCHANGER THERMAL-HYDRAULIC PREDICTIVE MODELS OVER EXPANDED BAFFLE SPACING AND REYNOLDS NUMBER RANGES ............ C.C. Gentry, W.M. Small | 103 |
| HEAT AND MOMENTUM TRANSFER PROCESSES THROUGH BANKS OF FLEXIBLE TUBES IN AIR CROSS-FLOW ............ E.E. Michaelides, Y. Chang, R.T. Bosworth | 109 |
| ANALYSIS OF HEAT TRANSFER IN PLATE HEAT EXCHANGERS ............ J.J. Marano, J.L. Jechura | 116 |
| PARAMETRIC STUDY OF AIR-COOLED HEAT EXCHANGER FINNED TUBE GEOMETRY ............ A. Ganguli, S.S. Tung, J. Taborek | 122 |
| THE GENERAL PREDICTION OF CONVECTIVE BOILING COEFFICIENTS IN PLATE-FIN HEAT EXCHANGER PASSAGES ............ J.M. Robertson, R.H. Clarke | 129 |

## THERMAL ANALYSIS OF STEAM GENERATORS
*W.H. Vance, Chairman, D.H. Cho, Co-chairman*

| | |
|---|---|
| A STUDY OF GEOMETRIC SCALING OF CURVED-ARM PRIMARY STEAM/WATER SEPARATORS ............ A.M. Eaton, W.P. Prueter, J.R. Wall | 135 |
| THERMIT-UTSG: A COMPUTER CODE FOR THERMOHYDRAULIC ANALYSIS OF U-TUBE STEAM GENERATORS ............ H.C. daSilva, N.E. Todreas, D.D. Lanning | 142 |
| VISUAL OBSERVATIONS OF FLOW PATTERNS THROUGH TUBE SUPPORT PLATES OF CIRCULAR HOLE AND TREFOIL DESIGN ............ G.W. Caille, E.R. Hosler, F.S. Gunnerson | 148 |

## GENERAL SOLAR HEAT TRANSFER
*D. Bharathan, Chairman, R.G. Nix, Co-chairman*

| | |
|---|---|
| MIXED LAMINAR CONVECTION IN TROMBE WALL CHANNEL ............ G.C. Huang, S.K. Chaturvedi | 156 |
| PERFORMANCE OF A FIXED MIRROR-DISTRIBUTED FOCUS SOLAR THERMAL SYSTEM: THE CROSBYTON SOLAR POWER PROJECT ............ L.D. Clements, K.L. Watson, J.D. Reichert | 163 |
| TIME DEPENDENT THREE DIMENSIONAL THERMAL ANALYSIS OF A SOLAR HEATING SYSTEM .. A.A. Arafa | 171 |
| NEW CORRELATIONS FOR CONVECTIVE HEAT TRANSFER COEFFICIENTS FOR FLAT PLATE SOLAR COLLECTORS ............ Said Shakerin | 178 |
| PERFORMANCE ANALYSIS OF HVAC SYSTEM FUELED BY SOLAR ENERGY OR A COMBINATION OF SOLAR, WOOD OR NATURAL GAS ............ O.J. Hahn, M.K. Richardson, R.C. Birkebak, W.R. Curtis | 185 |
| TEMPERATURE DISTRIBUTIONS AND HOT SPOTS IN A LAMINAR FLOW PHOTOCHEMICAL REACTOR (Abstract Only) ............ Falin Chen, A.J. Pearlstein | 190 |

## DIRECT FLUX SOLAR ENERGY PROCESS
*R.G. Nix, Chairman, D. Bharathan, Co-chairman*

| | |
|---|---|
| SOLAR RADIANT PROCESSING OF GAS PARTICLE SYSTEMS FOR PRODUCING USEFUL FUELS AND CHEMICALS .................. A.J. Hunt, J. Ayer, P. Hull, F. Miller, R. Russo, W. Yuen | 191 |
| MOMENTUM AND ENERGY EXCHANGE IN A SOLID PARTICLE SOLAR CENTRAL RECEIVER .................................................. J.M. Hruby, P.K. Falcone | 197 |
| EFFECTS OF HEATING RATE ON SOLAR THERMAL DECOMPOSITION OF ZINC SULFATE ........................................................ A.T. Raissi, M.J. Antal | 204 |
| GAS-PARTICLE FLOW WITHIN A HIGH TEMPERATURE SOLAR CAVITY RECEIVER INCLUDING RADIATION HEAT TRANSFER ................................ G. Evans, W. Houf, R. Greif, C. Crowe | 213 |
| VOLUMETRIC RECEIVER RADIATION HEAT TRANSFER ......................... M.K. Drost, J.R. Welty | 220 |

## NUMERICAL METHODS FOR MULTIPHASE FLOW SYSTEM AND COMPONENT ANALYSIS
*R.W. Lyczkowski, Chairman, J.H. Kim, Co-chairman*

| | |
|---|---|
| A STABLE NUMERICAL METHOD FOR ONE-DIMENSIONAL, TWO-PHASE FLOW ....... S.T. Free, A.L. Schor | 226 |
| BYPASS APPROACH OF FREE SURFACE MODELING (Abstract Only) ........................................ C.C. Miao, F.F. Chen, W.T. Sha, R.W. Lyczkowski | 233 |
| VARIATION IN THERMAL HISTORY DURING FREEZING DUE TO THE PATTERN OF LATENT HEAT EVOLUTION .................................... K.R. Diller, L.J. Hayes, M.E. Crawford | 234 |
| THE SAS4A/SASSYS-1 SODIUM BOILING MODEL FOR LMFBR WHOLE CORE ANALYSIS ........ Floyd Dunn | 240 |
| NUMERICAL MODELING OF THE PHASE SEPARATION PROCESSES IN BWR AND PWR STEAM SEPARATORS ............................................ S.S. Wang, G.S. Srikantiah | 246 |
| MULTIDIMENSIONAL TWO PHASE MODELING WITH THE COMMIX-2 COMPUTER PROGRAM .............. M. Bottoni, R.W. Lyczkowski, H.N. Chi, T.H. Chien, H.M. Domanus | 253 |

## NUCLEAR PLANT DEGRADED CORE COOLING
*M. Corradini, Chairman, M. El-Genk, Co-chairman*

| | |
|---|---|
| TRIGGERING AND ESCALATION BEHAVIOR OF THERMAL DETONATIONS ... C. Carachalios, M. Burger, H. Unger | 259 |
| ON THE PREDICTION OF MINIMUM VAPOR FILM BOILING ......................... M. Burger, H. Unger | 266 |
| ANALYSIS OF INFLUENCE OF STEAM SUPERHEATING ON PACKED BED QUENCH PHENOMENA ................................................................. T. Ginsberg | 273 |

# THERMALLY CONDUCTIVE SILICONE BASED MATERIALS FOR ATTACHING CONCENTRATOR SOLAR CELLS TO HEAT SINKS

L.C. Beavis ■ Photovoltaic Concentrator Research Division
M. Moss ■ Thermophysical Properties Division
Sandia National Laboratories, Albuquerque, New Mexico

We have application for low electrical conductivity-high thermal conductivity adhesives. The adhesive is to be used to attach a concentrator silicon solar cell to an aluminum heat sink. Early in our program we observed that commercially available adhesives did not achieve advertised thermal conductivities in minimum thickness layers. We ascribed the higher-than-expected thermal resistance to contact resistance. Therefore, we set about an experimental program to study the magnitude of contact resistance and the impact of varying the amount of thermally conductive additive to silicone-based adhesives. The thermal conductivity of freshly cured, unfilled silicone is about 0.05 W m$^{-1}$ K$^{-1}$. With thermal cycling this value increases to about 0.2 W m$^{-1}$ K$^{-1}$. The thermal conductivity of the material increases in a superlinear fashion with weight fraction of fill material. Thermal conductivities of 1 W m$^{-1}$ K$^{-1}$ are achievable with 0.75 alumina by weight.

## INTRODUCTION

A good adhesive for bonding silicon solar cells to the typical aluminum heat sink used in concentrator cell assemblies can improve the overall efficiency of photovoltaic concentrator arrays. Because the conversion efficiency of a good silicon concentrator solar cell--currently about 20% at 20°C--decreases as temperature increases, the electrically insulating adhesive used to bond the cell to the heat sink must have excellent heat transfer capabilities so that the remaining 80% of energy deposited on the cell (and converted to heat) can be efficiently conducted through the adhesive material to the heat sink and dissipated (Figure 1). In high-concentration systems currently being designed, the power to be dissipated is 20 to 80 W cm$^{-2}$. In addition to having a high thermal conductivity, to enable the cell to operate at as low a temperature as possible, the bonding layer formed by the adhesive must be resilient enough to accommodate the large thermal expansion mismatch between the cell mount and the heat sink. Finally, the adhesive must be capable of withstanding thousands of diurnal temperature cycles without mechanical failure.

In the past, thermal greases and silicones, enhanced to increase their thermal conductivity, have been used for

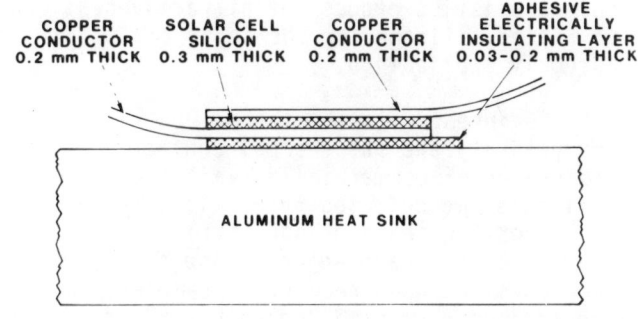

Figure 1. Schematic of cell mount.

this application. (Other polymeric materials have been tried, but the lack of resiliency during thermal cycling caused structural failure.) Thermal greases have the advantage of high thermal conductivity (1-2 W m$^{-1}$ K$^{-1}$) and facilitate the repair of cell assemblies, but require the use of mechanical constraints such as spring clips, used generally where easy disassembly of the cell from the heat sink is desired. Thermal greases also exhibit batch-to-batch variations in physical characteristics, and their long term stability is questionable.

Of the adhesives available commercially, silicone resins which cure to a semihard state show the most promise for use with photovoltaic systems. Because silicones remain flexible, they are capable of surviving thermal cycling, and they minimize the transfer of thermal-mechanical stress to the photovoltaic cell. Silicones also remain mechanically stable, but have relatively low thermal conductivity (0.05 to 0.2 W m$^{-1}$ K$^{-1}$). Filling these adhesives with thermally conductive, electrically insulating particles, e.g., boron nitride, magnesia, or alumina, can increase the conductivity above that of the unfilled adhesive while maintaining the desirable electrical and adhesive characteristics of the silicone. Some commercially available filled silicones have demonstrated certain undesirable characteristics: thermal contact resistance, electrical conductivity with copper, aluminum, or silicon carbide fillers, and a minimum bond thickness ranging from 0.15 to 0.4 mm, depending upon amount and type of filler. These characteristics reduce the attractiveness of filled silicone for the solar cell assembly application.

To investigate the thermal conductivity and the thermal contact resistance of commercially available silicones, we selected three silicone-based adhesives for testing--one prefilled with alumina by the manufacturer, and two which were unfilled when received. Samples of the silicones were filled with various amounts of alumina or magnesia particles. We then prepared sample stacks consisting of different particle-filled silicone adhesives between a high alumina-content ceramic disk and a 6061 aluminum alloy disk to simulate a cell assembly. After appropriate adhesive cure periods were determined, the thermal conductivity of the bonding layers was measured.

## MATERIALS AND EQUIPMENT

### Fill Material

Two fill materials--alumina and magnesia--were tested. The alumina particles ranged in size from 0.5 to 35 μm with a mean size of 10.5 μm. The magnesia particles ranged from 0.5 to 15 μm with a mean size of 4.8 μm. Both materials were stored in a dry, 90°C cabinet to prevent water absorption.

### Silicone-Based Adhesives

Three silicone-based adhesives were tested--one prefilled with alumina by the manufacturer, and two that we filled with various amounts of alumina or magnesia. The following were used:

McGhan NuSil Gel 8150--unfilled when received, very resilient after curing, and similar to Dow Corning 527 or General Electric 6159.

Dow Corning Sylgard 182--unfilled when received, less resilient than the McGhan NuSil 8150 after curing, and similar to General Electric 615.

McGhan NuSil R-2940--alumina-filled (one part adhesive to 20 parts filler, by weight) by the manufacturer, requiring no additional modification before use.

### Disks

Sample stacks were assembled with 18-mm diameter ceramic disks (96% alumina content) varying in thickness from 0.94 to 1.19 mm, and 6061 aluminum alloy disks of the same diameter varying in thickness from 2.31 to 2.34 mm.

### Colora Thermoconductometer

Thermal conductivity measurements at room temperature were made with a Colora Thermoconductometer, the operation of which has been described elsewhere.(1) The instrument permits rapid measurement--about 10 minutes per sample. It employs two fluids, with different boiling points, to heat and cool a small specimen of the material under investigation. The higher boiling point fluid is evaporated and condenses on the bottom of a silver plate in contact with the lower surface of the sample. Heat then passes through both the sample and a second silver plate in contact with the top of the sample, and finally boils a different fluid with a lower boiling point. The resulting vapor is condensed and collected in a graduated capillary tube.

The heat flow passing through the sample is derived from the rate of condensation of the upper fluid. If calibrated samples are measured with a selected liquid pair, the thermal resistance values of subsequent samples can be read directly from a calibration curve which plots resistance versus time. The accuracy is generally better than 10%, and repeatability on the same sample is within 2 to 4%.

## PREPARATION OF SAMPLES AND DETERMINATION OF CURE CYCLE

We first mixed the silicone with its hardener and deaerated the mixture at 1 torr absolute for 10 minutes. We then blended either alumina or magnesia in various amounts into samples of the silicone, and once again deaerated the material for 15 minutes.

Using the filled silicone samples, we prepared two sets of stacks to simulate cell assemblies: one set for evaluating appropriate fill ratios and cure cycles, and another for measuring thermal conductivity and contact resistance. To prepare the stacks, alumina ceramic disks and aluminum alloy disks were placed on opposite sides of a sample preparation mold (Figure 2), and the space between them was filled with particle-filled, deaerated silicone. The assembled stacks varied according to type of filler, ratio of amount of filler to adhesive, adhesive thickness and cure cycle. After assembly, the samples were deaerated for 25 to 40 minutes. Finally, the samples were cured at room or elevated temperature.

The first set of sample stacks was used to determine the optimum cure cycle for the silicone materials under consideration. In general, the magnesia-filled silicones were difficult or impossible to deaerate properly and very difficult to cure. As a result, only samples with high ratios of silicone to magnesia (5:1) were retained for further evaluation.

Table 1 summarizes our observations on curing McGhan NuSil Gel 8150 and Dow Corning Sylgard 182 containing various ratios of alumina to silicone. (Because McGhan NuSil R-2940 was prefilled by the manufacturer, it was not included in these cure-cycle tests.)

Though no samples to be used for thermal conductivity measurements were cured at temperatures over 60°C, some loaded adhesive was cured at 90°C and it, like the gels cured at ambient temperature, remained tacky. In general, gels remained tacky, as expected, whether filled or not.

## DETERMINATION OF CONTACT RESISTANCE AND THERMAL CONDUCTIVITY

The second set of sample stacks was used to measure the contact resistance and the thermal conductivity of the adhesive bond. Fourier's law for unidirectional heat conduction states

$$Q = \frac{kA\Delta T}{h} . \qquad (1)$$

Thermal resistance is defined by

$$R = \frac{\Delta T}{Q} . \qquad (2)$$

Hence,

$$R = \frac{h}{Ak} . \qquad (3)$$

The total thermal resistance along the cylindrical axis of a sample is the sum of the component resistances:

$$R_C = R_B + R_{A\ell} + R_{Cer} . \qquad (4)$$

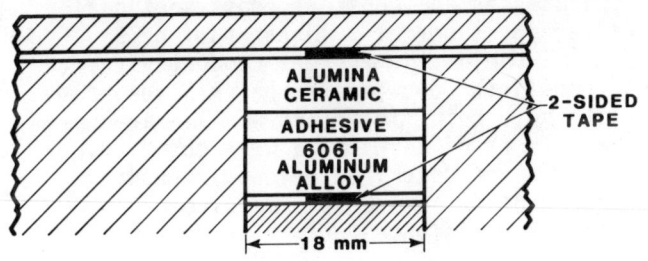

| | THICKNESS |
|---|---|
| ALUMINA CERAMIC | 1.0 mm |
| ALUMINUM ALLOY | 2.3 mm |
| ADHESIVE | 0.1 mm TO 2.0 mm |

Figure 2. Thermal conductivity sample schematic.

Table 1. Fill Ratios by Weight (Silicone: Alumina) and Description of Cures:

| Fill Ratio | Conditions | Cure Time | Results |
|---|---|---|---|
| McGhan NuSil Gel 8150 | | | |
| 1:5 | Vacuum, 25°C | 30 min | Cured |
| 1:3 | 25°C | 30 h | Cured |
| 1:2 | 25°C | 16 h | Hard, tacky |
| 1:1 | 60°C | 50 min | Hard, very tacky |
| Dow Corning Sylgard 182 | | | |
| 11:48 | 65°C | 16.5 h | Pliable, not tacky |
| 1:2 | 65°C | 7.5 h | Hard, not tacky |
| 1:1 | 65°C | 20.5 h | Pliable, not tacky |
| 2:1 | 65°C | 64 h | Pliable, not tacky |
| 1:1 | 150°C | 40 min | Less pliable, not tacky |

$R_C$ is the quantity measured in the Colora apparatus, and the value of bond resistance, $R_B$, is calculated using known values of $k_{A\ell}$ and $k_{Cer}$ which, with known dimensions, give $R_{A\ell}$ and $R_{Cer}$.

If we suppose that $R_B$ is composed of the resistance of the adhesive, $R_{Ad}$, and a possible contact resistance, $R_{Con}$, arising from incomplete wetting of the disks, then

$$R_B = R_{Ad} + R_{Con} \qquad (5)$$

where

$$R_{Ad} = \frac{h_{Ad}}{Ak_{Ad}} \approx \frac{h_B}{Ak_{Ad}} \qquad (6)$$

Equating $h_{Ad}$ and $h_B$ is justified by microscopic examination of the bonds. To a good approximation, therefore,

$$R_B = \frac{h_B}{Ak_{Ad}} + R_{Con} \qquad (7)$$

In this analysis, it is assumed that $R_{Con}$ does not change significantly with $h_B$ for a given adhesive. Contact resistance could conceivably vary in a system containing a relatively hard bonding agent whose adhesion to a substrate might be affected by thickness-dependent curing stresses. The adhesives in this study were pliable, and the assumption of a constant $R_{Con}$ seems reasonable.

For a number of samples of different thicknesses and constant cross-sectional area, the slope of a linear plot of $R_B$ vs. $h_B$ yields the thermal conductivity. A non-linear plot means that $k_{Ad}$ varies with $h_B$. Both linear and non-linear variations of $R_B$ with $h_B$ were observed, and will be described below. A positive intercept at $h_B = 0$ is a measure of any contact resistance that may exist.

Figure 3 shows the variation of bond thermal resistance, $R_B$, with bond thickness, $h_B$, for McGhan NuSil Gel 8150 filled with various amounts of alumina particles. A compression of the lightly particle-loaded adhesives was noted when the samples were clamped in the Colora apparatus. All reductions in thickness were measured in situ with a dial gauge, and the data reflect the adhesive thicknesses during measurement. Linear least-squares fits to the data are shown.

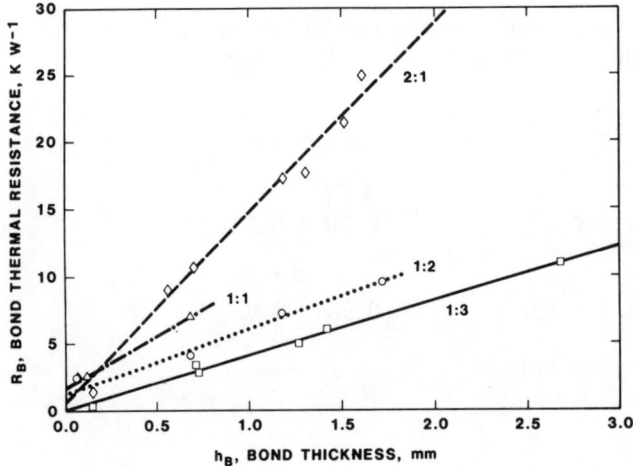

Figure 3. Bond thermal resistance vs. bond thickness for McGhan NuSil Gel 8150 adhesives filled with alumina particles. Adhesive/filler weight ratios indicated.

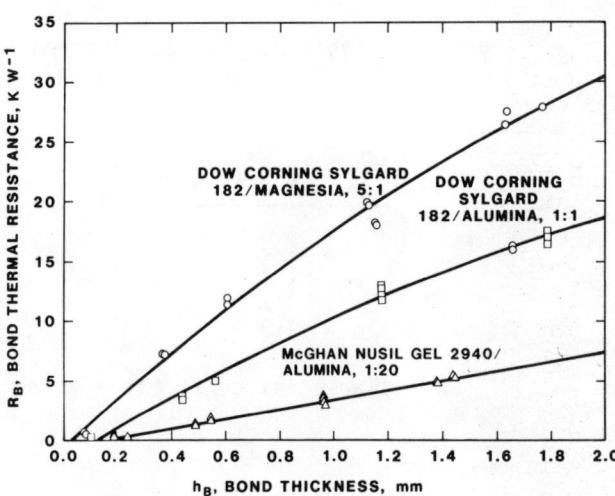

Figure 4. Bond thermal resistance vs. bond thickness for particle-filled adhesives. Adhesive/filler weight ratios indicated.

The contact resistances, $R_{Con}$, are tabulated in Table 2. $R_{Con}$ is zero for the 2:1 and 1:3 samples, within experimental error. The 1:2 and 1:1 samples show a positive $R_{Con}$, but only two data points were available for the 1:1 sample. The thermal conductivity, $k_{Ad}$, of the filled adhesives is also given in Table 2. Thermal conductivity increases with greater fractions of alumina filler, except that $k_{Ad}$ for 1:2 and 1:3 are the same within experimental error. The errors associated with $R_{Con}$ and $k_{Ad}$ are those computed in the least-squares fits combined with the estimated error of 6% in the measurement of sample area.

A number of the adhesives delaminated upon removal from the Colora apparatus. This was most pronounced for the 2:1 composition. The poor bond to the substrate in these cases did not result in a high contact resistance. Contact was evidently intimate, but the adhesion was weak.

Figure 4 shows the variation of bond thermal resistance with bond thickness for Dow Corning Sylgard 182 silicone containing either alumina or magnesia particles and McGhan NuSil R-2940 that was prefilled heavily with alumina (1:20) by the manufacturer. Within experimental error, the contact resistances between adhesives and the aluminum and alumina plates were nil ($R_B = 0$ at $h_B = 0$) as determined by least-squares fits to the data.

A thickness-independent thermal conductivity would produce a linear variation of $R_B$ with $h_B$. This was apparently the case for the R-2940 adhesive ($k_{Ad} = 0.97 \pm 0.07$ W m$^{-1}$ K$^{-1}$), but the plots for Sylgard 182/alumina, 1:1, and Sylgard 182/magnesia, 5:1 display curvature which implies that $k_{Ad}$ increases with $h_B$. The curve fits for the filled Sylgard adhesives were chosen arbitrarily to be quadratic. Based upon the fitted curve parameters and Equation (7), $k_{Ad}$ varies in the way shown in Table 3 where a few experimental values of $h_B$ were selected for illustration.

One possible explanation for $k_{Ad}$ increasing with $h_B$ is the presence of trapped air due to insufficient deaeration of the adhesive. If the amount of air remained constant, or nearly so, in all samples of a given adhesive type, it would represent a larger fraction of the volume in thinner samples and produce the observed result. Voids as large as 1 to 2 mm were indeed visible in a Sylgard 182/magnesia sample which had been sectioned and polished.

Table 2. Contact Thermal Resistance and Thermal Conductivity
of Alumina-Filled McGhan NuSil Gel 8150 Adhesives

| Adhesive:Filler Weight Ratio | $R_{Con}$(K W$^{-1}$) | $k_{Ad}$(W m$^{-1}$ K$^{-1}$) |
|---|---|---|
| 2:1 | 0.69 + 0.71 | 0.28 + 0.02 |
| 1:1 | 1.71* | 0.50* |
| 1:2 | 1.63 + 0.51 | 0.87 + 0.11 |
| 1:3 | 0.04 ± 0.23 | 0.96 ± 0.07 |

*Two-point curve fit - error indeterminate.

Table 3. Thermal Conductivity of Filled Sylgard 182 Adhesive

| Adhesive | $h_B$(mm) | $k_{Ad}$(W m$^{-1}$ K$^{-1}$) |
|---|---|---|
| Sylgard 182/alumina, 1:1 | 0.07 | 0.31 |
|  | 1.18 | 0.35 |
|  | 1.79 | 0.38 |
| Sylgard 182/magnesia, 5:1 | 0.08 | 0.20 |
|  | 1.13 | 0.22 |
|  | 1.77 | 0.24 |

## CONCLUSIONS

Magnesia-filled silicones, although of high potential value because of magnesia's high thermal conductivity, are so difficult to formulate and cure as to make them of no practical value in a production setting.

For the McGhan NuSil Gel 8150 silicone adhesives, addition of alumina particles monotonically increases the thermal conductivity for adhesive-to-filler weight ratios from 2:1 to 1:3. Maximum conductivity is 0.96 W m$^{-1}$ K$^{-1}$.

Insufficient deaeration of Dow Corning Sylgard 182 silicone adhesive containing alumina or magnesia particles is a possible cause of reduced conductivity. This would be more pronounced in thin samples in which bubbles represent a larger volume fraction of the adhesive than in thicker specimens. Such bubbles were observed.

Contact resistance was evident for some of the McGhan NuSil Gel 8150 adhesives, but not for the filled Sylgard 182 nor McGhan NuSil R-2940.

## NOTATION

Terms
- A = cross-sectional area, m$^2$
- ΔT = temperature difference, K
- h = thickness, m
- k = thermal conductivity, W m$^{-1}$ K$^{-1}$
- Q = heat flow, W
- R = thermal resistance, K W$^{-1}$

Subscripts
- Ad = filled adhesive
- Aℓ = aluminum disk
- B = bond
- C = composite
- Cer = alumina ceramic disk
- Con = contact

## LITERATURE CITED

1. W. E. Schwinkendorf and M. Moss, "Thermal Conductivity and Interface Resistance of Particle-Filled Silicone Resin," ASME Paper 84-HT-88, 22nd ASME-AIChE National Heat Transfer Conference, Niagara Falls, NY, August 6-8, 1984.

# THEORETICAL HEAT PUMP GROUND COIL ANALYSIS WITH VARIABLE GROUND FARFIELD BOUNDARY CONDITIONS

V.C. Mei ■ Oak Ridge National Laboratory, Oak Ridge, TN 37831

The operation of a single ground coil is described in detail mathematically. The mathematical model includes the effects of ground seasonal temperature variation, and the effect of coil fluid properties and flow characteristics, generally ignored by the traditional line source theory approach.

A computer code, based on the model, has been completed and validated satisfactorily by the field experimental results.

A parametric study indicates that the coil length, coil burial depth and soil thermal conductivity are important factors in determining the ground coil performance.

This computer code, one of the more advanced available so far, can be used for heat pump ground coil design if moisture migration is not a predominating factor in determining the soil thermal properties.

## INTRODUCTION

Ground-coupled heat pump systems have long been recognized for their energy conservation potential. Much theoretical and experimental work in this field has been accomplished. However, no comprehensive ground coil analysis has been derived so far. There are several major difficulties encountered: (a) the lack of knowledge of soil thermal conductivity and diffusivity, both highly dependent on the soil moisture content in a given location, (b) moisture migration, which results when a temperature gradient is imposed in the soil, (c) ice formation around the coil with attendant release of latent heat usually resulting in a step change in soil thermal conductivity in the frozen region, (d) the effect of the seasonal temperature variation at depths below the ground surface, (e) possible thermal resistance due to lack of intimate contact of the coil with soil, (f) the effect of the ground coil size and material, and (g) the effect of coil cyclic operations.

When the heat exchange rate between coil and soil is moderate, the soil thermal properties are relatively stable so that (a) and (b) may be neglected. Field experimental results (1) indicated that the latent heat released by the soil moisture freezing is small compared with total energy absorbed by the coil so that (c) can be discounted for a conservative design. Problem (e) is still under experimental study, (2) but the data have not been systematically published. Even if (a), (b), (c), and (e) can be justifiably ignored under certain conditions, the remaining difficulties still represent a complicated problem.

The most popular current theoretical approach utilizes line source (or cylindrical source) theory (3). The major drawbacks of this approach are (a) that one has to assume, or guess, the strength of the line source, which makes this approach more empirically dependent, (b) that the coil fluid-wall convective heat transfer resistance is generally ignored even when the fluid flow is in laminar region, (c) that the continuous coil fluid temperature change indicates that the strength of the line source will not be constant along the coil, and (d) that the effect of the seasonal temperature variation at depths is generally ignored so that the problem can be treated as radially symmetrical.

In this paper a three dimensional mathematical model based on energy conservation is formed to describe the theoretical analysis of a ground coil operation. The model considers the fluid flow inside the coil, coil material and size, and cyclic operation of the coil. The farfield conditions are specified with the empirical equation derived by Kusuda and Achenbach (4) so that they are a function of

---

Research sponsored by the Office of Building Energy Research and Development, U.S. Department of Energy under contract DE-AC05-85OR-21400 with Martin Marietta Energy Systems, Inc.

depth and time of the year. This model has been solved numerically. A simulation run of Brookhaven National Laboratory's field experimental data (5) for 44 days, by inputting experimental ground coil inlet fluid temperature as a function of time, indicated excellent prediction of test results. A parametric study was also performed to check the relative importance of different parameters.

This model can be used for ground coil design without presuming the coil-ground heat transfer rate as line source theory requires and yet considering more factors in analyzing the ground coil performance.

## MATHEMATICAL MODEL

A single pipe buried $r_F$ deep below the ground with fluid flow is shown in Fig. 1.

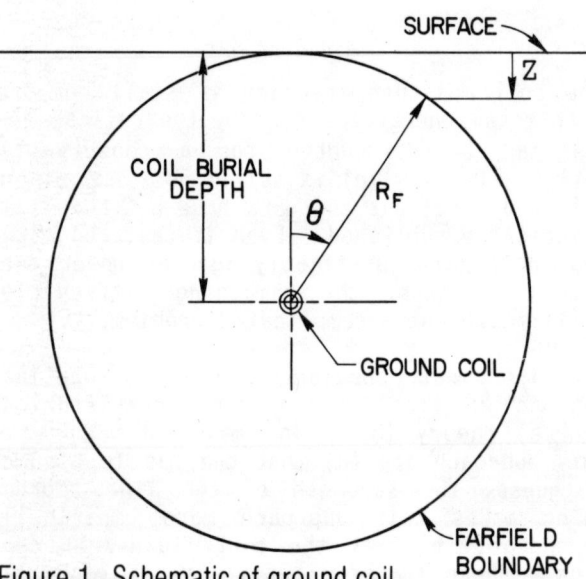

Figure 1. Schematic of ground coil.

The model is based on energy balance subject to the following assumptions: (1) the soil is homogeneous, (2) the soil thermal properties are constant, (3) the fluid temperature and velocity are uniform at any coil cross-section, (4) the coil is buried deep enough that the distance between ground surface and coil can be considered as farfield, (5) only single coil is used in the ground, and (6) heat transfer up to the coil outside wall is axially symmetrical.

With these assumptions, the following equations can be derived for the system shown in Fig. 1.:

(a) Heat flow in the fluid within the coil and coil wall

$$-V\frac{\partial T_f}{\partial X} + \frac{2K_p}{r_o \rho_f C_f}\frac{\partial T_p}{\partial r}\bigg|_{r_o} = \frac{\partial T_f}{\partial t} \quad (1)$$

(b) Energy balance in the coil wall

$$\frac{\partial^2 T_p}{\partial r^2} + \frac{1}{r}\frac{\partial T_p}{\partial r} = \frac{1}{\alpha_p}\frac{\partial T_p}{\partial T} \quad (r_o \leq r \leq r_1) \quad (2)$$

(c) Energy balance in the soil

$$\frac{\partial^2 T_s}{\partial r^2} + \frac{1}{r}\frac{\partial T_s}{\partial r} + \frac{1}{r^2}\frac{\partial^2 T_s}{\partial \theta^2} = \frac{1}{\alpha_s}\frac{\partial T_s}{\partial t} \quad (3)$$

$$(r_1 \leq r \leq r_F)$$

(d) Boundary conditions

At $r = r_o$ (fluid and coil inside wall)

$$h(T_p - T_f)\big|_{r_o} = K_p\frac{\partial T_p}{\partial r}\bigg|_{r_o} \quad (4)$$

At $r = r_1$ (coil outside wall)

$$T_p = T_s \quad (5)$$

$$2\pi K_p \frac{\partial T_p}{\partial r}\bigg|_{r_1} = K_s \int_0^{2\pi} \frac{\partial T_s}{\partial r}\bigg|_{r_1} d\theta \quad (6)$$

Equation (6) is due to assumption 6, where the soil temperatures away from the coil are not radially symmetrical.

At $r = r_F$ (farfield)

$$T_{SF} = TA - DT^*EXP\left(-Z\sqrt{\frac{\pi}{8766\alpha_s}}\right)*Cos\left(\frac{2\pi t_o}{8766}\right.$$
$$\left. - \theta_o - Z\sqrt{\frac{\pi}{8766\alpha_s}}\right) \quad (7a)$$

$$Z = r_F[1 - Sin(\frac{\pi}{2} - \theta)] \quad o \leq \theta \leq \pi \quad (7b)$$

where Eq. (7a) is Kusuda's correlation (4) and $\theta_o$ is the phase angle of the ground temperature variation at different depths due to soil thermal resistance.

(e) Initial conditions

$$T_f = T_p = T_s = T_i(x,r), t = o \quad (8)$$

where $T_f$, $T_p$, and $T_s$ are not necessarily initially equal but they have to be known functions of x and r.

(f) Fluid inlet condition

$$T_f(t,o) = T_{fi}(t) \quad (9)$$

where $T_{fi}$ is a known function of time t.

The model described so far is for the ground coil with fluid circulation. During the "off" cycle period, the fluid velocity, V, in Eq. (1) is zero and the boundary condition of Eq. (4) does not exist. The following equation replaces Eq. (4):

At $r = r_o$

$$T_f = T_p \quad (10)$$

Equation (10) is not exactly correct. Since the amount of fluid in the coil is relatively small compared with surrounding soil, the error will probably be small, particularly when the "off" period is long.

Ground coils usually are buried 1 to 1 1/2 m (3 to 5 ft) below ground surface. Past experience (1,6) indicated that thermal interference of the coil would not reach this far. It is, therefore, convenient to assume that the coil burial depth is the farfield radius.

## MODEL VALIDATION

The model was used to simulate the field test results provided by Brookhaven National Laboratory (BNL) (5). The fluid and ground temperatures were provided as daily averages. The heat pump total "on" time, daily energy absorbed from the ground, and average daily coil flow rates were also provided.

The following list gives the properties of the soil, coil, and fluid:

Coil length = 152.5 m (500 ft)
Coil burial depth = 1.2 m (4 ft)
Coil size = 3.8 cm (1.5 in.) nominal diameter
Coil material = medium-density polyethylene
Coil thermal conductivity = 0.46 W/m/°C (0.266 Btu/h/ft/°F)
Coil specific heat = 2174 J/kg/K (0.52 Btu/lb/°F)
Fluid = water-ethylene glycol (20 wt %) mixture
Soil = sandy with 10 vol % moisture content
Soil thermal conductivity = 1.731 W/m/°C (1 Btu/h/ft/°F)
Soil thermal diffusivity = 0.0036 $m^2$/h (0.039 $ft^2$/h)
Flow rate = 0.927 $m^3$/h (4.08 gpm) average during "on" time
Yearly average temperature, TA = 10.232°C (50.417°F)
Amplitude of yearly temperature variation, DT = 12.795°C (22.967°F)
Phase angle, $\theta_o$ = 0.352 radian

The thermal properties of the fluid were taken from the ASHRAE Handbook (7). The farfield temperature was allowed to vary as indicated by Eq. (7). The ground coil inlet fluid temperature from the test data was an input to the computer code.

The Reynolds number for the fluid flow, based on 0°C (32°F) fluid temperature and 0.927 $m^3$/h (4.08 gpm), is around 2300, which is in the transition region. The computer code can calculate variable Re along the coil. Donne and Bowditch (8) suggested, based on their experimental results, that Nusselt number of transitional flow between turbulent and laminar flow can follow two paths: one is the prolongation of the turbulent region and the other is the prolongation of the laminar region. However, the laminar transition region is less stable than the turbulent transition one. An increase in Reynolds number, air temperature (fluid used in their experiment) or pipe length over diameter ratio may lead to a change from laminar to turbulent flow. Since the ground coil length over diameter ratio is very high in this study, on the order of 3,000, we assumed that the flow was along the path of the prolongation of turbulent region. The Nusselt number was then calculated with Nu = $0.0155 Pr^{0.5} Re^{0.83}$ (9). The h value calculated as an input for the model validation was 710 W/$m^2$/°C (125 Btu/h/$ft^2$/°F).

A total of 44 days were simulated, starting on day number 329 (Nov. 26th), the day the heating season really began. Since only the fraction of "on" time per day was given in the experimental data, the computer code, which could handle the cyclic operation, was instructed to run the same fraction of "on" time per hour.

Figure 2 shows the simulation of the daily energy absorbed from the ground. The first nine days of simulation do not match the test results well due to experimental errors. However, after the first nine days, the computer simulation predicts the test results very well. Figure 3 shows the simulation of the coil exit fluid temperature during "off" cycle period. Again the first nine days are poorly simulated. The rest of the simulation indicates the same trend, but the computer calculated temperatures are, for the most part, about 1°C (2°F) higher than the test results. Part of the reason could be due to the lack of exact heat pump cycling schedule. Also a careful check on the experimental data reveals that the fluid temperatures are higher during the heat pump "on" cycle than during the "off" cycle period, which again could be caused by experimental errors. However, judging from the complicated nature of the problem, the model predicts the experimental results well.

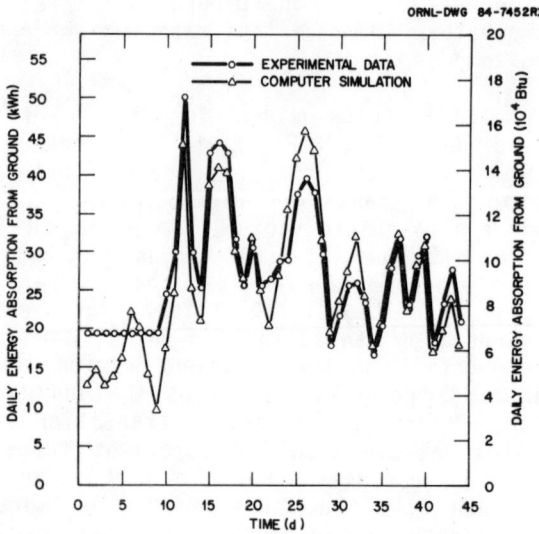

Figure 2. Computer simulation of 44 days of BNL field test results (5) on ground coil total energy absorption.

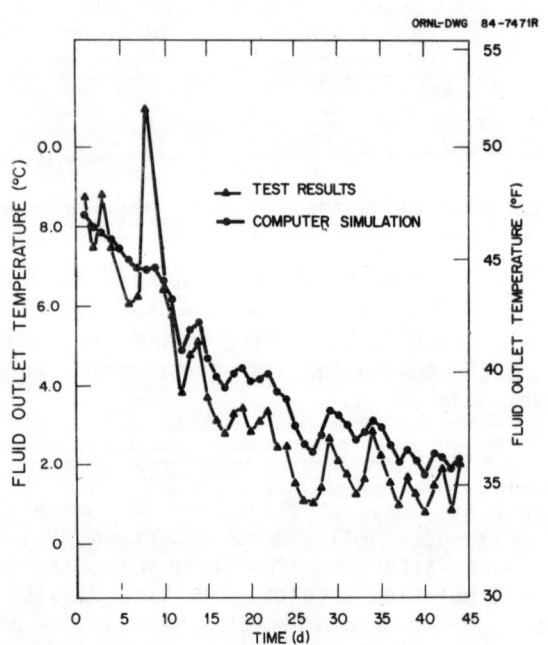

Figure 3. Computer simulation of 44 days of BNL field test results (5) on ground coil exit temperature at "off" cycle.

Figure 4 shows the calculated ground temperature distribution after 42 days' simulation. While there is no experimental result for comparison, the figure probably represents a realistic ground temperature distribution, which could be very useful for environmental impact studies if necessary.

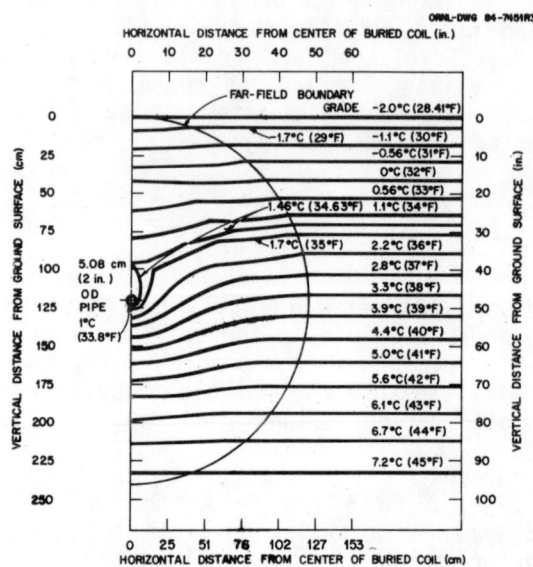

Figure 4. Calculated ground temperature distribution after 42 days simulation of BNL field test results (5).

## PARAMETRIC STUDY

The design of a ground coil involves many factors. A parametric study will provide the information to determine the relative importance of each factor and to improve the performance of the ground coil.

The parameters selected for investigation are: (a) soil thermal conductivity, (b) coil length, and (c) coil burial depth.

In order to provide a basis for comparison, a base run was completed with the following material and operating conditions:

15 day continuous operation,
Coil size = 4.4 cm (1.74 in.) nominal diameter,
Coil material = medium density polyethylene,
Coil thermal conductivity = 0.46 W/m/°C (0.266 Btu/h/ft/°F),
Coil length = 152 m (500 ft),
Fluid is 20% ethylene glycol and water mixture,
Fluid inlet temperature = 0°C (32°F),
Soil thermal conductivity = 1.731 W/m/°C (1.0 Btu/h/ft/°F)
Fluid rate = 0.927 m³/h (4.08 gpm),
Ground burial depth = 1.22 m (4 ft).

Figure 5 shows the effect of soil thermal conductivity. When the soil thermal conductivity is reduced by half from 1.731 to 0.866 W/m/°C (1.0 to 0.5 Btu/h/ft/°F), the total energy absorbed is reduced by 23%. When it is doubled from 1.731 to 3.462 W/m/°C (1.0 to 2.0 Btu/h/ft/°F), the total energy absorbed increases by 20%. This indicates that soil thermal conductivity is an important factor. For ground coil design, the soil thermal properties, particularly thermal conductivity, have to be defined first.

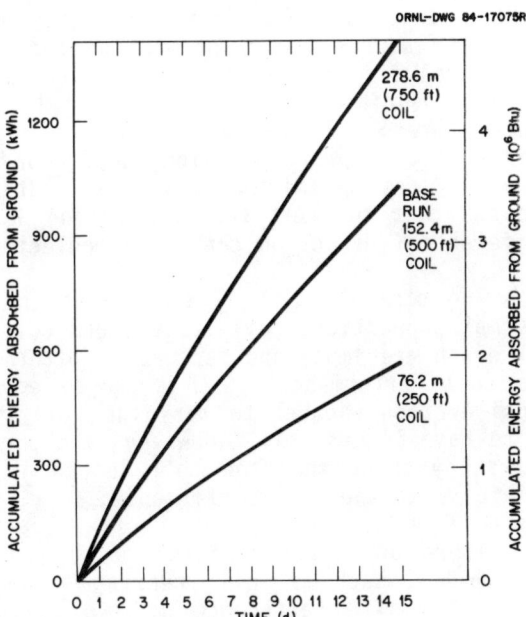

Figure 6. Effect of coil length on total energy absorption.

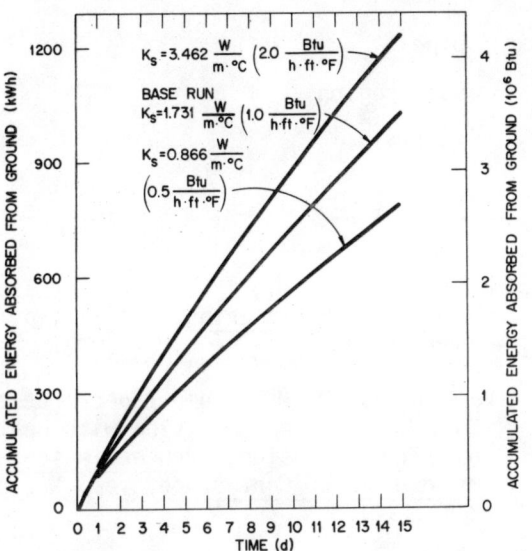

Figure 5. Effect of soil thermal conductivity on total energy absorption.

Figure 6 shows the effect of coil length. When the length is reduced to 76.2 m (250 ft) from 154.2 m (500 ft), the coil capacity is reduced by 44.5%. When the length is increased to 228.6 m (750 ft), the coil capacity increases 31%. These figures indicate that the coil length is a very important factor in determining the ground coil performance. Unlike soil thermal conductivity, coil length is a controllable factor.

Figure 7 shows the effect of coil burial depth. The deeper the coil is buried, the less fluctuation the surrounding soil temperature and the better the coil performs. Figure 7 indicates that when the coil burial depth increased from 1.22 m (4 ft) to 1.52 m (5 ft) and 1.83 m (6 ft), the coil capacity increased 13.1% and 23.4% respectively. However, the increased cost for deeper trenching has to be considered.

## DISCUSSION AND CONCLUSION

A set of partial differential equations describing the operation of ground coils for heat pump applications was solved numerically. It differs from other approaches by imposing a

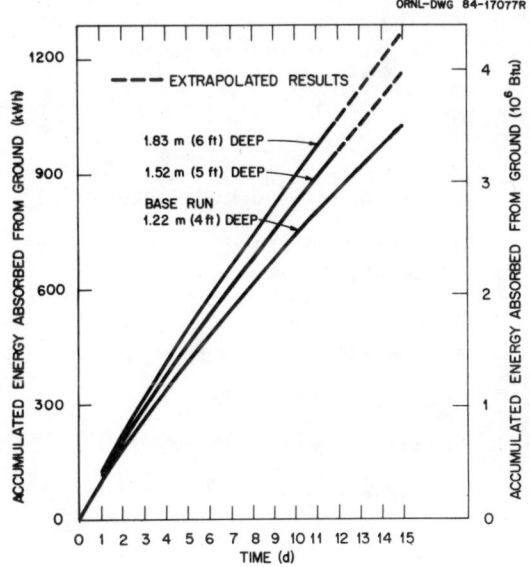

Figure 7. Effect of coil burial depth on total energy absorption.

ground farfield temperature as a function of the time of the year and depth below the grade.

The model was used to simulate the BNL's ground coil test results for 44 days with excellent match in total energy absorption from the ground. The calculated results indicated a deviation of 1°C (2°F) of the ground coil exit fluid temperature. The discrepancy is well within the acceptable experimental errors.

The model also provided a realistic ground temperature distribution as shown in

Fig. 4, which was far from the highly idealized line source theory prediction for buried cable analysis in a semi-infinite medium with constant surface and uniform initial medium temperatures. Figure 4 also shows that in coil vertical upward position, the ground temperature goes up and then goes down. This indicates that the heat transfer action in circumferential direction cannot be neglected.

The parametric study shows that the soil thermal properties, coil length, and coil burial depth are important factors in determining the coil performance. It is up to economic consideration whether to bury the coil deeper or to have longer coil. However, the parametric study shows that the coil length is most sensitive to the coil performance.

There are other factors not included in this model, such as soil freezing around the coil, thermal interference if two coils are buried very close to each other, and finally the moisture migration problem for ground coil summer operation. This model, nevertheless, is one of the more advanced ground coil theoretical analyses available so far.

## NOTATION

| | |
|---|---|
| C | specific heat, J/kg°C (Btu/lb°F) |
| DT | one half amplitude of annual surface temperature, °C (°F) |
| h | convective heat transfer coefficient, W/m²°C (Btu/h/ft²°F) |
| Nu | Nusselt number = $2hr_o/K_f$ |
| Pr | Prandtl number = $MC_f/K_f$ |
| Re | Reynolds number = $2V_o r_o \rho_f / \mu$ |
| r | radius, m (ft) |
| T | temperature, °C (°F) |
| TA | Annual average ground surface temperature, °C (°F) |
| t | time, min |
| $t_o$ | time of year, h |
| V | fluid velocity, m/min (ft/min) |
| x | distance along the ground coil, m (ft) |
| z | depth, m (ft) |
| P | density, kg/m³ (lb/ft³) |
| $\theta_o$ | phase angle, radius |
| A | thermal diffusivity = $K/PC_p$, m²/h (ft²/h) |
| M | fluid viscosity, kg/mh (lb/ft h) |

Subscripts

| | |
|---|---|
| 0 | pipe inside wall |
| 1 | pipe outside wall |
| f | fluid |
| i | initial |
| p | pipe |
| s | soil region |
| F | farfield |

## LITERATURE CITED

1. Coogan Jr., C. H., "Heat Transfer Rate," *Mechanical Engineering*, vol. 71, p. 495, June 1949.

2. Vestal Jr., D. M., "Some Aspects of the Soil Problem in Connection with Heat Pump Buried Coil Design," *Proceedings of Midwest Power Conference*, vol. XI, p. 279, 1949.

3. Svec, O., Goodrich, L. E., and Plamer, J. H. L., "Heat Transfer Characteristics of In-Ground Heat Exchangers," *J. of Energy Research*, vol. 7, pp. 263-78, 1983.

4. Carslaw, H. S., and Jaeger, J. C., *Conduction of Heat in Solids*, Oxford, Second Edition, 1959.

5. Personal communication with Dr. P. Metz about the ground-coupled test house. Brookhaven National Laboratory, Upton, New York, 1983.

6. *ASHRAE Fundamental Handbook*, Chapter 18 (1981).

7. Kusuda, T., and Achenbach, P. R., "Earth Temperature and Thermal Diffusivity at Selected Stations in the United States," *ASHRAE Trans.*, vol. 71, part 1, 1965.

8. Donne, M. D., and Bowditch, F. H., "High Temperature Heat Transfer," *Nuclear Engineering*, pp. 20-29, January 1963.

9. Kays, W. M., *Convective Heat and Mass Transfer*, pp. 109 and 173, McGraw Hill, New York (1966).

# FUEL PROCESSING ANALYSES FOR FUEL CELL POWER SYSTEMS

Gerard G. Elia ■ Westinghouse Electric Corp., Advanced Energy Systems Div., Pittsburgh, PA 15236

An analysis of Fuel Processing Reformers is presented with a simple and efficient digital computer code. A simulation is developed for a double counter-current heat exchange with steam/hydrocarbon fuel catalytic reactions producing hydrogen. Appropriate equations are presented for reforming reactions, rate kinetics, heat transfer, flow and pressure drop. Stream solution profiles illustrate two conceptual designs with discussion of important process variables. Short computing times enable the economic development of extensive parametric studies.

## INTRODUCTION

A digital simulation model is presented to illustrate key thermal-hydraulic and performance characteristics of the reforming process. This presentation highlights the simplicity of the digital reformer model. Typical model run times last on the order of several seconds for steady state convergence. Scoping analyses clearly show advantage over detailed multi-dimensional equation models which require orders of magnitude longer computing time. Results provide guidance to detailed design calculations.

## ANALYSIS MODEL

### Processes Description

The basics of hydrogen production by steam-hydrocarbon reforming can be described with a double counter-current process heat exchange shown in Figure 1. A reforming stream consists of a steam and hydrocarbon fuel mixture (methane will be employed in the example discussions) which passes through a catalyst bed where chemical kinetic reactions take place. The required heat flux for the catalyst reaction is provided by a counter-current pass combustion stream. Upon exiting the catalyst bed section of the reformer, the process stream enters a counter-current recuperation channel. The hydrogen rich recuperation stream transfers heat to the catalyst bed section.

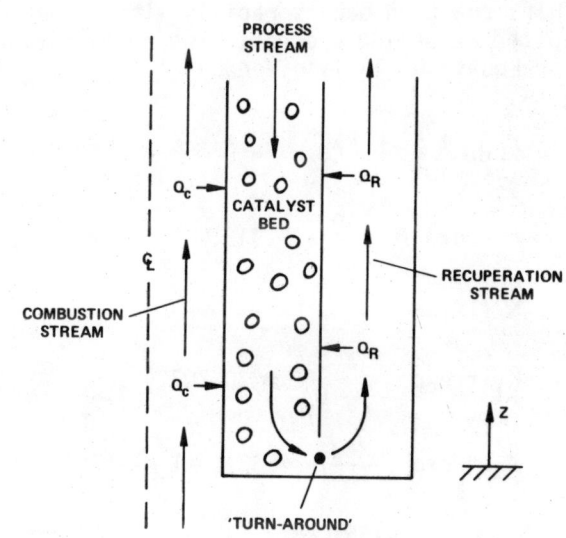

Figure 1. Internally-fired reformer basic model.

### Model Development

A simple analytical model employs the following assumptions:

1. All process streams have fully developed one dimensional (axial) flows with no radial gradients.
2. The streams possess ideal gas behavior.
3. The catalyst bed is homogeneous.

4. Heat transfer in the reformer tube is simulated with an equivalent convection mechanism.
5. External heat loss is ignored.
6. End effects are ignored for the fluid and thermal profiles.
7. The reformer tubes are independent of each other.

The following reactions are employed to simulate the hydrogen production:

$$CH_4 + 2H_2O \rightleftarrows CO_2 + 4H_2 \quad (1)$$

$$CO_2 + H_2 \rightleftarrows CO + H_2O \quad (2)$$

Equation (1) is the base reforming reaction while Equation (2) describes the accompanying shift reaction. This reaction mechanism was proposed by Moe and Gerhard (1) as a result of laboratory data.

The reforming reaction is the rate controlling process. It is assumed that the shift reaction occurs rapidly with respect to the reforming process. The rate relation for Equation (1) is given by:

$$r_{CH_4} = A [K\, P_{CH_4} P^2_{H_2O} - P^4_{H_2} P_{CO_2}] \quad (3)$$

$$A = \exp\left(\frac{31720}{T} - 7.912\right) \quad (4)$$

$$K = K_1 K_2 \quad (5)$$

$$K_1 = \exp\left(-\frac{49435}{T} - 30.707\right) \quad (6)$$

$$K_2 = \exp\left(\frac{8240}{T} - 4.33\right),\ T < 1560°R \quad (7a)$$

$$K_2 = \exp\left(\frac{7351.24}{T} - 3.765\right),\ T \geq 1560°R \quad (7b)$$

Another reaction mechanism is often presented in the literature:

$$CH_4 + H_2O \rightleftarrows 3H_2 + CO \quad (8)$$

$$CO + H_2O \rightleftarrows CO_2 + H_2 \quad (9)$$

Some authors have assumed the shift reaction at equilibrium or proposed a pseudo-equilibrium reaction at a temperature slightly below that of the process. Laboratory data in Reference (1) dictate the applicability of the present model for this discussion.

Note that Moe and Gerhard assumed that the shift reaction (Equation 2) occurs at a finite rate. The model uses empirical formulas to approximate equilibrium data for predicting the $CO_2$ to $CO$ ratio.

$$N_{CO_2}/N_{CO} = 1.0 \quad \text{for } T \leq 1460°R \quad (10a)$$

$$N_{CO_2}/N_{CO} = 1.0 - (T-1460)(0.0011 - 7.0 \times 10^{-5}\, S/G) \quad (10b)$$
$$\text{for } 1460°R \leq T \leq 2260°R$$

$$N_{CO_2}/N_{CO} = 0.12 + 0.056 \times S/G \quad (10c)$$
$$\text{for } T > 2260°R$$

where $N_{CO_2}/N_{CO}$ is the $CO_2$ to $CO$ molar ratio and $S/G$ is the $H_2O$ to $CH_4$ molar ratio with all $CO_2$ and $CO$ converted to $CH_4$.

In the model, the Hyman analysis (2) simulates the reforming process. Hyman's principal reference data were obtained for conditions where the methane conversion is greater than ninety percent of equilibrium at final process temperatures. Thus, the model should only be used when the process is near equilibrium conditions. For other process conditions, the results should be viewed only to establish trends.

The model is one dimensional and thus assumes only axial gradients. Results must be interpreted as indicative when a significant radial temperature gradient exists. Nevertheless, these conditions on the reformer model are not that restrictive. Many practical conventional reformers approach one dimensional operating conditions and equilibrium hydrocarbon conversion, especially when viewed in the overall performance perspective.

Typical mass (species), momentum and energy balances are employed for the three reformer streams: combustion, process and recuperation. These equations provide calculations for individual stream constituents, pressure and temperature. The expressions are classic in nature and will not be discussed here. Interested readers may consult any process text for further discussions.

Stream to stream energy transfer is represented by an overall heat transfer relation.

$$U = \frac{T_{s1} - T_{s2}}{\frac{1}{h_i A_i} + \frac{\ell}{Ak} + \frac{1}{h_o A_o}} \quad (11)$$

## Heat Transfer Correlations

The Beek correlation (3) is employed to calculate the effective convection film coefficient on the reforming side walls of the process flow channel.

$$h = \frac{k_g}{D_p}(2.58 \, Re_p^{0.33} \, Pr^{0.33} \\ + 0.094 \, Re_p^{0.8} \, Pr^{0.4}) \quad (12a)$$

$$h = \frac{k_g}{D_p}(0.203 \, Re_p^{0.33} \, Pr^{0.33} \\ + 0.220 \, Re_p^{0.8} \, Pr^{0.4}) \quad (12b)$$

Equation (12a) is employed for cylindrical shaped catalyst particles while (12b) holds for spherical particles.

The relation proposed by Leva (4) was also considered.

$$h = \frac{k_g}{D_p}[0.813 \, (\frac{D_p}{D_t}) \, e \, (\frac{-6D_p}{D_t})] \, Re_p^{0.90} \quad (13)$$

Experimental results (5) have shown that the Beek relationship adequately represents heat transfer in a packed bed and was therefore used in the reformer model. A key fact against the Leva relationship is the lack of Prandtl number effect.

Recent heat transfer measurements of low Prandtl number gas mixtures indicate that existing scaling laws for normal Prandtl number (~ 0.7) gases do not adequately represent heat transfer characteristics of these gas mixtures. In order to take this effect into account, the following correlation (6) was employed for the calculation of the film coefficient in the stream channels.

$$h = 0.0215 \, \frac{k}{D} \, (\frac{WD}{A\mu})^{0.8} \, Pr^{0.53} \quad (14)$$

Note that for certain conditions (e.g., combustion to process stream) radiation effects can be significant. In these cases the radiation may be approximated by an equivalent convective coefficient (7).

$$h_r = \mathcal{H}_{1-2} \, F_T \quad (15)$$

## Pressure Drop Correlations

Pressure drops are calculated in the combustion and recuperation streams based on typical relations for flow friction factor (8)

$$f = 0.0055 \, [1 + 2 \, (\frac{\epsilon}{D_t} \times 10^4 + \frac{10^6}{Re})^{0.33}] \quad (16)$$

The pressure drop is then calculated by the familiar momentum relation:

$$\Delta P = f \, \frac{L}{D} \, \frac{v}{2g} \, (\frac{W}{A})^2 \quad (17)$$

The pressure drop for the process stream (through the packed catalyst bed) is calculated with the Ergun (9) relationship.

$$\frac{\Delta P}{L} = \frac{150 \, (1-S)^2}{Re_p \, (S^3)} \, \frac{\rho \, v^2}{D_p} \\ + 1.75 \, \frac{\rho^2 v}{D_p} \, \frac{(1-S)}{S^3} \quad (18)$$

## Calculation Procedures

The calculation for reforming starts with the specification of the process stream inlet conditions (temperatures, pressures, flows) and the combustion and recuperation stream outlet conditions. A increment of reformer length (axial slice) is chosen. Methane conversion is calculated for this increment using the kinetic rate relations. The $CO_2$ to $CO$ ratio is computed from the empirical relationships (Equation 10). A stoichiometric balance is used to calculate product composition. Heats of reaction are calculated from the change in stream composition. Heat transfer coefficients for the tube (channel) are calculated with the overall conductance relation via Equation (11). Similar calculations are performed for the recuperation stream tube. The heat input from the combustion and recuperation streams to the process flow (catalyst bed) is then calculated, based on the temperatures at the beginning of the increment. The difference between heat input and heat absorbed by chemical reaction determines the temperature at the outlet of the increment. The process outlet conditions of the increment are the inlet conditions of the next axial slice, which steps the calculation along the channel. The combustion stream temperature change in the increment is obtained from a heat balance, and the temperature at the start of the next increment is calculated.

When the calculations reach the process stream (catalyst bed) exit, the heat transfer, the combustion stream temperature, and the recuperation inlet temperature are different from the true values. The recuperation inlet temperature should be equal to the process stream outlet temperature (i.e., 'turn-around' temperature). The recuperation outlet temperature and the combustion outlet temperature are then changed and the calculation iterates until the recuperation inlet and combustion stream outlet temperatures meet desired design values.

Note that there are two fluid flows being calculated upstream, the combustion and the recuperation streams. The process flow is calculated downstream, due to the irreversibility of the changing chemistry and physical properties. The code operator must guess the outlet conditions for the combustion and the recuperation streams. The code yields the temperatures at the process outlet after calculating incrementally along the entire tube. The calculated recuperation inlet temperature is then compared to the calculated process outlet temperature (they should be equal). The recuperation inlet temperature guess is then adjusted along with stream composition which is equal to the composition of the process stream outlet. The program iterates until the recuperation inlet - process outlet temperature difference meets the desired calculational accuracy.

At this point, the combustion inlet calculated and specified temperatures do not yet agree. Thus, the combustion outlet temperature is also changed in the code. This changes the balance which had been achieved on the process and recuperation streams. The program iterates, holding the new combustion outlet temperature, until the recuperation inlet and process outlet temperatures again agree. This process continues until all stream conditions have been satisfied.

## ANALYSIS RESULTS

The simple one dimensional code model (10) described above was used to provide scoping and guideline calculations for several conceptual designs. Two schemes are illustrated. The internally-fired or module concept (Figure 1) and the externally-fired or furnace concept shown in Figure 2. The major difference was the switch of physical location for the combustion and recuperation streams, inside to outside, respectively. The very short time required for the code calculations permitted reformer simulations with several hundred increments (axial slices). Extensive parametric studies were performed quite economically. These are discussed in the sections which follow.

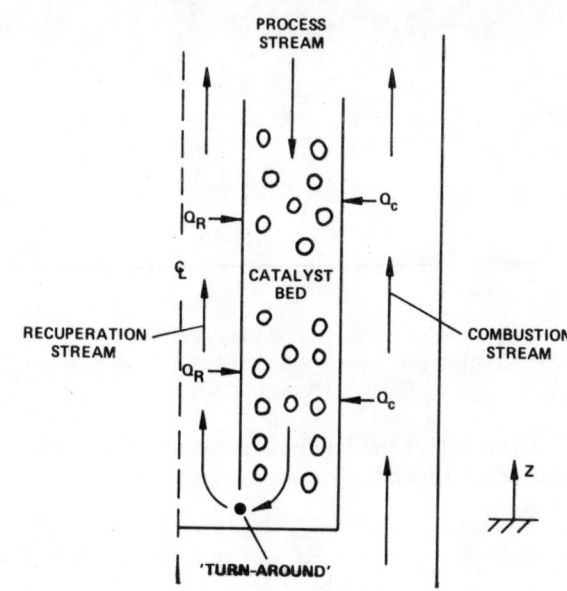

Figure 2. Externally-fired reformer basic model.

## Internally-Fired (Module) Concept

A parametric study was performed to determine the effect of reformer tube size (nominal diameter) on reformer capacity and required catalyst bed length. (11) For illustration consider the catalyst bed area to maintain a given process stream mass velocity which results in a flow velocity of approximately 3 ft/sec (91.4 cm/sec) based on the inlet conditions. This velocity results in a process gas pressure drop of approximately 12 psi (0.816 atm).

The size of the flow annulus for the combustion gas is now varied to maintain a combustion gas inlet velocity of approximately 110 ft/sec (3352.8 cm/sec). However, mechanical limitations required to fit the concentric pipes together restrict the minimum gap width of the annulus to approximately 1/4 inch (0.625 cm). In some of the shorter designs, this was further reduced. The limitation of a minimum gap width caused the combustion gas velocity to be lower in several of the large diameter, low capacity designs.

The results of the parametric study are shown in Figures 3 and 4. The reason the slope of the 16 inch (40.64 cm) curve decreases at lower capacity is the decreasing combustion gas velocity, caused by the limitation of a minimum gap size. The lower gas velocity results in a lower overall heat transfer coefficient and a longer catalyst bed. As an illustrative example the 10 inch (25.4 cm) nominal diameter, 24 tube case was selected. The resulting thermal profiles for the process fluid streams are given in Figure 5.

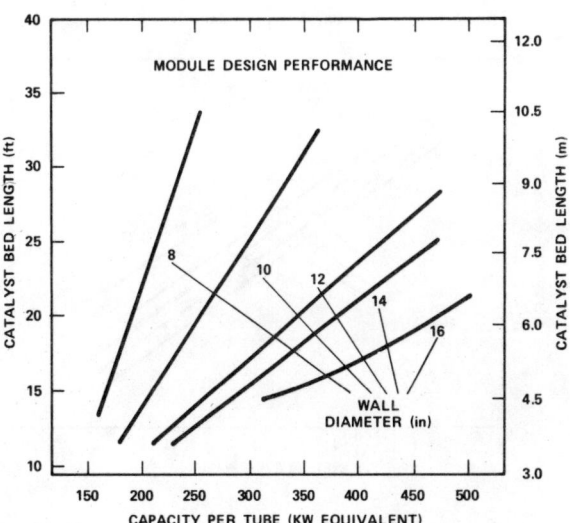

Figure 4. Modular reformer performance analysis.

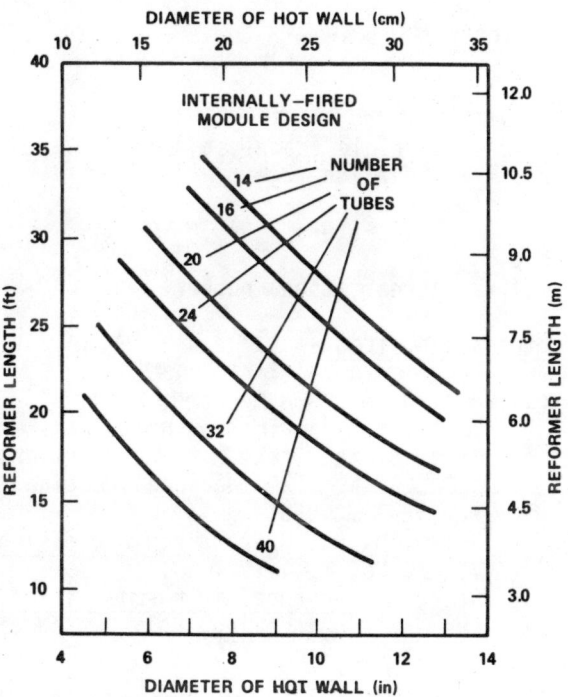

Figure 3. Internally-fired reformer design results.

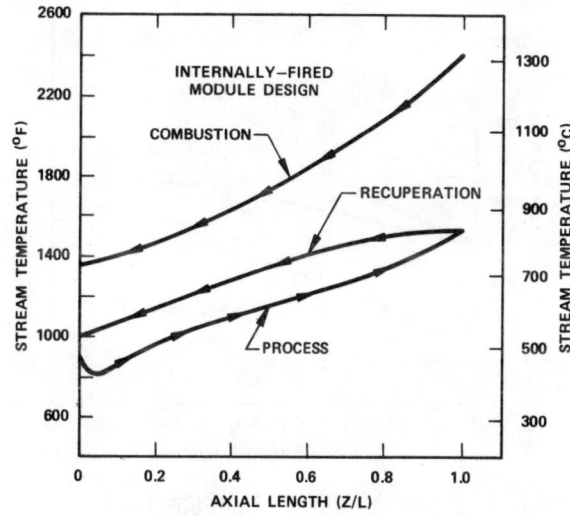

Figure 5. Modular reformer fluid stream temperatures.

### CONCLUSIONS

A simple digital computer code has been discussed for analysis of steam-hydrocarbon reformers. Assumptions for one-dimensional fully developed flows enable a fast and economical simulation of a complicated double-counter-current process heat exchange involving chemical kinetics. Short computation times permit numerous scoping calculation runs to develop extensive parametric studies. Applicability over several different kinds of reforming schemes has been demonstrated.

### Externally-Fired (Furnace) Concept

For comparative purposes an externally-fired design concept was analyzed. Again a series of code runs were made to produce a parametric study covering a range of design geometries. The results of the study are shown in Figure 6. Thermal profiles for an example case (10 inch (25.4 cm) nominal diameter, 24 tubes) is given in Figure 7.

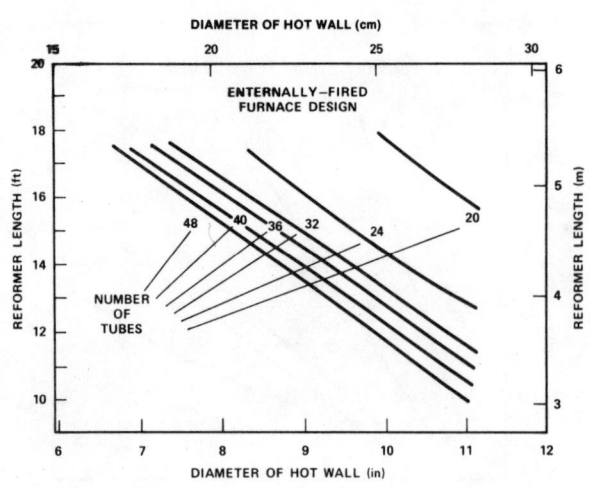

Figure 6. Externally-fired reformer design results.

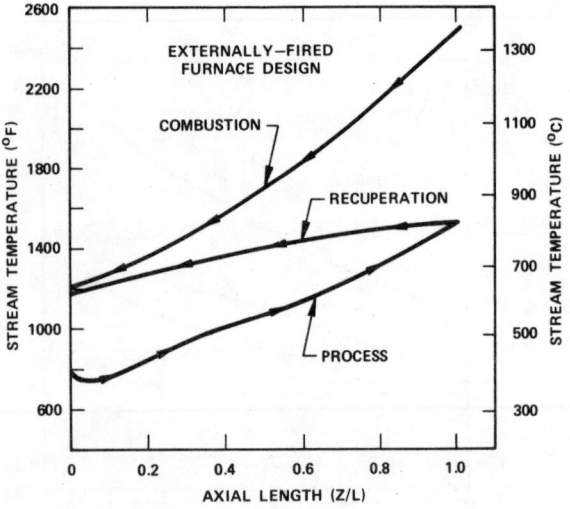

Figure 7. Furnace reformer fluid stream temperatures.

Acknowledgment

The author gratefully acknowledges guidance from B. L. Pierce (Westinghouse). A. P. Murray related many interesting discussions. The Electric Power Research Institute, D. M. Rastler, program manager, provided partial funding.

Notation

| | | |
|---|---|---|
| A | = | Area |
| $\bar{A}$ | = | Equivalent conduction area |
| D | = | Diameter |
| F | = | Temperature factor |
| f | = | Friction factor |
| g | = | Gravitational constant |
| h | = | Convective heat transfer coefficient |
| k | = | Thermal conductivity |
| L | = | Length |
| $\ell$ | = | Conduction length |
| N | = | Number of moles |
| Pr | = | Prandtl number |
| Q | = | Heat Transfer |
| Re | = | Reynolds number |
| r | = | Reaction rate |
| S | = | Porosity (catalyst bed) |
| T | = | Temperature |
| U | = | Overall heat transfer coefficient |
| V | = | Velocity |
| v | = | Specific volume |
| W | = | Mass flow rate |
| $\epsilon$ | = | Surface roughness |
| $\rho$ | = | Density |
| $\mu$ | = | Viscosity |
| $\mathcal{F}$ | = | Gray body shape factor |

Subscripts

| | | |
|---|---|---|
| g | = | gas |
| i | = | inside |
| p | = | particle (catalyst) |
| r | = | radiation |
| s | = | stream |
| T | = | temperature |
| t | = | tube |
| 1,2 | = | stream (tube) number |

Literature Cited

(1) Moe, J. M., and Gerhard, E. R., "Chemical Reaction and Heat Transfer Rates of the Steam-Methane Reaction," AICHE Fifty-Sixth National Meeting, May 16-19, 1965.
(2) Hyman, M. H., Hydrocarbon Processing, V. 47, No. 7, July 1968.
(3) Beek, J., Advances in Chemistry and Chemical Engineering, Vol. 3, Academic Press, New York, 1962.
(4) Leva, M., Industrial and Engineering Chemistry, July 1947.
(5) Van Dame, S. E., Smith, R. A., and Christner, L. G., "Experimental Steam Methane Reformer Heat Transfer Correlations," 1983 ASME Winter Meeting.
(6) Pierce, B. L., ASME Journal of Engineering for Power, V. 103, No. 1, January 1981.
(7) Krieth, F., Principles of Heat Transfer, 1958.
(8) Moody, L. F., ASME, V. 66, p. 671, 1944.
(9) Ergun, S., Chemical Engineering Progress, V. 48, p. 227, 1952.
(10) Pierce, B. L., Westinghouse Electric Corporation, personal communication, 1985.
(11) Parker, W. G. and Summers, W. A., Westinghouse Electric Corporation, personal communication, 1985.

# SYNTHETIC BRINE FOULING OF A GEOTHERMAL HEAT EXCHANGER

S.H. Chan, K.F. Neusen and C. DeBellis ■ University of Wisconsin—Milwaukee
Milwaukee, Wisconsin 53201

A geothermal brine loop has been designed and constructed to determine the fouling characteristics of a wide variety of synthetic brines on a 2.82 m (9.25 ft) long heat exchanger. The heat exchanger is heavily instrumented so that the local transient fouling behavior can be determined. A computerized data acquisition system is used to take temperatures and flow rates during experiments.

In the set of experiments reported the brine chemistry was the primary parameter of interest. Data for runs with different Reynolds numbers are also presented. The base chemistry was distilled deionized water saturated with silica. The NaCl concentration was varied from 0 to 1 molar and the pH was adjusted between 6.0 and 7.0 using NaAc. Each experiment ran for approximately 135 hours with continuous data acquisition. From the data, local and overall thermal fouling resistance rates were determined for the different brine chemistries.

## INTRODUCTION

Heat exchangers operating with geothermal brine are subject to scaling and an attendant increase in fouling resistance to heat transfer (1). The composition of natural geothermal brines varies widely but the presence of 10,000 ppm or more of total dissolved solids is common. Scaling with silica or calcium compounds can lead to a significant reduction in the overall heat transfer coefficient for the exchanger. While this problem has been recognized for a long time, it is largely unresolved and the predictive methods available are rather crude (2 to 5).

The lack of generalized fouling models is due to the fact that fouling takes place under many conditions and mechanisms, such as precipitation, particulate accumulation, chemical reaction, corrosion, biological organism attachment and component freezing (6). Despite the complexity of the process, some typical fouling behavior has been observed. The net fouling rate is expressed generally as the difference between $\phi_d$, a deposition, and $\phi_r$, a removal rate function (2, 5):

$$d(x_f/k_f)/dt = \phi_d - \phi_r \qquad (1)$$

Several previous investigations have reported experimental results for fouling heat transfer with geothermal brines in tubular heat exchangers. Tests of two ten foot (3.05 m) horizontal one inch (2.54 cm) titanium tubes were run for 92 hours and showed a 60% loss in heat transfer coefficient (7). Vertical stainless and carbon steel tubes of 1/2" O.D., 18" long using natural and synthetic brines were tested (8). Little scaling resulted when the brine was kept in the liquid phase. A test of four, 20 foot long shell and tube heat exchangers using 1" titanium tubes was reported using natural brine (9). In most instances, linearly increasing fouling resistance was observed due to a silica based scale. Experiments with twenty foot long horizontal double pipe heat exchangers were made and the observed linear fouling rates were used to design a 5 MW demonstration plant (10). Studies (11) conducted in Iceland were performed at two sites with heat exchangers of the same design but differing flow conditions. At one site, linear fouling was observed but an induction period was indicated at the other site.

An extensive study of fouling with synthetic brines on horizontal, six foot long, 1/4 inch titanium pipes was conducted at Oak Ridge National Lab (ORNL) and reported by Bohlmann et al. (12). The heat exchangers were operated in series and overall thermal resistances were reported which varied widely from exchanger to exchanger.

These results must be viewed as being far from conclusive and no general fouling model is yet available. The complexity of geothermal brine fouling usually necessitates resorting to experiments. Furthermore, no local thermal fouling resistances have been reported to date. The present purpose is to begin a comprehensive series of experiments to study parameters of importance before a general model can be derived. The number of parameters is staggering, including various chemical, material, thermal and hydraulic variables. The initial investigation focuses on effects of silica supersaturation, brine pH, Reynolds number and contents of sodium and chloride ions on formation of silica scale. The laboratory loop used is heavily instrumented to yield both overall and local fouling resistances for all experiments. Synthetic brines were used to allow for a systematic variation in brine chemistry while holding the other variables constant.

## FOULING THEORY

Based on the outside surface area, the overall heat transfer coefficient, $U_o$, of a tube fouled inside and outside of the tube is

$$\frac{1}{U_{of}} = \frac{1}{h_o} + \frac{A_o h_i}{A_i} + R_{so} + \frac{A_o R_{si}}{A_i} + R_w \quad (2)$$

Experimentally $R_{so}$ can be kept negligible by using a non-fouling liquid on the outside surface. Initially the tube is clean; thus $R_{si} = 0$. Therefore,

$$1/U_{of} - 1/U_{ocl} = (R_{si} + 1/h_{if} - 1/h_{ct})(A_o/A_i) \quad (3)$$

The "h" terms result from the fact that the scale can change the heat transfer coefficient on the fouling surface. This effect cannot be separated experimentally from the scale resistance; therefore, a fouling resistance, $R_f$, will be defined as all terms on the right-hand side. Then Equation (3) yields,

$$R_f = 1/U_{of} - 1/U_{ocl} \quad (4)$$

which represents an experimentally determinable thermal fouling resistance. This can be also expressed using mass per surface area, m, or scale thickness, x. The three quantities are interrelated by,

$$m = \rho_f x = \rho_f k_f R_f \quad (5)$$

The accumulation rate is then:

$$dR_f/dt = (1/k_f) \, dx/dt = (1/\rho_f k_f) dm/dt. \quad (6)$$

## EXPERIMENTAL SET-UP AND PROCEDURE

The experiments in the present study were conducted in UWM's Geothermal Test Laboratory using a facility whose design was reported previously (13). The major components of the fouling heat transfer loop are shown in Figure 1. All experiments were performed on the same co-current, double-tube heat exchanger with brine on the tube side and distilled deionized water on the shell side. The inner tube was a 9.25 foot (2.82 m) long, 1" (2.54 cm) diameter titanium tube with a 0.065 inch (0.165 cm) wall thickness. Thermocouples were also imbedded in 0.025 inch holes drilled through the titanium tube, using an E.D.M. machine, at one foot intervals along the tube length.

A calibration was performed on the brine tube prior to scaling runs with distilled water in the brine loop and a special bulk fluid temperature probe inserted in the center of the titanium tube. Its purpose was to obtain the thermal resistance, $R_{oi}$, between each of the eight thermocouples in the tube wall and the tube surface on the coolant side. From the results for a clean tube, it is possible to calculate the local overall heat transfer coefficients for a fouled heat exchanger. The method is that of Wilson as modified by Fisher et al. (14). Using a range of coolant velocities, an extrapolation $1/V_c$ to zero produces the desired tube wall resistance.

The thermal hydraulic conditions for each of the five runs can be found in Table 1 and the chemistry for each run is given in Table 2. Runs A to C used distilled deionized water at a pH of 6 and a variation in NaCl concentration. Runs D and DO used sodium acetate as a buffer to change the pH of the brine to above 7.0.

## EXPERIMENTAL RESULTS AND CONCLUSIONS

### Overall Thermal Fouling Resistance ($R_f$)

$R_f$ was defined in Equation (4) with $U_o$ evaluated by

$$U_o = q/A_o \Delta T_{LMTD} \quad (7)$$

The log mean temperature difference and q are calculated from the experimental brine mass flow rate, and the inlet and outlet temperatures of brine and coolant. $\rho_b$ and $C_{pb}$ of the brine are calculated (1). The $U_{ocl}$ values are listed in Table 1. The transient overall fouling resistance versus time curves shown in Figure 2 are discussed and compared below.

**TT1-A.** Very little fouling was found in this run. After 150 hours of operation the linear fouling rate ($dR_f/dt$) was very small (see Table 3). This indicates that although silica is the major constituent of the scale it takes more than just silica to form a significant deposit.

**TT1-B.** The scattering of data was caused by poor flow control. However, the data trends are still valid because immediately after each calibration (every 24 hours) the brine flow rate was known accurately. Therefore, if only the calibration points are considered (solid symbols on graph), the fouling curve begins with an induction period, followed by a linear increase in fouling with a fouling rate of $3.489 \times 10^{-6}$ (hr.ft$^2$.°F/Btu/hr)(Table 3).

**TT1-C.** As a result of the previous run the flow calibration technique was modified without total success (see Figure 2). The data between 100 and 122 hours is questionable and the jump in fouling resistance at 122 hours was unexplainable. The curve is characterized by an induction period of about 30 hours followed by a linear increase in fouling. Disregarding the jump at 122 hours, the linear fouling rate is the same for both sections (30 to 100 hours and 123 to 180 hours) at $1.314 \times 10^{-6}$ hr.ft$^2$.°F/Btu/hr.

**TT1-D.** The flow calibration procedure was modified one more time with excellent results. This curve starts with a drop in resistance and then increases asymptotically. The asymptotic value of resistance reached after 148 hours was $2.5 \times 10^{-4}$ hr.ft$^2$.°F/Btu.

**TT1-DO.** This was the only run with a variation in Reynolds number as seen in Table 1. In Figure 2 the silica concentration versus time graph is superimposed on the fouling curve. The unexpected drop in silica concentration below the saturation level caused a temporary repression of fouling. However, after the saturator was recharged at 98 hours the fouling process continued where it left off. This shows the extreme sensitivity of the fouling process to silica concentration. Overall, the fouling curve has the same shape as TT1-D if the hours between 22 and 98 are ignored, but the magnitude is much greater. After 193 hours the fouling resistance was $14.16 \times 10^{-4}$ (hr.ft$^2$.°F/Btu).

In comparing the results of these five runs the effect of sodium chloride concentration will be examined first. In runs A, B and C, the NaCl concentration varied from 0.0, 0.1 to 1.0 molar, respectively. The linear fouling rates (Table 3) indicate a nonlinear relationship between the NaCl concentration and fouling rate. This is unexpected. One would expect the fouling rate to increase with concentration. One explanation for the higher fouling rate in run B could be the higher pH of 6.8. Since pH is a major variable affecting fouling, it may overshadow the NaCl concentration. Another explanation might be that the tube was not chemically cleaned prior to running. Therefore, silica nucleation sites could have been present prior to the experiment, which would accelerate the fouling process. Run B may have to be repeated because of these factors.

The effect of pH is not completely clear at this time since runs D and DO (pH above 7) had asymptotic fouling rates while those with a pH below 7 had linear fouling rates. However, a comparison of run B to A and C would indicate that a higher pH increases the fouling rate as described earlier.

To demonstrate the effect of Reynolds number, runs D and DO can be compared. If the data between 20 and 100 hours is ignored in run DO, both curves have the same shape. However, the asymptotic fouling resistance in run DO is 466% higher than that of run D (see Table 3). This corresponds to a 32% lower Reynolds number in run DO. Therefore the lower Reynolds number increased the fouling rate in this brine chemistry. This would confirm Wahl's ([8]) conclusion and refute others ([7], [11]).

The asymptotic behavior of runs D and DO is probably a result of the particular chemistry used in these experiments. No other investigators have reported asymptotic fouling in real brines. An explanation, however, can be found in general fouling model Equation (1). The removal term $\phi_r$ could increase with time as the shear strength of the deposit decreases with thickness. The removal rate could also increase as a result of constant mass flow rate which causes a higher shear stress at the wall as the scale thickness increases.

The characteristic shapes of the fouling curves show an induction period for the runs with NaCl and an initial drop in resistance for the runs with NaAc. With the exception of TT1B (the induction period of which is not as clearly defined as others with a spoon shape), experiments C, D and DO show that the induction period is reduced as pH increases, in agreement with the batch type of experiments (15,16). The initial drop in resistance was reported in (11, 12). It is explained as an initial enhancement in heat transfer caused by scale induced turbulence which is eventually overcome by the increasing thermal resistance of the deposit. All curves then proceed with increasing fouling functions, i.e., linear or asymptotic.

## Variation in Local Thermal Fouling Resistance

With thermocouple readings in the wall and coolant ($T_{wi}, T_{ci}$) and the modified Wilson plot results ($R_{oi} + 1/h_{oi}$), it is possible to evaluate local overall heat coefficients, local brine temperatures and, therefore, local thermal fouling resistance at each thermocouple location.

The local thermal fouling resistance is calculated from

$$R_{fi}(t) = 1/U_{of}(t)_{i-1, i+1} - 1/U_{ocl}(t=0)_{i-1, i+1} \quad (8)$$

$U_{of}$ is evaluated from the energy balance.

$$\Delta Q_{i-1,i+1} = (T_w - T_c)_i \pi D_o \Delta x_i / (R_o + 1/h_o)_i \quad (9)$$

$$= U_{of\,i-1,i+1} \Delta T_{LMTD\,i-1,i+1}$$

$$\pi D_o \Delta x_{i-1,i+1} \quad (10)$$

with $\Delta T_{LMTD\,i-1,i+1} =$

$$\frac{(T_{b\,i+1} - T_{c\,i+1}) - (T_{b\,i-1} - T_{c\,i-1})}{\ln[(T_{b\,i+1} - T_{c\,i+1})/(T_{b\,i-1} - T_{c\,i-1})]}$$

The necessary $T_b$'s at thermocouple locations are evaluated as follows. The first $T_{b1}$ is calculated by a linear interpolation using the measured inlet and outlet brine temperatures, because a linear brine temperature profile is expected due to a high brine mass flow rate in the tube and the small $\Delta T$ drop in the coolant side. For other points, a combination of Equation (9) and $dQ = m_b C_{pb} dT_b$ yields

$$T_{bi} - T_{bj} = \frac{D_o \pi}{\dot{m}_b C_{pb}} \int_{x_i}^{x_j} \frac{T_w - T_c}{R_o + 1/h_o} dx$$

from which $T_{b2}$ is calculated using a two point numerical integration. For the third points (j = i+2) and beyond, Simpson's integration rule was used.

Figure 3 shows the transient local thermal fouling resistance at different locations in the heat exchanger for run TT1-D. This has the same shape as the TT1-D overall curve in Figure 2. In comparing the curves, there is a drop in the rate of change and in the asymptotically achieved value of resistance at 148 hours, as heat exchanger length (this refers to distance from the inlet of the heat exchanger) increases. Figure 4 is a graph of the asymptotically fouling resistance at the end versus heat exchanger length. The asymptotic value of resistance is reached faster at the exit than at the inlet locations. For other runs, the local curves have the same shape as the overall curves. For example, for the runs without NaAc (TT1A to TT1C), both local curves and overall curves appear to be linear. From a regression analysis undertaken to obtain the local fouling rate ($dR_{fi}/dt$), it is found that the fouling rate decreases with heat exchanger length. Figure 4 also shows the variation in fouling rate versus heat exchanger length for run TT1-C. This contradicts the findings that fouling was heavier at the exit (11).

Greater fouling at the inlet can be explained physically as follows. The difference between the saturation concentration of silica at the bulk brine temperature and the wall temperature is greatest at the inlet. A previously developed model (13) states that the fouling resistance rate is proportional to the square of this difference. Therefore, the largest amount of scale should form at the inlet provided the chemical reaction at the wall is rapid enough to deposit this excess silica there.

## Pressure Drop Across Heat Exchanger

The pressure drop across the heat exchanger was measured on an hourly basis by subtracting the pressure drop between B and E from the pressure drop between B and A (see Figure 1). The monotonic increase in pressure drop confirms the previous speculation that the surface is roughened by scale, even initially, which in turn induces turbulence and enhances the heat transfer, resulting in a spoon shape of fouling resistance curves. The percentage increases in the pressure drop for all runs are summarized in Table 3.

## Scale Deposition and Composition

A quantitative analysis of the scale was carried out by removing the deposit from the heat transfer surface with a high pressure water jet. The scale was removed from the jet water by passing it through a 20 micron filter. It was then dried and weighed. The mass of deposit per unit area (m) are given in Table 4, except for run A where insufficient scale was collected.

In comparing the mass deposited per surface area and fouling resistance in Equation (5), the density-thermal conductivity product of the scale can be calculated by dividing m by $R_f$ at the end of the experiment (Table 3). The values are all of the same magnitude which would tend to support the validity of Equation (5).

The general physical description of the scale was the same in all runs except A, where the deposit was a sparse dusting of fine white nodules much like those found in the shake down run. Take run TT1-DO, for example, there was no scale upstream from the heat exchanger; it started abruptly at the beginning of the heat transfer surface. The scale had a rippled appearance transverse to the flow direction with high pointed ridges and valleys that left the tube surface exposed. This is the same type of deposit observed by Gudmundsson (11) at Hversgeradi and all of the runs at ORNL (12). The thickness and density of the scale seemed to decrease from inlet to outlet. This was only visually confirmed in run DO when an adapter was available to take photographs with a borescope. This also confirms the drop in local fouling resistance with heat exchanger length caused by a thinner, less dense scale. Upon cleaning with the hydrolaser, the scale was harder to remove towards the inlet. Downstream from the instrumented tube the deposit continued without any change in shape, although the scale at locations where heat transfer was occurring was harder to remove.

The composition of the scales is given in Table 4. The major component of the scale in all runs is silica, in agreement with a survey of brine scales made by Wahl (1). The only time silica was not a major component was in scales from wells that had high calcium or iron content. The zinc found in runs B and C was a result of cleaning hydrolaser using city water. Upon drying the scale sample, the zinc from the pipe lines in the water was left in the scale sample. In runs D and DO the scale sample for analysis was taken prior to cleaning and thus the zinc was gone.

Even with the high sodium and chlorine content of runs B and C little of these constituents was found in the scale. However, their presence in the brine makes a great difference in the fouling rate as seen in Table 3.

## CONCLUDING REMARKS

Synthetic brines have been tested in a high pressure, high temperature, corrosion resistant loop to study silica fouling under different pH, NaCl concentrations, and thermal hydraulic conditions. Although some interesting results have been obtained, they are far from complete and continuation of the systematic measurement is desirable.

## ACKNOWLEDGEMENT

This research was sponsored by the National Science Foundation (NSF MEA-8311788).

## NOTATION

| | |
|---|---|
| A | area, $m^2$ (10.76 $ft^2$) |
| h | heat transfer coefficient, $W/m^2 k$ (0.176 $Btu/ft^2 hr F$) |
| $k_f$ | thermal conductivity of scale, $W/m k$ (0.578 $Btu/hr ft F$) |
| M | water mass (volume) flow rate, $kg/s$ (15.8 gal/min) |
| R | heat transfer resistance, $m^2 k/W$ (5.68 $ft^2 hr F/Btu$). |
| $U_o$ | overall heat transfer coefficient, $W/m^2 k$ (0.176 $Btu/ft^2 hr F$) |
| $x_f$ | fouling deposit thickness, m (3.28 ft) |
| $\rho_f$ | scale density, $kg/m^3$ (0.062 $lbm/ft^3$) |

Subscripts

| | | | | | |
|---|---|---|---|---|---|
| cl | cleaned | i | inside | s | scale |
| f | fouling | o | outside | w | wall |

## LITERATURE CITED

1. Wal, E.F. Geothermal Energy Utilization, John Wiley & Sons, New York (1978).
2. Taborek, J., Aoki, T., Rittes, K.B. and Palen, J.W., Chem. Eng. Prog., 68(2), 59 (1972).
3. Taborek, J., Aoki, T., Rittes, R.E., Palen, J.W. and Knudsen, J.G., Chem. Eng. Prog., 68(7), 69 (1972).
4. McCabe, W.L. and Robinson, C.S., Ind. & Eng. Chem., 16, 478 (1924).

5. Kern, D.Q. and Seaton, R.E., <u>British Chem. Eng.</u>, 4, 258 (1959).
6. Epstein, N., "Fouling: Technical Aspects," Univ. of British Columbia, Vancouver B.C. Canada, (1979).
7. Feisinger, D.E., Mulliner, D.K. and Bishop, H.K., "Geothermal Field Test Heat Exchanger Evaluation," San Diego Gas & Elec., (1973).
8. Wahl, E.F., Yen, I.K. and Bartel, W.J., "Silica Scale Control in Geothermal Brines," Office of Saline Water, U.S. Dept. of Int. Washington D.C. #14-30-3-41, (1974).
9. Lombard, G.L., "Heat Exchanger Tests with Moderately Saline Thermal Brines," San Diego Gas & Elect. Co., (1978).
10. Mines, G.L. and Neill, D.I., "Analysis of Raft River Heat Exchanger Fouling Data Using a Multivariate Regression Technique," EG & G, Inc., Idaho Falls, Idaho (1978).
11. Gudmundsson, J.S. and Bott, T.R., Desalination, 28, 125 (1979).
12. Bohlmann, R.G., Shor, A.J., Berlinski, P. and Mesmer, R.R., "Silica Scaling in Simulated Geothermal Brines," Oak Ridge Nat. Lab., (1981).
13. Chan, S.H., Neusen, K.F., Shadid, J. and Rau, H., "A Silica Deposition Model and a High Pressure, High Temperature Loop Design for Transient Fouling Heat Transfer Measurements of Geothermal Brines," proc. of ASME/JSME Thermal Eng. Joint Conf., Honolulu, 2, 139 (1983).
14. Fisher, P., Suitor, J.W. and Rittes, R.B., <u>Chem. Eng. Prol. Syp. Ser.</u>, 71(7), 66 (1975).
15. Rothbaum, H.P. and Rohde, A.G., <u>J. of Colloid and Interface Sci.</u>, 71(3), 533 (1979).
16. Weres, O., Yee, A. and Tsao, Leon, "Kinetics of Silica Polymerization," Lawrence Berkeley Lab., LBL-7033, UC-4, Berkeley, Calif. (1980).

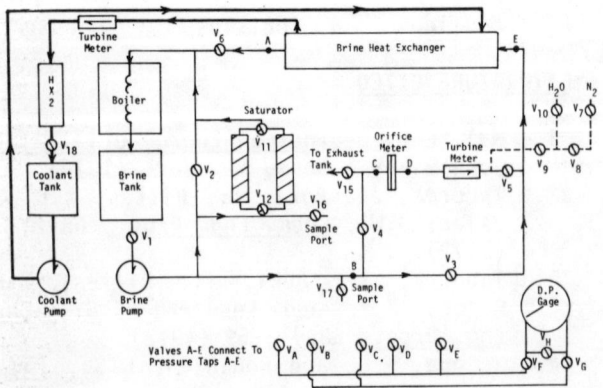

Figure 1 Schematic of Geothermal Fouling Loop

TABLE 1 - Thermal Hydraulic Conditions

| Run # | $Re_b$ | $T_b$ in F[1] | $T_b$ out[2] | $T_c$ in | $T_c$ out[2] | $\dot{M}_b$ gal/min. | $\dot{M}_c$ | $U_o(t=0)$ | $t_{tp}$ hr |
|---|---|---|---|---|---|---|---|---|---|
| TT1-A | 73,130 | 330.0 | 254.5 | 160.7 | 162.6 | 4.15 | 169.5 | 478.2 | 150 |
| TT1-B | 77,500 | 327.8 | 254.1 | 159.6 | 161.6 | 4.30 | 168.5 | 477.1 | 133 |
| TT1-C | 75,790 | 329.3 | 256.0 | 160.4 | 162.3 | 4.21 | 167.8 | 460.5 | 193 |
| TT1-D | 76,150 | 328.7 | 256.6 | 161.4 | 163.3 | 4.23 | 168.9 | 459.3 | 148 |
| TT1-D0 | 51,920 | 330.8 | 241.5 | 159.3 | 161.4 | 3.00 | 110.0 | 397.5 | 193 |

1. Subscripts: b = brine; c = coolant; 2. At beginning of run; 3. $Re_b$ based on $D_i$, $T_b$ and water properties. $U_o$ in Btu/hr ft$^2$F; $t_{tp}$ = test period.

TABLE 2 - Chemical Composition in Fouling Experiments

| Run # | $SiO_2$[1] | NaCl[2] | NaAc | pH | TDS[3] |
|---|---|---|---|---|---|
| TT1-A | 0.0111 (670) | 0 | 0 | 5.9 | 670 |
| TT1-B | 0.0098 (590) | 0.1 (58500) | 0 | 6.8 | 6440 |
| TT1-C | 0.0095 (570) | 1.0 (58500) | 0 | 5.8 | 59070 |
| TT1-D | 0.0109 (660) | 0 | 0.1 (13610) | 7.3 | 14250 |
| TT1-D0 | 0.0113 (680) | 0 | 0.1 (13610) | 7.4 | 14290 |

1. Saturation concentration. 2. Units of molarity and (ppm). 3. Total dissolved solids ppm.

TABLE 3 - Comparison of Fouling Behavior

| Run # | Characteristic Fouling Behavior and Value | Percent Increase in Δp Across Heat Exchanger | $\rho_f k_f$ Btu.lb hr.ft$^4$.hr |
|---|---|---|---|
| TT1-A | Linear 0.227[1] | 0 | --- |
| TT1-B | Linear 3.489[1] | 200 | 19.52 |
| TT1-C | Linear 1.314[1] | 64 | 12.79 |
| TT1-D | Asymptotic 2.50[2] | 193 | 54.24 |
| TT1-D0 | Asymptotic 14.16[2] | 317 | 24.78 |

[1] Linear Fouling Rate (hr.ft$^2$.F/Btu/hr) x $10^{-6}$
[2] Fouling Resistance at End of Run (hr.ft$^2$.F/Btu) x $10^4$

TABLE 4 - Scale Deposition and Composition

| Run # | TT1-A | TT1-B | TT1-C | TT1-D | TT1-D0 |
|---|---|---|---|---|---|
| M x $10^3$ | Neg. | 4.763 | 3.736 | 13.56 | 35.09 |
| $\dot{M}$ x $10^5$ | --- | 4.215 | 1.936 | 9.162 | 28.53[2] |
| $SiO_2$[1] | --- | 87.6 | 82.3 | 84.9 | 97.1 |
| $Na_2O$[1] | --- | --- | --- | 2.3 | 1.3 |
| Cl[1] | --- | --- | <1 | --- | --- |
| CaO[1] | --- | <1 | --- | 1.7 | --- |
| FeO[1] | --- | 4.7 | 9.3 | 5.6 | 1.6 |
| ZnO[1] | --- | 6.2 | 6.3 | 5.6 | --- |

[1] Composition in weight percent of total sample
[2] Time =.(193 hr) - (70 hr of zero fouling)
M and $\dot{M}$ in lb/ft$^2$ and lb/ft$^2$hr

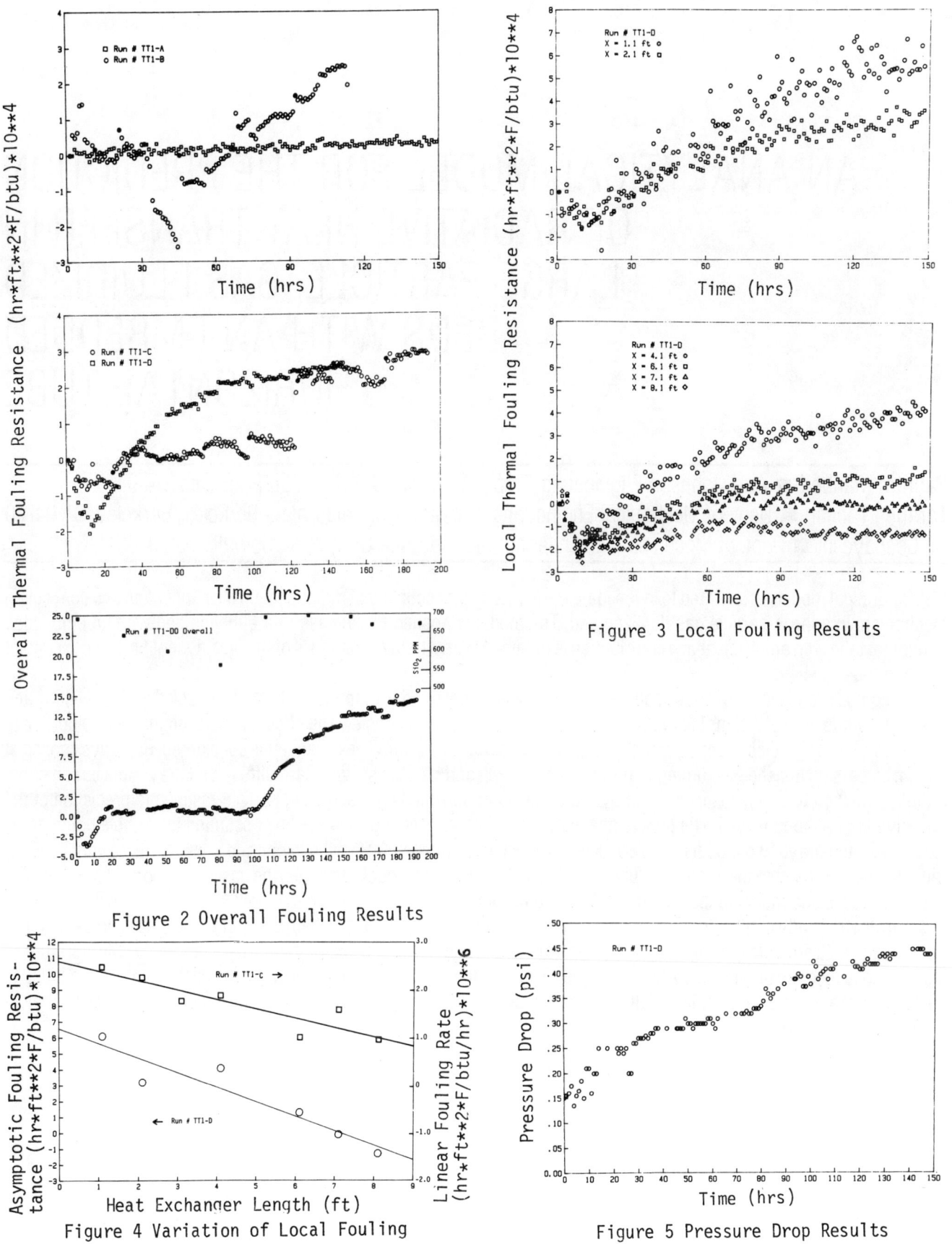

Figure 2 Overall Fouling Results

Figure 3 Local Fouling Results

Figure 4 Variation of Local Fouling

Figure 5 Pressure Drop Results

# AN ANALYTICAL MODEL FOR THE PREDICTION OF RADIATIVE HEAT TRANSFER IN LARGE-PARTICLE, GAS FLUIDIZED BEDS WITH AN EMBEDDED HORIZONTAL TUBE

Bahram Mahbod ■ Axel Johnson Engineering Corp., 666 Howard Street, San Francisco, CA 94105
Taraneh Tabesh ■ Dept. of Mechanical Engineering, University of California—Berkeley, Berkeley, CA 94720
Ali Goshayeshi ■ Dept. of Mechanical Engineering, Oregon State University, Corvallis, OR 97331

A theoretical model is presented to investigate the interaction or coupling of radiation with the conduction and convection mechanism in a large-particle gas fluidized bed. The model is applied to the case of a bubbling bed with an embedded horizontal tube. The emulsion phase contribution is obtained from a detailed analysis of the first layer of particles in contact with the tube.

The method proposed here is based on the single particle model of Adams and Welty [10] with coupled gas convection and unsteady conduction of Adams [11] with addition of the radiative heat transfer. The effect of microroughness of a portion of the particle surface in close contact with the heat transfer surface on overall heat transfer is considered. Radiative cooling of the particles is established assuming that the bed, tube wall and particle surfaces are diffuse and gray. Application of net radiative method to an enclosure formed by a gray cylinder surrounding a spherical particle resulted in prediction of the radiative flux to the tube wall. The bubble phase radiative heat flux is approximated by assuming the bubble boundary is isothermal and gray. Two models were considered to establish an estimate of the lower and upper limits of bubble phase radiation heat transfer to the immersed tube. The first of the foregoing models assumes the gas within the bubble is optically thin and the second model takes gas absorption into account.

A parametric study of the effects of particle thermal and physical properties on heat transfer to the tube is presented. The numerical results for both total and radiative heat transfer are compared with experimental data reported in the literature as well as those recently obtained by Alavizadeh [29]. The heat transfer coefficients calculated using the model were found to be within the range of experimental results obtained by others. The model is expected to be valid for mean particle diameters greater than 2.00 mm.

# A SINGLE CHAR PARTICLE COMBUSTION MODEL OF $SO_2$ RETENTION BY LIGNITE ASH DURING FLUIDIZED BED COMBUSTION

F.W. Cox, W.E. Genetti and Y.Y. Lee ■ Chemical Engineering Department
University of Mississippi, University, Mississippi 38677

A pseudo steady-state model was developed for combustion of a porous single char particle that included both boundary layer and intraparticle reaction and diffusion characteristics. The model includes the energy equation as well as four species equations. The four gas species represented by the model are $O_2$, $CO$, $CO_2$ and $SO_2$. The $SO_2$ species equation allows for the sulfur to exist as $H_2S$, and $COS$ in the char pores when a reducing environment exists. The sulfur gases are captured by $CaO$ in the ash through the following reactions:

$$CaO + SO_2 + 1/2\ O_2 \rightarrow CaSO_4$$
$$CaO + H_2S \rightarrow CaS + H_2O$$
$$CaO + COS \rightarrow CaS + CO_2$$

The results show that lignite char particles can exist in a fluidized bed combustor either in an ignited state or an unignited state depending on the particles radius, bed temperature, and oxygen concentration. The model reveals trends for the effects of bed temperature and oxygen concentration for both ignited and unignited particles.

Devolatilization occurs first in the combustion of coal. The volatiles consist of $H_2$, $CO$, $H_2S$, $CH_4$, $CO_2$ and vaporized tars. Since devolatilization occurs in about one-half to one second for a two millimeter particle in a fluidized bed, it is believed that these gases undergo combustion away from the char particle.

After devolatilization the primary forms of sulfur in the char should be ferrous sulfide and organic sulfur. The sulfur release during this stage of combustion can be represented by the following reactions:

$FeS + 1.5\ O_2 \rightarrow FeO + SO_2$    S1
$Organic\text{-}S + O_2 \rightarrow CO + SO_2$    S2
$Organic\text{-}S + CO_2 \rightarrow CO + COS$    S3

Burning of $FeS$ forms $FeO$ and $FeSO_4$ at 400°C. As the temperature is increased less $FeSO_4$ is formed. At 700°C and higher very little $FeSO_4$ is formed and the reaction proceeds as in reaction S1, Schwab and Philiais ([1]). Ferrous sulfide is very stable in a reducing environment and releases very little of its sulfur in practical temperature ranges, Attar ([2]).

Organic sulfur forms $SO_2$ and $CO$ when it is reacted with $O_2$ as in reaction S2. In a reducing environment the organic sulfur may react with $CO_2$ to form carbonyl sulfide. Hydrogen sulfide and carbonyl sulfide can be related by the following reversible reaction:

$CO + H_2S \rightleftharpoons COS + H_2$

The reactions of the basic minerals with $H_2S$, carbonyl sulfide and sulfur dioxide are the major reasons for retention of sulfur in char during combustion. $CaO$ and $MgO$ react with these gases and retain the sulfur in the forms of sulfide or sulfate. The major retention reactions are:

$MO + H_2S \longrightarrow MS + H_2O$    C1
$MO + COS \longrightarrow MS + CO_2$    C2
$MO + SO_2 + 1/2\ O_2 \longrightarrow MSO_4$    C3

M is the alkali metals Ca or Mg. Reations C1 and C2 have rate constants at least ten times smaller than C3. All three of the reaction rates are first-order with respect to sulfur species.

## A SINGLE PARTICLE CHAR COMBUSTION MODEL

A pseudo steady-state model was developed for combustion of a porous single char particle that included both boundary layer and intraparticle reaction and diffusion characteristics. The model consists of a spherical char particle, with conical pores, surrounded by a boundary layer. The boundary layer thickness is determined by using the Nusselt Number calculated for a sphere of diameter, $d_p$, with fluid flowing at a relative velocity, $U_0$, around it. The expression given by Ranz and Marshall ([3]) for the Nusselt Number is:

$$Nu_p = 2 + 0.6\ Re^{1/2}\ Pr^{1/3} \qquad (1)$$

The distance $r_f$ from the center of the particle to the outer edge of the boundary layer is calculated from Equation (2)

$$r_f = \frac{-Nu_p (r_p)}{(2 - Nu_p)} \qquad (2)$$

This equation is calculated by assuming an effective stagnant boundary layer for a given Nusselt Number.

The combustion reactions used in the model are:

$C + 1/2\ O_2 \longrightarrow CO$  Exothermic
$CO + 1/2\ O_2 \longrightarrow CO_2$  Exothermic
$CO_2 + C \longrightarrow 2CO$  Endothermic

The char carbon is oxidized to CO through the first reaction for a char particle that is reaction rate controlled. The CO is oxidized to $CO_2$ by the second reaction. These reactions are exothermic. In a reducing environment the char carbon may "burn" in a carbon dioxide atmosphere by the Boudouard reaction which is endothermic.

Sulfur is released in the model by oxidation of ferrous sulfide and organic sulfur with $O_2$ to form $SO_2$. The organic sulfur is also released in the model by reaction with $CO_2$ to form carbonyl sulfide. These reactions are:

$FeS + 3/2\ O_2 \longrightarrow SO_2 + FeO$
$(Organic)S + O_2 \longrightarrow CO + SO_2$
$(Organic)S + CO_2 \longrightarrow CO + COS$

In a reducing environment which may exist in the pores, the sulfur containing gases will exist as $H_2S$ or COS. In an oxidizing environment the sulfur containing gas will be $SO_2$.

Sulfur is captured in the model by the three reactions:

$SO_2 + 1/2\ O_2 + CaO \longrightarrow CaSO_4$
$H_2S + CaO \longrightarrow H_2O + CaS$
$COS + CaO \longrightarrow CO_2 + CaS$

Sulfur dioxide can react with CaO to form $CaSO_4$, $H_2S$ can react with CaO to form CaS, and carbonyl sulfide may react with CaO to form CaS. In the analysis the assumption is made that all $CaCO_3$ initially in the ash has been calcined to CaO during pyrolysis. This is a good assumption for particle temperatures greater than 800°C.

The char model differential equations are developed by mass and energy balances about a spherical shell of thickness dr. The resulting steady-state differential equations are:

### Energy Equation

$$R_h (-\Delta H_r)\ \theta r^2 + R_s (-\Delta H_s)\ 2\ \theta r\ \frac{\sqrt{1 + \alpha^2}}{\alpha}$$
$$= \frac{\partial}{\partial r}\left(-k\ r^2 \frac{\partial T}{\partial r} + \sum_{i=1}^{4} N_i\ \bar{C}_p\ (T - T_R^\circ)\ r^2\right)$$

### Species Equations

$$R_h\ r^2 + R_s\ 2\ \frac{\sqrt{1 + \alpha^2}}{\alpha}\ r =$$
$$-\frac{\partial}{\partial r}\left(r^2\ D_i\ \frac{\partial C_i}{\partial r}\right) + \frac{\partial}{\partial r}\left(r^2\ C_i\ \bar{U}\right)$$

where:  i  species
1  $O_2$
2  $CO_2$
3  CO
4  $SO_2$

There are four species equations; that is, one for each species. The boundary conditions are the known free stream temperature and species compositions at the outer edge of the boundary layer. At the center of the particle the temperature and concentration gradients are zero. Also radiant energy is lost at the particle surface.

The discretization equation for this system for node J is derived from the differential equations using a control volume finite-difference method. The discretization equation developed in this manner expresses the conservation of energy or mass species for the finite control volume, just as the differential equation expressed it for an infinitesimal control volume, Patankar (4). Figure 1 shows the symbols representation of the finite control volume.

### Energy Nodal Equation

$$a_J T_J = a_{J+1}\ T_{J+1} + a_{J-1}\ T_{J-1} + b$$

$$a_{J+1} = k\ \frac{r_{J+1}^2}{\Delta r_{J+1}} - \sum_{i=1}^{4} N_i\ \bar{C}_p\ \left(1 - \frac{T_o}{T_{J+1}}\right) r_{J+1}^2$$

$$a_{J-1} = k\ \frac{r_J^2}{\Delta r_J}$$

$$a_j = k \frac{r_{J+1}^2}{\Delta r_{J+1}} + k\frac{r_J^2}{\Delta r_J} - \sum_{i=1}^{4} N_i \overline{C}_p (1 - \frac{T^\circ}{T_J}) r_J^2$$

$$b = R_h(-\Delta H_r)\theta \left(\frac{r_{J+1}^3 - r_J^3}{3}\right)$$

$$+ R_s(-\Delta H_s)\theta \frac{\sqrt{1 + \alpha^2}}{\alpha} (r_{J+1}^2 - r_J^2)$$

## Species Nodal Equation

$$a_{i,J} C_{i,J} = a_{i,J+1} C_{i,J+1} + a_{i,J-1} C_{i,J-1} + b_i$$

$$a_{i,J+1} = D_{i_{J+1}} \frac{r_{J+1}^2}{\Delta r_{J+1}}$$

$$a_{i,J-1} = D_{i_J} \frac{r_J^2}{\Delta r_J} + \overline{U}_J r_J^2$$

$$a_{i,J} = D_{i_{J+1}} \frac{r_{J+1}^2}{\Delta r_{J+1}} + D_{i_J} \frac{r_J^2}{\Delta r_J} + \overline{U}_{J+1} r_{J+1}^2$$

$$b_i = R_h \left(\frac{r_{J+1}^3 - r_J^3}{3}\right) + R_s \frac{\sqrt{1 + \alpha^2}}{\alpha} (r_{J+1}^2 - r_J^2)$$

With a 50 node system there will be 50 energy discretization equations and 50 species discretization equations for each of the four species for a total of 250 equations and 250 unknowns.

The reaction rate expressions used are shown on Table 1. Undoubtedly, the kinetics and mechanisms are more complex than assumed; however, similar assumptions have been used by Sotirchos and Amundson (5). The sulfur release reaction kinetics are assumed to be proportional to Reaction 2 and 3 as shown in Table 1. The proportionality constants $\alpha$ and $\beta$ are the moles of ferrous sulfide per mole of carbon and moles of organic sulfur per mole of carbon, respectively. The sulfur capture reaction kinetics are taken from the literature as shown on Table 1. Since carbonyl sulfide capture kinetics is first order with respect to the sulfur species and the rate constant is very close to that for $H_2S$ capture, it was assumed that the sulfur capture kinetics of Reaction 8 and 9 are the same. The concentration of COS and $H_2S$ are lumped together. Since the development of the mathematical model is independent of the reaction rate expressions used, these parameters and expressions could be easily changed in the model.

The char data, physical data and bulk stream parameters used to represent lignite are given in Table 2.

## MODEL RESULTS

The model results show that lignite char particles can exist in a fluidized bed combustor either in an ignited state or an unignited state depending on the particle radius, bed temperature and oxygen concentration. Figure 2 shows particle temperature versus bed temperature for a 1 mm particle with the bulk stream $O_2$, $CO_2$ and CO at 13, 3.6 and 7.2 mole percent respectively. It should be noted that the particle temperature is only slightly higher than the bed temperature for bed temperature below 1100°K. When the bed and particle temperatures are only slightly different the particle is considered unignited (in the boundary layer). At bed temperatures greater than 1120°K, the particle temperature is much greater than the bed temperature. This condition exists in the ignited particle range. Multiple solutions exist between these states.

The transition from unignited to ignited occurs at the higher bed temperatures for smaller particles and at lower bed temperatures if humid air is used for combustion. Other workers have shown that particles undergo ignition for initial bed temperatures higher than a critical value and extinction after some conversion level, Sotirchos and Amundson (5).

Figure 3 shows the concentration profiles for an unignited particle. This Figure shows that our definition of an unignited particle actually means a particle that is unignited in the boundary layer. The $O_2$, $CO_2$ and CO profiles are relatively flat in the boundary layer, but strong concentrations gradients exist in the particles. This indicates the combustion reaction are occurring within the particle. The $SO_2$ concentration peaks at the particle surface at about 700 ppm.

Figure 4 shows the concentration profile for an ignited particle. This Figure indicates combustion of CO occurring in the gas film. Both the $CO_2$ concentrations and the temperature peak in the gas film surrounding the particle. The $SO_2$ profile peaks at the particle surface at about 3800 ppm. This is a much higher $SO_2$ concentration than that calculated for the unignited particle.

Figure 5 shows the fraction of the sulfur capture versus bed temperature for the same particle size and bulk steam conditions as shown on Figure 2. A decrease in fraction of sulfur capture as bed temperature is increased is shown for the unignited state. A slight increase in the fraction of sulfur capture as bed temperature is increased is shown for the ignited state.

Figure 6 shows the fraction of sulfur captured versus oxygen composition for a 3 mm particle at a bed temperature of 1000°K. In the unignited state, sulfur capture is calculated to be almost independent of $O_2$ bulk stream concentration. However, sulfur capture is calculated to increase with the $O_2$ bulk stream composition in the ignited state.

Figure 7 shows the effect of particle size on sulfur capture. Sulfur capture is shown to increase with particle radius. As the particle radius is increased the particle temperature decreases if the bed temperature is constant.

## CONCLUSIONS

1. The char model indicates that particles go from an unignited state to an ignited state in the normal operating range of fluidized bed combustors.

2. The temperature at which the particle changes from an unignited state to an ignited state increases with a decrease in particle size.

3. For unignited particles:

   a. The fraction of sulfur captured decreases as the bed temperature is increased.

   b. The fraction of sulfur captured is almost independent of bulk stream $O_2$ concentration.

4. For ignited particles:

   a. The fraction of sulfur captured increases as the bulk stream $O_2$ concentration is increased.

   b. The fraction of sulfur captured increases as the bed temperature is increased.

5. Sulfur capture increases with particle size.

## NOTATION

| | |
|---|---|
| $a_{i,J}$ | Coefficient in the Species Nodal Equation, $L^3 t$ |
| $a_J$ | Coefficient in the Energy Nodal Equation, $Q/tT$ |
| $b, b_i$ | Parameters in the Energy and Species Nodal Equations, $Q/tT$ and $M/t$ |
| $C_i$ | Concentration of gas species i, $M/L^3$ |
| $\overline{C_p}$ | Mean heat capacity, $Q/MT$ |
| $D_i$ | Diffusivity of gas species i, $L^2/t$ |
| $\Delta H_h$ | Heat of reaction for homogeneous reactions, $Q/M$ |
| $\Delta H_s$ | Heat of reaction for heterogeneous reactions, $Q/M$ |
| J | Nodal index, dimensionless |
| k | Thermal conductivity, $Q/LtT$ |
| K | Reaction rate constant |
| $N_i$ | Molar flux of species i, $M/L^2 t$ |
| $Nu_p$ | Particle Nusselt Number, dimensionless |
| Pr | Prandl Number, dimensionless |
| r | Radial position, L |
| $r_f$ | Radial position to outer edge of the boundary layer, L |
| $r_J$ | Radial position to the $J^{th}$ Nodal position, L |
| $r_p$ | Pore radius, L |
| $r_c$ | Char particle radius, L |
| $\Delta r_J$ | Nodal radial finite difference, L |
| $R_h$ | Homogeneous reaction rates, $M/L^3 t$ |
| $R_s$ | Heterogeneous reaction rates, $M/L^2 t$ |
| Re | Reynolds Number, dimensionless |
| T | Temperature, T |
| $T_R^{\circ}$ | Reference temperature, T |
| U | Convective velocity in the boundary layer, $L/t$ |

### Greek Letters

| | |
|---|---|
| $\alpha$ | $r_p/r_c$, dimensionless |
| $\theta$ | porosity, dimensionless |

## LITERATURE CITED

1. Schwab, G.M. and J.J. Philiais, *American Chemical Society*, Vol. 69, p. 2588 (1974).

2. Attar, A., *Fuel*, Vol. 57, (April 1978).

3. Ranz, W. and W. Marchall, *Chemical Engineering Progress*, 48, p. 141 (1952).

4. Patankar, S.V., *Numerical Heat Transfer And Fluid Flow*, Hemisphere Publishing Corporation, New York, NY (1980).

5. Sotirchos, S.V. and N.R. Amundson, *Ind. Engr. Chem. Fund.*, Vol. 23, p. 191 (1984).

6. Howard, J.B., Williams G.C. and D.H. Fine, 14th International Sympos. on Combus., The Combustion Institute, Pittsburgh, PA (1973)

7. Smith, I.W. and R.J. Tyler, <u>Fuel</u>, Vol. 51, (1972).

8. Dutta, S., Wen, C.Y. and R.J. Belt, Industrial Engineering Chemistry Process Design and Development, (1977).

9. Wen, C.Y. and M. Ishida, <u>Environmental Science and Technology</u>, Vol. 7, No. 8, (1973).

10. Westmoreland, P.R., Gibson, J.B. and D.P. Harrison, <u>Environmental Science and Technology</u>, Vol. 11, No. 5 (1977).

11. Yang, R.T. and J.M. Chen, <u>Environmental Science and Technology</u>, Vol. 13, No. 5 (1979).

Table 1. Reaction and Rate Equations

Reaction

$CO + 1/2\ O_2 \rightarrow CO_2$     Exothermic (Howard, (6))

$R_1 = k_1 \exp(-\frac{15,098}{T})\ C_{CO}\ C_{O_2}^{1/2}$

   K-mol $CO/m^3 \cdot sec$

$k_1 = 3.04 \times 10^9\ (\frac{m^{3/2}}{k\ mol^{1/2}\ sec})$

$-\Delta H_c(298) = 5.65 \times 10^8\ J/K\ mol\ O_2$

$C + 1/2\ O_2 \rightarrow CO$     Exothermic (Smith & Tyler (7))

$R_2 = k_2\ T \exp(-\frac{20,131}{T}) C_{O_2}$

   K-mol $C/m^2 \cdot sec$

$k_2 = 3.76\ (\frac{m}{°K\ sec})$

$-\Delta H_c(298) = 2.21 \times 10^8\ J/k\ mol\ O_2$

$C + CO_2 \rightarrow 2CO$     Endothermic  (Dutta, (8))

$R_3 = k_3 \exp(-\frac{29,790}{T})\ C_{CO_2}$

   K-mol $C/m^2 \cdot sec$

$k_3 = 1.47 \times 10^5\ (\frac{m}{sec})$

$-\Delta H_2(298) = -1.725 \times 10^8\ J/k\ mol\ CO_2$

$FeS + 3/2\ O_2 \rightarrow FeO + SO_2$     $R_4 = \alpha R_2$

$(Organic)S + O_2 \rightarrow SO_2 + CO$     $R_5 = \beta R_2$

$(Organic)S + CO_2 \rightarrow COS + CO$     $R_6 = \beta R_3$

$SO_2 + CaO + 1/2\ O_2 \rightarrow CaSO_4$     (Wen, (9))

$R_7 = k_7 \exp(-\frac{8807}{T})\ C_{SO_2}$

   K-mol $SO_2/m^2 \cdot sec$

$k_7 = 2.386\ m/sec$

$H_2S + CaO \rightarrow CaS + H_2O$     (Westmoreland, (10))

$R_8 = k_8 \exp(-\frac{2597}{T})\ C_{H_2S}$

   K-mol $H_2S/m^2 \cdot sec$

$k_8 = 3.732 \times 10^{-4}\ (m/sec)$

$COS + CaO \rightarrow CaS + CO_2$     (Yang, (11))

$R_9 = R_8$

Table 2. Char Data, Physical Data, and Bulk Stream Parameters

Porosity = 0.35

BET Area = $2.1 \times 10^5\ m^2/Kg$

Density = 900 $Kg/m^3$

Ash Content = 0.5 wt fraction

Emissivity = 0.9

Average Heat Capacity of Gas = 39.77 (KJ/Kg - mol °K)

Thermal Conductivity of Char = 0.157 (J/m sec °K)

Thermal Conductivity of Gas = $2.627 \times 10^{-4} \times T^{.8}$ (J/m sec K)

Diffusivity of $O_2$, CO in Gas Film = $5.123 \times 10^{-9} \times T^{1.5}(m^2/sec)$

Diffusivity of $CO_2$ in Gas Film = $4.211 \times 10^{-9} \times T^{1.5}(m^2/sec)$

Diffusivity of $SO_2$ in Gas Film = $3.538 \times 10^{-9} \times T^{1.5}(m^2/sec)$

Knudson Diffusion in Char Pores = $9700\ r_p\sqrt{\frac{T}{M}}\ (m^2/sec)$

   $r_p$: pore radius in cm; T: °K; M: Mol wt of gas

Superficial Gas Velocity = 2 m/sec

Pressure = 1 atmosphere

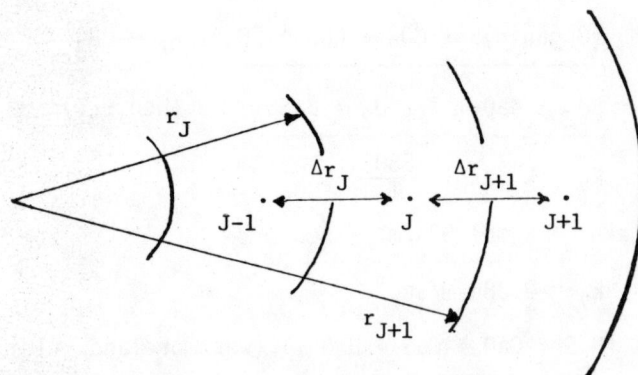

Figure 1. Finite control volume.

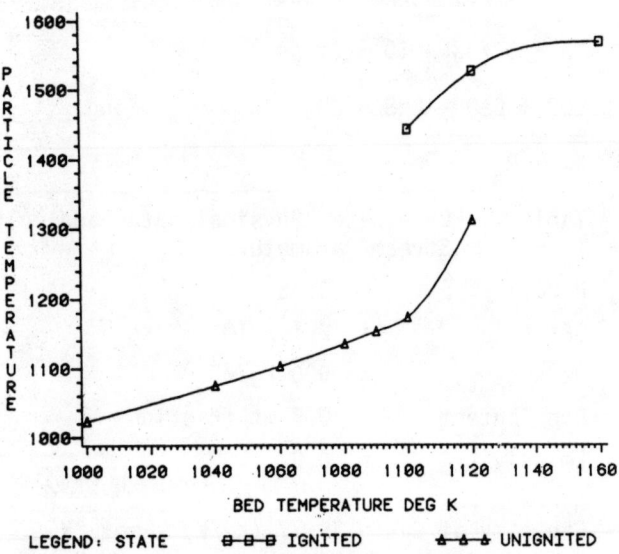

Figure 2. Sulfur capture vs. bed temperature.

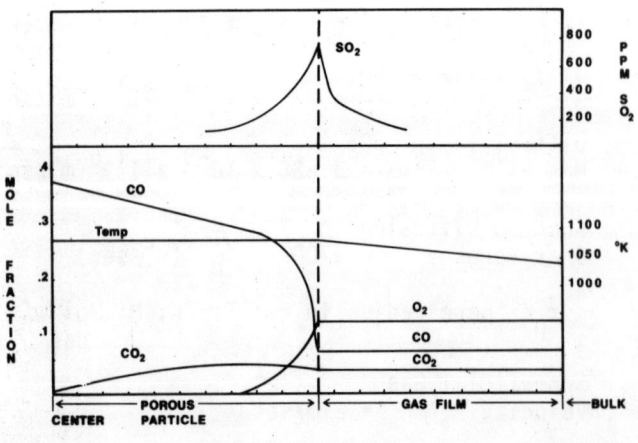

Figure 3. Unignited char particle concentration profile.

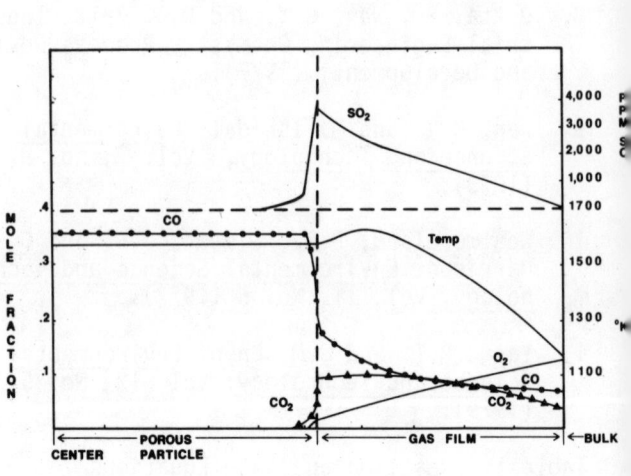

Figure 4. Ignited char particle concentration profile.

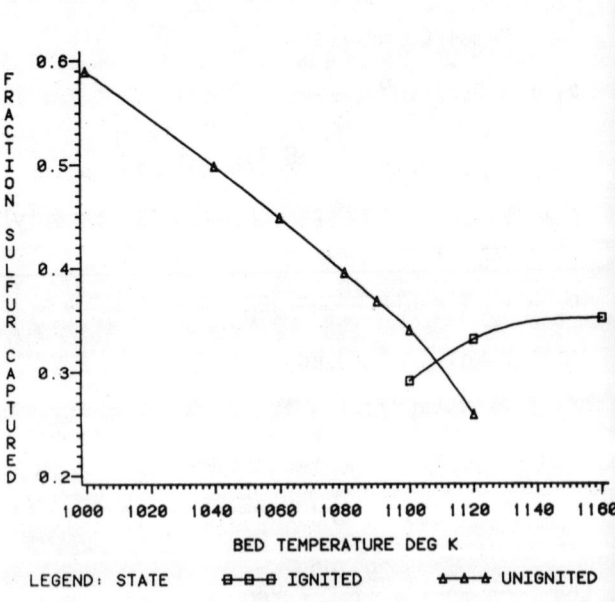

Figure 5. Sulfur capture vs. bed temperature.

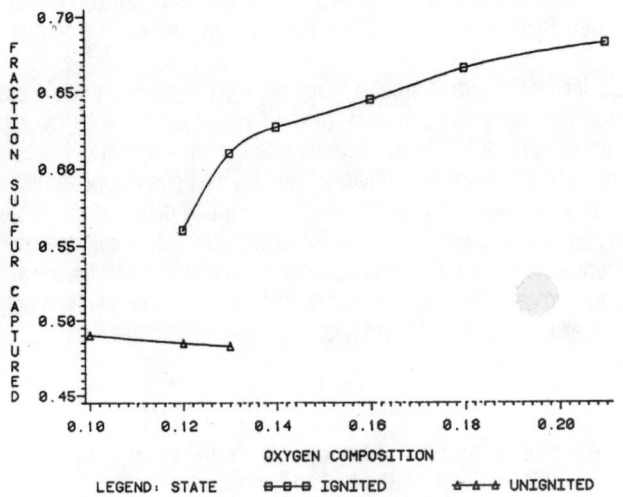

Figure 6. Sulfur capture vs. oxygen composition.

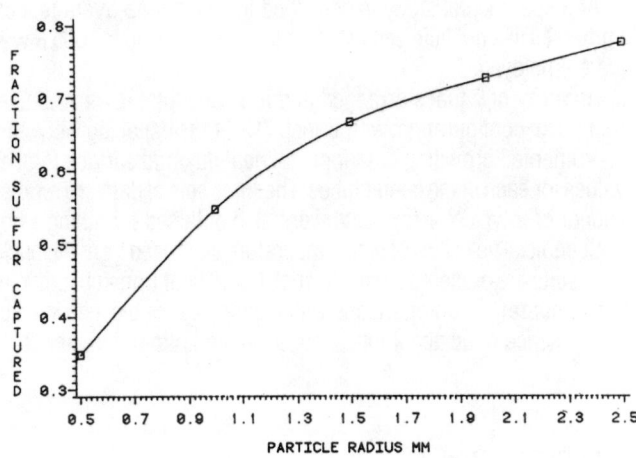

Figure 7. Sulfur capture vs. particle radius.

# LOCAL HEAT TRANSFER COEFFICIENTS FOR HORIZONTAL TUBE ARRAYS IN HIGH TEMPERATURE LARGE PARTICLE FLUIDIZED BEDS—AN EXPERIMENTAL STUDY

A. Goshayeshi, J.R. Welty, R.L. Adams and N. Alavizadeh ■ Department of Mechanical Engineering
Oregon State University, Corvallis, OR 97331

An experimental study is described in which time-average local heat transfer coefficients were obtained for arrays of horizontal tubes immersed in a hot fluidized bed. Bed temperatures up to 1005 K were achieved. Bed particle sizes of 2.14 mm and 3.23 mm nominal diameter were employed.

An array of 9 tubes arranged in three horizontal rows was used. The 2 inch (50.8 mm) diameter tubes were arranged in an equilateral triangular configuration with 6 inch (15.24 cm) spacing between centers. The center tube in each of the three rows in the array was instrumented providing data for local heat flux and surface temperature at intervals of 30° from the bottom to the top—a total of 7 sets of values for each of the center tubes. The three sets of data are representative of the heat transfer behavior of tubes at the bottom, top, and in the interior of a typical array. Data were also obtained for a single horizontal tube to compare with the results of a tube-bundle performance.

Superficial velocities of high temperature air ranged from the packed-bed condition through approximately twice the minimum fluidized level.

Results are presented in the forms of local heat transfer coefficients as functions of angular position, superficial velocity, particle size, and bed temperature. Comparisons with results for a single tube in a bubbling bed indicate only slight effects on local heat transfer resulting from the presence of adjacent tubes. Tubes in the bottom, top, and interior rows also exhibited different heat transfer performance.

## INTRODUCTION

Gas-fluidized beds are noted for their excellent heat transfer characteristics. It is generally acknowledged that high rates of heat transfer can be achieved between the bed and immersed surfaces or vessel walls. Fluidized bed combustion of coal in particular, is receiving attention of scientists and engineers throughout the world, especially in the United States.

Fluidized-bed combustion offers great potential for the utilization of all ranks of coal to meet air standards. Crushed coal is usually burned in a bed of dolomite or limestone in which bed particles are held in suspension by upward flowing air. The inert material reacts with the sulfur dioxide produced during combustion of high-sulfur coal forming a dry solid disposable waste. The heat produced in the bed is transferred to heat exchange tubes (which are usually horizontal) for steam generation in electric utility applications and for liquid or gas heating in process industries.

A large number of investigations reported in the literature have emphasized spatial-averaged heat transfer results. Local time-averaged heat transfer coefficients with immersed tubes provide additional information of importance to the combustor designer. A number of recent investigations (1 to 9) have reported local heat transfer results for cases involving small ($d_p$ < 1 mm) or large (> 1 mm) particles and cold or hot bed conditions. Most of the previous investigations [reviewed by Saxena, et al. (10)] were restricted to small particles and low bed temperatures where tubes were electrically heated and local heat fluxes were evaluated by measuring the power input along with measurements of tube wall temperatures for heat transfer coefficient calculations.

In the present work, tubes were instrumented by thin thermopile-type transducers for direct measurements of local heat fluxes at hot bed conditions. Also, lack of information of local heat transfer coefficients on tubes at different location in the array was the motivation for the study.

For the purpose of tube array design, the following considerations were made: In an array of horizontal tubes, it is known that the heat transfer to any tube, not affected by side walls [it is known that the solids motion is different for centrally located tubes and those adjacent to a side wall (11)], is affected only by the presence of the neighboring tubes. It is also known that the minimum spacing between tubes should not fall much below a pitch/diameter ratio of 2, to prevent severe restriction of particle motion and a sharp decrease in heat transfer (12,13). Typical heat exchanger designs employ tubes with diameters in the range 50.8 mm (2 inch) arranged horizontally (14).

To fulfill all of the above design aspects, an array of 9 tubes arranged in three horizontal rows, was used. The 50.8 mm diameter tubes were positioned in an equilateral triangular configuration with 15.24 cm (6 inch) spacing between centers. The center tube in each of the three rows was instrumented providing data for local heat fluxes and surface temperatures at intervals of 30° from the bottom to the top - a total of 7 sets of values for each of the center tubes. The three sets of data are representative of the heat transfer behavior of tubes at the bottom, top and the interior of a typical array. The particular equilateral triangular tube array was selected because this configuration seems to give better fluidization characteristics, and hence better heat transfer, than an in-line (square pitch) array (15). This tube arrangement is the most common in current industrial practice.

Data were also obtained for a single horizontal tube to compare with the results of tube bundle performance.

EXPERIMENTAL APPARATUS

Figure 1 shows a schematic illustration of an instrumented tube. Three tubes were instrumented by mounting 7 thermopile-type heat flux transducers side-by-side around half of each tube periphery (30° apart). The Micro-Foil Heat Flow Sensors, with copper-constantan surface temperature thermocouples, were bonded to the tubes with relatively high-temperature thermal conducting epoxy and were tightly covered by 0.127 mm (0.005 inch) thick stainless-steel shim stock.

Measurements were conducted in the Oregon State University high-temperature fluidized-bed facility. A schematic of the assembly is shown in Figure 2.

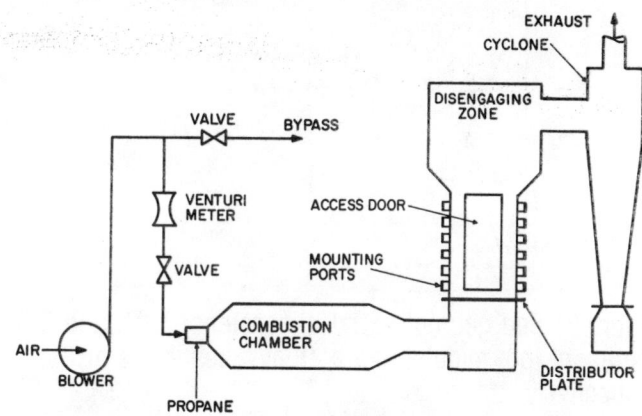

Figure 2. Schematic illustration of the Oregon State University high temperature fluidized bed facility.

Propane was burned in a refractory lined combustion chamber and the hot combustion gases directed into the 0.30 m x 0.60 m (1 ft x 2 ft) test section through a distributor plate. The tube array was positioned horizontally within the bed. A proportional type controller was used to regulate the propane flow rate and maintain the desired gas temperature.

A digital data acquisition system was used to record local heat fluxes, surface temperatures and computed local heat transfer coefficients at each position on the instrumented tubes.

EXPERIMENTS

Experiments were conducted at bed temperatures of 810 K (1000°F), 922 K (1200 °F) and 1005 K (1350°F). A granular refractory (Ione Grain) was used as bed material. Bed particle sizes of 2.14 mm and 3.23 mm nominal diameter were employed. The size distribution of this material remained stable after many hours of bed operation.

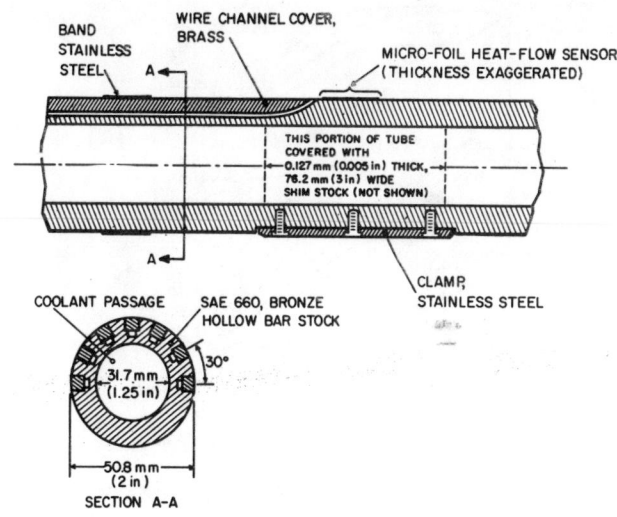

Figure 1. Instrumented tube.

An array of nine tubes arranged in three horizontal rows, as previously described, was used. Data were also obtained for a single horizontal tube for comparison with the results of tube-bundle performance. This tube was located where tube M was positioned in the array. Figure 3 is an illustration of the array geometry.

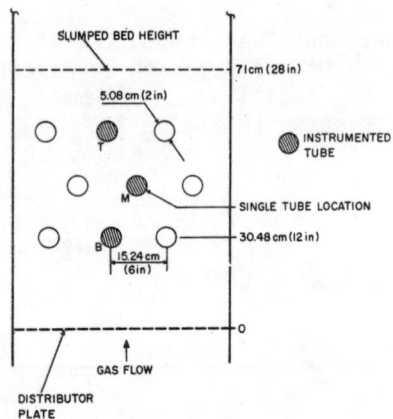

Figure 3. Bed geometry and dimensions; T, M, and B represent top, middle, and bottom tubes in the array, respectively.

A summary of test conditions, along with the thermal properties [measured by Gafourian (16)] and chemical compositions of Ione Grain particles, is given in Table 1.

Table 1. Particle properties.

|  | $d_p$ (mm) | $T_B$ (K) | $U_{mf}$ (m/s) | $U_o$ (m/s) | $\bar{T}_W$** (K) |
|---|---|---|---|---|---|
| Tube Array | 2.14 | 810 | 1.60 | 1.0-2.8 | 404 |
|  |  | 922 | 1.65 | 1.2-2.96 | 414 |
|  |  | 1005 | 1.60 | 1.24-3.5 | 431 |
|  | 3.23 | 810 | 2.20 | 1.42-2.76 | 392 |
|  |  | 922 | 2.25 | 1.43-3.2 | 409 |
|  |  | 1005 | 2.30 | 1.58-3.48 | 428 |
| Single Tube | 2.14 | 810 | 1.55 | 1.24-2.8 | 391 |
|  |  | 922 | 1.60 | 1.08-3.12 | 396 |
|  | 3.23 | 810 | 2.20 | 1.58-2.76 | 382 |
|  |  | 922 | 2.25 | 1.5-3.13 | 402 |

*53.5% silicon, 43.8% alumina, 2.3% titania, .4% other;
$\rho_s$ = 2700 kg/m³; k = 1.26 W/mK; $C_p$ = .22 cal/gK;
$\epsilon$ = .86 (estimated).

**Spatial - Average tube wall temperature for tube M in the array and single tubes at bubbling bed. The wall temperature for tubes T, M, and B were within a few degrees of each other at fluidized bed conditions.

Throughout this paper, the top, bottom, and middle tubes are designated as "T", "B", and "M", respectively.

## RESULTS AND DISCUSSION

Heat transfer results are presented both for tube arrays and single tubes as time-averaged local values. The effect of superficial gas velocity, particle size and bed temperature are shown. Results for the top, middle and bottom tubes are also compared with those for a single tube. The uncertainty in local heat transfer coefficient values is estimated at ±9%.

### Effect of Superficial Gas Velocity

The most obvious effect is the very large change in magnitude of the local heat transfer coefficient at the upper stagnation point on each tube as the superficial gas velocity is changed. In Figure 4, the local heat transfer coefficient, $h_\theta$, on the top of tube M, shows an increase from 6 to 185

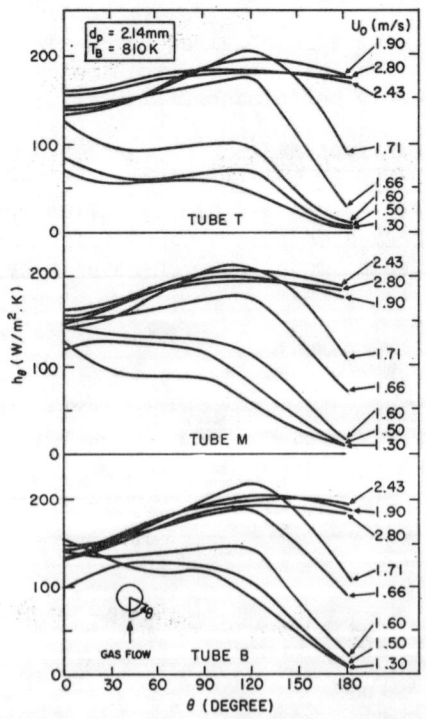

Figure 4. Time-averaged local heat transfer coefficient vs. angular position for top, middle, and bottom tubes in the array.

W/m².K with a superficial gas velocity increase from 1.3 to 2.43 m/s. Similar trends are also displayed for tubes T and B. This behavior is known to be related to the formation of a relatively cool stagnant cap of particles which remain on the top of the tubes at low superficial velocities. At higher velocities, however, the stack is removed. Available information in the literature (17 to 20) confirms this observation.

The distribution of the local heat transfer coefficient around the surface of the tubes seems to approach an asymptotic limit as the gas velocity is increased. This limit was reached at a slightly larger value of $U_o/U_{mf}$ for the smaller particles than for the larger particles at the same bed temperature when compared with data of our other work (21), which was carried out in the same facility for a single tube and particle sizes of .5 mm and 1.0 mm. Therefore, higher velocities and larger bed particle diameters are attractive from the standpoint of thermal stresses in tubes. This behavior is also consistant with the results of George (17) and Catipovic (20).

At the lower stagnation points on the tubes, the heat transfer coefficient varies with $U_o$ in an irregular manner for packed beds. For fluidized beds, stagnation point heat transfer rates are nearly constant. The increase of $h_\theta$ at any position (except at $\theta$ = 180°), for the range of superficial velocities considered, is higher for tube T than for tubes M and B for all bed temperatures and particle sizes considered.

As the gas velocity is increased the location of maximum local heat transfer coefficient, $h_{\theta_{max}}$, varies from the bottom to the side of the tubes (around 120° to 135°) and in some cases tends to shift to the top.

Local heat transfer coefficients are shown as a function of $U_o$ for all 7 locations on tube M in Figure 5. Values of the spatial-averaged heat transfer coefficient, $\bar{h}$, are also shown. Thermal stresses would, in general, be greatest at, or near, $U_{mf}$. Similar behavior was experienced by tubes B and T. For the smaller particles tested ($d_p$ = 2.14 mm), the maximum local heat transfer coefficient was observed to be highest at moderate superficial gas velocities (for all tubes, Figures 4 and 6). For the larger particle size, $h_{\theta_{max}}$ continued to increase with increasing gas velocity over the entire range (Figure 7).

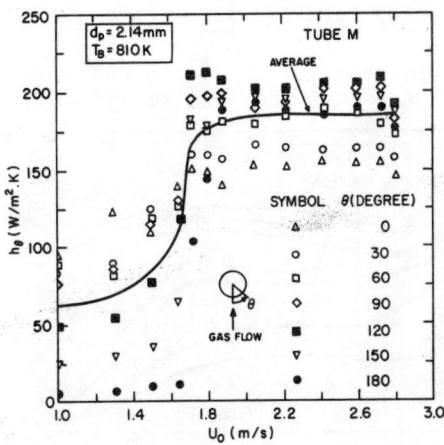

Figure 5. Time-averaged local heat transfer coefficient vs. angular position for single tubes.

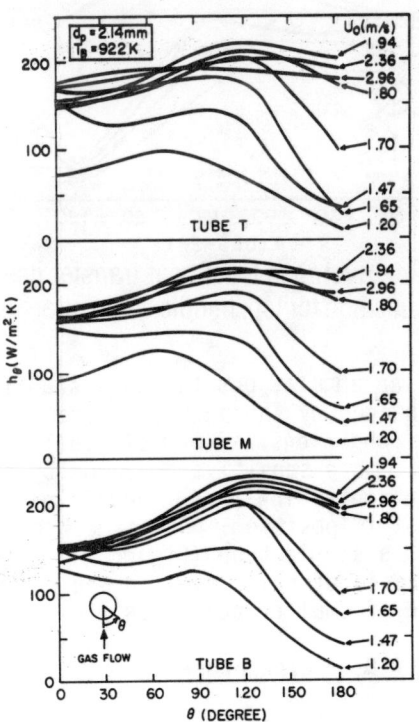

Figure 6. Time-averaged local heat transfer coefficient vs. superficial gas veloctiy on the middle tube in the array.

## Effect of Particle Size

An observation of Figures 4 and 7, indicates that the local coefficients at a

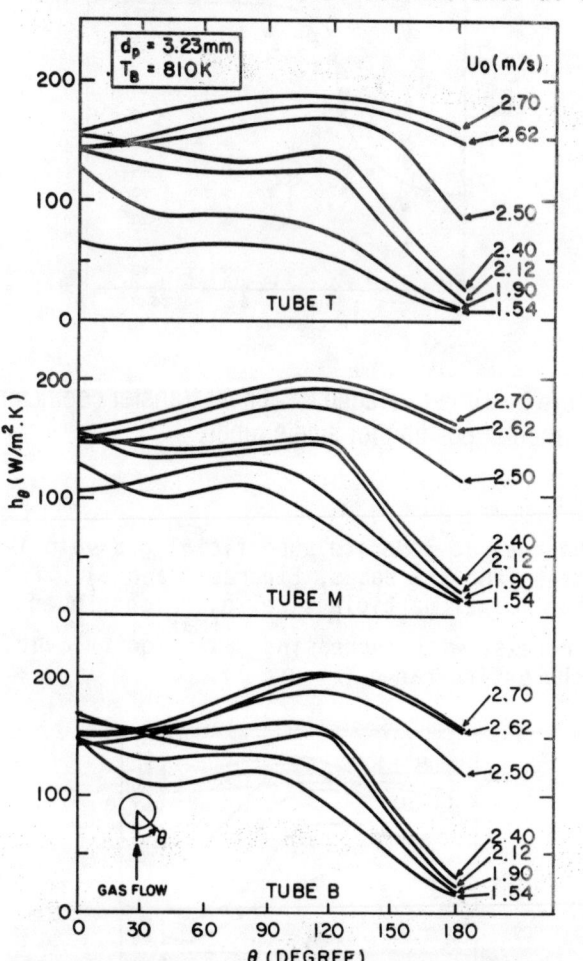

Figure 7. Time-averaged local heat transfer coefficient vs. angular position for top, middle, and bottom tubes in the array.

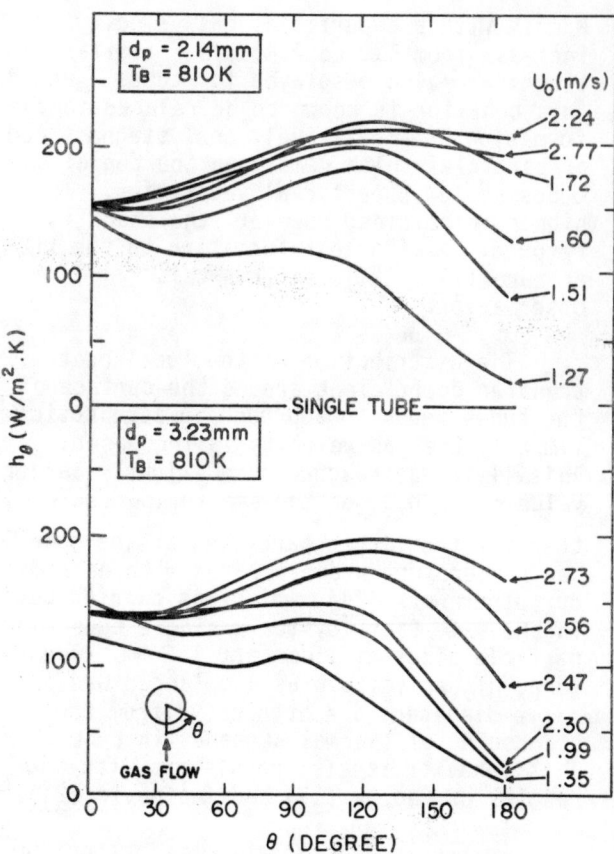

Figure 8. Time-averaged local heat transfer coefficient vs. angular position for top, middle, and bottom tubes in the array.

particular angular position are significantly influenced by the particle size under certain conditions. A bed of smaller particles is more sensitive to velocity changes than one containing larger particles, except for the 180° position. The same behavior is true for a single tube (Figure 8). Also, higher rates of heat transfer are obtained in a bed of smaller particles.

## Effect of Bed Temperature

Local heat transfer coefficients did show a moderate increase with increasing bed temperature. For tube B, the largest local coefficient, $h_{\theta_{max}}$ increased by less than 7% for an increase in bed temperature from 810 K to 922 K. (Smaller increases were observed for tubes M and T.) This is likely due to an increase in radiant heat transfer between the bed and the tubes. Alavizadeh et al (22) found, for $d_p$ = 2.14 mm, a radiation contribution of about 8% and 13% for bed temperatures of 810 K and 1050K respectively.

## Comparison of Tube Heat Transfer Performance

Figure 9 shows local heat transfer coefficients for tubes B, M, and T for different superficial gas velocities. At velocities well above $U_{mf}$, the difference in local values for the three tubes did not exceed 20%, whereas, in some cases at packed bed conditions, the differences were significantly greater. At slugging or near slugging cases (and all bed conditions), the values of local heat transfer coefficients for the side of the tubes ($\theta$ = 90°) nearly coincide. This suggests that, at high gas velocities, similar particle motion exists

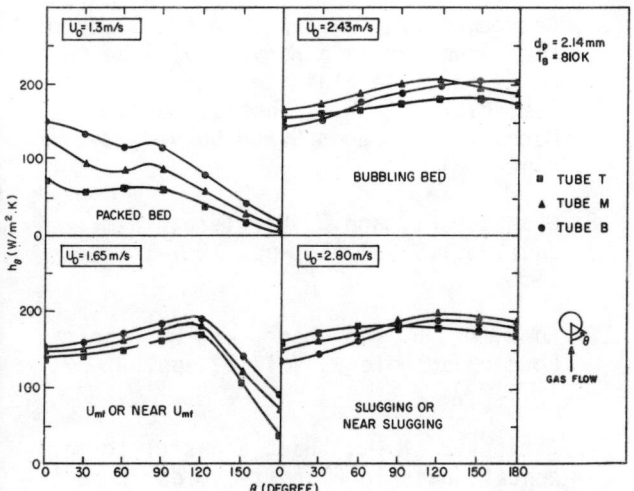

Figure 9. Time-averaged local heat transfer coefficient vs. angular position for top, middle, and bottom tubes in the array at different superficial gas velocities.

near the sides of the tubes. Except at velocities near $U_{mf}$, tube T experienced less variation in $h_\theta$ than tubes M and B. A comparison between Figures 4, 7, and 8 suggest that, in the case of bubbling beds, single tube studies are useful for predicting heat transfer with staggered-tube bundles with a pitch-to-diameter ratio of 3 or greater. Similar conclusions have been reached by other investigations (20,23) for a pitch-to-diameter ratio of 2 or greater.

ACKNOWLEDGEMENT

This work was supported by a grant from Battelle Memorial Institute, Pacific Northwest Laboratories, Richland, Washington.

NOTATION

| Symbol | Description |
| --- | --- |
| B | Bottom tube in array |
| $c_p$ | Specific heat |
| $d_p$ | Surface mean particle diameter |
| $\bar{h}$ | Spatial-average bed-to-tube heat transfer coefficient |
| $h_\theta$ | Time-average local bed-to-tube heat transfer coefficient at angular position $\theta$ |
| $h_{\theta max}$ | Maximum time-average local bed-to-tube heat transfer coefficient at angular position $\theta$ |
| k | Thermal conductivity |
| M | Middle tube in array |
| T | Top tube in array |
| $T_B$ | Temperature of fluidized bed |
| $\bar{T}_W$ | Spatial-average tube wall temperature |
| $U_o$ | Superficial gas velocity |
| $U_{mf}$ | Superficial gas velocity at minimum fluidization conditions |

Greek symbols

| $\theta$ | Angular position |
| --- | --- |
| $\rho_s$ | Density of solid particles |
| $\varepsilon$ | Emissivity |

LITERATURE CITED

1. Berg, B.V., and A.P. Baskakov, Int. Chem. Eng., Vol. 14, No. 3, pp. 440-43, 1974.

2. Gelperin, N.I. and V.G. Ainshtein, "Heat Transfer in Fluidized Beds," in Fluidization, ed. by J.F. Davidson and D. Harrison, p. 471, Academic Press, London, 1971.

3. Cherrington, D.C., L.P. Golan, and F.G. Hammitt, "Industrial Application of Fluidized Bed Combustion -- Single Tube Heat Transfer Studies," in the Proceedings of the Fifth International Conference on Fluidized Bed Combustion, Vol. III, pp. 184-209, Dec. 1977.

4. Chandran, R., J.C. Chen, and F.W. Staub, J. of Heat Transfer, Vol. 102, pp. 152-57, Feb. 1980.

5. Golan, L.P., G.V. Lalonde, and S.C. Weiner, "High Temperature Heat Transfer Studies in a Tube Filled Bed," Exxon Res. & Eng. Co., 6th Int. Conf. on FBC, Atlanta, GA, Vol. 3, pp. 1173-1185, April 1980.

6. George, A.H. and J.R. Welty, AIChE Journal, Vol. 30, 1984.

7. Baskakov, A.P., B.V. Berg, O.K. Vitt, N.F. Filippovsky, J.M. Goldobin, and V.K. Maskaev, V.K., Powder Tech., Vol. 8, pp. 273-82, 1973.

8. Vadivel, R. and V.N. Vedamurthy, "An Investigation of the Influence of Bed Parameters on the Variation of the Local Radiative and Total Heat Transfer Around an Embedded Horizontal Tube," Proceedings of the 6th Int. Conf. on Fluidized Bed Combustion, Vol. 3, Atlanta, GA, April 1980.

9. Noack, R., Chem. Ing. Tech., Vol. 42, No. 6, pp. 371-76, March 1970.

10. Saxena, S.C., N.S. Grewal, J.D. Gabor, S.S. Zabrodsky, and D.M. Galershtein, "Heat Transfer Between a Gas Fluidized Bed and Immersed Tubes," Advances in Heat Transfer, Vol. 14, 1978.

11. Peeler, J.P.K. and A.B. Whitehead, Che. Eng. Sci., Vol. 37, No. 1, pp. 77-82, 1982.

12. Grewal, N.S. and S.C. Saxena, Ind. Eng. Che. Process Des., Dev., Vol. 22, No. 3, pp. 367-376, 1983.

13. Borodulya, V.A. and V.L. Ganzha, Int. J. of Heat & Mass Transfer, Vol. 23, pp. 1602-1604, 1980.

14. Strom, S.S., T.E. Dowdy, W.C. Lapple, J.B. Kitto, T.P. Stanoch, R.H. Boll, and W.L. Sage, "Preliminary Evaluation of Atmospheric Pressure Fluidized-bed Combustion Applied to Electric Utility Large Steam Generators," EPRI Report No. RP 412-1, Electric Power Research Institute, Palo Alto, CA, Feb. 1977.

15. Gelperin, N.I., V.G. Ainshtein, and L.A. Korotyanskaya, Int. Che. Eng., Vol. 9, No. 1, pp. 137-42, Jan. 1969.

16. Ghafourian, M.R., "Determination of Thermal Conductivity, Specific Heat, and Emissivity of Ione Grain," M.S. project, Dept. of Mech. Eng., Oregon State University, June 1984 (unpublished).

17. George, A.H., "An Experimental Study of Heat Transfer to a Horizontal Tube in a Large Particle Fluidized bed at Elevated Temperature," Ph.D. Thesis, Dept. of Mech. Eng., Oregon State University, June 1981.

18. Glass, D.H., and D. Harrison, Chem. Engr. Sci., Vol. 19, pp. 1001-1002, 1964.

19. Loew, O., B. Schmutter, and W. Resnick, Powder Technology, Vol. 22, pp. 45-57, 1979.

20. Catipovic, N.M., "Heat Transfer to Horizontal Tubes in Fluidized Beds: Experiment and Theory," Ph.D. Thesis, Dept. of Che. Eng., Oregon State University, 1979.

21. Alavizadeh, N., Z. Fu, R.L. Adams, J.R. Welty, and A. Goshayeshi, "Radiative Heat Transfer Measurements for a Horizontal Tube Immersed in Small and Large Particle Fluidized Beds," to be presented at the International Symposium on Heat Transfer, Beijing, China, October 1985.

22. Alavizadeh, N., R.L. Adams, J.R. Welty, and Goshayeshi, A., "An Instrument for Local Radiative Heat Transfer Measurement in a Gas-Fluidized Bed at Elevated Temperatures," 22nd ASME/AIChE National Heat Transfer Conference, Niagara Falls, Aug. 1984.

23. Borodulya, V.A., V.L. Ganzha, A.I. Zheltov, S.N. Upadhyay, and S.C. Saxena, Letters in Heat and Mass Transfer, Vol. 7, pp. 83-95, 1980.

# EVOLUTION AND COMBUSTION OF VOLATILES FROM A SINGLE COAL PARTICLE

Pradeep K. Agarwal ■ Department of Chemical Engineering, The University of Adelaide, S.A., 5001

A single particle model is proposed for the evolution and combustion of coal volatiles. The model is based on a coupled heat transfer and chemical reaction limited description of the volatile evolution. The analysis is divided into pre-ignition and post-ignition stages. The flame temperature required in the post-ignition stage calculations is estimated from a modified Schvab-Zeldovich formulation. Approximate expressions are proposed for the estimation of the flame temperature as well as the volatiles burn out time. The model results are compared with the experimental data reported for single coal particles in stagnant as well as convective oxidizing environments. The application of the model to fluidized beds is also discussed. Model predictions are also compared with the volatile burnout times in fluidized beds.

## INTRODUCTION

Much research has been reported on the combustion mechanism of carbon/char in fluidized beds (1). Little has, however, been reported on the detailed process of evolution and combustion of coal volatiles, and the impact of these phenomena on the subsequent combustion of residual char and the importance of considering devolatilization in the overall modelling of fluidized bed combustion is now being rapidly recognized (2 to 5).

Experimental studies on the devolatilization of coal in combusting systems suggest that the devolatilization time, measured as the time between the ignition and extinction of volatiles, may be correlated by an expression of the form

$$\tau_v = k_v d^n \quad (1)$$

However, Essenhigh (6) noted that there does not appear to be a theoretical explanation for the values of $k_v$ obtained experimentally. Nor have the variations (or the lack of variation) with the type of coal and the operating conditions been understood. Experimental investigations in fluidized beds have been few and the reported results appear to be contradictory. Yates (7) and coworkers (8) studied the combustion of a single suspended bituminous coal particle (2 cm diameter) in a fluidized bed (200 μm sand bed particles) and observed that the volatiles evolved and escaped in the form of bubbles. They concluded that the combustion of volatiles would then depend on the hydrodynamics of the bed and on the degree of contact with the fluidizing air. Pillai (9) observed elongated diffusion flames around coal particles and suggested that the formation of bubbles (8) was probably due to the constraint in the movement of the coal particle imposed by the suspending wire.

The disparity between the assumptions for volatile release in the overall modelling of fluidized bed combustors has been pointed out (2 to 5, 10). Borghi et al (2), recognizing the importance of treating volatile release as a rate process, considered heat transfer to the coal particle and the isothermal kinetics of coal decomposition (11) as the possible rate limiting steps. The inadequacy of the isothermal particle assumption has been pointed out by Agarwal et al (4, 5). The mathematical formulation (2) of the diffusion flame has been questioned by Stubington (10).

## THE MODEL

In previous papers, Agarwal et al (4, 5) have considered the devolatilization of single coal particles under inert conditions. This paper extends the analysis to devolatilization under oxidizing conditions. The analysis is presented in two stages separated by ignition.

## Stage I.

In this stage, prior to ignition, the cold coal particle introduced into the hot bed/gas would heat up to the point where the volatiles would start evolving from the surface of the particle. This stage, corresponding to pyrolysis prior to ignition, may be treated mathematically as described earlier Agarwal et al (4, 5).

**Model I** In this general formulation (4), heat transfer (both to and through the coal particle) and chemical reaction are assumed to be the rate limiting steps for pyrolysis. Then

$$X_{avg} = [3\int_0^{R_o} X \, r^2 dr]/R_o^3 \quad (2)$$

where X is given (11) by,

$$X = \int_0^\infty \exp[-\int_0^t (k_o e^{-E/RT} dt)] \, g(E) dE \quad (3a)$$

$$g(E) = [\sigma(2\pi)^{1/2}]^{-1} \exp[-(E - E_o)^2/2\sigma^2] \quad (3b)$$

The particle temperature profile is calculated from the analytical solution (12) of the unsteady state heat conduction equation with a convective boundary and uniform initial temperature conditions

$$T(r,t) = T_a - \sum_{i=1}^\infty A_i \frac{\sin\beta_i \, r/R_o}{\beta_i r/R_o} \quad (4a)$$

$$A_i = 2(T_a - T_o) \frac{\sin\beta_i - \beta_i \cos\beta_i}{\beta_i - \sin\beta_i \cos\beta_i}$$

$$\exp[-\beta_i^2 \alpha t/R_o^2]$$

where $\beta_i$'s are the roots of the transcedental equation

$$\beta \cos \beta = (1 - Bi) \sin \beta \quad (4b)$$

Using Equations (2 to 4b) the integration was performed numerically (13).

**Model II** For the specific case of large particles (>1 mm), heat transfer was assumed to be the rate limiting mechanism with chemical reaction controlling only the residual amount of volatiles (5). Then, assuming a linear temperature dependence for volatiles retained between $T_1$ and $T_2$,

$$X = \begin{cases} 1.0 & T < T_1 \\ (T_2 - T)/(T_2 - T_1) & T_1 < T < T_2 \\ 0.0 & T > T_2 \end{cases} \quad (5)$$

Combining Equations (2), (4) and (5), analytical expressions were obtained for $X_{avg}$.

## Stage II

Since coal is a partially pyrolysing substance, the ignition could take place either homogeneously (in the gas phase) or heterogeneously at the solid surface. For particle sizes greater than 65 μm, homogeneous ignition is expected to take place as the surface flux of the volatiles would be large enough to prevent the oxygen from reaching the coal surface(14). Thomas et al(15) pointing out that the ignition phase would depend on coal reactivity showed that heterogeneous ignition could take place for particle sizes as large as 1 mm. A more detailed analysis would incorporate an appropriate criterion for the purpose of this simplified analysis it will be assumed that spontaneous ignition takes place in the gas phase once a minimum amount of pyrolysis (assumed to be 5% of the initial volatiles content) takes place.

Once a diffusion flame is formed within the boundary layer of the coal particle, the flame would increase the rate of devolatilization. This in turn would increase the flux of volatiles from the coal particle pushing the stoichiometric flame sheet further away from the coal particle. The flux may indeed be large enough to prevent oxygen from entering the boundary layer around the coal particle; that under such circumstances the flame would be extinguished. This extinction would result in the lowering of the volatiles flux and the flame would be reestablished. Within a fluidized bed, volatile combustion is expected to occur when the coal particle is in the bubble phase and be negligible when it is in the emulsion phase (17). After a stable phase of volatile combustion, the volatiles would be depleted from within the coal particle, the flame would decrease in intensity, move closer to the particle surface and collapse permitting oxygen to attack the residual char. The complexity of the problem would be compounded by the effect of finite combustion kinetics and the time dependent release of different gaseous and tarry species from within the coal particle. Moreover for high gas velocities, the flame would not be spherical but elongated due to forced convection. In view of the complexity of the movement of the flame front, it is assumed that the flame temperature remains constant and the flame radius is $(R_0 + \delta/2)$.

The apparent similarity between the gas phase combustion of coal volatiles and the problem of fuel droplet combustion has been pointed out by Essenhigh[6]. The solution of the liquid fuel droplet combustion [18] may be adapted to the present problem keeping in mind that the heat transmitted back from the flame raises the temperature of the whole solid with time and it is this changing temperature profile which results in the thermal breakup of bonds in the chemical structure of the coal. Consequently the 'heat of vaporization' as well as the surface temperature are functions of time. It may be shown that

$$T_f \cong 0.95[(\Delta H f Y_{oa}/C_{pg}) + T_a] \quad (6a)$$

Also, the boundary layer thickness may be estimated as

$$\delta \simeq 2 R_0 k_g / Bi \, k_s \quad (6b)$$

The devolatilization in the presence of the volatile flame may be now be characterized. Defining $t' = t - t_{ig}$, where $t_{ig}$ is the ignition delay, the temperature profile is estimated from the analytical solution of the heat conduction equation with the initial condition obtained from Equation 4 at $t = t_{ig}$. Thus,

$$T(r,t') = T_f - \sum_{j=1}^{\infty} N_j \frac{\sin \beta_j r/R_0}{\beta_j r/R_0} \exp\left(-\frac{\beta_j^2 \alpha t'}{R_0^2}\right) \quad (7a)$$

where

$$N_j = \frac{\sum_{i=1}^{\infty} A_i \Big|_{t_{ig}} \frac{\beta_j^2}{\beta_i} \left[ -\frac{\sin(\beta_i + \beta_j)}{(\beta_i + \beta_j)} + \frac{\sin(\beta_i - \beta_j)}{(\beta_i - \beta_j)} \right]}{\beta_j - \sin \beta_j \cos \beta_j} \quad (7b)$$

and $\beta_j$'s are the roots of Equation (4b) for $Bi_f = 2Bi$.

The particle temperature profile given by Equations (7 a-b) may now be used in conjunction with the volumetric average devolatilization expressions presented earlier. For the more general model (Model I), numerical integration is required.

For the large particle model (Model II), analytical expressions may be obtained as

$$X_{avg} = \left(\frac{r_1}{R_0}\right)^3 - \left(\frac{r_2^3 - r_1^3}{R_0^3}\right) \left(\frac{T_f - T_2}{T_2 - T_1}\right)$$

$$+ \frac{3}{R_0^3 (T_2 - T_1)} \left[ \sum_{j=1}^{\infty} \frac{N_j' \sin \beta_j \, r/R_0}{(\beta_j/R_0)^3} - \right. \quad (8)$$

$$\left. \sum_{j=1}^{\infty} \frac{r \cdot N_j' \cos \beta_j \, r/R_0}{(\beta_j/R_0^2)} \right]_{r_1}^{r_2}$$

where $N_j' = N_j \exp[-\beta_j^2 \alpha t'/R_0^2]$

## RESULTS AND DISCUSSION

### Application to Single Isolated Particles

The assumptions and the results of the model are verified by comparison between experimental and predicted volatile burn out times. Numerical calculations suggest that for d > 0.5 mm it is possible to approximate the volatile burnout time from the model calculations by the expression

$$\tau_v = \frac{0.0175}{\alpha} \left[ 1 + \ln\left(\frac{1}{\phi}\right) \frac{15}{\beta_{1f}^2} \right] d^2 \quad (9a)$$

where

$$\phi = (T_f - T_2)/(T_f - T_0) \quad 0 < \phi < 1 \quad (9b)$$

The heat transfer Biot number may be estimated from the single sphere correlation

$$Nu_p = 2.0 + 0.6 \, Re_p^{1/2} \, Pr^{1/3} \quad (10)$$

Equation 10 suggests that for lower $Re_p$ (corresponding to devolatilization under stagnant oxidizing conditions the devolatilization time would be proportional to $d^2$ because the $Nu_p$ would be approximately 2 (and hence $\beta_{1f}$ would be constant) for all particle diameters.

Considering the experimental conditions of Essenhigh (6), the flame temperature may be estimated from Equation 6a. Using appropriate values (18) and considering that radiation heating was employed leading to $T_a \simeq T_0$, $T_f$ may be estimated as $\simeq 2100$ K. For different types of coal, $T_2$ may vary from 900-1200 K. For $Nu \simeq 2$, $\beta_{1f}$ may be estimated as $\simeq 1$. With $\alpha = 0.1$ mm$^2$/s, it may be shown that Equation 9 reduces to

$$\tau_v = (1.13 - 1.2)d^2 \qquad (11)$$

Equation 11 is in excellent agreement with the data and correlation of Essenhigh (6). This equation also suggests that the effect of coal type under stagnant oxidizing conditions is not very strong. Also, over a wide range of ambient of ambient oxygen concentrations and gas temperatures the flame temperature $T_f$ is expected to vary within the range of 2000-3000 K. It is then possible to construct upper and lower limits for expected devolatilization times. In Figure 1, the reported experimental data (6, 19 to 22) are compared with the model predictions in terms of the upper and lower limits.

For larger $Re_p$ the value of $Nu_p$ (and consequently Biot number as well as $\beta_{1f}$) is expected to be larger according to the Equation 10. Consequently the value of the coefficient of $d^2$ is expected to drop as noted by Carabogdan (20). Additionally, since $Re_p$ would depend on the particle diameter, it is expected that the $d^2$ relationship would not be valid. To demonstrate this aspect the experimental conditions of Ragland and Weiss (23) were simulated. Once again upper and lower limits were constructed for the expected values of $\phi$. $\beta_{1f}$ was calculated from the actual particle diameter and flow conditions. The upper and lower limits along with the experimental data of Ragland and Weiss (23) are presented in Figure 2. $\tau_v'$ as well as $\tau_v$ are plotted against particle diameter. Once again, the model predictions appear to be in agreement with the data. The correlation of Essenhigh (6), reasonable only for stagnant oxidizing conditions, is seen to increasingly overpredict the dependence of devolatilization time on particle size for convective oxidizing conditions.

## Application to Fluidized Beds:

An accurate estimate of the devolatilization time in fluidized beds requires the knowledge of the relative amounts of time spent by the coal particle in the gas phase (where volatile combustion takes place) and in the emulsion phase (where the volatile flame is expected to be extinguished, however pyrolysis would still be taking place). Consequently, it may be expected that

$$\tau_v = p\,\tau_{vg} + q\,\tau_{ve} \qquad (12)$$

where p and q are relative phase residence time probabilities such that $p + q = 1$. Obviously, p and q depend on the hydrodynamics of the fluidized bed in terms of bed particle size, fluidizing velocity as well as the coal particle size. For large coal particles with much smaller bed particles and lower fluidizing velocities such as used by Yates et al (8) the value of p is expected to be small. Consequently the volatiles would form as detached bubbles. It may be noted that the bubble size of the gas phase plays an important role in such considerations. If the gas bubble size is less than the size of the coal particle at any specified height within the bed, then the particle would be expected to be predominantly in the emulsion phase. This would explain the contradictory results mentioned earlier obtained by Pillai (9) and Yates et al (8). This would also explain why volatile combustion appears to be greater at the top of the bed where the bubble sizes are expected to be much larger than within the bed.

The contribution of the pyrolysis like conditions in the emulsion phase during devolatilization under combustion conditions would explain the strong bed temperature dependence obtained in the experimental results of Pillai (24). In Figures 3 a - b, the experimental results are compared with model predictions using Equation 12 (using Model I) for three types of non swelling or moderately swelling coals. The results are seen to be good agreement. The thermophysical properties of the coals were taken to be the same as used in earlier simulations (4, 5). The kinetic parameters chosen are at best rough estimates as pyrolysis results for these coals do not appear to have been reported. The values of p and q have been used as adjustable parameters here. A rough justification for the value is obtained from the experimental results of Mickley et al (25) who found that for higher excess gas velocities, the fraction of immersed heater exposed to gas bubbles was about 0.5.

An important difference between the isothermal particle model formulations in the literature (2) and the present non-isothermal particle models arises in the estimation of the maximum temperature of the coal particle. The maximum temperature during devolatilization according to the model of Borghi et al (2), would be approximately $T_2$; the present model suggests that the surface temperatures may be substantially higher. These higher temperatures are expected to have an important influence on the subsequent combustion of char. Additionally the temperatures may be higher than the ash softening temperature which could lead to the formation of ash shells around the coal particles possibly prior to the char combustion stage. This has been observed experimentally for Mississippi lignite (26).

## ACKNOWLEDGEMENTS

The author is grateful to Ms. Julie Tonkin, Lyn Earnshaw, and Diane Ovens for typing the several drafts of the manuscript.

## NOTATION

$A_i$ (i = 1 to ∞) coefficients

Bi    heat transfer Biot number.

$Bi_f$    Biot number corresponding to the flame radius.

$C_{pg}$    specific heat of the gas.

d    particle diameter, mm.

E    activation energy, kJ/mol.

$E_0$    mean of the activation energy distribution, kJ/mol.

f    stoichiometric coefficient, gm fuel/gm oxygen.

$\Delta H$    heat of combustion.

$k_v$    constant, sec/mm$^n$.

$k_g$    thermal conductivity of the gas.

$k_s$    thermal conductivity of the coal.

$k_0$    preexponential factor, sec$^{-1}$.

$N_j'$, $N_j$ (j = 1 to ∞) coefficients.

$Nu_p$    Nusselt number.

n    constant.

p    probability of the coal particle being in the gas phase.

q    probability of the coal particle being in the emulsion phase.

R    Universal Gas Constant, kJ/mole K.

$R_0$    radius of the particle, mm.

$Re_p$    Reynolds number based on coal particle diameter.

r    radial position within particle, mm.

$r_1$    radial position corresponding to temperature $T_1$, mm.

$r_2$    radial position corresponding to temperature $T_2$, mm.

T    temperature, K.

$T_a$    gas temperature, K.

$T_f$    flame temperature, K.

$T_1$    temperature at which devolatilization begins, K.

$T_2$    temperature at which devolatilization is complete, K.

$T_0$    original temperature of the coal particle, K.

t    time, sec.

$t_{ig}$    ignition delay, sec.

t'    $t - t_{ig}$, sec.

X    fractional amount of volatiles retained.

$X_{avg}$    volumetric average fraction of volatiles retained.

$Y_{oa}$    oxygen mass fraction in the ambient.

$\alpha$    thermal diffusivity of coal, mm$^2$/s.

$\beta$    roots of Equation 4b.

$\delta$    boundary layer thickness, mm.

$\sigma$    standard deviation of the activation energy distribution, kJ/mol.

$\tau_v$    devolatilization time, sec.

$\tau_v'$    devolatilization time measured as the total lapse time to flame extinction, sec.

$\tau_{ve}$    devolatilization time if the coal particle was in the emulsion phase, sec.

$\tau_{vg}$    devolatilization time if the coal particle was in the gas phase, sec.

$\phi$    parameter defined in Equation 9.

## LITERATURE CITED

1. Ross, I.B. and Davidson, J.F., Trans.I.Ch.E., 1979, 57, 215 (1979).

2. Borghi, G., Sarofim, A.F. and Beer, J.M., paper presented A.I.Ch.E. 70th Annual Meeting, New York, (1977).

3. LaNauze, R.D., Fuel, 61 771 (1982).

4. Agarwal, P.K., Genetti, W.E. and Lee, Y.Y., Fuel 63, 1748 (1984).

5. Agarwal, P.K., Genetti, W.E. and Lee, Y.Y., Fuel, 63, 1748 (1984).

6. Essenhigh, R.H., J.Eng.Power, 85, 183. (1963).

7. Yates, J.G., Private Communication, (1983).

8. Yates, J.G., Macgillivray, M. and Cheesman, D.J., Chem.Eng.Sci., 35, 2361 (1980).

9. Pillai, K.K., J.Inst.Energy, 132, (1982).

10. Stubington, J.F., J.Inst.Energy, 191, (1980).

11. Anthony, D.B. and Howard, J.B., A.I.Ch.E.J., 4, 625, (1976).

12. Jakob, M., 'Heat Transfer', John Wiley and Sons, New York, (1959).

13. Agarwal, P.K., 'Fluidized Bed Combustion of Wet Low Rank Coals', Ph.D. Dissertation, University of Mississippi (1984).

14. Howard, J.B. and Essenhigh, R.H., Eleventh Symp. (Intl) Combustion, Combustion Institute, Pittsburgh, 399, (1967).

15. Thomas, G.R., Harris, J.J. and Evans, D.G., Combust Flame, 29, 193 (1977).

16. Annamalai, K. and Durbetaki, P., Comb Flame, 29, 193, (1977).

17. Cowley, L.T. and Roberts, P., Proc.Fluid.Combustion Conf, Capetown, 2(5.3), 443, (1981).

18. Kanury, A.M., 'Introduction to Combustion Phenomena', Gordon and Breach Publishers, New York, (1975).

19. Carabogdan, I., St.cerc.energ.electr., Bucaresti, 15(2), (1965).

20. Carabogdan, I., St.cerc.energ.electr., Bucaresti, 1, (1967).

21. Stambuleanu, A., 'Flame Combustion Processes in Industry', Abacus Press, Kent, (1976).

22. Ivanova, I.P. and Babii, V.L., Teploenergetika, 13(4), 54 (1966).

23. Ragland, K.W. and Weiss, C.A., Energy, 1979, 4, 341 (1979).

24. Pillai, K.K., J.Inst.Energy, 142, (1981).

25. Mickley, H.S., Fairbanks, D.F. and Hawthorn R.D., Chem.Eng. Progr. Symp. Series, 57 (32), 51, (1961).

26. Genetti, W.E., Personal Communication (1984).

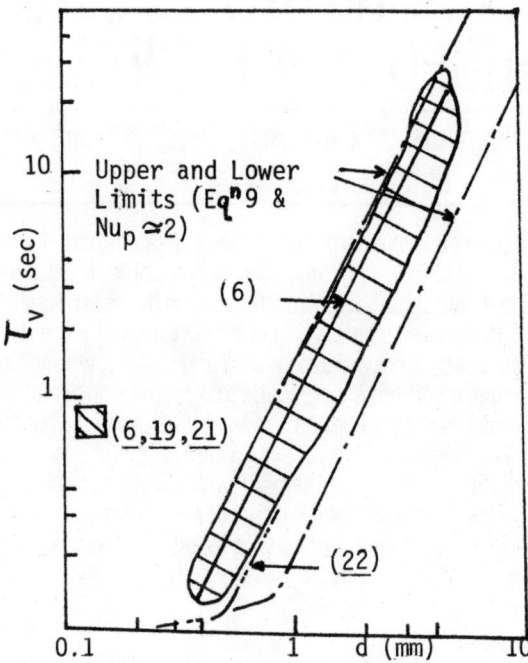

FIGURE 1: Data and Model Predictions; Stagnant Oxidizing Conditions.

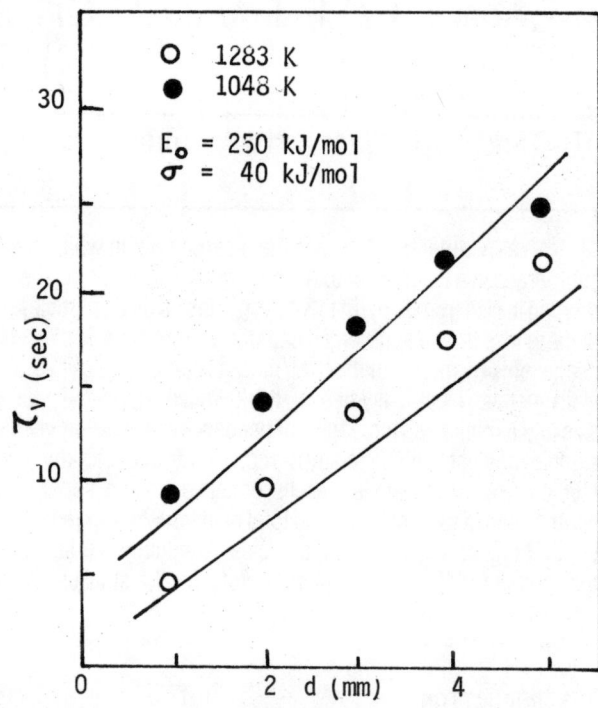

FIGURE 3a: Data (24) — Nostell Coal and Model Predictions; Fluidized Bed.

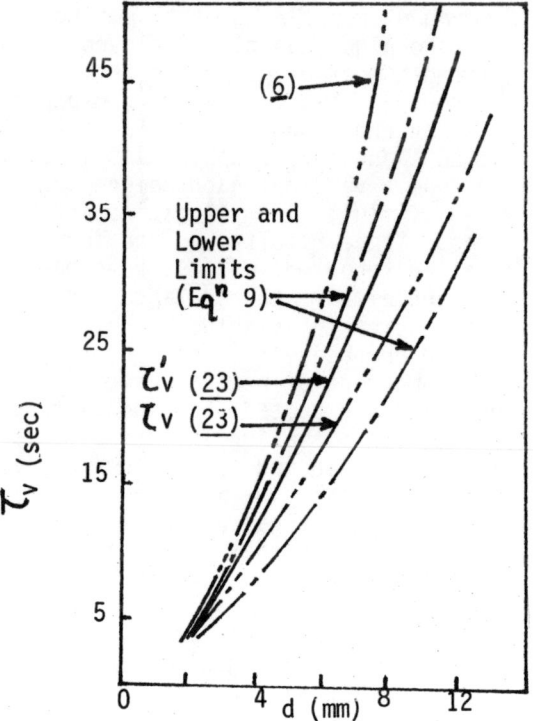

FIGURE 2: Data (23) and Model Predictions; Convective Oxidizing Conditions.

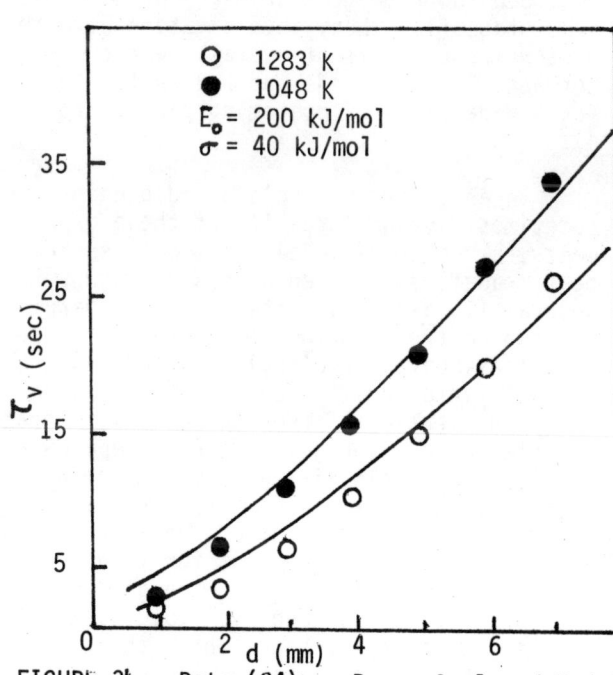

FIGURE 3b: Data (24) — Rexco Coal and Model Predictions; Fluidized Bed.

# THE CONTRIBUTION OF GAS CONVECTION TO TOTAL HEAT TRANSFER FOR A HORIZONTAL CYLINDER SUBMERGED IN A FLUIDIZED BED

J.E. O'Brien, M.L. Wade and A.M. Terpolilli ■ The Pennsylvania State University, Dept. of Mechanical Engrg. University Park, Pennsylvania 16802

An experimental study has been carried out in which the gas convective and total heat transfer were measured for a horizontal tube submerged in a fluidized bed. The experiments were performed in a 30.5 cm O.D, 1.32 m high transparent acrylic column, utilizing glass beads ranging in size from 215 $\mu$m to 3.4 mm as the fluidized particles. The gas convective contribution was determined by measurements of the rate of mass loss from a submerged naphthalene cylinder. The resulting mass transfer coefficients were converted to gas convective heat transfer coefficients by means of the heat/mass transfer analogy. Total heat transfer coefficients were measured under identical conditions using an instrumented, electrically heated, thick-walled copper cylinder. Gas convective coefficients were found to increase significantly with particle size and exhibited an increasing dependence on fluidizing velocity for the larger particles. The gas convective results of the present study are significantly underpredicted by the correlation of Baskakov and Suprun, indicating the importance of the differences in geometry for the two studies. The heat transfer results obtained from a separate set of experiments are in agreement with previous studies. The relative contribution of gas convection to total heat transfer ranged from 6.8 percent for the smallest particles at optimum fluidization for heat transfer to 100 percent for the largest particles. At minimum fluidization, the gas convective contribution was found to be much more significant, accounting for 40-100 percent of the total for the smallest and largest particles, respectively.

## INTRODUCTION

The rate of heat transfer to a tube submerged in a fluidized bed depends upon a number of complex factors, including the properties of the bed material and the fluidizing gas, bed and tube geometries, and the fluidized state ([1]). Many investigators have reported measurements of overall heat transfer between fluidized beds and horizontal tubes for a wide range of operating parameters ([2-4]).

In an effort to explain and predict the sometimes divergent results of these experiments, a number of models of bed-to-surface heat transfer have been proposed. Among these, several ([1], [5-7]) have attempted to separate the two controlling low temperature transport processes--particle circulation and gas convection--into additive components. The relative contribution of these processes to heat transfer at a given gas flow rate depends mainly on particle size and system operating pressure. Several experimental studies ([8-10]) have demonstrated the increasing importance of the gas convective component for large particle fluidized beds in which high gas velocities are required to achieve fluidization, and at high operating pressure where the gas heat capacity becomes significant. Other situations in which gas convection plays an important role in the heat transfer process include gas velocities near the minimum fluidizing velocity and in regions of stagnant particles, as on top of a submerged horizontal tube ([5]).

Despite the difficulty in resolving the separate contributions of particle and gas convection in a fluidized bed by means of heat transfer experiments, it is possible to independently measure the gas convection component by means of analogy to mass transfer from an evaporating or sublimating surface. This technique has been employed in a very limited number of geometries by several investigators. In an oft-referenced paper, Baskakov and Suprun report mass transfer coefficients due to gas convection in a fluidized bed from a vertical naphthalene surface for a range of fluidizing velocities and particle sizes. These results, in the form of a correlation reported in ([11]), have been extended, for lack of any other data, to various unrelated geometries (including horizontal tubes) for the purposes of predicting the gas convective component of heat transfer ([1],[7],[8]). However, it is expected that the distinctly different flow situation and the effects of bubble interaction and the top defluidized cap on a horizontal tube will generally result in a very different gas convective heat transfer coefficient than that for a vertical tube at any given fluidizing velocity.

The present experimental study has been undertaken with the goal of providing a complete set of basic data on the gas convective contribution of heat transfer to a horizontal tube in a fluidized bed as a function of particle size and gas velocity. This task has not been attempted previously. The results will also provide an opportunity to reevaluate the various hypotheses which appear in the literature concerning gas convection in fluidized beds.

## APPARATUS AND PROCEDURE

The experiments were performed in a 30.5 cm O.D., 1.32 m high transparent acrylic column, allowing continuous visual observation of flow patterns. Air was supplied to the bed by means of a large reciprocating, oil-free compressor capable of delivering a flow rate of 0.165 m³/s at a maximum pressure of 2.07 MPa. Two large pressure regulators were used to step down the line pressure leading to the bed, which was operated at essentially atmospheric pressure. The air flow rate was monitored by one of the three calibrated rotameters, depending on the magnitude of the flow rate which ranged from 9.93 kg/h to 520 kg/h during the course of the experiments. Air pressure at the bed inlet was measured by a mercury manometer. The base of the bed was flanged to a conical diffuser at the top of which the air distributor was located. The distributor consisted of a sandwich arrangement of two fine-mesh (250μm spacing) brass screens separated by a 7.6 cm thick packed bed of 2.05 mm glass beads. Exhause air was vented directly to the outside by means of a 25.4 cm diameter galvanized sheet metal duct.

Technical quality solid glass spheres were employed as the fluidized particles in these experiments. The five particle sizes chosen for the experiments (see Table 1) ranged from 215 μm to 3.40 mm in diameter. This size range was chosen in order to encompass a range of gas convective contribution to total heat transfer from less than 10% to virtually 100% (11).

Two sets of experiments were performed, both involving a submerged 2.54 cm O.D. horizontal cylinder. In the first set, the gas convective component of the total heat transfer coefficient was determined by the naphthalene sublimation technique. For these experiments, the submerged cylinder consisted of two solid naphthalene halves, each 8.9 cm long. These cylinders were formed by a casting procedure in which a hollow aluminum core was placed inside a highly polished cylindrical brass mold. Mass losses from the two halves were determined separately by careful before and after weighings on a Sartorius precision analytical balance, accurate to 0.1 mg. The two mass losses were then averaged and divided by the run duration time, yielding the average mass transfer rate, $\dot{m}_{Av}$. The bed temperature for the mass transfer experiments was measured by two NBS standard mercury-in-glass thermometers. These

Table 1  Properties of Test Particles

| Particle Designation | Mean Size (μm) | $U_{mf}$ (m/s) Exp. | $U_{mf}$ (m/s) (Eq. 2) | $U_{opt}$ (m/s) Exp. | $U_{opt}$ (m/s) (Eq. 4) |
|---|---|---|---|---|---|
| P-010 | 215 | 0.038 | 0.036 | 0.22 | 0.332 |
| P-023 | 513 | 0.219 | 0.218 | 0.423 | 0.560 |
| P-047 | 945 | 0.492 | 0.524 | 0.689 | 0.776 |
| A-205 | 2050 | 1.03 | 1.06 | 1.23 | 1.15 |
| A-340 | 3400 | 1.41 | 1.49 | 1.69 | 1.50 |

temperatures were used in calculating the naphthalene surface vapor density, $\rho_w$, from the Sogin (12) vapor pressure correlation. Mass transfer coefficients were then determined directly from the expression:

$$K = \frac{\dot{m}_{Av}}{A_{wn}(\rho_w - \rho_\infty)} \qquad (1)$$

where $A_{wn}$ is the "wetted" naphthalene surface area and $\rho_\infty$ is the free-stream naphthalene vapor density, which is zero.

In the second set of experiments, total heat transfer coefficients were measured by means of a 17.8 cm long instrumented thick-walled (0.32 cm wall thickness) copper cylinder (2.54 cm O.D.). The cylinder was heated internally by a cartridge heater and thermally isolated by the use of low conductivity nylon end supports. The measured power input to the cartridge heater yielded the cylinder average heat flux. Four type K thermocouples (0.25 mm diameter), circumferentially located 90° apart, were imbedded in the outer surface of the copper at the cylinder midspan. In addition, two thermocouples submerged in the bed allowed determination of the temperature difference between the bed and the cylinder surface.

The geometry and flow conditions for the two sets of experiments were identical. In addition, the uniform wall temperature thermal boundary condition provided by the thick-walled copper tube is completely analogous to the uniform naphthalene vapor density boundary condition of the mass transfer experiments. The test cylinder for both sets of experiments was located 38.1 cm above the distributor plate. The unfluidized depth of the bed was 61.0 cm.

## EXPERIMENTAL RESULTS

### Mass Transfer Results

Overall mass transfer results are presented in Figure 1. Since the hypothesis behind these mass transfer experiments is that the mass loss from the cylinders is due primarily to interstitial gas flow, independent of direct particle effects, the trends in the

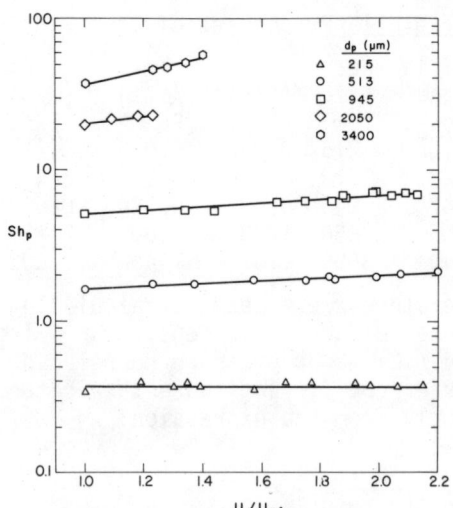

Figure 1. Dependence of Sherwood number on relative fluidization numbers.

mass transfer results can also be taken to represent trends in the gas convection contribution to overall heat transfer via the well established analogy between heat and mass transfer.

Several observations may be noted from inspection of Figure 1. The most obvious of these is the fact that as the particl size and corresponding fluidizing velocities increase, the $Sh_p$ values increase. This result is due partly to the fact that the characteristic dimension used in calculating the $Sh_p$ values is $d_p$, the particle diameter. In addition, however, the actual mass transfer coefficients increase with particle size, which is to be expected since these have resulted strictly from gas convective effects. For a specified particle size, Figure 1 indicates the dependence of $Sh_p$ on $U/U_{mf}$. This dependence is strongest for the larger particle sizes. The explanation for this behavior is that for the small particle sizes, small increases in air flow rate above minimum fluidization result in all of the excess air passing through the bed in the bubble phase, contributing little to gas convective mass loss from the cylinder. Thus, it appears to be a reasonable assumption for small particle sizes that the gas convective contribution to heat transfer is largely unaffected by excess air flow (1,5). For the large particle sizes, however, gas convection is the dominant transport mechanism at all air flow rates. Furthermore, the high porosity, large particle beds do not readily form well-defined bubbles at flow rates above minimum fluidization. Therefore, a strong dependence of $Sh_p$ on $U/U_{mf}$ results.

Minimum fluidizing velocities were determined experimentally by observation of pressure drop behavior. These experimental values for $U_{mf}$, which are listed in Table 1, compare favorably with corresponding predictions obtained from the Ergun equation (13):

$$Re_{mf} = 25.7\{[1 + 5.53 \times 10^{-5} Ar]^{1/2} - 1\} \quad (2)$$

The predicted values from Equation (2) are also shown in Table 1.

An oft-referenced correlation for predicting the gas convective contribution to total heat transfer in a fluidized bed is that attributed to Baskakov and Suprun (11). This correlation summarizes the results of a set of experiments in which a vertical naphthalene surface was submerged in a fluidized bed. The correlation has the form:

$Nu_{pconv} = 0.0175\ Ar^{0.46} Pr^{0.33}$; $(U \geq U_{opt})$

or, when $(U_{mf} < U < U_{opt})$ (3)

$Nu_{pconv} = 0.0175\ Ar^{0.46} Pr^{0.33} (U/U_{opt})^{0.3}$

The quantity $U_{opt}$, which appears in the correlation, is the optimum fluidizing velocity for heat transfer. Numerical values for $U_{opt}$ have been obtained from the Todes (13) correlation:

$$U_{opt} = \frac{\nu_f}{d_p} \left(\frac{Ar}{18 + 5.22\ Ar^{1/2}}\right) \quad (4)$$

and are listed in Table 1 along with experimentally determined values of $U_{opt}$ obtained from heat transfer experiments to be discussed later.

The Baskakov correlation has been applied to the conditions of the present experiments, and the results are summarized in Table 2. The calculated values are obtained by substituting the Schmidt number for the naphthalene-air system, $Sc = 2.56$, for the Prandtl number in accordance with the heat/mass transfer analogy. In addition, the experimentally determined values of $U_{opt}$ have been used in forming the ratio, $U/U_{opt}$.

Table 2 Comparison of Mass Transfer Results to the Baskakov and Suprun Correlation

| Particle Designation | $U = U_{mf}$ | | $U = U_{opt}$ | |
|---|---|---|---|---|
| | $Sh_p$ Exp. | $Sh_p$ Eq.(3) | $Sh_p$ Exp. | $Sh_p$ Eq.(3) |
| P-010 | 0.335 | 0.361 | ~0.39 | 0.492 |
| P-023 | 1.60 | 1.20 | 1.94 | 1.63 |
| P-047 | 5.08 | 3.43 | 5.27 | 3.80 |
| A-205 | 19.6 | 10.3 | 22.7 | 10.9 |
| A-340 | 42.1 | 18.0 | 52.4 | 19.0 |

A comparison of the values of $Sh_p$ obtained from the present experiments for the submerged horizontal tube and those obtained from the correlation of Equation (3) reveals generally poor agreement, especially for the larger particle sizes where the correlation significantly underpredicts the mass transfer rates. The level of agreement for the two smallest particle sizes in Table 2 is reasonable (<30% difference). It is felt that the distinct difference in the geometry between the submerged horizontal tube of the present study and the submerged vertical surface of Baskakov and Suprun is responsible for the diverging results for the large particle sizes. As the particle size increases, fluidizing velocities and bed porosities increase. Correspondingly, gas convection becomes more important and the particular geometry presented to the flow by the submerged object has a much greater influence on the convective heat transfer.

In addition to comparing the magnitudes of the experimental results presented in Table 2 to those calculated from Equation (3), it is worth noting that for a specified particle size, the dependence of $Sh_p$ on fluidizing velocity for the present experiments was found to be very different from the $(U/U_{opt})^{0.3}$ of Equation (3). As noted previously when discussing Figure 1, this dependence becomes much more pronounced for the larger particle sizes, as would be expected from physical reasoning. Furthermore, no "leveling off" of the gas convective data is apparent for velocities exceeding $U_{opt}$, as predicted by the Baskakov correlation.

## Heat Transfer Results

The results of the heat transfer experiments are displayed in Figure 2, which is a dimensional plot of the overall heat transfer coefficient, h, versus the fluidizing mass velocity, G. As indicated in the Figure, the data for all five particle sizes are shown. Inspection of Figure 2 reveals that, on the average, heat transfer coefficients decrease with increasing particle size, despite the much higher flow velocities required for fluidization of the large particles. This trend is due to the fact that as the particle size is increased, the relative contribution of the more effective particle conduction heat transfer mechanism decreases, while the gas convection mechanism becomes dominant.

Also evident in Figure 2 is that for a specified particle size, an abrupt rise in heat transfer coefficient occurs at flow rates

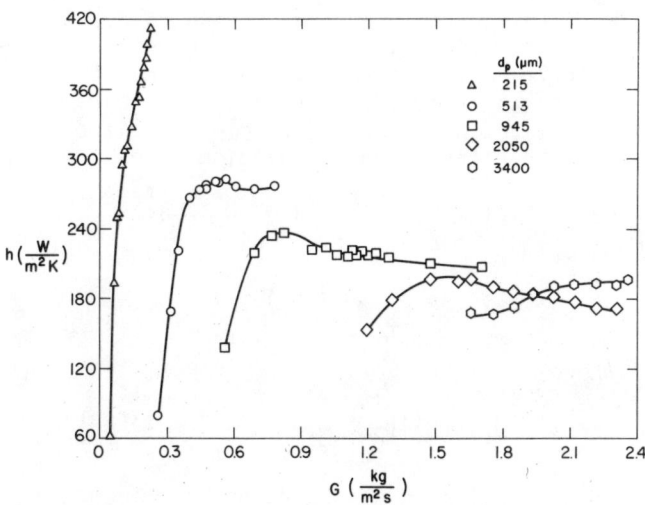

Figure 2. Total heat transfer coefficiencies.

just above that required for minimum fluidization. This abrupt rise is generally followed by a leveling off or even a decrease in heat transfer coefficient, indicating that an optimum fluidizing velocity for heat transfer exists. Experimentally determined optimum fluidizing velocities are listed in Table 1, along with values for $U_{opt}$ calculated from the previously discussed Todes (13) correlation, Equation (4). Agreement between the experimental values and the correlation is reasonable. The value of $U_{opt}$ for the smallest particle size was never quite achieved in the present experiments; thus the experimental entry listed in the Table is an estimate based on extrapolation.

Heat transfer results for the two smallest particle sizes are compared to the correlation of Grewal and Saxena (4) in Figure 3.

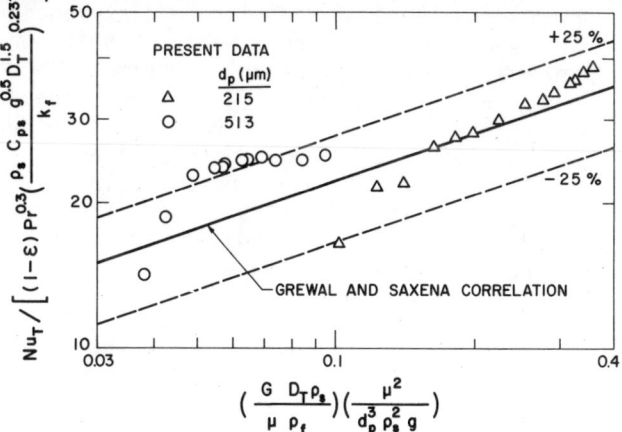

Figure 3. Comparison of present small particle heat transfer results with the correlation of Grewal and Saxena.

This correlation was chosen as a basis for comparison since it was developed very recently and very carefully, and since it attempts to include all the important parameters associated with small particle fluidized bed heat transfer. The correlation takes the form:

$$Nu_T = 47(1-\varepsilon)\left(\frac{G}{\rho_f}\frac{D_T}{\mu}\frac{\rho_s}{d_p^3 \rho_s^2 g}\right)^{0.325}$$

$$\times \left(\frac{\rho_s}{k_f}\frac{C_{ps}}{k_f}D_T^{3/2} g^{1/2}\right)^{0.23} Pr^{0.30} \quad (5)$$

where (6)

$$\varepsilon = \frac{1}{2.1}\left[0.4 + \left\{4\left(\frac{\mu G}{d_p^2 \rho_f (\rho_s - \rho_f)\phi_s^2 g}\right)^{0.43}\right\}^{1/3}\right]$$

This correlation was based on experiments in which particle sizes ranged from 167 to 504 μm. Therefore, comparison with the data of the present experiments for the two smallest particle sizes, 215 μm and 513 μm, is appropriate.

The level of agreement between the present experimental data and the correlation is very good, generally within ±25%. Furthermore, the present data agree with the actual data of (4) for glass beads even more closely than comparison with the correlation indicates, since the glass bead data of (4) also tends to be about 20% high with respect to the correlation.

No such correlation is available for large particle heat transfer results. In fact, there is a shortage of reliable heat transfer data for large particle beds. The gas convective model for heat transfer in large particle beds, formulated by Adams and Welty (6) and experimentally validated by Catipovic et al (14) does provide some basis for comparison. Direct application of the Adams and Welty model requires knowledge of the voidage distribution surrounding the cylinder. This quantity was not measured in the present experiments. However, Catipovic et al. (14) provide values for the relative fluidizing velocities at which an average voidage, $\varepsilon$, around the cylinder was found to be 0.57 for several particle sizes. Average heat transfer results were then measured at this voidage condition and compared to the predictions of the model, resulting in excellent agreement. In order to allow for a comparison of the present data with the Adams-Welty model, $Nu_p$ values are presented in Table 3 at relative fluidizing velocities corresponding to $\varepsilon = 0.57$. Both the experimental values of the present study and the predictions of the Adams-Welty model, obtained from Figure 3 of (4), are listed. The comparison is made for the three largest particle sizes used in the present study. The model is seen to underpre-

Table 3 Comparison of Measured Values of the Overall Nusselt Number with Predictions of the Adams-Welty Model

| Particle Diameter $d_p$ (mm) | $U/U_{mf}$ [$\varepsilon$=0.57] | U (m/s) | $Re_p$ | $Nu_p$ Exp. | $Nu_p$ Model |
|---|---|---|---|---|---|
| 0.945 | 1.19 | 0.58 | 35.1 | 7.8 | 6.7 |
| 2.05 | 1.08 | 1.11 | 145.2 | 13.7 | 10.5 |
| 3.40 | 1.05 | 1.48 | 320.6 | 21.4 | 15.9 |

dict the experimental values by an average of about 25%. It should be noted that the model also underpredicts the data of Canada and McLaughlin (15) by about 15%. The discrepancies are probably due to an inexact knowledge of the voidage surrounding the cylinder and its dependence on relative fluidization for the conditions of the present study. With these factors in mind, the level of agreement between the model and the present data is quite reasonable.

Combined Heat and Mass Transfer Results

The primary goal of this work was the determination of the relative contribution of gas convection to total heat transfer for a submerged horizontal cylinder in a fluidized bed. Thus far in this paper, the total heat transfer results and the gas convective mass transfer results have been presented and discussed separately. In order to compare the results of the heat and mass transfer experiments on the same basis, the gas convective heat transfer Nusselt numbers, $Nu_{pconv}$, can be calculated by multiplying the mass transfer Sherwood numbers by the ratio $(Pr/Sc)^n$, where the exponent n takes on values ranging from 0.3 to 0.4, depending on the nature of the flow and the correlation chosen. For the purposes of the following comparisons, n = 0.4 was chosen.

The gas convective Nusselt numbers and the total Nusselt numbers at two representative fluidizing conditions, minimum fluidization and optimum fluidization, are plotted as a function of Reynolds number based on particle diameter, $Re_p$, in Figure 4. In this Figure, the highest $Nu_p$ values represent total heat transfer results at the optimum fluidizing velocity for each of the five particle sizes. The second highest data set through which a curve is drawn are the total heat transfer results at minimum fluidizing velocities. The next two data sets represent the gas convective contributions at optimum and minimum fluidizing velocities, respectively.

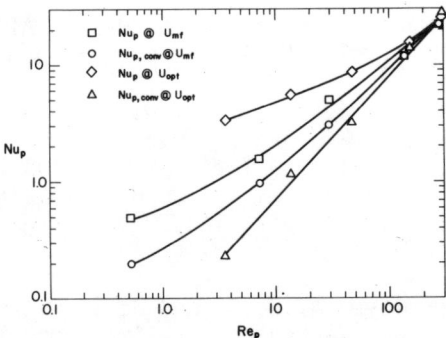

Figure 4. Relative contribution of gas convection to heat transfer at minimum and optimum fluidization.

The data shown in Figure 4 indicate the increasing relative importance of gas convection as particle size increases. In addition, for a specified particle size, the relative contribution of gas convection is seen to be much stronger at minimum fluidization than at optimum fluidization due to the lack of vigorous particle mixing at minimum fluidization. The assumption that the gas convective heat transfer contribution is 100% at minimum fluidization is not validated exactly for the smallest particle sizes. However, the assumption still appears to be reasonable in view of the high sensitivity of the heat transfer data near $U_{mf}$ to very slight increases in air flow.

All four data sets converge to nearly equal values for the largest particle size. The total and gas convective $Nu_p$ values approach one another since the relative contribution of gas convection is virtually 100% for these large particles. The $Re_p$ values corresponding to minimum and optimum fluidization approach one another because the value for $U_{opt}$ is very close to the value for $U_{mf}$ for large particles.

## CONCLUDING REMARKS

The experiments described herein represent the first definitive study of the gas convective contribution to total heat transfer for a horizontal cylinder submerged in a fluidized bed. Mass transfer coefficients obtained by the naphthalene sublimation technique were converted to gas convective heat transfer coefficients via the heat/mass transfer analogy. Total heat transfer coefficients were measured using an instrumented, electrically heated, thick-walled copper cylinder. Technical quality glass beads ranging in size from 215 μm to 3.40 mm were employed as the fluidized particles.

Mass transfer coefficients were found to increase with particle size due to the increased bed porosity and higher velocities required for large particle fluidization. In addition, for a specified particle size, the dependence of mass transfer rate on superficial velocity was found to be strongest for the largest particle sizes and nearly absent for the smallest particles. The gas convective results of the present study were found to be significantly underpredicted by the correlation of Baskakov and Suprun (11), indicating the importance of the differences in geometry for the two studies.

The results of the heat transfer experiments were found to agree both qualitatively and quantitatively with the work of previous investigators. Heat transfer coefficients were highest for the smallest particles and generally decreased with increasing particle size. For a specified particle size, heat transfer rates were found to increase rapidly at superficial velocities just above minimum fluidization and to subsequently level off and even decrease with further increases in flow rate, with the existence of a distinct optimum fluidizing velocity apparent for the intermediate particle sizes. The experimentally determined values of $U_{opt}$ were found to agree reasonably well with the correlation of Todes (13).

The relative contribution of gas convection to total heat transfer was determined by comparing the mass transfer results, scaled by the ratio $(Pr/Sc)^{0.4}$, to the heat transfer results. For the particle size range chosen, the ratio of gas convective heat transfer to total heat transfer ranged from 6.8% for the smallest particles at optimum fluidization to virtually 100% for the largest particles. At minimum fluidization, the gas convective contribution was found to be much more significant, ranging from 40% for the smallest particles to 100% for the largest particles.

## NOTATION

$A_n$ = Surface area of naphthalene
$Ar$ = Archimedes number
$C_{ps}$ = Particle specific heat
$d_p$ = Mean particle diameter
$D$ = Binary diffusion coefficient for naphthalene/air
$D_T$ = Cylinder diameter
$g$ = Acceleration due to gravity
$G$ = $\rho_f U$ = Fluidizing mass velocity
$h$ = Total heat transfer coefficient
$k_f$ = Air Thermal conductivity

K = Mass transfer coefficient
$\dot{m}_{Av}$ = Average rate of mass loss from naphthalene cylinder
n = Scaling exponent
$Nu_p$ = $hd_p/k_f$ = Nusselt number based on particle diameter
$Nu_{pconv}$ = $h_{conv}d_p/k_f$ = Gas convective Nusselt number
$Nu_T$ = $hD_T/k_f$ = Nusselt number based on cylinder diameter
Pr = Air Prandtl number
$Re_{mf}$ = $U_{mf}d_p/\nu_f$ = Particle Reynolds number at minimum fluidization
$Re_p$ = $Ud_p/\nu_f$ = Particle Reynolds number
Sc = Schmidt number
$Sh_p$ = $Kd_p/D$ = Sherwood number based on particle diameter
U = Superficial fluidizing velocity
$U_{mf}$ = Superficial velocity at minimum fluidization
$U_{opt}$ = Superficial velocity at optimum fluidization for heat transfer
$\varepsilon$ = Porosity
$\mu$ = Absolute air viscosity
$\rho_f$ = Air density
$\rho_s$ = Particle density
$\rho_w$ = Naphthalene vapor density at the cylinder surface
$\rho_\infty$ = Naphthalene vapor density away from the cylinder
$\nu_f$ = Kinematic viscosity of air
$\phi_s$ = Sphericity of fluidized particles

## LITERATURE CITED

1. Botterill, J. S. M., *Fluid-Bed Heat Transfer*, Academic Press, New York (1975).

2. Chen, J. C., "Heat Transfer to Tubes in Fluidized Beds," ASME Paper No. 76-HT-75, ASME-AIChE Heat Transfer Conference, St. Louis, MO (1976).

3. Andeen, B. R. and L. R. Glicksman, "Heat Transfer to Horizontal Tubes in Shallow Fluidized Beds," ASME Paper No. 76-HT-67, ASME-AIChE Heat Transfer Conference, St. Louis, MO (1976).

4. Grewal, N. S. and S. C. Saxena, *International Journal of Heat and Mass Transfer*, Vol. 23, pp. 1505-1519 (1980).

5. Saxena, S. C., N. S. Grewal, J. D. Gabor, S. S. Zabrodsky and D. M. Galershtein, *Advances in Heat Transfer*, Vol. 14, pp. 149-247 (1978).

6. Adams, R. L. and J. R. Welty, *AIChE Journal*, Vol. 25, No. 3, pp. 395-405 (1979).

7. Catipovic, N. M., G. N. Jovanovic, T. J. Fitzgerald and O. Levenspiel, "A Model for Heat Transfer to Horizontal Tubes Immersed in a Fluidized Bed of Large Particles," *Fluidization*, Plenum Press, pp. 225-234 (1980).

8. Denloye, A. O. O. and J. S. M. Botterill, *Powder Technology*, Vol. 19, pp. 197-203 (1978).

9. Xavier, A. M., D. F. King, J. F. Davidson and D. Harrison, "Surface-Bed Heat Transfer in a Fluidized Bed at High Pressure, *Fluidization*, Plenum Press, pp. 209-216 (1980).

10. Borodulya, V. A., V. G. Ganzha and A. I. Podbereysky, "Heat Transfer in a Fluidized Bed at High Pressure," *Fluidization*, Plenum Press, pp. 201-207 (1980).

11. Baskakov, A. P. and V. M. Suprun, *International Chemical Engineering*, Vol. 12, No. 2, pp. 324-326 (1972).

12. Sogin, H. H., *ASME Transactions*, Vol. 80, pp. 61-71 (1958).

13. Davidson, J. F. and D. Harrison, *Fluidization*, Academic Press, New York (1971).

14. Catipovic, N. M., T. J. Fitzgerald, A. H. George and J. R. Welty, *AIChE Journal*, Vol. 28, No. 5, pp. 714-720 (1982).

15. Canada, G. S. and H. H. McLaughlin, *AIChE Symposium Series*, Vol. 74, No. 176, pp. 27 (1978).

# DENSE BED, SPLASH ZONE, AND FREEBOARD HEAT TRANSFER IN A FLUIDIZED BED COMBUSTOR

S. Adibhatla and M. Boggs ■ Kentucky Center for Energy Research Laboratory
University of Kentucky, Institute for Mining and Minerals Research
P.O. Box 13015, Lexington, Kentucky 40512-3015

Heat transfer coefficients for vertical tube heat exchange surfaces in a coal fired fluidized bed combustor have been calculated from operating data. The data have been obtained from a pilot plant sized fluidized bed with a bed cross section of 2'8" x 2'5" (0.81m x 0.74m). Bed height information obtained from a series of pressure taps as well as measured coolant flows and temperatures have been used to calculate the bed to tube and the above-bed to tube heat transfer coefficients (two zone model) or the dense bed to tube, splash zone to tube, and the above-bed to tube coefficients (three zone model). These heat transfer coefficients are then compared to predictions using existing theoretical and empirical correlations.

Data were obtained for a number of steady-state test runs at different loads and bed heights, while superficial velocity, bed temperature, and freeboard temperature were kept approximately constant. Analysis of heat transfer to vertical 2" (5 cm) O.D. tubes in water-wall panels was performed. Using the two zone model, bed to tube heat transfer coefficients in the 41-66 Btu/hrft$^2$ °F (233-375 W/m$^2$K) range and above-bed to tube heat transfer coefficients in the 14-19 Btu/hrft$^2$ °F (79-108 W/m$^2$K) range were obtained. Using the three zone model, the dense bed to tube, splash zone to tube, and the above-bed to tube heat transfer coefficients were found to be 67-73, 26-40, and 18-24 Btu/hrft$^2$ °F (380-414, 148-227, and 102-136 W/m$^2$K), respectively.

## INTRODUCTION

Heat transfer to horizontal and vertical tubes has been studied by a number of researchers. Most of the studies were performed in the laboratory, using silica sand, glass beads, or other non-attriting materials rather than limestone, the most common bed material for fluidized bed coal combustors. Based on these studies, a number of theoretical and empirical correlations have been proposed for the estimation of heat transfer coefficients for surfaces immersed in the bed. These correlations attempt to model the effect of parameters such as particle diameter, solids concentration, solid and gas densities and thermal conductivities, etc. to varying extents. However, the implicit assumption in these models is that there are only two zones of interest, a bed zone and a freeboard zone.

Cold models, photographs of actual fluidized beds, and erosion measurements on immersed surfaces all indicate the presence of a splash zone in the region between the bed and the freeboard. This zone is characterized by a lower solids concentration and a more violent solids motion than the deeper part of the bed (the dense bed). As a rule, the ratio of the splash zone height to the dense bed height increases with decreasing bed height (decreasing load). The present investigation attempts to measure the heat transfer coefficients in the dense bed, splash zone, and the above-bed regions of a coal fired fluidized bed combustor using dolomites and limestones as the sulfur sorbents. The intent is to see whether the difference in dense bed and splash zone heat transfer coefficients is due to difference in solids concentration alone, and whether the two zones can be considered to be one for the purposes of heat transfer calculations.

## AFBC SYSTEM AND TEST CONDITIONS

The AFBC system at the Kentucky Center for Energy Research Laboratory (KCERL) consists of a 2.7 million Btu/hr (0.79 MW thermal) combustor with a 2'8" x 2'5" (0.81m x 0.74m) bed. The heat transfer surface consists of two removable water-wall panels and an external flue gas heat exchanger downstream of a hot cyclone. There is no freeboard heat exchange surface, but part of the water walls are exposed to the freeboard for bed heights less than 50" (1.27 m). This part of the water walls is being refered to as the "above-bed" heat transfer surface in this paper. The coolant is a water/glycol mixture and there is no steam generation. The entire system is well-instrumented, with a microprocessor-based data acquisition and control system.

All the test runs used in this study were conducted at a superficial velocity of 5.1-5.8 ft/s (1.55 to 1.77 m/s), bed temperature of 1500 to 1600°F (815 to 871°C), and freeboard temperature of 1600 to 1800°F (871 to 982°C). There was substantial freeboard burning, as indicated by the fact that freeboard temperature was higher than bed temperature. Overall carbon combustion efficiency was between 95% and 99%, whereas thermal efficiency was between 73% and 76%. Coal and limestone feedrates changed with changes in loads from one test run to another, as did the bed height. Coal feed size was 1/4" x 0" (6300 μm x 0 μm), limestone feed size was No. 6 x 0" (3350 μm x 0 μm), and typical bed material size was No.8 x No.100 (2360 μm x 150 μm). The specific gravity of the bed material (partially sulfated limestone), found by measuring the volume of water displaced by a known weight of the bed material, was 2.5. Its skeletal density was 2.88. The bulk density of the bed material, measured by weighing a known volume of the bed material, was 87 lb/ft$^3$ (1400 kg/m$^3$).

## IN BED HEAT EXCHANGE SURFACES

The IMMR AFBC in-bed heat transfer equipment consists of two water wall units made up of two inch (5.1 cm) outside diameter **vertical** tubes spaced 3.25" (8.3 cm) center to center. Each water wall has nine tubes three feet (0.91 m) in length, which are located 15" (38 cm) above the bottom of the bed. The side of the tube facing the wall is somewhat shielded from the bed material, since the tubes are less than 1/2" (13 mm) away from the wall.

## CALCULATION OF DENSE BED AND SPLASH ZONE HEIGHTS

The bed height measurement system consists of six pressure taps, located 14", 20", 26", 32", 38" and 44" (36, 51, 66, 81, 97 and 112 cm, respectively), above the bottom of the combustor.

First, the overall bed height is calculated on the basis of each pressure tap reading and the bed pressure gradient between that tap and the previous one. This is done only for those pressure taps for which the pressure difference between the pressure tap reading and the freeboard pressure is greater than 2" WC (0.5kPa). The overall bed height used in the two zone heat transfer model described in the next section is the mean of all the bed heights calculated using this procedure.

Next, the dense bed height is defined as the height of the last pressure tap for which the bed pressure gradient is greater than 0.37 psi per foot of bed (8.5 kPa per meter of bed). Note that this value is somewhat arbitrary, and is approximately 2/3 the typical bed pressure gradient for the dense part of the bed.

Finally, the splash bed height is defined as the height of the last pressure tap for which the difference between the pressure tap reading and freeboard pressure is greater that 2" WC (0.5 kPa), plus the pressure difference times the pressure gradient in the splash zone.

When overall bed height, dense bed height, and splash bed height are calculated as described above, the overall bed height will be larger than the dense bed height but less than the splash bed height. The overall bed height can be interpreted as the height of the bed that is obtained if the splash zone is compressed until its density is the same as that of the dense bed.

## TWO-ZONE AND THREE-ZONE HEAT TRANSFER MODELS

The overall heat transfer coefficient was calculated using the inlet and outlet coolant temperatures along with the bed temperatures and heat transfer area:

$$U_{ov} = Q_{tot}/(A \cdot \Delta T_{lm}) \tag{1}$$

For the IMMR/KCERL fluidized bed combustor, the tubes were very close to the combustor walls. Hence the effective area was taken to be only 90% of the actual tube surface area.

The tube side coefficients, $H_i$, were calculated according to the flow regime (laminar, transition, or turbulent) by means of the following empirical correlations ($\underline{1}$) to ($\underline{3}$)

$$H_i D_i/k_g = 1.86[Re_t Pr(L/D_i)]^{1/3}(CF)^{0.14} \tag{2}$$

$$H_i D_i/k_g = 2[(D_i/4L)Re_t Pr]^{1/3}(CF)^{0.14} \tag{3}$$

$$H_i D_i/k_g = (f/8)(Re_t Pr/X)(CF)^{0.11} \tag{4}$$

$$X = 1.07 + 12.7(f/8)^{1/2}(Pr^{2/3} - 1) \tag{5}$$

$$f = (1.82 \log_{10} Re_t - 1.64)^{-2}, \quad CF = \mu_g/\mu_{gw} \tag{6}$$

Equations (2), (3), and (4) are for Re < 2200, 2200 < Re < 10,000, and Re > 10,000, respectively.

After calculating $U_{ov}$ and $H_i$, $H_o$ was calculated using the series resistance equation:

$$H_O = 1/[1/U_{ov} - D_o/(D_i H_i)] \quad (7)$$

$H_O$ was then correlated to the fraction of the tube in the bed $X_b$ (calculated from the overall bed height), to determine the bed-to-tube coefficient, $H_{bt}$, and the gas to tube or above-bed coefficient, $H_{ab}$. A simple linear relationship of the form

$$H_O = H_{bt}X_b + H_{ab}(1-X_b) = (H_{bt}-H_{ab})X_b + H_{ab} \quad (8)$$

was used. A straight line fit to data for $H_O$ plotted against $X_b$ will therefore have a slope ($H_{bt} - H_{ab}$) and intercept $H_{ab}$, which facilitates calculation of $H_{bt}$ and $H_{ab}$.

The heat transfer analysis was then carried one step further. In Equation (8), it was assumed that the tubes were exposed to only two distinct heat transfer zones. In reality, three zones of different heat transfer and physical properties exist in a fluidized bed. These are the dense zone, splash zone, and above-bed (lower part of the freeboard) zone. The overall outside heat transfer coefficient $H_O$ calculated using Equation (7) is related to the the heat transfer coefficients in the three zones as follows:

$$H_O = (H_{bt} - H_{ab})X_b + (H_{st} - H_{ab})X_s + H_{ab} \quad (9)$$

which is an equation of the general form

$$y = ax_1 + bx_2 + c \quad (10)$$

Knowing the dense and splash zone heights for eight test runs, the heat transfer coefficients in each of the three zones, $H_{bt}$, $H_{st}$, and $H_{ab}$, can be calculated from Equation (10) above by means of a linear regression analysis.

## EXPERIMENTAL RESULTS

Experimental results using the two zone and three zone models are presented in Table 1 for five test series. Each test series involved eight of four hour steady-state test runs using the a different coal and limestone combination. The five test series consist of five different combinations of coal and limestone. As the numbers in the table indicate, measured values of heat transfer coefficient are reasonably consistent. Differences in the measured heat transfer coefficients for the different test series are partly

Table 1. Measured Values of Heat Transfer Coefficients using the Two-zone and Three-zone Models.

| Test Series No. | Two-zone Model | | Three-zone Model | | |
|---|---|---|---|---|---|
| | $H_{bt}^*$ | $H_{ab}^*$ | $H_{bt}^*$ | $H_{st}^*$ | $H_{ab}^*$ |
| 1 | 53 | 14 | 67 | 17 | 18 |
| 2 | 41 | 17 | 46 | 26 | 18 |
| 3 | 61 | 18 | 67 | 40 | 20 |
| 4 | 57 | 19 | 69 | 28 | 23 |
| 5 | 66 | 18 | 73 | 30 | 24 |

*Values are in Btu/hr ft$^2$ °F (Multiply by 5.678 to obtain W/m$^2$ °K)

statistical, but are probably due to inaccuracies in bed height measurement and slight differences in particle size, bed density, and composition of bed material.

Figure 1 illustrates the increase in splash zone height with decrease in overall bed height (decrease in load).

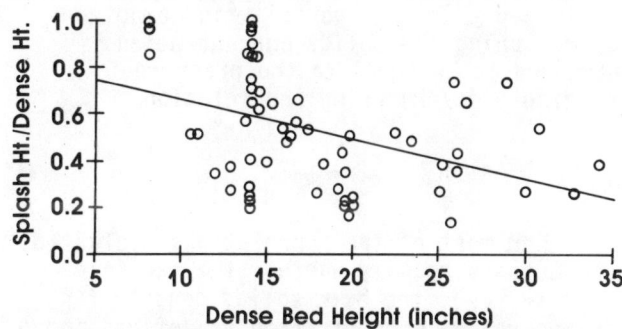

Figure 1. Variation in splash zone height with dense bed height.

## GENERAL REMARKS REGARDING HEAT TRANSFER MODELS

We follow the general lines of Verma & Saxena's(4) comparison of their experimental data to a number of correlations for heat transfer to vertical tubes in fluidized beds.

However, we also include a couple of correlations for horizontal tubes, and a correlation for heat transfer in the freeboard.

Prior to comparison of the measured heat transfer coefficients to the empirical and theoretical correlations, the following points need to be made:

(i) Some of the correlations are based

on experiments involving heat transfer to immersed (internal) surfaces, whereas others are for heat transfer to walls (external heat transfer surfaces).

(ii) Since bed material with a wide size distribution is usually less sensitive to velocity, and since the manner in which the average particle size, dp, is calculated for use in the correlation can make a substantial difference, the wide range of particle sizes used in our experiments needs to the emphasized. For instance, for the bed material collected in the pilot plant, the arithmetic mean ($d_p = \Sigma w_i d_{pi}$) is 0.00205 ft (625 μm) whereas the mean diameter based on the surface to volume ratio ($1/d_p = \Sigma w_i/dpi$) is 0.00143 ft (436 μm). Both these diameters have been used in evaluating the various correlations.

(iii) The bed porosity, assumed to be proportional to the pressure drop across unit height of the bed, is an important parameter for some of the correlations evaluated. The bed porosity is computed by measuring the solids concentration $\varrho_s$ (assumed to be equal to the pressure gradient, dp/dh) using the relation

$$1 - \varepsilon = (dp/dh)/(\varrho_p - \varrho_g) = \varrho_s/(\varrho_p - \varrho_g) \quad (11)$$

For most of the correlations evaluated, the authors specify whether the particle density ($\varrho_p$), the bulk solids density ($\varrho_b$), or the solids concentration as defined above ($\varrho_s$) was used. When the distinction was not clear, or if it was felt that the correlation might work better with $\varrho_s$ instead of $\varrho_p$, heat transfer coefficients were calculated using both values of density.

## THEORETICAL AND EMPIRICAL CORRELATIONS EVALUATED

Mickley and Trilling(7) fluidized glass spheres ($d_p$=40-450μm) in 2.875 inch (7.3 cm) I.D. internally and externally heated fluidized beds. They studied the effect of superficial velocity, particle size, and solids concentration on heat transfer. For the internally heated bed, they found that the gas velocity did not exert any effect independent of concentration, and proposed the following correlation:

$$h = 0.0433 \, (\varrho^2/dp^3)^{0.238} \quad \text{[English Units]}$$
$$= 0.028 \, (\varrho_s^2/dp^3)^{0.238} \quad \text{[S.I. Units]} \quad (12)$$

Miller & Logwinuk(8) measured heat transfer coefficients in a 2 inch (5.1 cm) bed. They fluidized silicon carbide, aluminum oxide, and silica gel using air, helium, and $CO_2$ as the fluidizing gas. They found that the heat transfer coefficient was independent of bed temperature in the range studied, almost independent of thermal conductivity of particles, and only slightly dependent on gas velocity. Effect of varying solids concentration was not considered. The correlation they proposed, with thermal properties being calculated at actual bed temperatures rather than at mean film temperature, is:

$$h = 1.5 X^{1.585} \quad \text{[English Units]}$$
$$= 6500 X^{1.585} \quad \text{[S.I. Units]} \quad (13a)$$

where, $X = G^{0.2} k_s^{0.045} k_g^{1.5}/dp^{0.6} C_{pg} \mu_g^{0.5}$ (13b)

Wender & Cooper(5) conducted a broad empirical study of nine independent sets of data on fluidized bed heat transfer. They proposed

$$y/F = 8.0 \, Re_p^{0.8} \quad (14a)$$

where, $y = (hd_p/k_g)/[(1-\varepsilon)(C_{ps}\varrho_p/C_{pg}\varrho_g)]$ (14b)

F = Correction Factor

= 1.0 for $L_H/D_{bed} > 7$

= 1 + 7.5 exp[-0.44($L_H/D_T$)/($C_{ps}/C_{pg}$)]

for $L_H/D_{bed} < 7$ (14c)

Zabrodsky, et al(9), while commenting on some of the correlations appearing in Russian publications, suggest the use of the following two correlations (S.I. units)

$$h_{max} = 35.7 \, k_g^{0.6} dp^{-0.36} \varrho_p^{0.2} \quad (15)$$

$$h_{max} = 0.88 \, Ar^{0.213} \quad (16)$$

In addition, they suggest that $h = 0.7 h_{max}$ to $0.8 h_{max}$ be used for design purposes.

Verma and Saxena(4) performed a number of studies using sand and millet seeds to study the influence of gas velocity, tube diameter, particle diameter, and bed size on heat transfer coefficient. They compared their data to a number of correlations and concluded that the correlation in Equation (15) above best described their data. They also suggested a modified form of the correlation proposed by Baerg, et al(10) (S.I. units):

$$h_{max} = 105.8 \ln [7.05 \times 10^{-6} \varrho_s/d_p] \quad (17)$$

but found that Baerg's correction for velocity, given as

$$h = h_{max} - 312 \exp[-8.85 (G - 6 \times 10^{-5} \varrho_b)] \quad (18)$$

did not describe their data adequately.

For horizontal tubes, Grewal and Saxena(11) proposed the following correlation, based on their experiments using electrically heated horizontal tube bundles in square beds of silica sand and alumina. They proposed (S.I. units)

$$h = 0.9 (Ar \cdot 0.0127/D_T)^{0.21}$$
$$(Cps/Cpg)^{0.2} (k_g/d_p)(CF) \quad (19)$$

where CF is a correction factor for tube pitch of the bundle.

Lapple, Viliamas, and Wenderoth(12) used data collected on the Babcock and Wilcox 36-square foot (3.34 m$^2$) fluidized bed facility and used the following relationship (Reynolds number corrected for bed voidage):

$$(Nu_f/Pr^{0.3} - 0.35)/(0.47 Re_f^{0.52}) = Y \quad (20)$$

For the zero dust loading limit, Y has the value of unity. In the presence of solids, they reasoned that the heat transfer coefficient is improved and Y should have a value greater than unity. They provided a correlation for Y based on their experiments, as well as correction terms for tube geometry and radiation, but the details will not be presented here.

Osman et al(13) used a similar argument to model heat transfer in the freeboard region. They modified well known correlations for heat transfer to vertical tubes by multiplying by a constant to account for elutriated solids in the gas stream. Their correlation was (S.I. units)

$$Nu = A Re^{0.55} Pr^{1/3} (\mu_g/\mu_{gw})^{0.14} \quad (21)$$

By dropping the viscocity correction term ($\mu_g/\mu_{gs}$), they found that A=3.95 for $d_p$=222μm and A=2.6 for $d_p$=488 and 778μm described their data best. For the no dust loading case, A=0.36.

## COMPARISON OF MEASURED VALUES TO EMPIRICAL & THEORETICAL CORRELATIONS

Heat transfer coefficients calculated by each of the models described above are compared to the measured values in Table 2. Values calculated using the theoretical and empirical correlations include two definitions of mean particle diameter, and, where appropriate, two definitions of density of bed material.

The Mickley and Trilling correlation, equation (12), seems to be best suited for prediction of h in the splash zone. For the dense bed, it underpredicts heat transfer coefficient values.

Table 2. Comparison of Measured Values to Values Calculated using Empirical and Theoretical Correlations.

| | | $d_p = (\Sigma w_i/d_{pi})^{-1}$ = 436μm | | $d_p = \Sigma w_i d_{pi}$ = 625μm | |
|---|---|---|---|---|---|
| | | $h_{bed}$* | $h_{splash}$* | $h_{bed}$* | $h_{splash}$* |
| Measured | | 67-73 | 26-40 | 67-73 | 26-40 |
| Mickley & Trilling | (Eqn. 12) | 36 | 29 | 28 | 22 |
| Miller & Logwinuk | (Eqn. 13) | 125 | - | 125 | - |
| Wender & Cooper | (Eqn. 14) | 34 | 21 | 32 | 20 |
| Zabrodsky | (Eqn. 15) | | | | |
| $\varrho = \varrho_p$ | | 84 | - | 74 | - |
| $\varrho = \varrho_s$ | | 73 | 66 | 64 | 58 |
| | (Eqn. 16) | | | | |
| $\varrho = \varrho_p$ | | 80 | - | 71 | - |
| $\varrho = \varrho_s$ | | 69 | 62 | 61 | 55 |
| Verma & Saxena | (Eqn. 17) | | | | |
| $\varrho = \varrho_p$ | | 69 | - | 62 | - |
| $\varrho = \varrho_s$ | | 56 | 47 | 49 | 40 |
| Grewal & Saxena | (Eqn. 18) | | | | |
| $\varrho = \varrho_p$ | | 59 | - | 52 | - |
| $\varrho = \varrho_s$ | | 51 | 46 | 44 | 40 |
| Lapple, et al. | (Eqn. 19) | 154 | 127 | 115 | 95 |

* Values are in Btu/hr ft$^2$ °F (Multiply by 5.678 to obtain W/m$^2$ °K)

The Miller and Logwinuk correlation, Equation (13), overpredicts the heat transfer coefficient, although its predictions seem to be better if the arithmetic mean particle diameter is used.

The Wender and Cooper model, Equation (14), is similar to the Mickley and Trilling correlation, in that it predicts splash zone heat transfer coefficients reasonably well while underestimating dense bed heat transfer coefficient.

The first of the Zabrodsky correlations, Equation (15), overestimates the dense bed h if $\varrho = \varrho_p$ is used. However, if $\varrho = \varrho_s$ is used, the prediction is very good. The second Zabrodsky correlation, equation (16), is equally good. Neither of the correlations does well in predicting the decrease in h with decrease in solids concentration observed in the splash zone.

The Verma and Saxena correlation, Equation (17), predicts the dense bed heat transfer coefficient very well, with $\varrho = \varrho_p$.

With $\varrho = \varrho_s$, the predicted heat transfer coefficient matches the measured overall in-bed heat transfer coefficient. However, it overestimates the splash zone heat transfer coefficient.

As expected, the horizontal tube bundle correlations of Grewal and Saxena (Equation 19) and Lapple et al. (Equation 20) do not lead to good predictions. Interestingly, one underestimates heat transfer coefficients while the other overestimates them.

The correlation of Osman et al., Equation (21), is incapable of predicting the above-bed heat transfer coefficient. This is to be expected, since the solids concentration near the tubes just above the bed is likely to be higher than the solids concentration in the higher regions of the freeboard. A value of A=7.8 in Equation (21) describes our data well.

## CONCLUSIONS

(a) Typical values for heat transfer to vertical tubes near the combustor wall (water wall tubes) in a coal fired AFBC pilot plant are:
Dense bed to tube = 71 Btu/hrft$^2$°F (403 W/m$^2$K)
Splash zone to tube = 26 to 40 Btu/hrft$^2$°F
                     (148 to 227 W/m$^2$S)
Above-bed to tube = 18 Btu/hrft$^2$°F (102 W/m$^2$K)

Note that, as detailed in earlier sections, the definition of splash zone used in this paper is somewhat arbitrary. However, the differences in measured heat transfer coefficients are large enough to warrant distinguishing between the dense bed and splash zones. Alternately, an overall bed height that includes both the dense and splash zones (see (e) below) can be used. In such a case, the typical bed to tube heat transfer coefficient is 53 to 66 Btu/hrft$^2$°F (300 to 375 W/m$^2$K).

(b) Dense bed heat transfer is best predicted by the correlations of Zabrodsky (Equations 15 and 16, with $\varrho = \varrho_s$) and the correlation of Verma and Saxena (Equation 17, with $\varrho = \varrho_p$), with mean particle diameter based on the surface to volume ratio, $d_p = (\Sigma w_i / dp_i)^{-1}$.

(c) Splash zone heat transfer coefficient is best predicted by the correlations of Mickley and Trilling (Equation 12) and Wender and Cooper (Equation 14). All other correlations overestimate this heat transfer coefficient. This indicates that there may be a need to correct for differences in bed solids concentration in the different zones by including a term such as $(\varrho_s/\varrho_p)^n$ or $(\varrho_s/\varrho_b)^n$ in the correlations for heat transfer coefficient, where n is empirically determined.

(d) Above-bed heat transfer coefficient can be estimated using Equation (21), with the constant in the equation, A, equal to 7.8.

(e) If the distinction between dense and splash zones is not made, the definition of bed height will not usually take into account the gradual decrease in bed density. That is, the bed height can be calculated as the equivalent height of bed that is obtained by compressing the splash zone until its density is the same as the density of the dense, lower part of the bed. In such a case, the two-zone model of heat transfer can be used. Heat transfer coefficient is then best predicted by the equations of Zabrodsky (Equations (15) and (16)), (Verma and Saxena (Equation 17), and Grewal and Saxena (Equation 19). See Table 2.

The advantage of the two zone model over the three zone model is that it is easier to calculate overall bed height than it is to calculate dense and splash zone heights. The disadvantage is that the variation in splash height zone with bed height, which can be an important factor in designing the placement of heat transfer surfaces for operation over a wide range of loads (bed heights), is not modeled accurately by the two zone model. Perhaps a combination of the two approaches, with the results of one being used to check the results from the other, is the best design philosophy.

## ACKNOWLEDGEMENTS

This work was funded by the Kentucky Energy Cabinet. The authors would also like to thank the other members of the AFBC group for their assistance.

## LITERATURE CITED

1. E.N.Sieder and G.E.Tate, Ind. Eng. Chem., Vol.28, p.1429, (1936).
2. McCabe, L.Warren, and J.C.Smith, Unit Operations of Chemical Engineering, 3rd ed., McGraw-Hill, Inc., New York, 1976.
3. Petukhov, B.S., Advances in Heat Transfer, (J.P. Hartnett and T.F. Irvine, eds.), Academic Press, New York, pp. 504-564, 1970.
4. R.S.Verma and S.C.Saxena, Energy, Vol.8, No.12, pp. 909-925, 1983.
5. L.Wender and G.T.Cooper, A.I.Ch.E.J., Vol.4, No.1, pp. 15-23, 1958.
6. H.A.Vreedenberg, J. Appl. Chem., Vol.2, Supplementary Issue No.1, pp. S26-S33, 1952.
7. H.S.Mickley and C.A.Trilling, Ind. Eng. Chem., Vol.41, No.6, pp. 1135-1147, June 1949.
8. C.O.Miller and A.K.Logwinuk, Ind. Eng. Chem., Vol.43, No.5, pp. 1220-1226, May 1951.
9. S.S.Zabrodsky, N.V.Antonishin, and A.L.Parnas, Can. J. Chem. Eng., Vol. 54, pp. 52-58, Feb./Apr., 1976.
10. A.Baerg, J.Classen, and P.E.Gishler, Can. J. Res., Vol. 28F, p287, 1950.
11. N.S.Grewal and S.C.Saxena, Ind. Eng. Chem. Proc. Des. Dev., Vol. 22, No. 3, pp. 367-376, 1983.
12. W.C.Lapple, V.K.Viliamas, and F.H.Wenderoth, Fuel Proc. Tech., Vol. 7, pp. 239-260, 1983.
13. M.I.Osman, S.N.Upadhyay, and S.C.Saxena, Energy, Vol.7, No.5, pp. 465-472, 1982.

## NOTATION

| | |
|---|---|
| A | Heat transfer area, $ft^2$ ($m^2$) |
| Ar | Archimedes number, $Gd_p^3 \rho_g(\rho - \rho_g)/\mu_g^2$; $\rho = \rho_p$ or $\rho_s$ |
| $c_{pg}$ | Specific heat of fluidizing gas, Btu/lb°F (kJ/kgK) |
| $c_{ps}$ | Specific heat of bed material, Btu/lb°F (kJ/kgK) |
| $d_p$ | Mean particle diameter, ft (m) |
| $d_{pi}$ | Diameter of particle in the i-th size range, ft (m) |
| $D_{bed}$ | Equivalent diameter of fluidized bed, ft (m) |
| $D_i$ | Inside diameter of tube, ft (m) |
| $D_T$ | Outside diameter of tube, ft (m) |
| f | Friction factor, dimensionless |
| G | Mass gas velocity, $\rho_g \times$ superficial velocity, lb/ft²hr (kg/m²s) |
| h | Heat transfer coefficient in the bed, Btu/hr ft²°F (W/m²K) |
| $h_{max}$ | Maximum in-bed heat transfer coefficient, Btu/hrft²°F (W/m²K) |
| $H_{bt}$ | Bed-to-tube heat transfer coefficient, Btu/hrft²°F (W/m²K) |
| $H_{ab}$ | Above-bed or gas-to-tube heat transfer coefficient, Btu/hr ft²° (W/m²K) |
| $H_i$ | Tube side heat transfer coefficient, Btu/hr ft²°F (W/m²K) |
| $H_o$ | Outside heat transfer coefficient, Btu/hr ft²°F (W/m²K) |
| $H_{st}$ | Splash zone-to-tube heat transfer coefficient, Btu/hr ft²°F (W/m²K) |
| $k_g$ | Thermal conductivity of fluidizing gas, Btu/hr ft°F (W/mK) |
| $k_s$ | Thermal conductivity of bed solids, Btu/hr ft°F (W/mK) |
| L | Length of tube, ft (m) |
| $L_H$ | Heated tube length, ft (m) |
| $Nu_p$ | Nusselt number based on particle size, $hd_p/k_g$ |
| $Nu_t$ | Nusselt number based on tube diameter, $hD_T/k_g$ |
| Pr | Prandtdl number, $\mu_g c_{pg}/k_g$ |
| $Q_{tot}$ | Total heat transfer rate, Btu/hr (W) |
| $Re_p$ | Reynolds number based on particle diameter, $Gd_p/\mu_g$ |
| $Re_t$ | Reynolds number on tube diameter, $GD_T/\mu_g$ |
| $U_{ov}$ | Overall heat ransfer coefficient, Btu/hr ft²°F (W/m²K) |
| $X_b$ | Fraction of tube in bed |
| $X_s$ | Fraction of tube in splash zone |
| $T_{lm}$ | Log mean temperature difference, °F (°K) |

## GREEK SYMBOLS

| | |
|---|---|
| $\varepsilon$ | Bed Voidage, dimensionless |
| $\mu_g$ | Viscocity of fluidizing ghas, Ns/m² (lb/ft hr) |
| $\mu_{gw}$ | Viscocity of gas at tube wall temperature, Ns/m² (lb/ft hr) |
| $\rho_b$ | Bulk density of bed solids, lb/ft³ (kg/m³) |
| $\rho_g$ | Density of fluidizing gas, lb/ft³ (kg/m³) |
| $\rho_p$ | Particle density of bed material, lb/ft³ (kg/m³) |
| $\rho_s$ | Solids concentration (Density of expanded bed), lb/ft³ (kg/m³) |

# HEAT TRANSFER IN TURBULENT FLUIDIZED BEDS

P. Basu ■ Center for Energy Studies
R. Dieh ■ Dept. of Chemical Engineering, Technical University of Nova Scotia, Halifax, Canada B3J 2X4

Experiments were carried out in a 102.2 mm diameter and 2000 mm tall fluidized bed of 75 micron sand and 50 micron FCC particles. An inverted cone section was attached to the top of the bed to reduce entrainments. Pressures were measured at three levels along the height of the bed. A heat transfer probe was located flushed with the bed wall. The heat transfer probe was a 25mm diameter x 100 mm long copper cylinder. Four thermocouples were located along the length of the cylinder. The insulated probe was heated by hot water from one end and temperatures were measured along its length. These measurements enabled calculation of heat flux along the probe and its wall temperature.

The bed was operated in the turbulent regime of fluidization. Heat transfer coefficients were calculated from above data and these were plotted against the velocity and voidage. The heat transfer coefficient in the transition zone between the turbulent and bubbling regime is of the same order of magnitude as that in the dense bubbling fluidized bed even though the voidage in this zone is as high as .75-0.9. The heat transfer coefficient increases with velocity but at a rate slower than that found in the dense bubbling beds. This has been attributed to the enhanced particle mobility in this regime. Experimental results are found to correlate well with the nondimensional plot of Wender and Cooper.

# HEAT TRANSFER IN GLASS

R. Viskanta ■ School of Mechanical Engineering, Purdue University, West Lafayette, IN 47907

This paper reviews heat transfer processes in glass which are relevant to glass manufacturing. The concept of radiative conductivity (i.e., diffusion approximation for radiation transfer) for modeling radiative transfer in glass is reexamined. Results of calculations show that the approximation yields reasonably accurate temperature distribution predictions in 10 cm or thicker glass layers, but not of temperature gradients at the opaque and transparent boundaries. The emphasis in the review is on modeling of heat transfer and particularly of radiation transfer in glass.

## INTRODUCTION

Glass manufacturing (melting, forming, annealing and tempering) requires knowledge of temperature distribution and depends on controlled heat addition and removal from glass. The importance of heat transfer in modern glass technology has been well recognized and discussion of problems is available [1-3]. The purpose of this review is to reexamine some old concepts and to discuss recent advances in heat transfer related to glassmaking.

Perhaps the first question that should be raised is why should we be concerned about heat transfer in glass? Manufacturers are capable of producing good quality glass without fully understanding all of the complexities associated with heat transfer. The main motivation to do this is that the better is heat transfer understood, the more effective can the knowledge of heat transfer be applied to optimize the individual processes involved, lead to insights of other physiochemical phenomena in glass manufacturing and may produce improvements in energy efficiencies.

A review process is a somewhat arbitrary activity, because of the decision one must make on what to include, what to omit and where to start. This paper should be considered as an update of previous reviews on the subject [1-3]. Space limitations do not permit thorough discussion of an extremely complex and diverse topic; therefore, the focus will be on more fundamental problems. No claim is made of the completeness of the review. In these days of many journals and other publications, it is possible that some relevant works may have been overlooked and others may have been purposely left out because space constraints did not permit their discussion.

## MODES OF HEAT TRANSFER IN GLASS

It has been recognized for some time [4] that glass is a semitransparent (i.e., to the visible and infrared parts of the spectrum) material. The first and the most important characteristic of this semitransparency is that absorption and emission of thermal radiation are volume and not surface phenomena, as with opaque materials. The property of glass which characterizes its opacity is the absorption coefficient. This coefficient has a profound effect on the nature of energy transport by radiation in glass. The radiant energy transfer in glass is governed by the radiative transfer equation (RTE). On a phenomenological level, this equation represents conservation of radiant energy on a pencil of radiation confined to an elementary solid angle $d\Omega$, in the wavelength interval $d\lambda$ and time interval $dt$. A detailed discussion of the fundamentals of radiative transfer in glass is available [5].

As in ordinary fluids, heat in glass is transferred by conduction, radiation and convection if glass is at sufficiently high temperature so that it flows. Conduction and convection in glass obey the laws and principles of ordinary fluids. These two modes of heat transfer are discussed in elementary textbooks [6] and need not be reviewed here. For a virtually incompressible liquid such as glass, the transient conservation of energy equation is

$$\rho c \left(\frac{\partial T}{\partial t} + \overline{v} \cdot \nabla T\right) = -\nabla \cdot \overline{q} + \dot{q} \qquad (1)$$

where the total (conductive plus radiative) heat flux vector $\overline{q}$ is given by

$$\overline{q} = -k\nabla T + \overline{F} \qquad (2)$$

and $\dot{q}$ denotes the volumetric rate of heat production (or absorption) due either to chemical reactions and/or Joulean heat dissipation. The conductive flux vector $(-k\nabla T)$ is given by Fourier-Biot law, with k being the thermal (phonon) conductivity. The radiative flux vector $(\overline{F})$ can be determined from the knowledge of the radiation field [5]. Of course, the velocity field $(\overline{v})$ would have to be given or computed in order to solve for the temperature distribution from Eq.(1).

Because of the "long range" nature of radiant energy transport, the expressions for the radiative flux vector or its components are in terms of integrals over space and spectrum [5]. They are usually quite complex and are very tedious to evaluate numerically, particularly for multidimensional problems. Radiative transfer in a two-dimensional rectangular geometry has been formulated [7-9]. An up to date review of literature on radiation transfer and combined conduction and radiation as well as convection and radiation is available [10], and there is no need to repeat this survey. Suffice it to mention that as a result of the complexity of the expressions for the radiative flux in multidimensional geometries, various approximate methods have been devised to predict radiation transfer and are amply detailed in radiation heat transfer literature [5,11,12].

The author has been able to identify only a single study in which the transient temperature distribution was predicted in a two-dimensional semitransparent solid (i.e., optical window) [13]. Most of the other analyses of heat transfer are one-dimensional, in which either rigorous integral or approximate formulations of radiative transfer are employed. The two- and three-dimensional problems arising in modeling of heat transfer in glass melting tanks and forehearths employ the diffusion approximation for radiative transfer.

## TRANSIENT HEAT CONDUCTION AND RADIATION IN A PLATE OF GLASS

In order to illustrate the salient features of heat transfer in glass and of integral formulation of radiation transfer in particular, we examine a geometrically simple physical situation. Consider transient heating or cooling (i.e., heat treating, annealing) of a sheet (plate) of glass. Suppose a one-dimensional planar layer of glass is heated (or cooled) by convection and radiation. The glass is assumed to be homogeneous, absorbing and emitting material, and scattering is neglected in comparison to absorption. The free surfaces of the glass plate are assumed to be optically smooth and their reflection and transmission characteristics are predicted from classical electromagnetic theory [12]. Diffuse radiation fluxes are assumed to be incident on the free surfaces of the layer from some external radiation sources.

Based on the above model, the transient, one-dimensional energy equation reduces to

$$\rho c \frac{\partial T}{\partial t} = \frac{\partial}{\partial z}\left(k \frac{\partial T}{\partial z}\right) - \frac{\partial F}{\partial z} \qquad (3)$$

For a plane layer of glass the local radiative flux can be formulated using the analysis given elsewhere [5]. Omitting the details, one can show that the local flux can be expressed as

$$F(z,t) = \int_0^\infty F_\lambda(z,t) d\lambda$$

$$= \int_0^\infty [F_\lambda^+(z,t) - F_\lambda^-(z,t)] d\lambda \qquad (4)$$

where the radiative fluxes $F_\lambda^+$ and $F_\lambda^-$ in the forward and backward directions (see Fig.1), respectively, are given by

$$F_\lambda^+(z,t) = F_{1\lambda}^0 \int_0^{\pi/2} [\tau_{1\lambda}(\theta^0) e^{-\tau_\lambda/\cos\theta}$$
$$+ \tau_{2\lambda}(\theta^0) \rho_{1\lambda}(\theta) e^{-(\tau_{H\lambda} + \tau_\lambda)/\cos\theta}]$$
$$\gamma(\tau_{H\lambda},\theta) \cos\theta \sin\theta \, d\theta$$
$$+ 2\int_0^{\tau_\lambda} n_\lambda^2 E_{b\lambda}[T(\xi)] E_2(|\tau_\lambda - \xi|) d\xi$$

$$+ 2 \int_0^{\pi/2} \{ \int_0^{\tau_{H\lambda}} n_\lambda^2 E_{b\lambda}[T(\xi)][e^{-(\tau_\lambda+\xi)/\cos\theta}$$

$$+ \rho_{1\lambda}(\theta)e^{-(2\tau_{H\lambda}-\tau_\lambda+\xi)/\cos\theta} ]$$

$$\rho_{2\lambda}(\theta)\gamma_\lambda(\tau_{H\lambda},\theta)d\xi\}d\theta \quad (5)$$

and

$$F_\lambda^-(z,t) = F_{2\lambda}^o \int_0^{\pi/2} [\tau_{1\lambda}(\theta^o)\rho_{2\lambda}(\theta)$$

$$+ e^{-(2\tau_{H\lambda}-\tau_\lambda)/\cos\theta}$$

$$+ \tau_{2\lambda}(\theta^o)e^{-(\tau_{H\lambda}-\tau_\lambda)/\cos\theta}$$

$$\gamma(\tau_{H\lambda},\theta)\cos\theta\sin\theta d\theta$$

$$+ 2 \int_\xi^{\tau_{H\lambda}} n_\lambda^2 E_{b\lambda}[T(\xi)]E_2(|\xi-\tau_\lambda|)d\xi$$

$$+ 2 \int_0^{\pi/2} \{ n_\lambda^2 E_{b\lambda}[T(\xi)][e^{-(2\tau_{H\lambda}-\xi-\tau_\lambda)/\cos\theta}$$

$$+ \rho_{2\lambda}(\theta)e^{-(2\tau_{H\lambda}-\tau_\lambda+\xi)/\cos\theta} ]$$

$$\rho_{1\lambda}(\theta)\gamma(\tau_{H\lambda},\theta)\sin\theta d\xi\}d\theta \quad (6)$$

where the function $\gamma(\tau_{H\lambda},\theta)$ accounts for the interreflections between the two faces of the plate and is given by

$$\gamma = [1 - \rho_{1\lambda}(\theta)\rho_{2\lambda}(\theta)\exp(-2\tau_{H\lambda}/|\cos\theta|)]^{-1} \quad (7)$$

In Eqs. (5) and (6) the optical depth $\tau_\lambda$ and the optical thickness (opacity) $\tau_{H\lambda}$ of the plate are defined as

$$\tau_\lambda = \int_0^z \kappa_\lambda(z)dz \quad \text{and} \quad \tau_{H\lambda} = \int_0^H \kappa_\lambda(z)dz \quad (8)$$

The physical meaning of the terms in Eqs.(5) and (6) is clear. For example, the first term on the right-hand-side of Eq.(5) represents the contribution to the forward flux due to the external radiation field incident on surface 1 and the second and third terms account for emission and absorption of radiation by glass between boundary 1 and plane z taking into account all multiple interreflections between the two interfaces.

The boundary condition on temperature on face 1 at z=0 is assumed to be of the convective type,

$$[-k\frac{\partial T}{\partial z}]_{z=0} = h_1[T - T(z,t)]_{z=0} \quad (9)$$

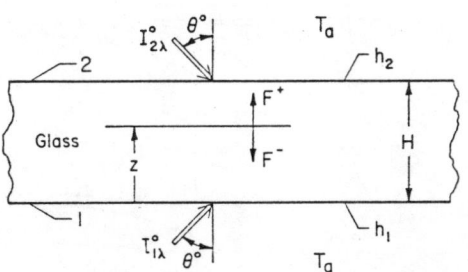

Fig.1. Schematic diagram and coordinate system.

and a similar one on face 2 at z=H. To complete the mathematical statement of the problem, the initial temperature of the plate (at t=0) is assumed to be specified.

A rigorous formulation of the expression for the radiative flux in a somewhat different form was reported over twenty five years ago by Gardon [14]. It was given here to show that mathematically correct formulations of radiative transfer problems in simple geometries are possible, but the expressions for the flux are very complex. The equations are time consuming to evaluate numerically even on fast digital computers.

The foregoing analysis has been applied to predict the transient temperature distribution in glass plates undergoing various heat treatments [14]. The results presented in Fig. 2 are intended to illustrate the effect of transparency on the thermal history for a plate being cooled by forced or natural convection. Comparisons are therefore made between two plates in which heat transfer is only by conduction and also by both conduction and radiation. The results clearly illustrate strong interaction between conduction and radiation and reveal the importance of radiation on the rate of cooling and on the temperature gradients in the glass near the free surface for the conditions considered. Neglect of radiation in comparison to conduction greatly underestimates the rate of cooling.

Even for large plates, near the edges heat transfer cannot be treated as one-dimensional. However, formulation of radiation transfer in two- or three-dimensions is rather formidable and does not appear to have been attempted.

## RADIATION DIFFUSION IN GLASS

The expression for the local radiative flux even for a planar geometry is complicated, and a much simpler expression for

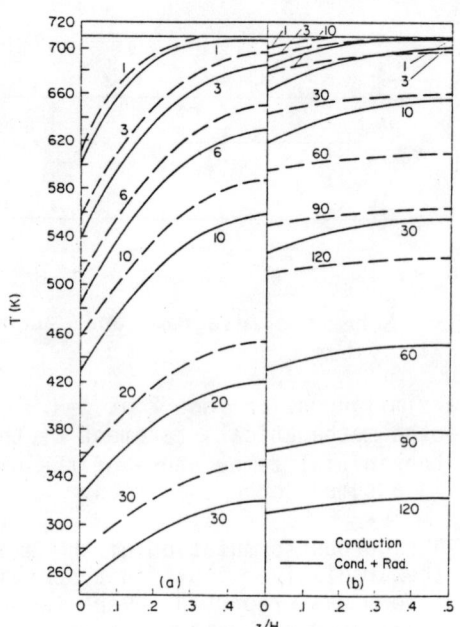

Fig.2. Temperature distribution in a 6 mm float glass plate initially at $707^\circ$C: (a) cooled by forced convection, $h_1 = h_2 = 210$ W/m$^2$ K and (b) cooled by natural convection, $h_1 = h_2 = 21$ W/m$^2$K. The numbers on the figure refer to time in seconds.

predicting the flux would be highly desirable. Consider the special case when there is no external radiation incident on the boundaries ($F_{1\lambda}^0 = F_{2\lambda}^0 = 0$), and the radiation characteristics of the two interfaces of the plate are the same [$\tau_{1\lambda}(\theta^0) = \tau_{2\lambda}(\theta^0) = \tau_\lambda$ and $\rho_{1\lambda}(\theta) = \rho_{2\lambda}(\theta) = \rho_\lambda^1$]. If it is assumed that glass is strongly absorbing (i.e., the spectral absorption coefficient $\kappa_\lambda$ is large), expansion of Planck's function in a Taylor series about z and retention of first order terms yields the following expression for the local radiative flux,

$$F = -\frac{4}{3} \int_0^\infty \frac{n_\lambda^2}{\kappa_\lambda} \frac{\partial E_{b\lambda}}{\partial T} \{1 - \rho_\lambda[E_3(\tau_\lambda) + E_3(\tau_{H\lambda} - \tau_\lambda)]\} d\lambda \frac{dT}{dz} \quad (10)$$

The above expression is valid only at optical distances far away from the boundaries, i.e., for which $\exp(-\int_0^\delta \kappa_\lambda dz) \ll 1$ and $\exp(-\int_{H-\delta}^z \kappa_\lambda dz) \ll 1$, where $\delta$ is the distance from the boundary. Comparison of Eq.(10) with the Fourier-Biot law for heat conduction reveals that the local radiative flux can be expressed as

$$F = -k_{rad}(T) \frac{dT}{dz} \quad (11)$$

where $k_{rad}(T)$ can be interpreted as a "radiative conductivity,"

$$k_{rad} = \frac{4}{3} \int_0^\infty \frac{n_\lambda^2}{\kappa_\lambda} \frac{\partial E_{b\lambda}}{\partial T} \{1 - \rho_\lambda[E_3(\tau_\lambda) + E_3(\tau_{H\lambda} - \tau_\lambda)]\} d\lambda \quad (12)$$

In the vicinity of the boundaries, $\tau_\lambda = \tau_{H\lambda} - \tau_\lambda \simeq O(1)$, the radiative conductivity $k_{rad}$ depends on the distance and therefore is not a property of glass. At distances optically far away from the boundaries, $\tau_\lambda \gg 1$ and $\tau_{H\lambda} - \tau_\lambda \gg 1$, Eq.(12) reduces to the Rosseland radiative conductivity,

$$k_R(T) = \frac{4}{3} \int_0^\infty \frac{n_\lambda^2}{\kappa_\lambda} \frac{\partial E_{b\lambda}}{\partial T} d\lambda = \frac{16 n^2 \sigma T^3}{3 \kappa_R} \quad (13)$$

where $\kappa_R$ is the Rosseland mean absorption coefficient and is defined as

$$n_\lambda^2 / \kappa_R = \int_0^\infty (\frac{n_\lambda^2}{\kappa_\lambda}) \frac{dE_{b\lambda}}{dT} d\lambda / \int_0^\infty \frac{dE_{b\lambda}}{dT} d\lambda \quad (14)$$

The index of refraction of glass $n_\lambda$ is usually a weak function of wavelength and can readily be averaged over the spectral range of interest.

If radiation can be treated as a diffusion process, the Rosseland radiative conductivity can be considered as a true property of glass, because it is only a function of chemical composition and temperature of glass. Once the spectral absorption coefficient $\kappa_\lambda(T)$ is known, the Rosseland radiative conductivity can be evaluated from Eq.(13). For common commercial glasses the radiative conductivities have been determined [15,16].

### EFFECTIVE CONDUCTIVITY OF GLASS

Heat transfer across a layer of glass heated on one face and cooled on another has been studied both analytically [17-19] and experimentally [18,19]. The interferometrically measured temperature distribution was found to be in good agreement with experimental data [18,19]; however, the extent of the interaction between conduction and radiation was relatively small because the temperatures were in general below 950 K. A priori knowledge of the spectral infrared radiative properties of glass permitted determination of the true thermal (phonon) conductivity [19].

Experimental measurements [19-21] have clearly established that the effective (phonon plus radiative) conductivity of glass, $k_{eff} = k + k_{rad}$, is not only a function of temperature but also of the sample thickness and of the radiative properties of the sample faces. This finding is not surprising and should have been expected from Eq.(12). Unfortunately, the dependence of $k_{rad}(T)$ on thickness and radiative surface properties causes problems in the use of $k_{eff}(T)$ information in practical heat transfer and temperature distribution calculations.

The concept of the effective conductivity is very simple and allows easy treatment of heat transfer in glass. Unfortunately, the concept of radiative conductivity breaks down near opaque and/or transparent boundaries and when the opacity of glass is not sufficiently large. In order to examine the range of validity of the diffusion approximation in glass, some calculations have been performed for a simple problem [22]. Steady-state heat transfer by combined conduction and radiation through a plane layer of glass on an opaque, diffuse substrate heated at the free surface by convection and radiation from the combustion space has been modeled. The free, optically smooth surface was exposed to a diffuse radiation flux incident on it.

Some temperature distributions calculated for container glass are illustrated in Fig.3. The results show that the diffusion approximations for radiative transfer in glass and the concept of radiative conductivity yields reasonable predictions for the temperature distribution, but large errors may result in the predicted temperature gradients in comparison to the profiles obtained from a model based on the integral formulation of radiative transfer in glass [22]. This is even true for a glass layer which is 30 cm thick. The modified diffusion approximation, Eq.(11), yields improved predictions for the temperature distribution near the boundaries, but the improvement is marginal. The results confirm that the concept of the radiative conductivity and the diffusion approximation for radiative transfer in glass should be used with extreme caution if correct temperature distribution and particularly temperature gradient predictions are to be expected near opaque and transparent boundaries in glass. The approximations should not be used to predict temperatures and heat transfer in glass if the opacity of glass is not sufficiently large.

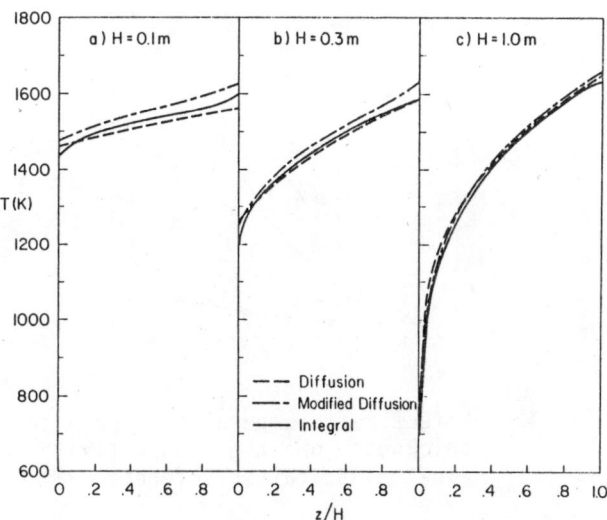

Fig.3. Comparison of temperature distributions in a layer of float glass heated by convection and radiation (after Reference 20).

In the past, the dependence of $k_{eff}$ on the sample thickness and radiative boundary conditions has not been recognized, and the differences in the measured conductivities have been attributed to the experimental techniques employed [23]. Recent measurements have clearly demonstrated that there is no single general method that would be equally applicable for all glasses and sample thicknesses [24]. The experimental results depend strongly on the sample thickness and show that the radiative conductivity can be greater than the effective conductivity (Fig.4). The findings suggest that the experimentally determined effective conductivity data for colorless glass of insufficient optical thickness obtained by some methods should be used with extreme caution. It is recommended to evaluate the published data from the standpoint of the experimental method and the sample thickness used in the measurements. For strongly absorbing glasses the measured $k_{eff}$ values are physical properties and can be generally used. However, for weakly absorbing colorless glasses the measured values of $k_{eff}$ are not appropriate for the calculation of the temperature distribution in glass [24]. For these glasses $k_{rad}$ is suitable for calculation of the temperature distribution only in the interior of thick glass layers.

## HEAT TRANSFER IN NONHOMOGENEOUS GLASSMELT

In the preceding discussion of heat transfer it was implied that glass is a homo-

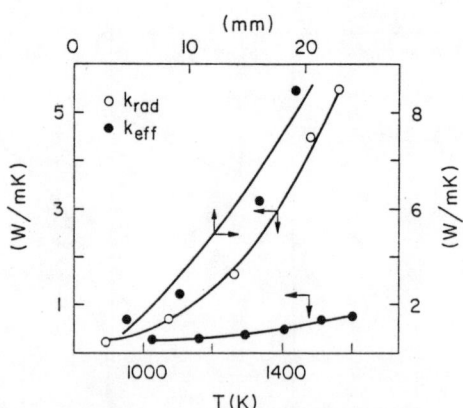

Fig.4. Effect of temperature and layer thickness on the conductivity of float glass (after Reference 22).

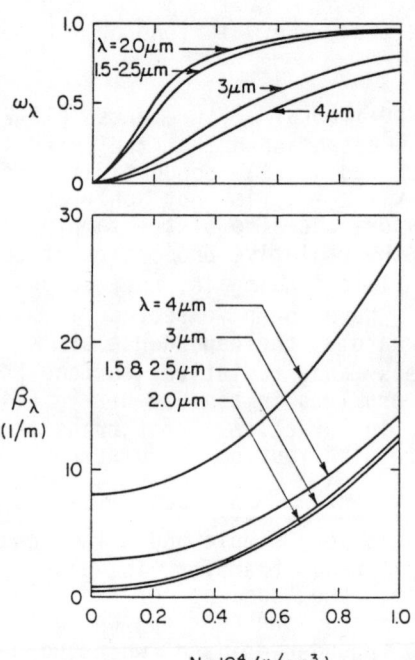

Fig.5. Effect of wavelength on the spectral extinction coefficient $\beta_\lambda$ and scattering albedo $\omega_\lambda$ for a gas-glass mixture: $T = 1400°C$, $D = 200$ μm, glass containing 0.1% $Cr_2O_3$.

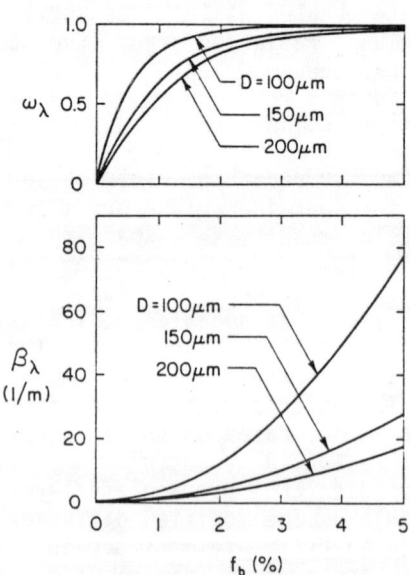

Fig.6. Variation of the spectral extinction coefficient $\beta_\lambda$ and spectral single scattering albedo $\omega_\lambda$ for a gas-glass mixture with volume fraction of gas $f_b$ in the mixture: $T = 1400°C$, $\lambda_b = 2.5$ μm, glass containing 0.1% $Cr_2O_3$.

geneous material. Presence of sufficient number of seeds in glassmelt such as gas bubbles or undissolved solid particles (particularly under the batch in a glass melting tank) is expected to affect not only the thermophysical and radiative properties of glassmelt but also flow and heat transfer.

Presence of inhomogenities such as gas bubbles in glass will cause scattering of radiation [11]. In order to assess the effect of the gas bubbles on the radiative properties of glassmelt and radiation heat transfer, the spectral absorption ($\kappa_\lambda$) and scattering ($\sigma_\lambda$) coefficients can be calculated for a typical glass using the analsis developed elsewhere [25]. The spectral radiative properties of glass and gas bubbles were weighted with respect to the volume of each to obtain volume averaged properties of the gas-glass mixture. The gas bubbles were assumed to have a diameter D and be distributed uniformly in the glass. Figures 5 and 6 show the results of calculations for the spectral extinction (absorption plus scattering) coefficient $\beta_\lambda$ and the single scattering albedo $\omega_\lambda$ ($=\sigma_\lambda/\beta_\lambda$). It is clear from Fig. 5 that as the bubble number density increases the extinction coefficient of the glassmelt also increases, and the melt becomes highly scattering as evidenced by the single scattering albedo $\omega_\lambda$. For a given gas bubble volume fraction $f_b$, the smaller diameter bubbles are more effective scatterers of radiation and increase both the extinction coefficient and single scattering albedo rather dramatically (Fig.6). Also, the calculations show that the gas-glass system is highly forward scattering. The phase function has a peak value of about 10,000 in the forward direction. These results suggest that the presence of gas bubbles in glassmelt and their effect on thermophysical and radiative properties would need to be properly accounted for if circulation and heat transfer in a glass melting tank (particularly below the batch where the bubble number density is large) is to be realistically simulated.

## MATHEMATICAL MODELING OF HEAT TRANSFER

There has been considerable amount of work done on the modeling of heat and momentum transfer related to glass manufacturing such as batch melting, glass circulation in glass melting tanks and forehearths, and glass forming. However, because of space limitations these topics cannot be discussed here, and reference is only made to available reviews [26-30].

## CONCLUDING REMARKS

During the past twenty five or so years significant advances have been made in understanding heat transfer in glass. However, the progress has not been as rapid as had been previously hoped for [4], and modern glass manufacturing is based as much on art as on engineering principles. In view of energy content of glass per unit mass leaving the furnace and the energy required to produce a unit mass of glass in most efficient furnaces, there appears to be considerable room for improving energy efficiency of glass manufacturing. There is not only a need for improved understanding of heat transfer in glass but also to apply and integrate what is known into mathematical models simulating glass making processes. This integration would contribute to the improvement of energy efficiency and productivity.

## ACKNOWLEDGEMENT

The author wishes to thank Mr. M.P. Menguc for predicting the radiative properties of gas-glass mixture given in the paper.

## LITERATURE CITED

1. Gardon, R., J. Am. Ceram. Soc., 44, 305-312 (1961).
2. Cooper, A.R., J. Non-Cryst. Solids, 26, 28-37 (1977).
3. Gardon, R., "Some Heat Transfer Problems in Glass Technology", in Proceedings of the Eighth International Congress of Glass 1968, pp. 85-95.
4. Condon, E.U., J. Quant. Spectrosc. Radiat. Transfer, 8, 369-385 (1968).
5. Viskanta, R. and Anderson, E.E., Heat Transfer in Semitransparent Solids, in Advances in Heat Transfer, vol.11, edited by T.F. Irvine, Jr. and J.P. Hartnett, Academic Press, New York (1975), pp.317-441.
6. Incropera, F.P. and DeWitt, D.P., Fundamentals of Heat Transfer, John Wiley, New York (1981).
7. Razzaque, M.M., Howell, J.R., and Klein, D.E., J. Heat Transfer, 106, 613-619 (1984).
8. Yuen, W.W. and Wong, L.W., J. Heat Transfer, 106, 433-440 (1984).
9. Fiveland, W.A., J. Heat Transfer, 106, 699-706 (1984).
10. Viskanta, R., Prog. Chem. Eng., 24A, 51-81 (1984).
11. Ozisik, N.M., Radiative Transfer and Interactions with Conduction and Convection, Wiley and Sons, New York (1973).
12. Siegel, R., and Howell, J.R., Thermal Radiation Heat Transfer, Second Edition, McGraw-Hill Book Co., New York (1981).
13. Amlin, D.W. and Korpella, S.A., J. Heat Transfer, 101, 76-80 (1979).
14. Gardon, R., J. Am. Ceram. Soc., 41, 200-207 (1958).
15. Genzel, L., Glastechn. Ber., 26, 69-71 (1953).
16. Blazek, A., Endrys, J., Kada, J., and Stanek, J., Glastechn. Ber., 49, 75-81 (1976).
17. Chui, G.K. and Gardon, R., J. Am. Ceram. Soc., 52, 548-553 (1969).
18. Anderson, E.E., Viskanta, R. and Stevenson, W.H., J. Heat Transfer, 95, 179-186 (1973).
19. Kunc, T., Lallemand, M., and Sauliner, J.B., Int. J. Heat Mass Transfer, 27, 2307-2319 (1984).
20. Anderson, E.E. and Viskanta, R., J. Am. Ceram. Soc., 56, 541-546 (1973).
21. Men, A.A., and Checkel'nitsky, A.Z., High Temp., 11(6), 1309-1312 (1973).
22. Viskanta, R. and Song, T.H., Glastechn. Ber. (in press).
23. Neuroth, N., Glastechn. Ber., 32, 197-198 (1959).
24. Blazek, A., Endry's, J. and Ederova, J., Glastechn. Ber., 56K, 524-529 (1983).
25. Menguc, M.P., and Viskanta, R., Comb. Sci. Tech. (in press).
26. Rawson, R., J. Non-Crystal Solids, 26, 3-25 (1977).
27. Mase, H., Ceramics, 14, 528-534 (1979).
28. Carling, J.L., Glass Tech., 23, 201-222 (1982).
29. Viskanta, R., J. Japan Soc. Mech. Engs., 87, 1258-1266 (1984) (in Japanese).
30. Rawson, H., Paper presented at 1984 Beijing International Symposium on Glass, September 3-7, 1984.

# EFFECT OF AIR BUBBLING ON CIRCULATION AND HEAT TRANSFER IN A GLASS MELTING TANK

A. Ungan and R. Viskanta ■ Heat Transfer Laboratory, School of Mechanical Engineering
Purdue University, West Lafayette, IN 47907

This paper presents numerical methodology to simulate the effect of air bubblers on glassmelt circulation and heat transfer in a glass melting tank. Due to high velocity and short residence time of the rising air bubbles, heat and mass transfer between air bubbles and glassmelt are neglected. Using existing empirical relations, the momentum exchange between a stream of air bubbles and glassmelt are calculated by considering bubble shape, size, relative velocity and air-glassmelt thermophysical properties. Some representative results of three-dimensional numerical simulations for flow and heat transfer with and without air bubblers are presented for a container glass melting tank. The results show significant effect of air bubblers on glass circulation patterns and the enhancement of heat transfer from the combustion space to the molten glass.

## 1. INTRODUCTION

Air bubblers are used in glass melting tanks to produce desirable glassmelt flow patterns, to insure glass uniformity and to improve fining [1]. A more detailed discussion of the advantages provided by the air bubbling in glass manufacturing is available and need not be repeated here [2-4].

Bubbling of air through viscous liquids and glass to simulate the conditions in glass melting tanks has been studied in small-scale laboratory experiments [3,5-7]. These studies indicate that for a specified air flow rate the average volume of bubbles was larger in the liquid of higher viscosity, but the number of bubbles formed show the opposite trend [5]. Flow visualization experiments [5,6] clearly show that air bubbling causes recirculation of the fluid (or molten glass) in the vicinity of the rising column of air bubbles. An experimental investigation of the effect of air bubbling in a small continuous glass melting furnace of two-ton has been reported [7]. The conclusions of the study were that the flow of molten glass induced by bubbling of air caused a continuous draw of molten glass of low temperature from near the bottom to the surface. As a result, there was an increase in heat transfer to the molten glass and a decrease in the lengthwise temperature difference.

The purpose of this paper is to develop a methodology for simulating the effects of air bubblers in glass melting operations. A three-dimensional mathematical model is used to predict the circulation and heat transfer in a glass tank with and without the bubblers. The results obtained are used to assess the effectiveness of the bubblers in altering flow patterns in the tank.

## 2. ANALYSIS

The system model considered in this study consists of a main model for the melt circulation and heat transfer and two submodels for the batch and the combustion space. The air bubbler model has been combined into the melt convection model.

### 2.1 Melt Convection Model

The behavior of the glassmelt within the tank can be simulated by solving the conservation equations of mass, momentum, and energy in three-dimensional space. It is well established in the literature that the radiative transport within the glass bath can be modeled by the Rosseland (or diffusion) approximation, and it can be imbedded into the diffusion term of the energy equation by use of the effective thermal conductivity concept.

The resultant governing equations can then be expressed in a general vector form,

$$\text{div}(\rho \bar{u} \phi) = \text{div}(\Gamma \, \text{grad} \phi) + S \qquad (1)$$

where $\phi$, $\Gamma$ and $S$ denote the dependent variable, the diffusion coefficient, and the source term, respectively. The corresponding quantities and symbols for the each conservation equation considered are given in Table 1.

Table 1. Values of $\phi$, $\Gamma$, and S.

| Equation | $\phi$ | $\Gamma$ | S |
|---|---|---|---|
| Continuity | 1 | 0 | 0 |
| x-momentum | u | $\mu$ | $-\frac{\partial P}{\partial x}+\frac{\partial}{\partial x}(\mu\frac{\partial u}{\partial x})+\frac{\partial}{\partial y}(\mu\frac{\partial v}{\partial x})+\frac{\partial}{\partial z}(\mu\frac{\partial w}{\partial x})$ |
| y-momentum | v | $\mu$ | $-\frac{\partial P}{\partial y}+\frac{\partial}{\partial y}(\mu\frac{\partial v}{\partial y})+\frac{\partial}{\partial z}(\mu\frac{\partial w}{\partial y})+\frac{\partial}{\partial x}(\mu\frac{\partial u}{\partial y})$ |
| z-momentum | w | $\mu$ | $-\frac{\partial P}{\partial z}-\beta\rho_0 g(T-T_0)+K(w_a-w)$ $+\frac{\partial}{\partial z}(\mu\frac{\partial w}{\partial z})+\frac{\partial}{\partial y}(\mu\frac{\partial v}{\partial z})+\frac{\partial}{\partial x}(\mu\frac{\partial u}{\partial z})$ |
| Energy | h | $\frac{k_{eff}}{c}$ | 0 |

According to Leibson et al.[8], when a gas is forced to flow through a submerged nozzle, very different behavior can be observed at high and low gas flow rates. At the former, a more or less continuous jet is formed, while at low flow rates, the formation of discrete bubbles is observed. Based on the work of Leibson et al. [8] for air-water systems and considering observations in an actual glassmelting furnace, it was assumed that the bubbles forming under operational conditions of a glass melter are discrete. The other pertinent assumptions employed in the air bubbler model are as follows:

i) The distance between the discrete air bubbles is sufficiently large and the melt is a continuous phase so that the correlations given for single gas bubble-liquid system are appropriate for the train of air bubbles in glassmelt.

ii) Due to high velocity and short residence time of the rising discrete bubbles, the heat and mass exchange between the bubbles and the melt are assumed to be negligible. The pressure within each bubble is assumed to be variable and equal to that of glassmelt so that air bubbles expand slowly as they rise.

iii) Ideal gas relations are assumed to be valid for the air contained in each of the bubbles.

In light of the physical phenomena and assumptions, the behavior of the single bubbles can be represented by the Newton's second law of motion. However, the inertia effects of the bubbles can be readily ignored compared to the magnitudes of buoyancy and drag forces in the glassmelt and the bubble velocity is basically determined by the balance of these forces.

A map of the different gas bubble domains in terms of Morton, Eotvos, and bubble Reynolds numbers to describe the bubble shape and behavior in a liquid has been given by Clift et al. [9]. This map has been used to meet the needs of the present application. Under operational conditions in a glassmelt the bubbles are expected to assume ellipsoidal-cap and/or spherical shapes. The associated drag coefficient and terminal velocities for these shapes are given in the literature [9]. Some relations for the detachment radius of the air bubble, which may be appropriate for the needs of the present study, have been reported [8,10-13]. However, consideration of the physics of bubble growth indicates that the growth and detachment of gas bubbles in liquids are affected by several factors, including liquid thermophysical properties, orifice shape and dimensions, air flow rate, liquid inertia and upstream pressure [10] but detailed experiments on an air-glassmelt system have not been reported in the literature. Also, the effect of antechamber, or volume between last large pressure drop (i.e., a valve) and actual nozzle or orifice, on the detachment radius has been found to be important in liquid metal-gas systems [14,15].

The drag force on each air bubble may be calculated from

$$F_D = \frac{1}{2} C_D \rho \pi r_a^2 (w_a - w)^2 \qquad (2)$$

This equation also represents the upward force exerted on the glassmelt by an ascending single bubble. For a train of bubbles, once the velocity of a single bubble is known, it is a simple matter to find the void ratio (or the number of bubbles per unit volume) in the (descritized) region directly above the air bubbler nozzles. Thus, the total force exerted on the glassmelt by the air bubbles immediately above the bubblers may be determined and accounted for in the z-momentum equation by a source term,

$$S_a = K(w_a - w) \qquad (3)$$

where, K represents the volumetric momentum exchange coefficient and may be expressed as

$$K = \frac{3}{8} \frac{\alpha}{r_a} C_D \rho (w_a - w) \qquad (4)$$

The system of equations (Table 1) was solved numerically using the SIMPLER algorithm

[16]. Modified hybrid scheme [17] was employed for modeling of convection and diffusion terms in the model equations.

## 2.2 Combustion Space and Batch Models

As a first approximation, heat transfer to the sink (batch-glass bath) was estimated by considering only radiation exchange between the sink and surrounding walls of the furnace (crown, sidewalls and endwalls), and the combustion products. In the analysis, Hottel's zonal method [18] was employed.

Since the details about the batch modeling have been reported earlier [19], only a brief summary seems to be appropriate here. The batch is assumed to be charged axially, and consists of 75% $SiO_2$, 15% CaO and 10% $Na_2O$ by weight. During the transformation, the chemical reactions are simulated by using modified form of Arrhenius equation [17]. The gas percolation within the blanket is considered and formulated by using Darcy's law, and heat exchange between gas-solid phase are accounted for. The system of equations are solved using the marching method of Patankar and Spalding [20].

## 2.3 Boundary Conditions and Coupling of the Models

Unless some of the ambiguity in the thermophysical data and other batch parameters are clarified and the model predictions are verified experimentally, at present, extension of the batch melting model from two to three dimensions appears to be impractical. However, with approximations 2D batch melting model can be coupled to 3D glassmelt circulation and heat transfer as well as the simplified combustion space heat transfer models. Accordingly, the interface conditions involving batch may be averaged in the spanwise direction [17].

The specifications of the other boundary conditions are as follows. At the refractories surrounding the glassmelt, conventional "no-slip" boundary conditions were employed for the flow field. The heat losses from these refractories were expressed in terms of the overall heat transfer coefficients. At the throat, a slug-flow assumption was employed while neglecting the construction details of the throat outside the tank and/or presence of a refining section. At the refractories surrounding the combustion space, the temperatures were specified and thus the combustion space heat transfer model was further simplified. The curvature of the glassmelt-combustion gas interface (free-surface of the glassmelt) was neglected even though it is recognized that this is an oversimplification of the reality, especially with the air bubblers. The shear stresses at the combustion gas-glassmelt interface in the horizontal directions (x- and y- directions) were neglected, while the melt velocity in the vertical (z) direction was assumed to vanish at the interface. The energy balance equates the net radiant heat transfer at the interface.

## 3. RESULTS AND DISCUSSION

In this section, the utility of the models is demonstrated by simulating a container glass tank. Some results of representative three-dimensional numerical simulations of flow and heat transfer with and without air bubblers are presented.

### 3.1 Model Parameters

The tank size considered (Fig.1) was 2 m wide and 6 m long, and it may be classified as small in glass industry. The melter depth was 0.75 m, and the average height from glassmelt to the crown was 1.0 m. The front, side and back wall refractory constructions were considered to be the same, and a value of 8.5 $W/m^2K$ was assigned to their overall heat transfer coefficients. The overall heat transfer coefficient at the bottom, however, was 4.0 $W/m^2K$. Heat loss calculations from these walls were performed by using ambient temperature of 320 K.

The raw materials are fed axially to the furnace from the back wall at a rate of 0.35 kg/s, in the form of a loose blanket having an uniform thickness of 0.125 m and an uniform velocity of 0.1 cm/s over the entire width of the tank. The inlet temperature of raw batch is 320 K. The conditioned glass is withdrawn

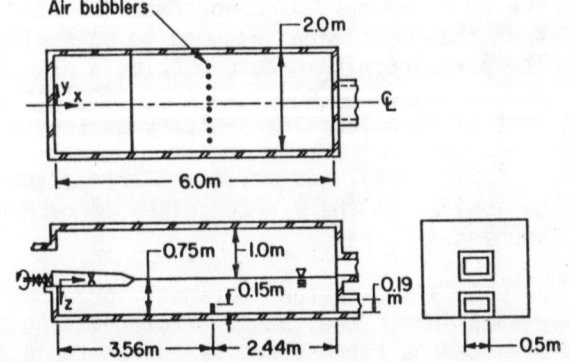

Figure 1. Schematic of the glass melting furnace and coordinate system.

through a submerged throat having a cross-sectional area of 0.5 x 0.19 m. The temperature distribution within the combustion space and at the surrounding refractories are assumed to be the same for the simulations considered. The values employed are listed in Table 2.

Table 2. Temperature distribution in the combustion space employed for the simulations. (Note: the temperature of gas volume elements are assumed to be 50K higher than those of the crown surface elements).

| Side wall | | | | | | | | | |
|---|---|---|---|---|---|---|---|---|---|
| 1693.1 | 1724.2 | 1757.2 | 1803.9 | 1820.0 | 1840.3 | 1834.2 | 1824.2 | 1810.6 | 1792.8 |
| 1661.1 | 1724.3 | 1752.2 | 1781.1 | 1824.0 | 1830.0 | 1846.9 | 1844.0 | 1835.6 | 1821.3 | 1801.5 | 1797.2 |
| 1670.0 | 1732.0 | 1760.0 | 1789.0 | 1825.0 | 1835.0 | 1848.0 | 1848.0 | 1839.0 | 1824.0 | 1803.0 | 1800.0 |
| 1686.8 | 1745.1 | 1771.1 | 1796.7 | 1830.0 | 1840.0 | 1851.9 | 1851.0 | 1843.3 | 1829.3 | 1807.9 | 1802.8 |

(Back wall, Front wall; 1.0 m — Crown 6.0 m — 1.0 m)

The thermophysical properties for the glassmelt and batch were taken from the literature. Although the complete list of values can be found elsewhere [17], the thermophysical properties of glassmelt employed in the simulations are as follows:

$\rho$ = 2500 (kg/m$^3$)

$\beta$ = 5.10$^{-5}$ (1/K)

$\log \nu = -6.21 + \dfrac{4570.98}{T - 525.6}$ ($\nu$ : m$^2$/s)

$k_{eff}$ = 5.386 − 0.021676 × T + 0.00002058 × T$^2$ (W/m K; T : K)

c = 1256 (J/kg K)

The air bubbler nozzles are assumed to be placed at x = 3.56 m, which is directly above the hottest crown surface and gas elements of the combustion space (Table 2). Only a single row of eight bubbler orifices are considered. The tubes protrude through the bottom blocks of the furnace 15 cm into the glass bath, and are spaced 25 cm apart, except in the immediate vicinity of the symmetry plane, where they are only 12.5 cm apart. The flow rate of air through each bubbler is 0.08 m$^3$/hr.

Assuming there exists a symmetry plane at y=0, half of the melt domain was descritized by 50x10x10 nodes.

## 3.2 Sample Simulations

The three-dimensional temperature and velocity fields generated by the models are difficult to display, and the best graphical representations were obtained by plotting the results in selected cross-sections. Figure 2 illustrates the flow field and temperature distributions within the glass bath for the baseline simulation without air bubblers. Those with the air bubblers are depicted in Fig. 3. Note that for the presentation of flow field with arrows, a two scales had to be used in order to accommodate the very high velocities in the vicinity of the air bubblers and to retain visual clarity in other sections of the tank. Also, the batch is shown schematically in the figures.

For the baseline simulation, the glass bath in the horizontal-longitudinal direction is partioned by two roll cells (Fig.2.a). The axial motion of the batch creates a recirculation underneath. The rest of the cross section is enveloped by the primary roll cell, which is basically the consequence of the sharp temperature gradients at the tip (or the

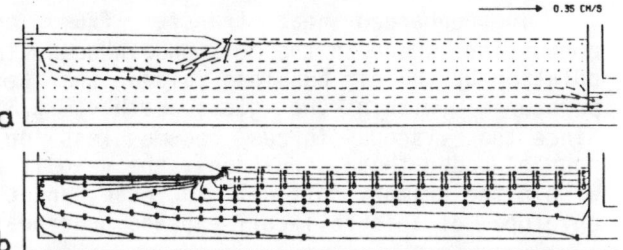

Figure 2. Glass velocity vectors (a) and isotherms (b) for the longitudinal (x-z) plane at y=0.06 m for the baseline simulation without air bubblers) (contour values: 1-1200 K; ΔT = 50 K)

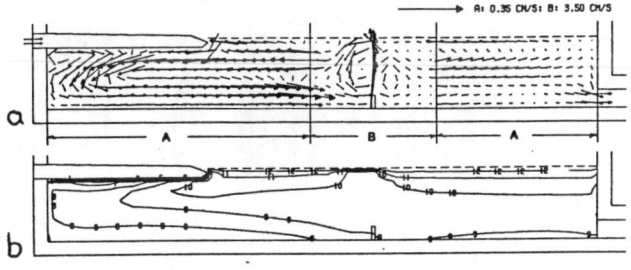

Figure 3. Glass velocity vectors (a) and isotherms (b) for the longitudinal (x-z) plane at y=0.06 m for the simulation with air bubblers (contour values 1-1200 K; ΔT=50 K).

leading edge) of the batch blanket. The temperature distribution in the vertical-longitudinal cross sections (Fig.2.b) can be characterized by nearly parallel isotherms below the combustion gas-glassmelt interface and the extension of these isotherms underneath the batch due to glass currents within the primary roll cell.

In the presence of air bubblers, the bottom cold glass currents are turned upward (not shown for the case of brevity). The melt velocities directly above the bubbler tubes acquire magnitudes as high as 3.5 cm/s. The velocities of the ellipsoidal-cap shaped air bubbles released from the bubbler nozzles are predicted to be as high as 12 cm/s. As shown in Fig.3, the cold glass is "lifted" to the free surface by the rising bubbles. The air bubblers enhance heat transfer from the combustion space to the glassmelt (Fig.4). As a result, total heat transferred to the sink (batch-glassmelt) from the combustion space is predicted to be 878.9 kW, while that without air bubblers is 791.7 kW. Table 3 summarizes the energy quantities on the batch-glassmelt for the two cases considered in this paper.

The enhanced heat transfer from the combustion space to the melt and induced circulation by the air bubblers raises the temperatures within the glass bath (Fig.3). Since the viscous forces become less pronounced under these conditions, the velocities within the primary roll cell increase, and the envelope of this roll cell expands underneath

Table 3. Energy balance on the batch and glassmelt system.

|  | Without air bubblers | With air bubblers |
|---|---|---|
| INPUT FROM:<br>Combustion gas-glassmelt and batch-glassmelt interfaces | Q(kW)<br>791.67 | Q(kW)<br>878.87 |
| OUTPUT FROM:<br>Bottom | 51.20 | 59.74 |
| Front wall | 15.05 | 16.41 |
| Back wall | 12.61 | 14.16 |
| Side walls | 89.66 | 98.56 |
| Throughput | 626.95 | 689.46 |

the batch at the expense of the counter rotating recirculation induced by the axial motion of the batch.

The batch length for the baseline case is predicted to be 1.98 m and the length with the air bubblers is only 1.70 m. Figure 5 shows the laterally "averaged" heat flux distributions at the combustion gas-batch and at the glassmelt-batch interfaces for both simulations. Although the heat flux distributions from the combustion space are approximately the same for batch, the flux from the glassmelt is considerably enhanced by the

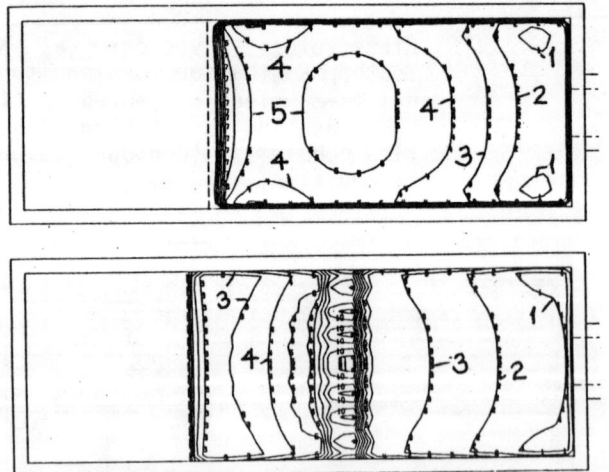

Figure 4.  Heat flux distribution at the combustion gas-glass melt interface (Top panel: without air bubblers; Bottom panel: with air bubblers) (contour values, top: 1-25 kW/m$^2$, $\Delta q$=2.5 kW/m$^2$, bottom: 1-30.0 kW/m$^2$, $\Delta q$=7.5 kW/m$^2$)

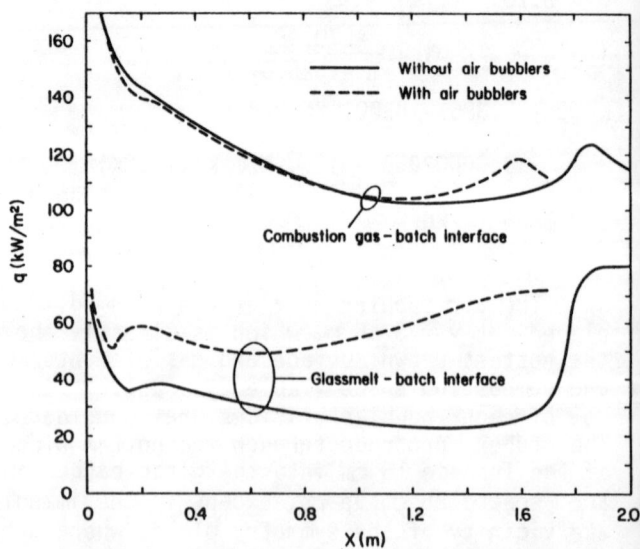

Figure 5.  Distribution of laterally averaged heat flux at the combustion gas-batch and glassmelt-batch interfaces for simulations with and without air bubblers.

presence of air bubblers. This is due to modification of the roll cells underneath the batch, which is caused by the air bubblers. As indicated in Table 4, the predicted increase with the air bubblers is 11% for the total heat input, and 70% for the average heat flux.

Table 4. Energy balance on the batch blanket. The numbers in paranthesis are average heat fluxes, i.e., $Q/(L_b W)$.

| | Without air bubblers | With air bubblers |
|---|---|---|
| INPUT FROM: | Q(kW) | Q(kW) |
| Combustion gas-batch interface | 524.51 (131.85) | 457.93 (134.69) |
| Glassmelt-batch interface | 144.08 (36.22) | 208.73 (61.39) |
| OUTPUT: | | |
| Chemical reactions | 197.75 | 197.75 |
| Sensible heat | 470.84 | 468.91 |

The three-dimensional flow characteristics, and perhaps the performance of a glassmelter, can be visualized somewhat better by a streakline, i.e., the path traced by a particle (having infinitesimally small volume) of the glassmelt. Figure 6 illustrates the streaklines of two particles in the melting tank, with and without air bubblers. The symbols on the streaklines denote the clock time such that between two sequential symbols (or stations) one hour of time has elapsed.

Even though it is an extreme and not a typical case, the particle of the baseline simulation spends only 9 hours within the glass bath, and without recirculating it is withdrawn from the tank through the throat. With the air bubblers (Figs.6.c and 6.d), the melted glass may recirculate several times while they are homogenized and the seeds can be disengaged or dissolved at the hot free surface glass layers.

## 4. CONCLUDING REMARKS

The results of numerical calculations presented show that the air bubblers are indeed capable of creating favorable effects on the glassmelt circulation, heat transfer from the combustion space to the melt and from glassmelt to the batch.

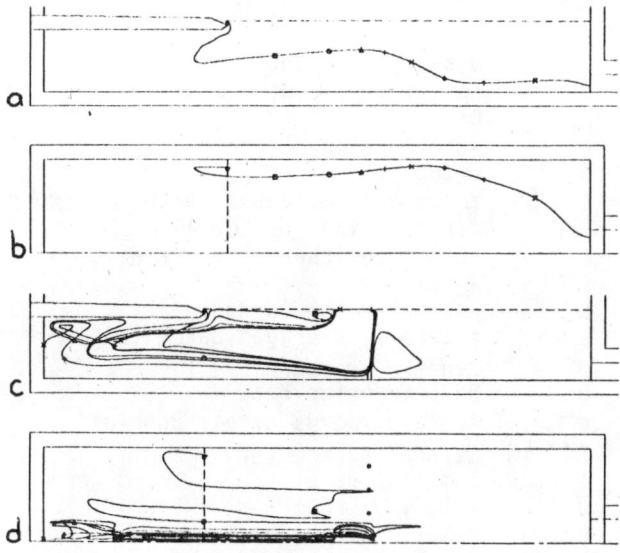

Figure 6. Longitudinal (a,c) and horizontal (b,d) views of the streaklines of the particles without (a,b) and with (c,d) air bubblers. (Starting locations: without air bubblers: x=1.98 m, y=0.9 m, z=0.01 m; with air bubblers: x=1.7 m, y=0.9 m, z=0.01).

The air bubbler model described should be considered as a first approximation of the complex momentum exchange processes between a train of air bubbles and glassmelt. It needs improved empirical relations for representing the momentum exchange of polydispersed air bubbles and detachment radius. The model must be verified using experimental data before it is employed for design purposes.

Since the solution procedure employed in the simulations is an iterative one, the computer time requied can be reduced significantly with the availability of a good initial guess for the field variables (i.e., from a previous, somewhat similar case). The computer time required for the simulations presented here was 20 to 30 minutes on Cyber 205 computer. Vectorization of the code is expected to reduce this time significantly.

ACKNOWLEDGEMENTS

The authors wish to acknowledge numerous helpful discussions with Mr. Warren H. Turner of the Fiber Glass Technical Center, PPG INDUSTRIES, INC., regarding processes simulated in the paper. Computer facilities were made available by the Purdue University Computer Center.

NOMENCLATURE

| | |
|---|---|
| $C_D$ | Drag coefficient |
| $c$ | Specific heat of glassmelt |
| $F_D$ | Drag force |
| $g$ | Acceleration due to gravity |
| $h$ | Enthalpy |
| $K$ | Volumetric momentum exchange coefficient defined by Eq.(4). |
| $k_{eff}$ | Effective (thermal and radiative) conductivity |
| $L$ | Length of the furnace |
| $L_b$ | Laterally averaged batch length |
| $p$ | Dynamic pressure |
| $Q$ | Heat transfer rate |
| $r_a$ | Average radius of air bubbles |
| $S$ | General source term |
| $T$ | Temperature |
| $\bar{u}$ | Velocity vector |
| $u,v,w$ | Axial, lateral, and vertical velocity components, respectively |
| $W$ | Width of the furnace |
| $x,y,z$ | Axial, lateral, and vertical coordinates, respectively. |

Greek Letters

| | |
|---|---|
| $\alpha$ | Void ratio |
| $\beta$ | Thermal expansion coefficient |
| $\Gamma$ | General diffusion coefficient |
| $\mu$ | dynamic viscosity |
| $\rho$ | Density |
| $\phi$ | General dependent variable |

Subscripts

| | |
|---|---|
| a | air bubbles |
| b | batch |
| o | reference |

REFERENCES

1. K.A. Pchelyakov, Glass and Ceramics 28, 402-405 (1971).
2. K.A. Pchelyakov, Glass and Ceramics 37, 228-231 (1980).
3. M. Kunugi, Ceramics 14, 520-529 (1979) (in Japanese).
4. R. Viskanta, Journal of the Japan Society of Mechanical Engineers 87, 1258-1266 (1984) (in Japanese).
5. M. Suzuki and K. Nagaoka, Osaka Industrial Engineering Research Institute Report 32, 49-55 (1981) (in Japanese).
6. M. Suzuki and K. Nagaoka, Osaka Industrial Engineering Research Institute Report 32, 215-221 (1981) (in Japanese).
7. M. Suzuki and N. Nagaoka, Osaka Industrial Engineering Research Institute Report 33, 395-402 (1982) (in Japanese).
8. I. Leibson, E.G. Holcomb, A.G. Gacoso and J.J. Jasmic, AIChE J. 2, 296-306 (1956).
9. R. Clift, J.R. Grace, and M.E. Weber, Bubbles, Drops and Particles, Academic Press, London (1978).
10. J. Szekely, Fluid Flow Phenomena in Metal Processing, Academic Press, New York (1979).
11. D.W. Van Krevelen and P.J. Hofijzer, Chem. Eng. Prog. 46, 29-35 (1950).
12. R. Kuman and N.R. Kuloor, Adv. Chem. Eng. 8, 256-368, (1970).
13. M. Ihara, T. Yamamoto and T. Arimari, J. Ceram. Assoc. 71, 63-71 (1963) (in Japanese).
14. G.A. Irons and R.I.L. Guthrie, Metall. Trans. 9B, 101-110 (1978).
15. M. Sano and K. Mori, Trans. Japan Inst. Met. 17, 344-352 (1976).
16. S.V. Patankar, Numerical Heat Transfer and Fluid Flow, Hemisphere Publishing Corp., Washington (1980).
17. A. Ungan, "Three-Dimensional Numerical Modeling of Glass Melting Process", Ph.D. Thesis, Purdue University, West Lafayette, Indiana (1985).
18. H.C. Hottel and A.F. Sarofim, Radiative Transfer, McGraw-Hill Book Co., New York (1967).
19. A. Ungan and R. Viskanta, Heat Transfer-Niagara Falls 1984, AIChE Symposium Series, No.236, Vol.80, 446-451, (1984).
20. S.V. Patankar and D.B. Spalding," "Heat and Mass Transfer in Boundary Layers" Intertext, London (1970).

# HEAT TRANSFER SIMULATION DURING FORMING OF GLASS CONTAINERS

Ihab H. Farag and Michael J. Beliveau ■ Chemical Engineering Department
University of New Hampshire, Durham, NH 03824

Richard L. Curran ■ Owens-Illinois, Inc., One Sea Gate (1-NTC), Toledo, OH 43666

An important step in the optimization of a glass container production cycle is the determination of the glass temperature distribution during heat treatment. The ideal approach to this problem is to formulate a theoretical model for comparison against experimental data measured in a well-determined system. Discrepancies between theory and experiment may then give further direction for model improvement. This approach, however, is limited because of the difficulties in measuring glass temperature distribution during forming.

Another approach is to use the model to predict glass surface heat fluxes during the forming cycle and test the computed results against published values of glass to metal heat fluxes measured during glass container production on an individual section (I.S.) Blow-and-Blow bottle machine.

A computer model has been developed to calculate the temperature distribution in a glass plate. The model includes the three modes of heat transfer: conduction, convection and radiation. The process is complicated by the simultaneous internal emission and reabsorption of radiant energy within the glass in the non-opaque region of the spectrum. With the problem being simplified to a one dimensional case, the glass plate is divided into several slices. An energy balance on each slice yields a system of integro-partial differential equations which are solved to obtain the temperature distribution.

Results in the form of time-temperature, time-distance and time-surface heat flux plots are presented and compared with published data.

Knowledge of the temperature history of glass during the bottle molding process is necessary because of the strong temperature dependency of the glass properties important to the molding process, especially the glass viscosity. Within the forming range, a few degrees change in glass temperature can change the glass viscosity by a factor of 1000, thus drastically affecting the molding process. This is especially true of the thin region near the glass surface, "enamel". This "chilled" layer maintains the shape of the bottle as it is removed from the mold. Being able to predict the glass temperature profile, especially the enamel temperature, should make it possible to optimize the dwell time in the mold and therefore increase production rates as well as product quality.

The results of this work is a computer model of the glass system for cases of heating or cooling by any combination of convection, conduction, or radiation. Temperature profiles as well as heat fluxes can be calculated and plotted. The energy density (heat content) of the glass can also be calculated.

Review of the literature reveals some calorimetry data [1,2], as well as modeling approaches [4] to account for heat loss from glass during forming operations, but few experimental data are available concerning the actual temperature profile in the glass. McGraw [1] gives some such data, but found it impossible to obtain data near the surface of the glass. In general the glass temperature data are difficult and time consuming to obtain. McGraw [1] extrapolated his data to include the glass surface, but admittedly had doubts about their accuracy. Since the "enamel" is the controlling region for glass forming, and given the difficulties in obtaining data, a model was needed to predict the glass temperature distribution. Discarding the earlier concept of apparent radiative conduction, which was dependent upon the glass thickness and which did not take into consideration the effects of the glass surface or multiple reflections, Gardon [7,8,9,10] was the first to develop a model incorporating radiation in a comprehensive manner. Farag and Curran [11,12] presented and discussed a model to interpret pyrometric temperatures of a glass plate.

McGraw [2] used the Gardon type model to interpret his experimental data for thin glass sections and found that radiation played an unimportant role in the heat transfer during normal pressing operations. He did note the effect that radiation would have in slower cooling of glass as well as the surface reheat effect. He also noted a large decrease in the surface heat transfer coefficient with time, from an initial value greater than 11 kw/m$^2$°K to about 2 kw/m$^2$°K at the end of the normal dwell period [3]. He credited this change to the glass pulling away from the mold with an accompanying shift from free molecule conduction to bulk conduction in the increasing air gap.

Jones [6] also used a Gardon type model including radiation to model a single glass plate after he found that a cylindrical model without radiation could not account for the experimental temperature profiles found in glass sections slightly thicker than those of McGraw [2]. He concluded that radiation was important in thicker sections.

Given the lack of data and the difficulties involved in obtaining good experimental data, a need for a computer model including all forms of heat transfer was obvious. The Gardon [8] type model seemed to be the best available, and was thus chosen as a starting point.

## MODEL OVERVIEWED

The first step in analyzing the symmetrical, one-dimensional heat transfer in a glass plate is to divide the plates into a number of slices as shown in figure 1.

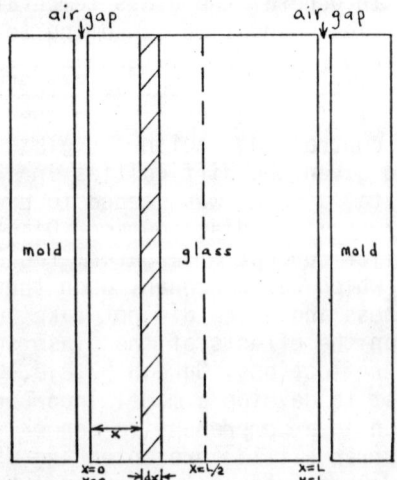

Figure 1  Energy Balance on Elemental Slice

The modes of heat transfer considered include conduction, convection, and opaque as well as volumetric radiation. The latter is due to the semitransparent nature of the glass. Multiple reflections that occur between all the glass, air, and mold interfaces need to be considered - (See references 11 & 12). As a consequence of multiple relfections and because the reflectivities of the normal and parallel components of the radiation are different, the radiation is partially polarized. Thus, when dealing with reflected radiation, it is necessary to treat the normal and parallel polarized components separately.

The volumetric nature of the radiative energy exchange makes it necessary to consider: (a) absorption of external radiation, (b) internal emission of radiation, and (c) reabsorption of internally emitted radiation.

The temperature profile in the glass is calculated by performing an energy balance on each slice. This results in a system of second order partial integro-differential equations. The finite difference method is then used to reduce this result to a system of first order ordinary differential equations which are solved using a Runge-Kutta method.

## ASSUMPTIONS

(1) The glass plate is homogeneous and isotropic in all directions and are being heated or cooled symetrically.

(2) One-dimensional net heat transfer.

(3) The attenuation of monochromatic radiation obeys the Bouguer-Lambert Law.

(4) Reflection and refraction at the surface of the glass obeys Fresnel's and Snell's equations respectively for dielectric materials.

(5) emitted radiation is initially diffuse and unpolarized.

## ENERGY BALANCE

Figure 1 shows an elemental slice of thickness dx at level x within a slab of glass for the 1-plate model. Performing an energy balance upon this slice results in:

$\rho C_p (\partial T/\partial t)dx$ = net heat flux. The net heat flux is composed of two terms; conduction, which where appropriate also includes surface effects (convection and surface opaque radiation), and volumetric radiation.

## 1. Conduction

The net conductive flux for an interior plane, written in a finite difference form is $k(T_{x+dx} - 2T_x + T_{xdx})/dx$. At the center plane this becomes $2k(T_{xdx} - T_x)/dx$. Conduction as well as convection and surface opaque radiation are included in the net "conductive" flux at the surface:

$$2[k(T_{x+dx} - T_x)/dx + h_{ex}(T_{Aex} - T(x)) + \int_{\lambda opaq}^{\infty} \frac{(W_{m\lambda} - W_{B\lambda}(x))}{1/\varepsilon_{m\lambda} + 1/\varepsilon_{g\lambda} - 1} d\lambda ]$$

## 2. Radiation

When dealing with the radiation term, it is helpful to consider three volumetric radiative terms, the absorption of external radiation, $QA_x$, the internal emission of radiation, $QE_x$, and the reabsorption of internally emitted radiation, $QR_x$. Since these are volumetric quantities it is necessary to multiply them by the thickness of the slice, $dx$, before calculating the radiative flux. Thus,

net radiative flux $= (QA_x + QE_x + QR_x)dx$

where

$$QA_x = 2n^2 \int_0^{\lambda opaq} \gamma_\lambda W_{m\lambda} \phi_m \, d\lambda$$

$$QE_x = 4n^2 \int_0^{\lambda opaq} \gamma_\lambda W_{B\lambda}(x) \, d\lambda, \text{ and}$$

$$QR_x = 2n^2 \int_0^{\lambda opaq} \gamma_\lambda \sum_{y=0}^{L/2} W_{B\lambda}(y) \psi(y) \, d\lambda$$

The $QA_x$ term contains the auxiliary function $\phi_m$ and the $QR_x$ term contains the auxiliary function $\psi$. These auxiliary functions which are not presented here, account for direct radiation and multiply internally reflected radiation. Unlike Gardon's auxiliary functions, $\phi_m$ and $\psi$ also take into account radiation that reflects from the surface of the mold, which is believed to be an important consideration for real mold surfaces. Further details of auxiliary functions and of the model are given in Beliveau [13].

## RESULTS

### A. Comparison With Gardon's Results for Radiant Heating

#### 1. Gardon Type Model

In an effort to duplicate the results of Gardon [8] several runs were made using the auxiliary functions $\phi$ and $\psi$ as defined by Gardon for a plate of glass 0.6 cm thick. The glass had the same physical properties as used by Gardon, and will hence be called Gardon type glass.

The present model was used to calculate temperature profiles for the case of radiant heating of the Gardon type glass with an initial isothermal glass temperature of 27°C and a mold temperature of 707°C. The emissivity of the mold is taken as 0.9. The computed temperature profile is shown on Figure 2. This result nearly duplicates the results of Gardon.

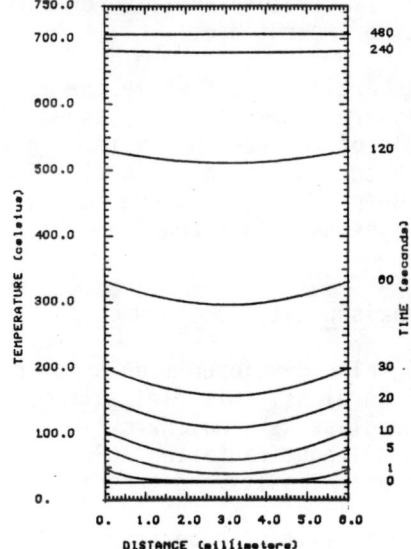

Figure 2 Temperature Profile for Gardon Type Glass using Gardon Model

#### 2. 1-Plate Model

In the Gardon type model it is assumed that any radiation that reflects from the outside of the glass back to the mold is completely absorbed by the mold. Since the mold reflectivity is not necessarily zero, it is clear that reflection from the mold will occur. Therefore multiple reflection will take place between the glass and the mold as it does internally in the glass. The results of the present work, the 1-plate Model, for the

Gardon type glass are shown in Figure 3.

Comparing this result to that for the Gardon type model, it is seen that the temperature

"Surface Chill" effect has been reduced and the center of the glass has begun to cool.

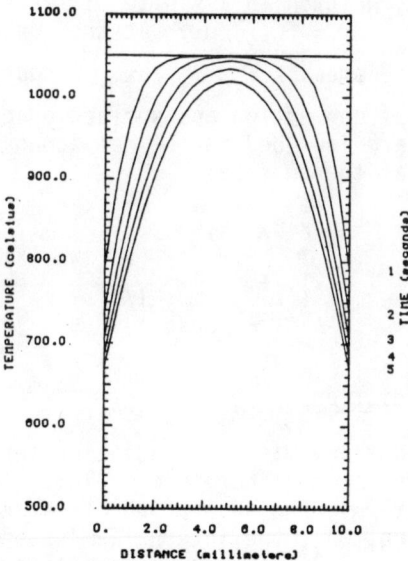

Figure 4  Temperature Profile for Jones Type Glass

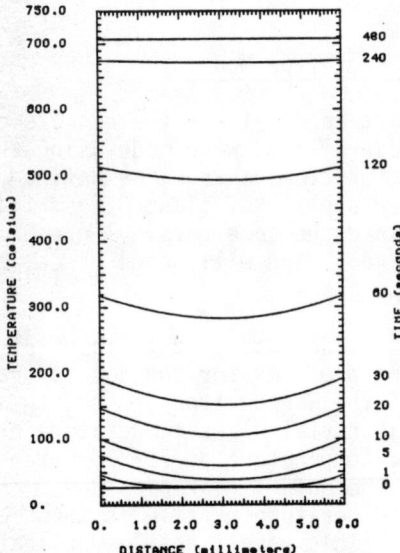

Figure 3  Temperature Profile for Gardon Type Glass using Present Model

gradients are identical while the actual temperatures are somewhat lower, especially near the middle of the heating cycle. This is caused by the increased radiactive interaction between glass slices and the decreased radiation exchange between the glass and the hot mold.

### B. Comparison With Jone's Models

Jones [6] also developed models for heat transfer in glass. His first attempt was to model the glass as a hollow cylinder in the mold, but without radiation. Because of unexplainable results, most notably the "surface chill effect" (rapid cooling of the surface with little temperature drop at the center), Jones [6] realized that some other form of energy transfer, must be responsible for removal of heat from the center of the glass. Due to the difficulties in handling reflected radiation in the cylindrical model, Jones developed a single plate model. The present model is used to calculate the heat transfer in a Jones type glass plate of 0.01 meter thick, heat transfer coefficient of 1568 watts/sq. m.-C, an initial glass temperature of 1050°C, a mold temperature of 488°C, and a mold emissivity of 0.9. The temperature profiles at 0, 1, 2, 3, 4, and 5 seconds are shown in Figure 4. It can be seen that the

### C. Comparison With McGraw's Data For Glass Pressing Using 1-Plate Model

McGraw's data [2] allows a direct comparison with actual experimental data. Using the present model and the properties for McGraw's type glass, the temperature distribution was predicted. The geometry for this case is: Thickness of plate is 0.00711 meter, initial isothermal glass temperature is 1141°C, temperature of the mold is 488°C, heat transfer coefficient is 6247 W/m$^2$-C, and the mold emissivity is taken as 0.9. The final temperature after 2 seconds is shown in Figure 5. Comparing this to

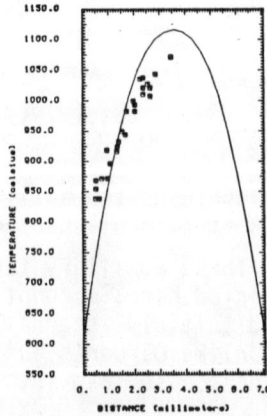

Figure 5  Temperature Profile for McGraw Type Glass

McGraw's result, some differences appear. First, the center temperature is about 45°C higher than that reported by McGraw [2]. However, there seems to be a discripancy in McGraw's result. His data show a rather large temperature gradient at the centerline, not zero as should be the case. The true temperature might very well be somewhat higher than he has measured. Second, the model predicts a surface temperature that is approximately 120°C lower than given by McGraw [2]. Given his lack of data near the surface and the need to extrapolate in a region of a steep temperature gradient, 7 and 8 respectively. McGraw's results, admittedly, may not be accurate.

### D. Comparison With O-I Calorimetric Data

Owens-Illinois also provided calorimetric data as well as surface temperature measurements on plate glass samples. This data is easily compared to results from present model, as as well as for slow cooling of actual glass plates, in this case 1/2" thick or 0.00127 meter. The initial glass temperature is 571°C and the ambient temperature is 27°C with a convective heat transfer coefficient of 7 W/sq.m.-C. Due to the lack of a mold surface, the "mold" emissivity (ambient environment) is taken as unity. For a cooling time of 345 seconds, the total flux is shown in Figure 6, and the integrated energy density is 4.01 mega joules/sq. meter. This value falls comfortable in the range of 3.3 to 4.4 mega joules/sq. meter given by O-I.

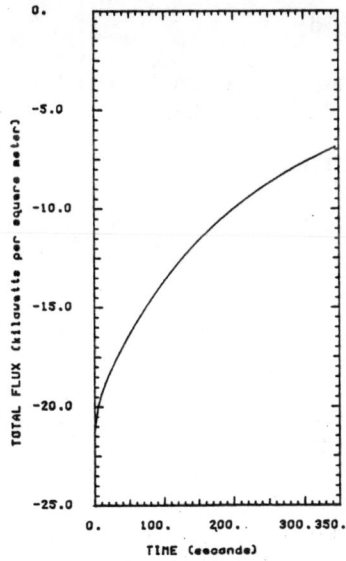

Figure 6  Total Flux History for 1/2" Plate of O-I Type Glass.

### E. Comparison With O-I Surface Temperature Data

The surface temperature data provide by O-I is for slow cooling of single glass plates under the same conditions as the calorimetric data. With an assumed heat transfer coefficient of 7 W/sq.m.-C, the present model was used to simulate the cooling process for a number of the glass thicknesses as supplied by O-I. For two of these thicknesses 1/4", and 3/4", comparison of the computed results with the experimental data is shown in Figures 7 and 8 respectively.

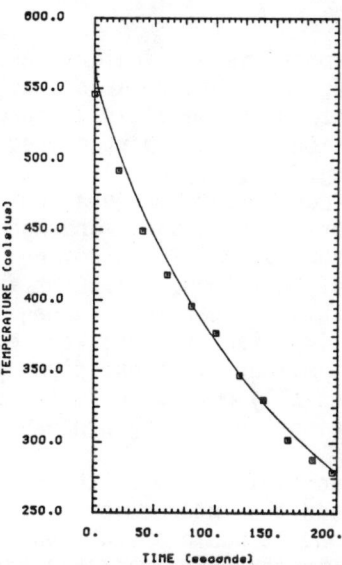

Figure 7  Surface Temperature History for 1/4" Plate of O-I Type Glass

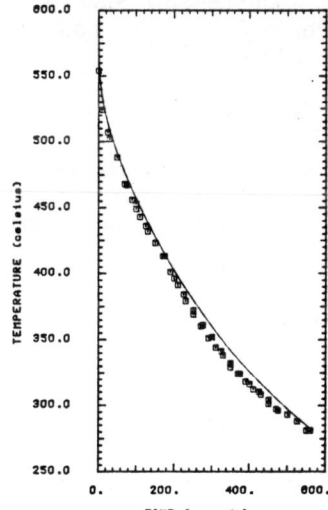

Figure 8  Surface Temperature History for 3/4" Plate of O-I Type Glass

The results is an almost perfect match between calculated and experimental surface temperatures. The only departure is in the low readings in the experimental data at the initial conditions for the thinner glass plates. Since the glass was allowed to reach thermal equilibrium within the oven at 571°C, this should be the temperature measured at time equal to zero. The low reported values are indicative of the difficulties in measuring glass temperatures in thin sections. The good agreement between the experimental data and the computed results indicates the appropriateness of the model for the case of glass plates.

CONCLUSIONS

The present model is particularly useful for investigating such phenomena as the "surface chill" effect where the glass surface or "enamel", which is the important region in glass forming, is at a much lower temperature and has a steep temperature gradient compared to the center of the glass. The model has also been shown useful for studying the importance of radiation, which for thin glass sections with high heat flux is negligible, but becomes important for thicker sections or those with a milder flux condition applied. This is especially noticeable in the surface "reheat" during the cooling process.

Due to the complex nature of the radiation component, and the interaction of conduction and radiation, it would appear that a computer model is the only available comprehensive solution to the problem. The model developed here has many practical applications in the glass industry, in such processes as pressing, blowing, heating, cooling, and tempering of glass.

ACKNOWLEDGEMENT

The authors wish to thank Owens-Illinois, Inc. for permission to publish this work as well as supplying data of temperature profiles and heat flux.

NOTATION

Symbols

$C_p$    Specific heat of glass, (J/Kg-°K).

$h$    Convective heat transfer coefficient (w/m²-°K).

$k$    Thermal conductivity of glass, (w/m-°K).

$L$    Total glass thickness, (m)

$n$    Refractive index.

$t$    Time, (s)

$T$    Absolute temperature, (°K).

$W$    Emissive power for radiat flux (w/m²).

$x,y$    Distance from plane to glass surface, (m).

Greek Symbols

$\gamma$    Absorption coefficient, (m⁻¹).

$\varepsilon$    Emissivity.

$\rho$    Density of glass, (kg/m³).

$\rho$    Reflectivity of glass surface.

$\phi$    Auxiliary function for external radiation.

$\psi$    Auxiliary function for internal radiation.

Subscripts

B    blackbody

ex    external radiation

g    glass

m    mold

opaq    lower wave length limit of opaque band

LITERATURE CITED

1. Babcock, C.L. and McGraw, D.A. Glass Ind., 38, 3(1957) 137-42, 144-46, 148-51, 161.

2. McGraw, D.A., J. Amer. Ceram. Soc., 35, 7(1961) 353-363.

3. McGraw, D.A., Glastechnische Berichte, 46, 5(1973) 89-91.

4. Henriette, J., Meunier, H., Capurso, T. and Mariani, R., AIChE Symposium Series 79, 225(1983) 401.

5. Belentepe, Y.C., AIChE Symposium Series

80, 236(1984) 297-302.

6. Jones, S.P., Basnett, P. and Parker, G.C., "A Theoretical Processes Investigation of Heat Transfer Processes in Glass Forming", British Glass Industry Research Association, Rep. No. 20, 1966.

7. Gardon, R., J. Amer. Ceram. Soc. 39, 8 (1956) 278-287.

8. Gardon, R., J. Amer. Ceram, Soc., 41, 7(1958) 200-209.

9. Gardon, R., J. Amer. Ceram, Soc. 44, 7 (1961) 305-312.

10. Chui, G.D. and Gardon, R., J. Amer. Ceram. Soc., 52, 10(1969) 548-553.

11. Farag, I.H. and R.L. Curran, Glastechnische Berichte, Vol. 56K, Bd. 1, pp 319-324, also 13th Int. Congress on Glass, July 4-9, 1983, Hamburg, W. Germany.

12. Farag, I.H., and R.L. Curran (1984), "Application of Radiation Pyrometry to Glass Temperature Measurements", 22nd ASME/AIChE National Heat Transfer Conf., Aug. 5-8, Niagara Falls, N.Y., also AIChE Symposium Series, "Heat Transfer-Niagara Falls 1984", 236, Vol. 80, pp. 291-296

13. Beliveau, M.J., M.S. Thesis, Chemical Engineering Department (advisor I.H. Farag), UNH, Durham, N.H. (1985).

# HEAT TRANSFER ANALYSIS OF A BINARY COMPUND (CHROME ORE-MAGNESIA) IN SOLIDIFICATION THROUGH A TEMPERATURE RANGE

Yilmaz C. Belentepe ■ Corning Glass Works, Corning, New York

A method for analyzing the solidification process through the Finite Difference technique was developed for a material whose latent heat liberation, as well as thermal properties, are temperature dependent. This method was applied to the Corning Glass Works' fused cast refractory manufacturing process. The findings are presented in this paper. The proposed method is general in nature and can easily be adapted to other materials.

## Introduction

This study was conducted on a Corning Glass Works' fused cast refractory (commercially known as Corhart C-104 Refractory). This refractory has a composition of 45% chrome ore and 55% magnesia, and it solidifies through a temperature range between 2550 and 2060°C. These refractory oxides are melted in an electric arc furnace at about 2550°C and then cast into a mold. The resulting billets are then cut into various sizes and used in lining steel making furnaces.

An analysis of the solidification of this material was made and is presented in this paper. The fundamental conduction heat transfer equation was modified to represent transient internal energy. The internal energy consists of the sensible heat and may also include the latent heat of solidification. The standard temperature scale was modified according to Sarjan and Slack's (10) method to accommodate the temperature-dependent thermal conductivity. The internal energy equation was then solved through a computer using a system of finite difference equations. Finally, the standard temperature at each network point was obtained by successive backward substitution of internal energy into the temperature-internal energy relationship.

---

Yilmaz C. Belentepe is presently with General Electric Company, Cleveland, Ohio.

The major advantage of this approach is that the solidification analysis can be made for a material whose latent heat liberation, as well as its thermal properties, are temperature dependent.

## Heat Transfer During Solidification

The change from liquid to solid phase during the solidification of a multi-component system does not take place at a constant temperature, as in the case for a pure metal. The liquid first starts to freeze at the liquidus temperature, it then freezes partially and gradually thru a temperature range until its temperature drops to the eutetic temperature; then the remaining liquid freezes isothermally. The temperature-dependent physical properties, (especially the latent heat liberation) in this transition region introduce difficulties in obtaining an analytically exact solution. Therefore, a numerical method is used for this analysis.

The differential equation of transient heat conduction in three-dimensions is given by:

$$\rho c \frac{dT}{d\tau} = \frac{\partial}{\partial x}(k\frac{\partial T}{\partial x}) + \frac{\partial}{\partial y}(k\frac{\partial T}{\partial y}) + \frac{\partial}{\partial z}(k\frac{\partial T}{\partial z}) \quad (1)$$

The technique reported by Sarjan and Slack (10), shown in Equation 2, is used to introduce a modified temperature scale to overcome the temperature-dependent thermal conductivity.

$$\phi = \int_{T_o}^{T} \frac{k}{k_o} dT \qquad (2)$$

Here $k$ and $k_o$ are the thermal conductivity at temperature $T$ and $T_o$ respectively, (the subscript o refers to a reference point).

Equation 1 with the help of Equation 2 becomes:

$$\rho c \frac{dT}{d\tau} = k_o \left( \frac{\partial^2 \phi}{\partial x^2} + \frac{\partial^2 \phi}{\partial y^2} + \frac{\partial^2 \phi}{\partial z^2} \right) \qquad (3)$$

The definition of specific heat yields:

$$c = \frac{dH}{dT} \qquad (4)$$

Here, $H$ is the internal energy of the material. Substituting Equation 4 into Equation 3.

$$\rho \frac{dH}{dT} \frac{dT}{d\tau} = k_o \left( \frac{\partial^2 \phi}{\partial x^2} + \frac{\partial^2 \phi}{\partial y^2} + \frac{\partial^2 \phi}{\partial z^2} \right) \qquad (5)$$

Making use of Equation A-1 (Appendix A) for density, Equation 5 becomes:

$$\rho_o f \frac{dH}{dT} \frac{dT}{d\tau} = k_o \left( \frac{\partial^2 \phi}{\partial x^2} + \frac{\partial^2 \phi}{\partial y^2} + \frac{\partial^2 \phi}{\partial z^2} \right) \qquad (6)$$

A new term "U", which is related to the internal energy is defined as:

$$\frac{dU}{dT} = f \frac{dH}{dT} \qquad (7)$$

Then Equation 6 can be rearranged to:

$$\frac{dU}{d\tau} = \frac{k_o}{\rho_o} \left( \frac{\partial^2 \phi}{\partial x^2} + \frac{\partial^2 \phi}{\partial y^2} + \frac{\partial^2 \phi}{\partial z^2} \right) \qquad (8)$$

With these above transformations the temperature field becomes a function of only two parameters, internal energy and the special (modified) temperature scale. This temperature field is much easier to solve than Equation 1 with several temperature dependent properties.

## The Special Temperature Scale

The temperature scale $\phi$ is defined by Equation 2. By taking the reference temperature as zero °C and substituting Equation A-4 into Equation 2,

$$\phi = \int_{T_o=0}^{T} \{1 + (b_k/a_k)T\} dT$$

and performing the integration, the relation between the special temperature scale $\phi$ and the standard scale $T$ becomes:

$$\phi = T + (b_k/2a_k)T^2 \qquad (9)$$

The inversion of the above equation is:

$$T = \{-1 + \sqrt{1 + 2\phi b_k/a_k}\}/(2b_k/a_k) \qquad (10)$$

## Internal Energy Parameter "U"

This parameter is defined for two states of the material. The first one is the solidus state, which is from zero to eutectic temperature (2060°C). The second one is the solidus-liquidus state, which is from the eutectic temperature to the liquid temperature (2550°C).

A. <u>Solidus State ($0 \leqslant T \leqslant 2060°C$)</u>: In this region, the material provides only sensible heat. Therefore, Equation 7 can be reduced to:

$$\frac{dU}{dT} = fc \qquad (11)$$

Solving for U

$$U = \int_0^T fc \, dT \qquad (12)$$

By substituting Equation A-2 and Equation A-3 into Equation 12 and performing the integration a fourth order binomial expression was obtained. A plot of this equation indicates that a straight line approximation can be used with good accuracy. The new equation for U is then defined as:

$$U = b_{us} T \qquad (13)$$

Substituting the above equation into Equation 9 and rearranging yields

$$\phi = b_{\phi s} U + c_{\phi s} U^2 \qquad (14)$$

The numerical values of the coefficients in the above and the foregoing equations are given in Table A-1 in the Appendix A.

B. <u>Solidus - Liquidus State ($2060 < T \leqslant 2550°C$)</u>: This region is the freezing or melting zone of the material. The material contains a fraction of the latent heat of fusion as well as the sensible heat. The internal energy "H" can be expressed as:

$$H = \int_0^T c \, dT + L \qquad (15)$$

Here, $L$ is the temperature-dependent latent heat of fusion.

Substituting Equation 15 into Equation 7

$$\frac{dU}{dT} = f \frac{d}{dT} \left( \int c \, dT + L \right)$$

Solving for U

$$U = \int_0^T fc \, dT + \int_{T=2060}^{T} f \left( \frac{dL}{dT} \right) dT \qquad (16)$$

The first term in the above equation is identical to Equation 12, therefore, substituting the result of Equation 12 into Equation 16 and Equation A-2 for f and the derivative of Equation A-5 for ($dL/dT$);

$$U = a_{u1} + b_{u1}T + c_{u1}T^2 + d_{u1}T^3 + e_{u1}T^4 \quad (17)$$

Using Equation 17 and the relations between U, T and $\phi$ following two binomial equations for T and $\phi$ as functions of U are obtained:

$$T = a_{t1} + b_{t1}U + c_{t1}U^2 + d_{t1}U^3 \quad (18)$$

and

$$\phi = a_{\phi 1} + b_{\phi 1}U + c_{\phi 1}U^2 + d_{\phi 1}U^3 \quad (19)$$

Plots of these equations are shown in Figures 1 and 2. Note in particular that the relation between U and $\phi$ is a nonlinear one, and so is the relation between U and T in the solidification zone.

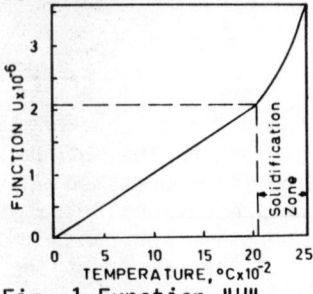

Fig. 1 Function "U" vs. Temperature

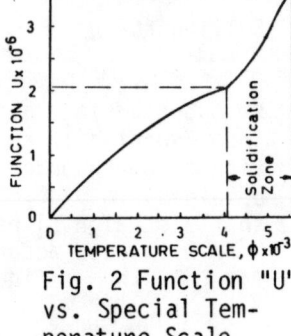

Fig. 2 Function "U" vs. Special Temperature Scale

## Mold

For this study a small experimental mold, similar to the production mold, was constructed. A cross section of this mold is shown in Figure 3. Basically the mold was made of graphite plates backed up by alumina powder contained in a steel can. Cooling of the mold was inititally controlled by the thermal storage capacity of the mold. A casting made in this mold freezes and cools down to room temperature under the natural cooling condition imposed on the mold by the environment. In the normal production the billets, upon cooling, are removed from the mold and cut into various size bricks and used in lining steel making furnaces. Molds are then recycled for new castings.

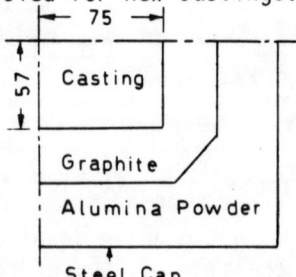

Figure 3 - Quarter Section of the Casting and the Mold (Dimensions are in millimeters.)

## Computational Technique

A set of finite difference equations for the casting and mold structure were defined according to the central difference formulation. After the new values of U were calculated throughout the newtwork, the corresponding special temperatures were determined from Equation 14 and Equation 19 and finally standard temperatures T, were obtained from Equation 10.

The boundary interface temperature between the casting and the mold was first calculated by the contact temperature relation. However, contact ceased within a few seconds after the casting was made. The boundary temperatures on the casting and the mold surfaces were then defined separately by extrapolating the temperatures in each domain toward the boundaries and matching the heat flux across the boundaries. This was accomplished by using Newton's forward interpolation scheme. The heat flux in this case predominantly consisted of radiation and a small amount of conduction which occurred thru the gas in the narrow gap between the casting and the mold.

## Analysis of Computer Result

For comparison purposes a constant temperature solidification, though hypothetical, is included here. This analysis indicates that the temperature distribution, solidification time and rate are drastically different for these two cases. Temperature vs. distance relations for both cases are shown in Figure 4. The solidification of a binary material (Case A) resulted in a higher degree of curvature in the temperature distribution thru the thickness than the constant temperature solidification (Case B). As expected, for the binary or higher order compositions, the liquidus temperature at the center of the casting drops steadily, as well as at the other locations. This indicates that the material solidifies not only from the boundaries toward the center but at the interior points too. However, for the case of solidification at constant temperature the liquid temperature stays constant until the material solidifies completely, as in pure metals. Knowing the exact temperature distribution in the casting is very important. Because, excessive thermal stresses in the casting cause cracks which under develops the inherent strength of the billets.

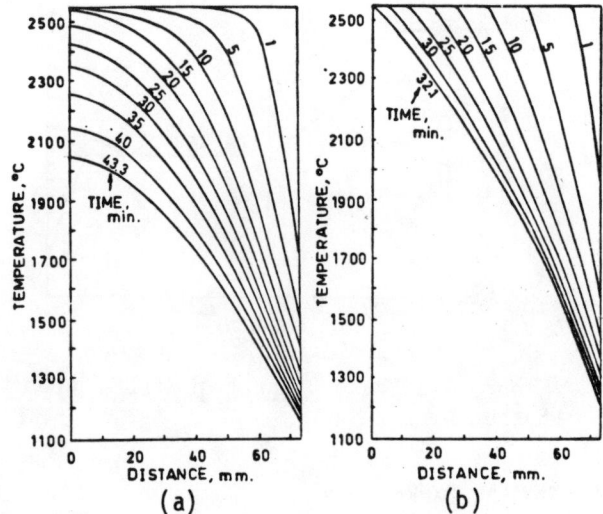

Figure 4 - Solidification vs. Thickness (a) Through Temperature Range, (b) - Constant Temperature

The temperature distribution of the casting for various elapsed times are shown in Figure 5. Initially the casting was at 2550°C and the mold was at 25°C. The largest temperature drop between various points of the casting occurred about 6 minutes after the melt was poured. The 2060°C isothermal line in Figure 5 (a) separates the solidified material zone from the liquid-solid mixture. The temperature distribution at the end of solidification is shown in (b).

## Conclusion

The approach presented here, is rather easy and helpful in performing the solidification analysis for a material whose latent heat liberation as well as its thermal properties are temperature dependent.

Knowing the exact temperature distribution in the casting during the solidification and cooling period helps to reduce thermal stresses in the casting and improves the strength of the material produced.

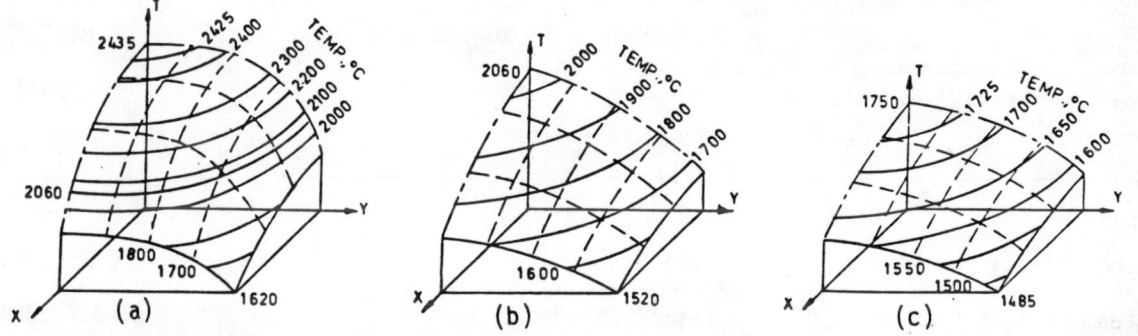

Figure 5 - Temperature distribution of the casting. (a), (b) and (c) are for 5; 21 and 30 minutes after the casting is made respectively.

## APPENDIX A

**Thermal Properties:** Thermal properties of C-104 (Corning Glass Works' product number) refractory brick are taken partly from internal reports and partly from the published data. The accuracy of foregoing property equations are very good in the range of 800 to 2550°C.

**Density:** The maximum attainable density of C-104, at room temperature, is given as 4180 kg./m$^3$. Using this value and the linear thermal expansion characteristics of the material, the following binomial equation is set up for the density:

$$\rho = \rho_o f \quad \text{(A-1)}$$

here

$$f = a_\rho + b_\rho T + c_\rho T^2 \quad \text{(A-2)}$$

**Specific Heat:** The specific heat of C-104 brick material as a function of temperature for the composition of 55% Magnesia and 45% Chrome Ore, is expressed with very good accuracy as linear:

$$c = a_c + b_c T \quad \text{(A-3)}$$

**Thermal Conductivity:** The thermal conductivity of C-104 brick is defined with reasonable accuracy by the following equation:

$$k = a_k + b_k T \quad \text{(A-4)}$$

**Latent Heat of Fusion:** The amount of latent heat liberated from the C-104 composition during the solidification process is determined by applying the "lever rule" on the phase diagram of the MgO-Chromite System (1). A reproduction of this diagram is shown in Figure A-1. The line CB on this figure represents the C-104 composition. As the material cools, the percent solidified material dictates the liberated amount of latent heat of fusion.

A binomial expression is fitted to the latent heat data and given below:

$$L = a_l + b_l T + c_l T^2 \qquad (A-5)$$

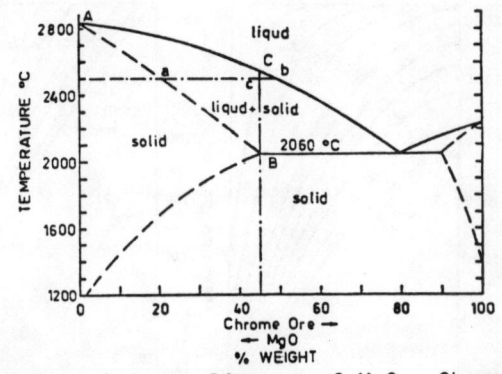

Figure A-1 Phase Diagram of MgO - Chromite System

TABLE A-1 COEFFICIENTS IN BINOMIAL EXPRESSIONS

| Function | Coefficient | Value | Function | Coefficient | Value |
|---|---|---|---|---|---|
| Density ($\rho$) | $a_\rho$ | 1.0 | Internal Energy parameter | $a_{ul}$ | 13963371.0 |
|  | $b_\rho$ | $-0.211 \times 10^{-4}$ |  | $b_{ul}$ | $-13544.17$ |
|  | $c_\rho$ | $-0.113 \times 10^{-7}$ | Solidus-Liquidus | $c_{ul}$ | 3.855250 |
| Specific Heat (c) | $a_c$ | 957.0 | State U(T) | $d_{ul}$ | $0.272943 \times 10^{-5}$ |
|  | $b_c$ | 0.0862 |  | $e_{ul}$ | $-0.209149 \times 10^{-7}$ |
| Thermal Conductivity (k) | $a_k$ | 2.873 | T(U) | $a_{tl}$ | $-819.7888$ |
|  | $b_k$ | $0.266 \times 10^{-2}$ |  | $b_{tl}$ | $0.25411 \times 10^{-2}$ |
| Latent Heat (L) | $a_l$ | 14,274,260.0 |  | $c_{tl}$ | $-0.702364 \times 10^{-9}$ |
|  | $b_l$ | $-14,548.9$ |  | $d_{tl}$ | $0.709616 \times 10^{-16}$ |
|  | $c_l$ | 3.70176 | $\phi(U)$ | $a_{\phi l}$ | $-3973.4692$ |
| Special Temperature Scale ($\phi$) | $a_\phi$ | 1.0 |  | $b_{\phi l}$ | $0.694253 \times 10^{-2}$ |
|  | $b_\phi$ | $0.46293 \times 10^{-3}$ |  | $c_{\phi l}$ | $-0.188086 \times 10^{-8}$ |
| Internal Energy Parameter, Solidus State U(T), $\phi$(U) | $b_{us}$ | 1004.73 |  | $d_{\phi l}$ | $0.190159 \times 10^{-15}$ |
|  | $b_{\phi s}$ | $0.9953 \times 10^{-3}$ |  |  |  |
|  | $c_{\phi s}$ | $0.45858 \times 10^{-9}$ |  |  |  |

## NOTATION

| Symbol | Description | Unit |
|---|---|---|
| c | Specific Heat | J/kg-°C |
| H | Internal Energy | J/kg |
| k | Thermal Conductivity | W/m.-°C |
| L | Latent Heat | J/kg |
| T | Temperature | °C, °K |
| x, y, z | Coordinate System | |

Greek Symbols

| | | |
|---|---|---|
| $\rho$ | Density | kg./m$^3$ |
| $\tau$ | Time | Second |
| $\phi$ | Special Temperature Scale | |

## LITERATURE CITED

1. Alper, A. M.; Doman, R. C. and McNally, R. N., "High Temperature Oxides" V. 1, Academic Press Inc.

2. Bathelt, A. G.; Viscanta, R., "ASME Journal of Heat Transfer", V. 101, p. 453, (1979).

3. Bathelt, A. G. and Viscanta, R., "International Journal Heat and Mass Transfer", V. 23, p. 1493, (1980).

4. Belentepe, Y. C., "Thermal Studies of the Solidification of C-104 Refractory Material", Corning Glass Works Internal Reports, No. E-68-75, (1968) and E-69-16 (1969).

5. Cho, S. H. and Sunderland, J. E., "ASME Journal of Heat Transfer", p. 214, (1974).

6. Cravalho, E. G.; Huggins, C. E. and O'Callaghan, M. G., "ASME Journal of Heat Transfer", V. 102, p. 673, (1980).

7. Ozisik, M. N. and Uzzell, J. C. Jr., "ASME Journal of Heat Transfer", V. 101, p. 331, (1979).

8. Ramsey, J. W.; Sparrow, E. M. and Varejao, L.M.C., "ASME Journal of Heat Transfer", V. 101, p. 732, (1979).

9. Rohsenow, W. M., Development in Heat Transfer. MIT Press (1964).

10. Ruddle, R. W., The Solidification of Castings, The Institute of Metals, Richard Clay and Co. Ltd., Bungay, England.

11. Shamsunder, N. and Sparrow, E. M., "ASME Journal of Heat Transfer", p. 333, (1975).

12. Saitoh, T., "ASME Journal of Heat Transfer", V. 100, p. 294, (1978).

13. Siegel, R., "ASME Journal of Heat Transfer", V. 100, p. 3, (1978).

# AN IMPROVED HEAT TRANSFER CORRELATION FOR LAMINAR FLOW OF HIGH PRANDTL NUMBER LIQUIDS IN HORIZONTAL TUBES

J.W. Palen and J. Taborek ■ Heat Transfer Research, Inc., Alhambra, California

Classical theoretical solutions for laminar heat transfer in horizontal tubes often fail to predict actual data, due to the inability to account for physical property variations between the tube wall temperature and the fluid bulk temperature.

Seven presently available empirical laminar heat transfer correlations were tested against over 600 horizontal tube data points (over 400 unpublished) on hydrocarbon oils. These ranged in Prandtl number from 15 to 19,000; in Grashof number from 0.5 to 27,000,000, and in bulk-to-wall viscosity ratio from 0.001 to 55. Most of the data were for conditions more closely approximating constant tube wall temperature than the other possible limiting condition, constant heat flux.

An improved horizontal tube correlation was developed for average Nusselt numbers which gives more consistent prediction over the entire data set; and, in addition, extrapolates in extreme cases to forms compatible with theory. The bulk of the data used were for conditions more closely approximating constant wall temperature than constant heat flux.

## INTRODUCTION

Many industrial fluids such as heavy oils or polymer solutions have viscosities high enough to prevent turbulent flow even at moderately high design velocities. Heat transfer rates to such laminarly flowing fluids are relatively low. The cost of the large heat transfer surfaces required for such fluids, therefore, dictates careful, accurate thermal sizing with a minimum of overdesign.

In spite of voluminous work on laminar flow heat transfer, now spanning nearly a century, many questions are still unanswered and much potential for improvement still remains.

The availability of a wider range of laminar flow data than ever previously correlated, as well as computer-aided evaluation techniques, offered a potential for much needed further improvement.

## IDEAL LAMINAR FLOW - CONSTANT PHYSICAL PROPERTIES

Complications introduced by varying physical properties and entrance effects make the real laminar flow problem significantly different from the ideal case adaptable for analytical solution. However, as with other topics, analysis of the ideal case provides a theoretical groundwork from which correlational extensions can be made.

Ideal laminar flow may be defined as that for which fluid particles move only along streamlines parallel to the surfaces containing the flow. This restriction requires a fully developed velocity profile and excludes superimposed natural convection currents. Since molecules move only parallel to the walls, no convection in a perpendicular direction is possible, and all heat transfer from the walls to the fluid must be by conduction alone. This fact, plus the low thermal conductivity of most liquids (except liquid metals), explains the relatively low heat transfer rates obtained for laminar flow.

The basic approach to laminar flow problems is to write equations for conservation of mass, momentum, and energy for a differential fluid element, and then make assumptions required to simplify the equations sufficiently for solution.

Although a general analytical solution in closed form is not possible, even for the simplified equations, a series type solution was obtained in 1885 by Graetz and later simplified by Leveque.

The Graetz solution is discussed to some extent in most heat transfer texts. A relatively recent discussion is given in Ref. 1.

# REAL LAMINAR FLOW - VARYING PHYSICAL PROPERTIES

The theoretical ideal laminar flow equations are strictly valid only for the limiting condition of zero temperature driving force for which the physical properties are constant across the tube cross sections. It was early recognized, in comparing theoretical solutions with actual data, that variation of physical properties can have a significant effect in practical applications.

## VARYING VISCOSITY

The viscosity of heavy oils and other viscous materials often varies greatly with temperature. Under normal operating conditions the viscosity at the wall temperature may be an order of magnitude or more different from the viscosity at the bulk temperature, causing distortion of the velocity profile. Figure 1 illustrates the relative effects for the extremes of the bulk-to-wall viscosity ratios of 0, 1, and ∞.

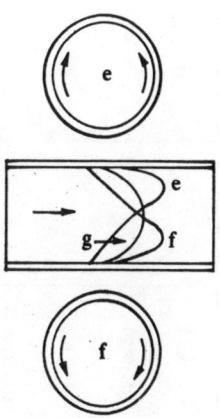

e) horizontal heating
f) horizontal cooling

Figure 2  Schematic Velocity Profiles Combined Natural and Forced Convection

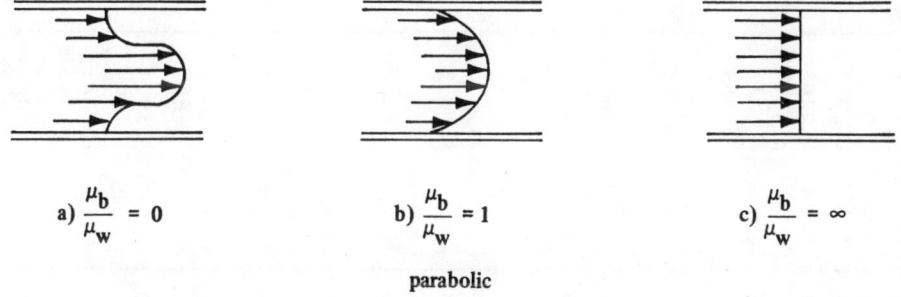

a) $\frac{\mu_b}{\mu_w} = 0$    b) $\frac{\mu_b}{\mu_w} = 1$    c) $\frac{\mu_b}{\mu_w} = \infty$

parabolic

Figure 1  Velocity Profiles as Functions of Viscosity Ratio

An early attempt to account theoretically for varying viscosity was made by Yamagata (2). Although the mathematical treatment was quite elegant, the resulting equations had to be further corrected empirically to agree with existing data.

## VARYING DENSITY

Density differences between the fluid at the tube wall and in the bulk cause natural convection currents which significantly affect heat transfer rates. The extent of natural convection effects and the methods of accounting for them depend on the vector relationship between natural convection flow and forced convection flow.

An illustration is shown in Figure 2.

## DATA

Existing correlations (2 to 9) give widely differing results under certain combinations of extreme parameters. In order to determine the overall ranges of applicability of the various equations, a set of over 600 laminar heat transfer data points was assembled. Of these, 297 were unpublished proprietary data and 99 were unpublished data obtained from the University of Tulsa.

Table 1 shows the wide range of parameters covered by the data. Much care was taken to assemble the best available data set which would completely cover, and even exceed, the range of parameters normally encountered in commercial applications.

TABLE 1  Ranges of Parameters and Conditions of the Data Set

| DATA SET | Source | Fluid | Process | D, in. | L/D | Re | Pr | $C_p$ | $\mu_b/\mu_w$ | Gr |
|---|---|---|---|---|---|---|---|---|---|---|
| 1 | HTRI | ALTA-VIS 530 | COOLING | 0.406–0.834 | 226–464 | 15–1600 | 630–2500 | 27–134 | 0.001–0.568 | 6–808 |
| 2 | HTRI | ALTA-VIS 530 | HEATING | 0.406–0.834 | 226–464 | 7–2100 | 970–6300 | 60–560 | 1.9–8.7 | 4–430 |
| 3 | HTRI | SAE 250 | COOLING | 0.406–0.834 | 226–464 | 50–1100 | 500–2600 | 26–160 | 0.007–0.97 | 98–4600 |
| 4 | HTRI | SAE 250 | HEATING | 0.406–0.834 | 226–464 | 1–1200 | 450–4400 | 35–460 | 1.6–11 | 0.3–2900 |
| 5 | HTRI | SAE 30 | HEATING | 0.406–0.834 | 226–464 | 12–2100 | 80–970 | 5–68 | 1.1–9.8 | 24–86,300 |
| 6 | SIEDER AND TATE | 16° API | COOLING | 0.620 | 98.7 | 4–1100 | 490–13,900 | 25–870 | 0.006–0.069 | 0.8–2700 |
| 7 | SIEDER AND TATE | 21° API | COOLING | 0.620 | 98.7 | 100–2100 | 150–570 | 8–35 0.15 | 0.05 | 490–16,000 |
| 8 | SIEDER AND TATE | 24.5° API | HEATING | 0.620 | 98.7 | 60–1300 | 360–570 | 25–40 | 3.9–5.4 | 474–1150 |
| 9 | UNIV. OF TULSA | 29.2° API | HEATING | 0.495 | 19.4–480 | 110–540 | 500–600 | 42–55 | 9.1–10.9 | 290–390 |
| 10 | UNIV. OF TULSA | 31.7° API | COOLING | 0.495 | 19.4–480 | 14–2000 | 170–260 | 11–15 | 0.11–0.52 | 240–4040 |
| 11 | UNIV. OF TULSA | 35° API | HEATING | 0.495 | 19.4–480 | 300–1100 | 90–110 | 7.9–8.1 | 3.2–3.6 | 14,000–18,000 |
| 12 | UNIV. OF TULSA | 37° API | HEATING | 0.495 | 19.4–480 | 70–2100 | 15–17 | 0.99–1.1 | 0.53–0.63 | 97,000–311,000 |
| 13 | UNIV. OF TULSA | 39° API | COOLING | 0.495 | 19.4–480 | 160–2100 | 44–47 | 2.9–3.1 | 1.5–2.5 | 64,000–89,000 |
| 14 | KERN AND OTHMER | CYLINDER OIL | HEATING | 0.62–2.5 | 49–193 | 20–750 | 760–5600 | 50–370 | 8.5–55 | 450–70,000 |
| 15 | KERN AND OTHMER | CORE OIL | HEATING | 0.62–2.5 | 49–193 | 80–460 | 350–1500 | 23–100 | 5.7–44 | 900–$1.9 \times 10^6$ |
| 16 | KERN AND OTHMER | TRANSFORMER OIL | HEATING | 0.62–2.5 | 49–193 | 190–2100 | 90–300 | 6–20 | 4–20 | $0.23 \times 10^6$–$27.7 \times 10^6$ |
| 17 | OLIVER | GLYCERINE | HEATING | 0.50 | 72 | 0.2–0.6 | 9000–14,000 | 1100–1700 | 1.9–3.5 | 0.5–1.5 |
| 18 | OLIVER | GLYCERINE | COOLING | 0.50 | 72 | 0.1–0.3 | 17,000–19,000 | 2100–2300 | 0.42–0.44 | 0.3 |

## EVALUATION OF PRESENT CORRELATIONS

A complete listing of all fluids investigated, as well as the percentage of points predicted by each literature correlation tested within the ±30 percent error band, is given in Table 2.

In general, the best correlation predicted about 70 to 100 percent of the data used in its own development within ±30 percent, but all equations tended to break down in some part of the extreme range of conditions of the entire data set.

As would be expected, the equations which did not include a natural convection correction predicted very low on all data with appreciably high Gr.

Of the correlations, including natural convection effects, overall agreement tended to improve with the chronological date of the correlations as a result of the increased amount of data used for development.

Two additional references (10,11) were brought to our attention after completion of the paper, and should receive further consideration.

## NEW CORRELATIONS

Since none of the literature correlations tested gave satisfactory results over the entire data range, it was necessary to develop an improved, more general correlation.

The new correlations were intended to have the following characteristics:

1. Show no serious discrepancies over the entire range of data.

2. Predict conservatively on data for which discrepancies are largest.

3. Extrapolate logically to extreme conditions, giving results consistent with theoretical limiting cases. For example, the heat transfer coefficient should become a constant as tube length becomes very large to agree with the limiting logarithmic mean Graetz solution.

4. Be the simplest form which holds to condition 3 and gives acceptable accuracy.

5. Be based on arithmetic mean physical properties and result in an average heat transfer coefficient for the tubeside fluid.

## VARIABLE GRAETZ NUMBER EXPONENT

Since the Pigford analysis (3) shows that the exponent on Gz should be a function of the velocity profile, it was reasonable to try to correlate this exponent as a function of the viscosity ratio, $\mu_b/\mu_w$. In order to be compatible with the limiting case of Gz = 0 for a log mean driving force, the following form was considered for the low natural convection region:

$$Nu_{\ell m} = 3.66 + A\ (Gz)^m \quad (1)$$

$$Gr < 200$$

Values of A and m were back-calculated from the data, giving A = 1.45 and m = $f\ (\mu_b/\mu_w)$. A curve relating m to $\mu_b/\mu_w$ gave values very close to those of Pigford in the range in which the Leveque assumption was fairly valid, $\mu_b/\mu_w > 0.1$. The data in the deep cooling range, $\mu_b/\mu_w < 0.1$, were scattered, but the best curve through the entire data set for Gr < 200 was represented by:

$$m = \frac{0.5}{1 + 0.5 \left(\frac{\mu_b}{\mu_w}\right)^{(-0.25)}} \quad (2)$$

## EMPIRICAL METHOD

Another logical approach was to determine empirical constants using dimensionless parameters. This approach is similar to methods employed in turbulent flow, where the inability to predict velocity and temperature profiles accurately by theory led to abandonment of theoretical methods in favor of simple, easily used design equations relating pertinent dimensionless groups by means of empirically determined constants.

The following basic format, which was successfully used in the past for both laminar and turbulent correlations, was selected:

$$Nu_{\ell m} = C + A(Re)^n\ (Pr)^p\ (D/L)^r\ (\mu_b/\mu_w)^s$$

where constant C accounts for the condition of Nu → C if L → ∞.

In order to account for natural convection, the Grashof number must be included. Since the effect of Gr is to impart an additional velocity component to the flow, it is proposed to simply modify the Reynolds number by adding axial and tangential velocity components.

TABLE 2  Percentage of Data Points Predicted Within ± 30 Percent For Each Correlation

| DATA SET | Source | Fluid | Process | 2 | 4 | Literature Reference 5 | 6 | 7 | 8 | 9 | Eq. 3 | No. of Points |
|---|---|---|---|---|---|---|---|---|---|---|---|---|
| 1 | HTRI | ALTA VIS | COOLING | 43.4L | 47.6H | 90.5 | 76.1 | 90.5 | 90.5 | 90.5 | 81 | 67 |
| 2 | HTRI | ALTA VIS | HEATING | 20.9H | 28.4 | 28.4H | 35.6H | 42.6H | 35.6H | 71.4H | 92.9 | 43 |
| 3 | HTRI | SAE 250 | COOLING | 43.8L | 33.3S | 71.4L | 76.2S | 61.8L | 76.2S | 57.1L | 76.2L | 32 |
| 4 | HTRI | SAE 250 | HEATING | 91.3 | 86.3 | 86.3 | 63.6H | 91.0 | 95.0 | 86.3L | 86.4L | 80 |
| 5 | HTRI | SAE 30 | HEATING | 81.4S | 84.0H | 70.6L | 75.5L | 60.0L | 93.4 | 60.0L | 91.0 | 75 |
| 6 | SIEDER AND TATE | 16° API | COOLING | 0L | 12.5H | 100 | 62.0L | 100 | 100 | 100 | 100 | 8 |
| 7 | SIEDER AND TATE | 21°API | COOLING | 40L | 20H | 80L | 100 | 80L | 100 | 80L | 100 | 5 |
| 8 | SIEDER AND TATE | 24.5° API | HEATING | 100 | 100 | 100 | 100 | 100 | 100 | 100 | 100 | 6 |
| 9 | UNIV. OF TULSA | 29.2° API | HEATING | 83.6 | 83.6 | 83.6 | 83.6 | 83.6 | 83.6 | 83.6 | 83.6 | 6 |
| 10 | UNIV. OF TULSA | 31.7° API | COOLING | 40.8S | 54.2L | 28.8L | 44.1 | 6.8L | 52.3L | 13.5L | 69.3L | 42 |
| 11 | UNIV. OF TULSA | 35° API | HEATING | 100 | 80.0L | 60.0L | 40.0L | 20.0L | 80.0L | 20.0L | 100 | 5 |
| 12 | UNIV. OF TULSA | 37° API | HEATING | 20.5S | 44.8L | 0L | 0L | 5.0L | 50.0S | 0L | 88.3 | 20 |
| 13 | UNIV. OF TULSA | 39° API | COOLING | 65.5L | 34.6L | 0L | 7.7L | 3.8L | 50.0L | 0L | 96.6 | 26 |
| 14 | KERN AND OTHMER | CYLINDER OIL | HEATING | 33.3H | 85.3 | 49.4L | 54.6H | 85.3 | 88.0 | 45.3L | 96.1 | 75 |
| 15 | KERN AND OTHMER | CORE OIL | HEATING | 48.8H | 81.9 | 26.8L | 87.7 | 65.9L | 96.3 | 19.6L | 83.9 | 82 |
| 16 | KERN AND OTHMER | TRANSFORMER OIL | HEATING | 20.0H | 57.5H | 20.0L | 92.5 | 62.5L | 92.5 | 12.5L | 98.7 | 40 |
| 17 | OLIVER | GLYCERINE | HEATING | 7.1H | 100 | 21.4H | 21.4H | 92.9 | 78.5H | 100 | 100 | 14 |
| 18 | OLIVER | GLYCERINE | COOLING | 0H | 100 | 100 | 100 | 100 | 100 | 100 | 100 | 3 |
| | TOTAL DATA | | | 5.03 | 66.1 | 52.3 | 63.2 | 64.3 | 81.9 | 51.7 | 88.1 | 629 |

Note: L = low, H = high, S = scattered

A modified velocity parameter, $Re^*$, may be defined as:

$$Re^* = [Re^2 + B(Gr)]^{0.5} \quad \text{or} \quad Re^* = Re + B\,Gr^{0.5}$$

for vectorial and for arithmetic addition of the velocity components respectively.

The correlating form then uses $Re^*$ substituted for $Re$. Regression analysis using both forms of $Re^*$ showed that the simpler arithmetic form gave better correlational results. The quality of the regression correlation was quite good for such a wide range of variables as included in our data set. A multiple correlation coefficient of 0.9 was obtained on a scale of 0 for no correlation to 1.0 for perfect correlation.

The resulting best equation was:

$$Nu_{\ell m} = 2.5 + 4.55\,(Re^*)^{0.37} \left(\frac{D}{L}\right)^{0.37} (Pr)^{0.17} \left(\frac{\mu_b}{\mu_w}\right)^{0.14} \quad (3)$$

The effective Reynolds Number is calculated as follows

$$Re^* = Re + 0.8\,Gr^{0.5} \exp\left(-42/Gr^2\right) \quad (4)$$

The exponential term in Equation (4) was added to the original definition as it was found necessary to suppress the Gr- effect below the value of about 2.0. For $Gr > 2$, the exponential term is approximately equal 1.0.

It is of interest that the power on the viscosity ratio term as determined by regression analysis was exactly the same as that determined by Sieder and Tate.

The following general limitations must be imposed upon Equation (3).

$0 < \mu_b/\mu_w < 55$

$20 < Pr < 10,000$ *(higher or lower Pr ranges may not be applicable)*

$0.1 < Re < 2000$

$0 < Gr < 30,000,000$

$40 < L/D < \infty$

None of the data of Refs. 10 and 11 were within the above ranges. However, the glycerine-water data of Ref. 11 ($Pr \simeq 300$) were also well predicted by Equation (3).

## CONCLUSIONS AND RECOMMENDATIONS

Velocity and temperature profiles in horizontal tubes with natural convection are complicated and difficult to predict with classic laminar flow theoretical concepts.

All previously available laminar flow correlations tested give poor predictions in some ranges of the large data set.

Of the previously published methods, the Oliver equation (8) gives the best overall results, but predicts low for light oils and high for very viscous oils in heating.

An empirical correlation similar in form to those used for turbulent flow was developed as a practical and relatively very accurate design equation for laminar flow. *However, it must be recognized that an empirical equation should not be used outside the range of data for which it was developed (see Table 1).*

The recommended correlation is based on arithmetic mean physical properties and gives an average value of the inside heat transfer coefficient for the tube. Since most data were for conditions approximating constant wall temperature (not electrically heated), the correlation should be better suited to constant wall temperature than to constant heat flux cases.

## NOTATION

A,B,C    constant

D    inside tube diameter, m

Gr    Grashof number, with properties evaluated at average bulk temperature

Gz    Graetz number, with properties evaluated at average bulk temperature

h    average tubeside heat transfer coefficient for tube of length L, W/m$^2$ K

k    thermal conductivity at average bulk temperature, W/m K

L    tube length, m

m    exponent on the Graetz number

$Nu_{\ell m}$    Nusselt number based on the logarithmic mean temperature difference = h D/K

| | |
|---|---|
| n | constant |
| Pr | Prandtl number with properties evaluated at average bulk temperature |
| ṗ | constant |
| Re | Reynolds number with properties evaluated at average bulk temperature |
| Re* | Reynolds number corrected for natural convection, with properties evaluated at average bulk temperature |
| r | constant |
| $\mu_b$ | viscosity evaluated at average bulk temperature, $Ns/m^2$ |
| $\mu_w$ | viscosity evaluated at average temperature, $Ns/m^2$ |

## LITERATURE CITED

1. Seller, J. K., Tribus, M., and Klein, J. S., Trans. ASME, Feb. 1956, 441.

2. Yamagata, K., Mem. Fac. Eng., Kyushu Imp. Univ., 8, 365 (1940).

3. Pigford, R. L., CEP Symposium Series 51, No. 17, 1955.

4. Colburn, A. P., Trans. AIChE 29, 174 (1933)

5. Sieder, E. N. and Tate, G. E., Ind. Eng. Chem. 28, 1429 (1936)

6. Kern, D. Q. and Othmer, D. F., Trans. AIChE 2, 517 (1933)

7. Eubank, O. C. and Proctor, W. S., S. M. Thesis in Chemical Engineering, M.I.T.

8. Oliver, D. K., Chem. Eng. Sci. 17, 335-350 (1962)

9. Hausen, H., Zeit. V. D. I., Beiheft Verfahrenstechnik, No. 4, 91-98 (1943)

10. Brown, A. R. and Thomas, M. A., "Combined Free and Forced Convection Heat Transfer for Laminar Flow in Horizontal Tubes," Journal of Mechanical Engineering Science, Vol. 7, 440-448 (1965)

11. Depew, C. A. and August, S. F., "Heat Transfer Due to Combined Free and Forced Convection in a Horizontal and Isothermal Tube," ASME Journal of Heat Transfer, Vol. 93, 380-384 (1971)

# DESIGN AND OPTIMIZATION OF HEAT EXCHANGERS FOR BATCH HEATING BY THE NTU-EFFECTIVENESS METHOD

Roger Crane and Rene Arrazola ■ University of South Florida, Tampa, Florida

A design method has been developed for sizing external heat exchangers for batch processes based on the effectiveness-NTU method. The method is applicable to any heat exchanger configuration and design. Based on the specified process conditions a solution is obtained which directly specifies the required batch circulation rate and the associated heat exchanger effectiveness. This approach is particularly useful for those types of heat exchanger designs which permit explicit sizing.

## Introduction

The use of external heat exchangers for batch heating has increased significantly as newer low cost units such as plate and frame have gained acceptance. Such units offer several advantages over jacketed vessels or immersion heaters including increased heat transfer area and elimination of large internal agitators. In many cases, it is possible to both reduce batch heat up time and power consumption.

The sizing of heat exchangers will generally depend on the batch size, circulation rate, initial and final design temperatures, utility flow rate and temperature, external feed rate and location, and the magnitude of associated chemical reactions. These parameters may be combined to size a heat exchanger using a trial and error procedure (1, 2, 3). An alternative is to relate each of these parameters to the heat exchanger effectiveness. The heat exchanger effectiveness is related to the thermal size of the unit through the appropriate effectiveness-NTU relations for specific pass arrangements. The advantage of this approach is that for certain heat exchanger types explicit expressions have been obtained for unit size as a function of the required NTU (4) and the trial and error procedure is avoided. The NTU method is already well established as a technique for steady state applications. This article discusses applications of the method to batch process.

## Basic Concepts

The solution will be limited to well mixed vessels having negligible evaporative losses. It is assumed that the product of heat capacity and circulation rate remain constant, that the vessel and its contents are initially at uniform temperature, that the effectiveness-NTU relations apply and that all pipes are well insulated. All chemical reactions are considered to be proportional to either the reactant input rate (fast reaction) or to the accumulated reactant mass (slow reaction).

Referring to Figure 1, fluid is drawn from the vessel and pumped through an external heat exchanger. Reactant feed during the transient may be added either directly to the vessel or through the circulation line. Under the above assumption, an overall energy balance would appear as:

$$\frac{dE}{dt} = Q_{hx} - Q_l + Q_r + Q_{in} \qquad (1)$$

The terms on the right side of the equation refer respectively to the energy input to the system through the external heat exchanger, losses to the environment, heat of reaction and the input feed.

Detailed expressions may be obtained for each of the terms in Equation (1). For a steady fluid addition to the system, the equation becomes:

$$\frac{d[(m_s t + M)C_p T_p]}{dt} + M_v C_v F_v \frac{dT_p}{dt} = \lambda \eta (T_u - T_{pi})$$
$$- U_v A_v (T_p - T_a) + Q_r + m_s C_p T_s \quad (2)$$

where the heat exchanger process side inlet temperature, $T_{pi}$, is equal to the batch temperature, $T_p$, if the inlet is directly from the vessel. For the case in which process feed is mixed with a stream from the vessel in the heat exchanger inlet line:

$$T_{pi} = [m_s T_s + (m_p - m_s)T_p]/m_p$$

Equation (2) is a simple first order, non-homogeneous differential equation which may now be solved for each of the cases of interest. The solution for cases involving slow reactions with process feed is given in Equation (3b). Equation (3a) applies to all other cases:

$$(T_f - \alpha \beta)/(T_i - \alpha \beta) = e^{-K/\alpha} \quad (3a)$$

$$\frac{T_f + \alpha(T_f C_p m_s - \beta - Q_r m_s t) + \alpha^2(Q_r C_p m_s M_1 - C_p m_s \beta)}{T_i + \alpha(T_i C_p m_s - \beta) + \alpha^2(Q_r C_p m_s M_1 - C_p m_s \beta)}$$
$$= e^{-K/\alpha} \quad (3b)$$

where $\alpha = 1/(\lambda \eta \gamma + U_v A_v + m_s C_p)$

$\gamma = 1 - m_s/m_p$ for feed to line,
otherwise $\gamma = 1$

$\beta = \lambda \eta \phi + U_v A_v T_a + m_s C_p T_s + Q_r m_s + Q_r M$

$\phi = (T_u m_p - T_s m_s)/m_p$ if feed is to line,
otherwise $\phi = T_u$

$K = [\ln(1 + m_s t/M_1)]/(m_s C_p)$ with feed,
otherwise $K = t/(M_1 C_p)$

## Linearizing Results

These equations provide implicit expressions for the process circulation rate, heat capacity and heat exchanger effectiveness as a function of batch time. While it is possible to work with the implicit equations directly, it is much more convenient to deal with an explicit form. The authors have linearized the exponential term using the method of weighted residuals presented by Villadsen and Michelsen (5). Then $\exp(-X) = (0.993 - 0.289 X)/(09.36 + X)$. This approximation holds within 5% over the interval $0.05 < X < 2.50$. This is equivalent to a transient in which the batch undergoes a temperature change ranging between 10 and 95% of the ultimate temperature change.

Introducing the linearized exponential term into Equations (3a) and (3b), rearranging terms and simplifying, the following second order polynomial equation is obtained for all cases except for those involving slow reactions:

$$C1\ \lambda^2 \eta^2 + C2\ \lambda\ \eta^2 + C3\ \eta^2 + C4\ \lambda\ \eta + C5\ \eta + C6$$
$$= 0 \quad (4)$$

where for cases expecting those involving line feed:

$C1 = K(1.289\ T_u - 0.289 T_i - T_f)$
$C2 = 0$
$C3 = 0$
$C4 = 0.993\ T_i - 0.936\ T_f - 0.057\ T_u + K(1.289\ T_u K1 + 1.289\ K2 - 0.578\ T_i K1 - 2\ T_f K1)$
$C5 = 0$
$C6 = -K\ K1^2(0.289\ T_i + T_f) + K1(1.289\ K\ K2 + 0.993\ T_i - 0.936\ T_f) - 0.057\ K2$

for cases involving no reaction and with reactant feed to the circulation line:

$C1 = K(1.289\ T_u - 0.289\ T_i - T_f)$
$C2 = C_p m_s K[0.578\ T_i + 2\ T_f - 1.289(T_s + T_u)]$
$C3 = K\ m_s^2\ C_p^2(1.289\ T_s - 0.289\ T_i - T_f)$
$C4 = 1.289\ T_u K\ K1 + 1.289\ K\ K2 + 0.993\ T_i - 0.936\ T_f - 0.057\ T_u - 2\ K\ K1\ (0.289\ T_i + T_f)$
$C5 = C_p m_s[0.057\ T_s - 1.289\ K(T_s K1 + K2) - 0.993\ T_i - 0.936\ T_f + 2\ K\ K1(0.289\ T_i + T_f)]$
$C6 = -K\ K1^2(0.289\ T_i + T_f) + K1(1.289\ K\ K2 + 0.993\ T_i - 0.936\ T_f) - 0.057\ K2$

For cases involving slow reactions and a reactant feed to the circulation line the linearized equation becomes:

$$C1\ \lambda^3 \eta^3 + C2\ \lambda^2 \eta^3 + C3\ \lambda\ \eta^3 + C4\ \eta^3 + C5\ \lambda^2 \eta^2$$
$$+ C6\ \lambda\ \eta^2 + C7\ \eta^2 + C8\ \lambda\ \eta + C9\ \eta + C10 = 0 \quad (5)$$

where the constants are defined as follows for reactant feed to the vessel:

$C1 = K\ (0.289\ T_i + T_f - 1.289\ T_u)$
$C2 = 0$
$C3 = 0$
$C4 = 0$

$$C5 = K\{(3\,K1 + C_p m_s)(0.289\,T_i + T_f) - Q_r m_s t$$
$$- 1.289[(2\,K1 + C_p m_s)T_u - K2]\}$$
$$+ 0.993\,T_i - 0.936\,T_f + 0.057\,T_u$$

$$C6 = 0$$
$$C7 = 0$$
$$C8 = K1^2 K(0.867\,T_i + 3\,T_f - 1.289\,T_u)$$
$$+ K1[1.986\,T_i - 1.872\,T_f - 2\,K\,Q_r m_s t$$
$$- 2.579\,K\,K2 + 2\,C_p m_s K(0.289\,T_i + T_f)$$
$$+ 0.057\,T_u - 1.289\,C_p m_s T_u K] + (C_p m_s)$$
$$[1.289\,K(Q_r M_1 - K2) + 0.057\,T_u + 0.936\,T_f$$
$$- 0.993\,T_i] + 0.936\,Q_r m_s t + 0.057\,K2$$

$$C9 = 0$$
$$C10 = K1^3\,K(0.289\,T_i + T_f) + K1^2[0.993\,T_i$$
$$- 0.936\,T_f + C_p m_s(0.289\,T_i + T_f)$$
$$- 1.289\,K\,K2 - K\,Q_r m_s t] + K1[0.057\,K2$$
$$+ 0.936\,Q_r m_s t + 1.289\,K\,C_p m_s(Q_r M_1 - K2)$$
$$- C_p m_s(0.993\,T_i - 0.936\,T_f)]$$
$$- 0.057\,C_p m_s(Q_r M_1 - K2)$$

and for slow reactions with reactant feed to the circulation line:

$$C1 = K(0.289\,T_i + T_f - 1.289\,T_u)$$

$$C2 = K\,m_s C_p(1.289\,T_s + 2.578\,T_u - 0.867\,T_i - 3\,T_f)$$

$$C3 = K\,m_s^2\,C_p^2(0.867\,T_i + 3\,T_f - 2.578\,T_s - 1.289\,T_u)$$

$$C4 = 1.289\,K\,m_s^3\,C_p^3\,T_s$$

$$C5 = K\{(3\,K1 + C_p m_s)(0.289\,T_i + T_f) - Q_r m_s t$$
$$- 1.289[(2\,K1 + C_p m_s)T_u - K2]\}$$
$$+ 0.993\,T_i - 0.936\,T_f + 0.057\,T_u$$

$$C6 = K\,m_s C_p[6\,K1\,(-0.289\,T_i - T_f) + 2.578\,K1$$
$$(T_s + T_u) + 2\,Q_r m_s t + 2.578\,K2 + 1.289\,m_s C_p$$
$$(T_s + T_u) - 2\,m_s C_p(0.289\,T_i + T_f)] - 1.986\,T_i$$
$$+ 1.872\,T_f - 0.057(T_s + T_u)$$

$$C7 = K\,m_s^2\,C_p^2[K1(0.867\,T_i + 3\,T_f - 2.578\,T_s)$$
$$- 1.289(K2 + m_s C_p T_s) + m_s C_p(0.289\,T_i + T_f)$$
$$- Q_r m_s t] + 0.993\,T_i - 0.936\,T_f + 0.057\,T_s$$

$$C8 = K1^2 K(0.867\,T_i + 3\,T_f - 1.289\,T_u)$$
$$+ K1[1.986\,T_i - 1.872\,T_f - 2\,K\,Q_r m_s t$$
$$- 2.579\,K\,K2 + 2\,C_p m_s K(0.289\,T_i + T_f)$$
$$+ 0.057\,T_u - 1.289\,C_p m_s T_u K] + (C_p m_s)$$
$$[1.289\,K(Q_r M_1 - K2) + 0.057\,T_u + 0.936\,T_f$$
$$- 0.993\,T_i] + 0.936\,Q_r m_s t + 0.057\,K2$$

$$C9 = K\,m_s C_p[K1^2(0.867\,T_i + 3\,T_f - 1.289\,T_s)$$
$$- K1(2\,Q_r m_s t + 2.578\,K2 - 0.578\,T_i m_s C_p$$
$$- 2\,T_f m_s C_p + 1.289\,T_s m_s C_p + 1.289\,m_s C_p$$
$$(Q_r M_1 - K2)] + m_s C_p[(2\,K1 - C_p m_s)$$
$$(0.993\,T_i - 0.936\,T_f) + 0.057(K1\,T_u + K)$$
$$+ 0.936\,Q_r m_s t]$$

$$C10 = K\,K1^3(0.289\,T_i + T_f) + K1^2[0.993\,T_i$$
$$- 0.936\,T_f + m_s C_p(0.289\,T_i + T_f) - 1.289\,K$$
$$K2 - K\,Q_r m_s t] + K1[1.289\,K\,m_s C_p(Q_r M_1 - K2)$$
$$- m_s C_p(0.993\,T_i - 0.936\,T_f) + 0.057\,K2$$
$$+ 0.936\,Q_r m_s t] - 0.057\,m_s C_p(Q_r M_1 - K2)$$

In the four preceding equations;

$$K1 = U_v A_v + m_s C_p$$

$$K2 = U_v A_v T_a + m_s C_p T_s, \text{ with no reactions}$$

$$K2 = U_v A_v T_a + m_s C_p T_s + Q_r M, \text{ slow reactions}$$

$$K2 + U_v A_v T_a + m_s C_p T_s + Q_r m_s, \text{ fast reactions}$$

Solution of these equations for either process circulation rate or heat effectiveness will require obtaining the roots of either a quadratic or a cubic equation. The methods are well known, so the final equations are omitted for brevity.

Experience has shown that most processes will have a negligible heat of reaction or can be treated as a fast reaction. Solving for the positive roots of Equation (4) it is found that the product of the x efficiency and the process side thermal capacity ($\lambda\eta$) is constant. In effect knowing the process circulation rate and the required heating time it is possible to calculate directly the required average heat exchanger effectiveness. Alternately having specified the heat exchanger effectiveness it is possible to solve directly for the required circulation rate. In cases in which a noncondensing utility is utilized, it is convenient to divide the product $\lambda\eta$ by the utility thermal capacity to obtain $\eta R = \text{Constant}/(m_u C_u)$.

Lines of constant $\eta R$ are shown in Figures 2, 3 and 4 developed from the TEMA (6) temperature efficiency charts for the cases of counter flow, 1-2 and 2-4 configurations respectively. One advantage of this approach is that the results are applicable to any flow arrangement for which effectiveness-NTU relations are available. Thus it is possible to immediately compare the performance of any unit for which effectiveness-NTU curves exist.

Example

A batch of 75 m³ of corn syrup is to be heated from 10 to 30°C in one hour by withdrawing 40 1/s and circulating it through an external hx. A 40°C hot water stream is available with a flow rate of 40 1/s. The process is to be conducted in a room maintained at 15°C in a vessel with a total surface area of 35 m² and with an overall heat transfer coefficient estimated at 10 W/m K. The thermal conductivity, specific heat, density and viscosity may be taken at 3.81 & 1.25 W/m K, 270 & 4.19 kJ/kg-K, 1350 & 1000 kG/m³, 175 & 0.70 cp for the corn syrup and the hot water respectively.

Substituting the process parameters into Equation (4) and solving for the positive root it is found that the product of the process thermal capacity and the hx efficiency is 129182 J/sec-K. This number is independent of the circulation rate of the utility or process stream. The process side thermal capacity is calculated to be 145800 J/sec-K. Dividing this value into the product gives a required hx efficiency, $\eta$, of 0.885. If a condensing utility were specified, the intercept of the $\eta = 0.885$ and the $R = 0$ lines on Figure 2, 3 or 4 would specify the unit NTU. In this example a noncondensing utility is specified so that an alternate approach is used. Note that the maximum effectiveness is 1 so that in the limit the process thermal capacity must exceed 129182 J/sec-K. Using a maximum available utility of 40 1/s, both sides of this equation can be divided by the utility thermal capacity to obtain the product of the thermal capacities and the hx efficiency, $R \eta = .77$. Look for an intercept of the $\eta = 0.885$ and the $R \eta = .77$ lines in Figures 2, 3 and 4. No such intercept is found for the 1-2 or the 2-4 configurations indicating that they will not work with the specified conditions. Note that no effort was made to actually design such a unit. This is in direct contrast to previous methods. For the counterflow arrangement, it is found that a unit with an effectiveness of 0.885 and a thermal size of 5.32 NTU is adequate. Using this value, the exit temperatures are known as functions of the inlet conditions for any time in the heating process. A detailed thermal design of a practical arrangement may then proceed using these temperatures. This quasistatic approach permits the use of standard steady state design methods beyond this point.

An alternate problem occurs in which no process side circulation rate is specified. In this case, a 1-2 unit may be specified. Design to operate near the temperature cross will provide moderate circulation rates and moderate hx thermal size. In Figure 3 find the intercept of the temperature cross line and the $R \eta = 0.77$ line. Extrapolation indicates that a thermal capacitance ratio of about 0.2 corresponding to a process flow rate of 177 1/s.

CONCLUSION

The authors have demonstrated a method to design external heat exchangers for batch operations based on the effectivness-NTU method. The method is able to accommmodate a wide range of batch configurations and varying chemical reaction rates. It avoids much of the iteration associated with other methods and is readily adaptable to standard computational methods.

NOTATION

Parameters

A - Surface area
C - Specific Heat
$F_v$ - Ratio of vessel to batch temp rise

M - Initial process fluid mass
$M_1$ - Total initial mass $[M+(M_v F_v C_v/C_p)]$

$m_p$ - Process mass flow rate

$m_s$ - Side stream mass flow rate

$M_v$ - Vessel Mass
Q - Energy input to process

R  - $(C_p m_p)/(C_u m_u)$
t  - Batch time
T  - Temperature
$U_v$ - Vessel heat transfer coefficient
η  - Hx effectiveness based on process side
λ  - $C_p m_p$

Subscripts

a    ambient conditions
f    final conditions
i    initial conditions
in   input feed
l    loss of environment
p    process fluid
pi   hx inlet at process stream
s    side stream
u    utility stream
v    vessel

LITERATURE CITED

1. Kern, D. Q. *Process Heat Transfer*, McGraw Hill, New York, NY, 1950.

2. Clasen, L. J., *Proc. 1976 Lec. Ser. NJ, AIChE*, p. 26, 1978.

3. Malone, R. J., *Chem. Eng.*, 87 (24), 95, 1980.

4. Crane, R. A., *Modern Heat Transfer Technology*, McGraw Hill Seminar, New York, NY, 1981.

5. Villadsen, J. & Michelsen, M. L., *Solutions of Differential Equation Models by Polynomial Approximation*, Prentice-Hall, Englewood Cliffs, NJ, 1978.

6. *Standards of Tubular Exchanger Manufacturers Assoc.*, TEMA, 6th ed., 1978.

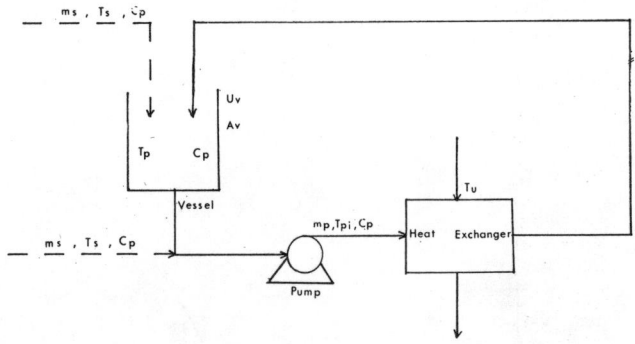

Figure 1. Schematic of the batch heating system.

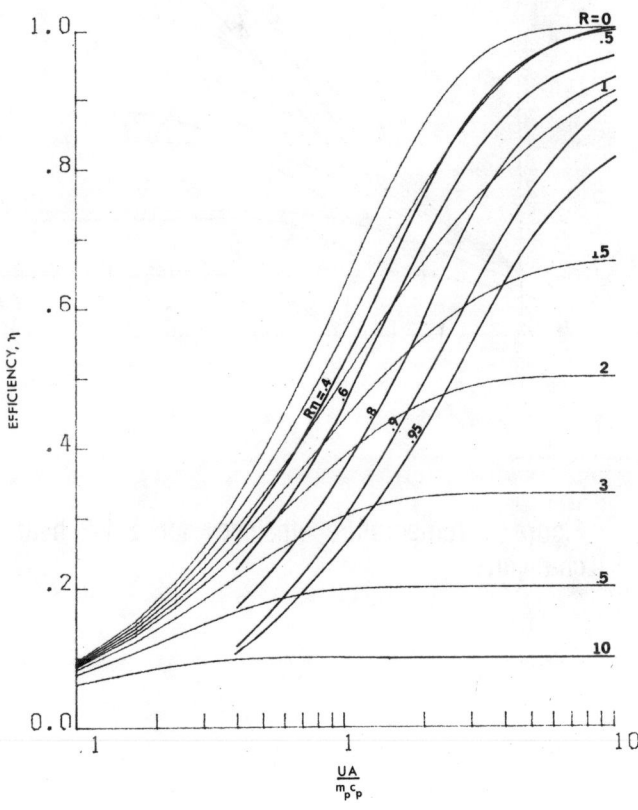

Figure 2. Temperature efficiency relation for a counter flow heat exchanger.

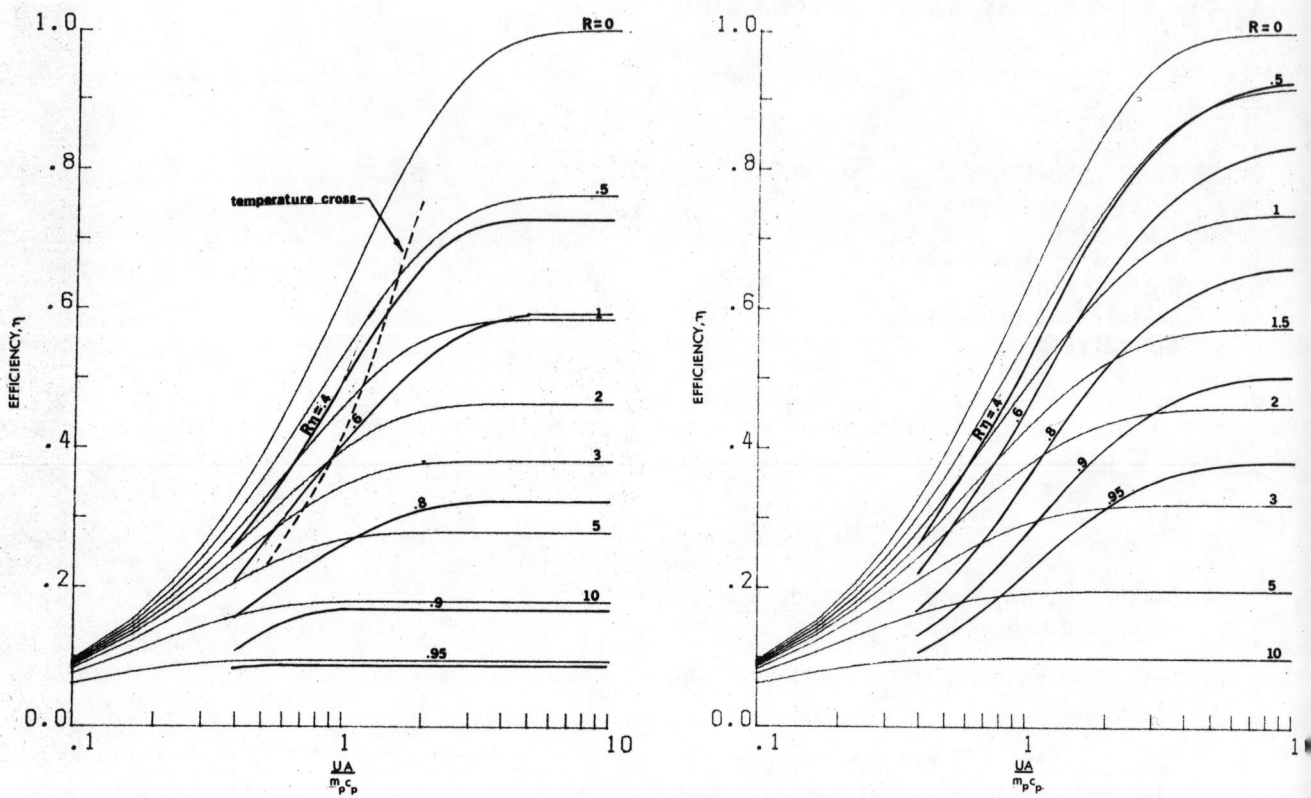

Figure 3. Temperature efficiency for a 1-2 heat exchanger.

Figure 4. Temperature efficiency relation for a 2-4 heat exchanger.

# RODbaffle EXCHANGER THERMAL-HYDRAULIC PREDICTIVE MODELS OVER EXPANDED BAFFLE-SPACING AND REYNOLDS NUMBER RANGES

C.C. Gentry and W.M. Small ■ Phillips Petroleum Company, Bartlesville, Oklahoma 74004

Heat transfer and pressure loss predictive models for RODbaffle heat exchangers have been developed over an expanded baffle spacing and Reynolds number range. Test results have been obtained from eleven RODbaffle bundle configurations and a single unbaffled test bundle, using water and light oil as test fluids. Heat transfer results are presented in conventional Nusselt, Reynolds, and Prandtl number form, with exchanger geometric parameters accounted for in geometric coefficient functions. The present geometric coefficient functions for both laminar and turbulent flow are valid over the expanded baffle spacing range from 76 mm (3.00 in) to 457 mm (18.00 in) and approach unbaffled exchanger values as asymptotic limits. Recently obtained test data, using a 152 mm (6.00 in) diameter, 1511 mm (59.5 in) long RODbaffle test exchanger, have been combined with previously reported results to extend the Reynolds number range from 300 to 150,000.

## INTRODUCTION

In the present investigation, experimental results from eleven different RODbaffle test exchangers and one unbaffled test exchanger have been utilized to extend the range of applicability of the RODbaffle heat exchanger thermal-hydraulic predictive models previously reported by Gentry, Young, and Small (1,2,3,4). The thermal-hydraulic predictive models presented herein are applicable over a wider baffle spacing range from 76 mm (3.00 in) to 457 mm (18.0 in) and over an increased Reynolds number range from 300 to 150,000. The present, more general predictive methods are also seen to represent test results for water and oil with the same degree of accuracy as the previous models. Features of the RODbaffle heat exchanger, which utilizes a parallel array of support rods for positive tube support, are depicted in Figure 1, with detailed constructional characteristics presented by Gentry (1,2,3,4,).

## TEST APPARATUS

The test apparatus used in the present investigation is similar to that previously described by Gentry (3,4,). Pertinent dimensions for the RODbaffle test exchangers used in the development of the present thermal-hydraulic predictive models are summarized in Table I, with a sketch of the two basic shell configurations shown in Figure 2.

As is evident in Table I, configurations ARA through ARH consist of bare tube RODbaffle bundles, having a bundle diameter of 260 mm (10.25 in.) and a bundle length of 3023 mm (118.7 in.). The unbaffled test bundle, configuration UNB, was identical to RODbaffle configurations ARC and ARD,

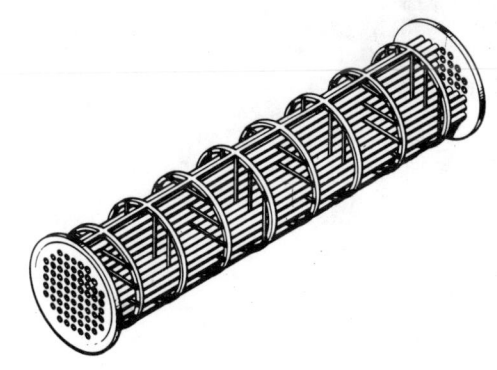

Figure 1 RODbaffle Heat Exchanger Details

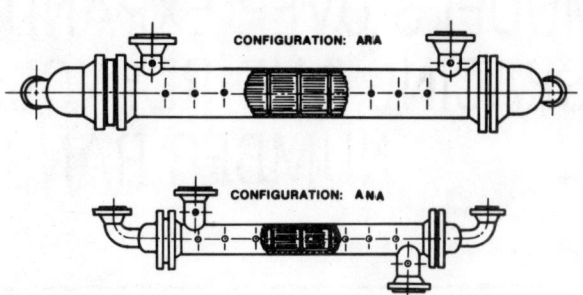

Figure 2 RODbaffle Test Exchangers

Table I  Test Exchanger Dimensions

| BUNDLE ID | BUNDLE DIA (mm) | BUNDLE LENGTH (mm) | TUBE TYPE | NUMBER TUBES | TUBE OD (mm) | ROD DIA (mm) | BAFFLE SPACE (mm) |
|---|---|---|---|---|---|---|---|
| ARA | 260 | 3023 | Bare | 137 | 12.7 | 4.76 | 124 |
| ARB | 260 | 3023 | Bare | 137 | 12.7 | 4.76 | 248 |
| ARC | 260 | 3023 | Bare | 134 | 12.7 | 4.76 | 76 |
| ARD | 260 | 3023 | Bare | 134 | 12.7 | 4.76 | 152 |
| ARE | 260 | 3023 | Bare | 112 | 15.9 | 3.18 | 76 |
| ARF | 260 | 3023 | Bare | 112 | 15.9 | 3.18 | 152 |
| ARG | 260 | 3023 | Bare | 112 | 15.9 | 3.18 | 152 |
| ARH | 260 | 3023 | Bare | 112 | 15.9 | 3.18 | 76 |
| ARJ | 260 | 1994 | Bare | 137 | 12.7 | 4.76 | 152 |
| ARL | 260 | 1040 | Bare | 137 | 12.7 | 4.76 | 152 |
| ANA | 152 | 1511 | Bare | 89 | 9.53 | 3.18 | 152 |
| UNB | 260 | 3023 | Bare | 134 | 12.7 | 4.76 | — |

except for the absence of baffling. Test configurations ARC, ARD, and UNB have been used to develop the baffle spacing relationship in the bundle heat transfer geometric coefficient functions. Configuration ANA is a smaller test unit, having a bundle diameter of 152 mm (6.00 in.) and a bundle length of 1511 mm (59.5 in.), and has been used in the present study to obtain high Reynolds number results. Bundle length to diameter effects have been determined using configurations ARD, ARJ, and ARL. Finned tube surface configuration effects have been established using low-finned tube configurations, as previously described in detail by Gentry (3,4).

## EXPERIMENTAL METHODS

For all RODbaffle test bundles, a series of isothermal runs were first conducted to establish isothermal pressure loss results. Following isothermal runs, heat transfer and pressure loss data were simultaneously obtained. Thermal duties for each test exchanger configuration were taken as the mean of shellside and tubeside duty values, which generally agreed to within ± 5%. Overall conductance values for bare and low finned tubes were determined from the mean thermal duty and the log mean temperature difference. Tubeside water convective coefficients for water and oil test runs were obtained using the Lawrence and Sherwood (5) correlation for turbulent tubeside water flow. Dimensionless Nusselt and Reynolds number groups used in the heat transfer correlations employ a characteristic diameter ($D_h$), which is defined in terms of the net flow area and wetted perimeter for a single tube layout in Equation (1). The characteristic diameter ($D_p$) used in the pressure loss Reynolds number was based on the net shellside flow area and total wetted perimeter of the shell and all tubes, as expressed in Equation (2). RODbaffle test parameters for the present study are shown in Table II.

$$D_h = \frac{4[P_t^2 - \pi/4 D_t^2]}{\pi D_t} \quad (1)$$

$$D_p = \frac{4[\pi/4 (D_s^2 - n_t D_t^2)]}{\pi (D_s + n_t D_t)} \quad (2)$$

## THERMAL-HYDRAULIC RESULTS

Previous thermal-hydraulic predictive models developed by Gentry, Young, and Small (3,4) represent extensive test data obtained over the recommended range of RODbaffle exchanger design conditions quite well. However, recent interest has been expressed by Taborek (6) in RODbaffle exchangers having baffle spacings greater than twice the maximum spacing upon which the previous thermal prediction models were based. At baffle spacings greater than 350 mm, the previous heat transfer predictive models were excessively conservative, predicting results which were lower than normally expected in unbaffled exchangers. In the present study, heat transfer data from an unbaffled test bundle are combined with results from RODbaffle exchangers having identical bundle geometry to develop bundle geometric coefficient functions which are valid over a much wider baffle spacing range. In addition, recent RODbaffle test data have been obtained which extends the Reynolds number range from 300 to 150,000.

Table II Range of RODbaffle Test Parameters

| Parameters | Range |
|---|---|
| Reynolds Number, $Re_h$ | 300 to 150,000 |
| Prandtl Number, $Pr$ | 2.9 to 45 |
| Viscosity Ratio, $\mu_b/\mu_w$ | 0.36 to 0.95 |
| Baffle-to-Shell Area Ratio, $A_b/A_s$ | 0.52 to 0.748 |
| Leakage-to-Shell Area Ratio, $A_\ell/A_s$ | −0.0899 to 0.219 |
| Length-to-Diameter Ratio, $L_t/D_{bo}$ | 4.02 to 11.58 |
| Fin Pitch-to-Height Ratio, $p/\epsilon$ | 0.965 to 2.99 |

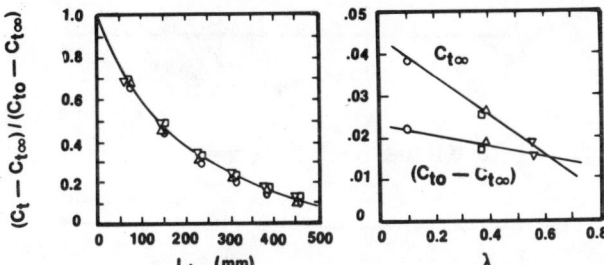

Figure 3 Geometric Coefficient Functions

## HEAT TRANSFER PREDICTIVE MODEL

The heat transfer predictive models for RODbaffle heat exchangers in the present investigation have been expressed in dimensionless Nusselt, Reynolds, and Prandtl number form, as in previous correlations by Gentry (3,4). Effects of the tube bundle geometry and tube surface configuration are accounted for in the geometric coefficient functions ($C_L$) and ($C_T$), for both laminar and turbulent flow.

In developing the present bundle geometric coefficient functions ($C_\ell$) and ($C_t$), test results from configurations ARC, ARD, and UNB each having the same bundle leakage-to-shell area ratio ($A_\ell/A_s$) were utilized. The functional relationship for ($C_\ell$) and ($C_t$) as a function of baffle spacing at constant ($A_\ell/A_s$) was found to be exponential rather than linear as in the previous heat transfer correlation. Similar relationships for ($C_\ell$) and ($C_t$) were developed for configurations ARA-ARB, ARE-ARF, and ARG-ARH, which were constructed to produce three different leakage-to-shell area ratios. Geometric coefficient function results for turbulent flow for the four leakage-to-shell area ratios are present in normalized form as a function of baffle spacing in Figure 3. Also shown in Figure 3, are plots of ($C_{t\infty}$) and ($C_{to} - C_{t\infty}$) as a function of the transformed leakage-to-shell area ratio variable $\lambda$.

Asymptotic conditions for both laminar and turbulent flow, established using unbaffled configuration UNB results, were assumed to occur at a baffle spacing of 457 mm (18.0 in). The present heat transfer predictive models for both laminar and turbulent flow are expressed in basic form in Equations (3) and (4), with the transition between regimes occurring at a Reynolds number value of 2000.

$$Nu = C_L (Re_h)^{0.6} (Pr)^{0.4} \phi \qquad (3)$$

$$Nu = C_T (Re_h)^{0.8} (Pr)^{0.4} \phi \qquad (4)$$

Present RODbaffle exchanger geometric coefficient functions ($C_L$) and ($C_T$) for laminar and turbulent flow are defined as the product of the bundle geometric function ($C_\ell$)($\xi_\ell$) and ($C_t$)($\xi_t$) and the tube surface configuration functions ($C_{s\ell}$) and ($C_{st}$), in Equations (5) and (6).

$$C_L = (C_{s\ell})(C_\ell)(\xi_\ell) \qquad (5)$$

$$C_T = (C_{st})(C_t)(\xi_t) \qquad (6)$$

The bundle geometric coefficient ($C_\ell$) for laminar flow, developed in the present study, is expressed as an exponential function of both the baffle spacing ($L_b$) and the bundle leakage to shell area ratio ($A_\ell/A_s$) in Equation (7), with the correlating parameters ($C_{\ell\infty}$) and ($C_{\ell o} - C_{\ell\infty}$) defined in Equations (8) and (9).

$$C_\ell = C_{\ell\infty} + (C_{\ell o} - C_{\ell\infty}) \exp[-0.00615(L_b)] \qquad (7)$$

$$C_{\ell\infty} = 0.182 - 0.172(\lambda) \qquad (8)$$

$$(C_{\ell o} - C_{\ell\infty}) = 0.120 - 0.0533(\lambda) \qquad (9)$$

For turbulent flow, the bundle geometric coefficient ($C_t$) is defined in Equation (10), with similar correlating parameters ($C_{t\infty}$) and ($C_{to} - C_{t\infty}$) expressed in Equations (11) and (12).

$$C_t = C_{t\infty} + (C_{to} - C_{t\infty}) \exp[-0.00496(L_b)] \qquad (10)$$

$$C_{t\infty} = 0.042 - 0.0417(\lambda) \qquad (11)$$

$$(C_{to} - C_{t\infty}) = 0.023 - 0.0117(\lambda) \quad (12)$$

The transformed variable ($\lambda$) utilized in both laminar and turbulent flow correlating parameters is defined in terms of the bundle leakage-to-shell area ratio ($A_\ell/A_s$), in Equation (13).

$$\lambda = [(A_\ell/A_s) + 0.1]^{0.5} \quad (13)$$

Shellside flow area ($A_s$), appearing in the expression for $\lambda$, is defined in Equation (14) as the net flow area inside the shell, less the cross sectional area of all tubes. The leakage flow area ($A_\ell$), also appearing in the expression for $\lambda$, is defined in Equation (15) as the peripheral area existing between the shell inner diameter and the bundle outer tube limit, minus the projected area of the baffle ring.

$$A_s = \pi/4 (D_s^2 - n_t D_t^2) \quad (14)$$

$$A_\ell = \pi/4 [(D_s^2 - D_o^2) - (D_{bo}^2 - D_{bi}^2)] \quad (15)$$

Effects of bundle length-to-diameter ratio ($L_t/D_{bo}$) are accounted for in the geometric coefficients ($\xi_\ell$) and ($\xi_t$) for laminar and turbulent flow, based on test results from configurations ARD, ARJ, and ARL, as defined in Equations (16) and (17).

$$\xi_\ell = \left[ 0.9600 + 0.2697 \exp\left\{-0.01705 (L_t/D_{bo}-1)^2\right\} \right] \quad (16)$$

$$\xi_t = \left[ 0.9600 + 0.2437 \exp\left\{-0.01614 (L_t/D_{bo}-1)^2\right\} \right] \quad (17)$$

Surface configuration coefficients ($C_{s\ell}$) and ($C_{st}$) are defined in Equations (18) and (19) for low finned tubes.

$$C_{s\ell} = 0.519 + 0.146 (P/\epsilon - 0.965)^{2.117} \quad (18)$$

$$C_{st} = 0.503 + 0.318 (P/\epsilon - 0.965)^{1.155} \quad (19)$$

In Figure 4, a comparative plot is provided which indicates that the present thermal predictive model represents test data for configurations UNB, ARC, and ARD, covering a baffle spacing range from 76 mm

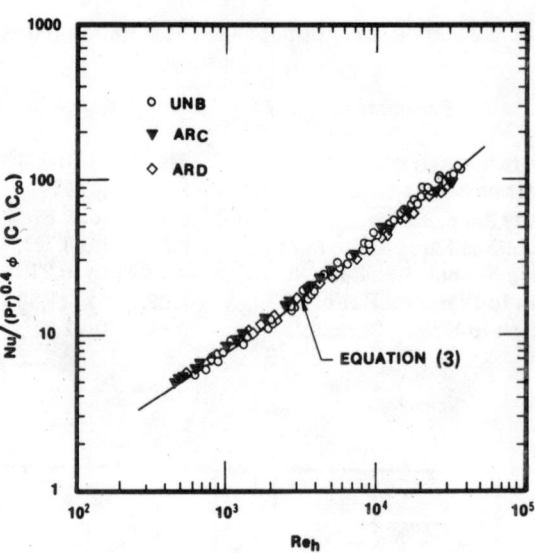

Figure 4 RODbaffle Heat Transfer Comparison

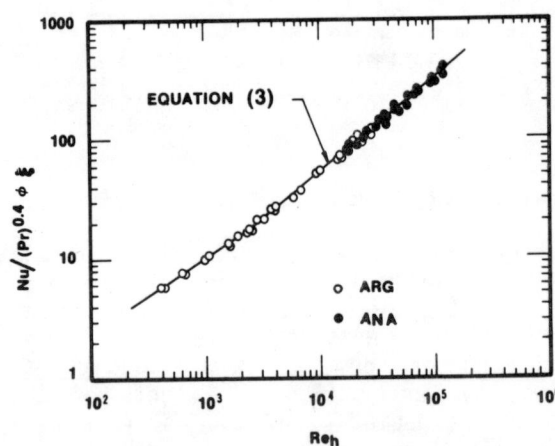

Figure 5 Thermal Results Over Wide Re Range

(3.00 in.) to 457 mm (18.0 in.), with a high degree of accuracy.

In addition to expanding the baffle spacing range for the heat transfer predictive models, experimental data have been obtained for configuration ANA, having a bundle diameter of 152 mm (6.00 in) and a bundle length of 1511 mm (59.5 in.), to expand the maximum Reynolds number limit from 40,000 to 150,000. Test results for configurations ARG and ANA are shown in Figure 5 and indicate that the present predictive model accurately represents RODbaffle test data over the expanded Reynolds number range from 300 to 150,000.

## PRESSURE LOSS PREDICTIVE MODEL

Pressure loss for flow through the shellside of a RODbaffle heat exchanger, excluding inlet and exit nozzles, has previously been defined by Gentry, Young, and Small (3,4) as the sum of an unbaffled, longitudinal flow component ($\Delta P_L$) and a baffle flow contribution ($\Delta P_b$), in Equation (20).

$$\Delta P = \Delta P_L + \Delta P_b \qquad (20)$$

In the defining equation, the unbaffled, longitudinal flow contribution is expressed in Fanning form in Equation (21), with the appropriate bare tube or low finned friction factors being employed. The baffle flow component for both bare and low finned tubes is expressed in terms of the baffle flow coefficient ($K_b$) and the baffle velocity ($U_b$) in Equation (22).

$$\Delta P_L = \frac{2 \rho f_F L_t U_s^2}{D_p} \qquad (21)$$

$$\Delta P_b = K_b n_b \frac{\rho U_b^2}{2} \qquad (22)$$

The baffle flow coefficient ($K_b$) used in Equation (22) is found to be a function of the baffle Reynolds number and geometric coefficients ($C_1$) and ($C_2$) in Equation (23).

$$K_b = \Psi [C_1 + C_2/Re_b] \qquad (23)$$

Coefficients $C_1$ and $C_2$ appearing in the baffle flow coefficient model are expressed as exponential functions of the baffle-to-shell area ratio ($A_b/A_s$) in Equations (24) and (25). Effects of bundle length-to-diameter ratio were accounted for by the variable $\Psi$, which is defined as a function of ($L_t/D_{bo} -1$) in Equation (26).

$$C_1 = 1.2053 \exp\{-1.6229 (A_b/A_s)\} \qquad (24)$$

$$C_2 = 48,732 \exp\{-6.8915 (A_b/A_s)\} \qquad (25)$$

$$\Psi = [1.00 + 0.22 \exp\{-0.02015(L_t/D_{bo}-1)^2\}] \qquad (26)$$

Baffle flow area ($A_b$) used in the baffle coefficient model is defined as the shellside flow area ($A_s$) minus the projected

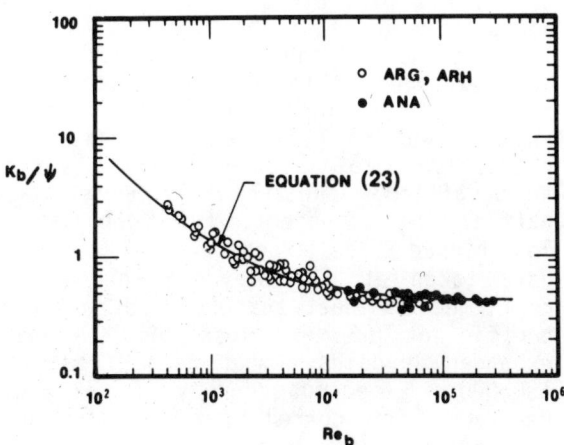

Figure 6 Flow Coefficient Over Wide Re Range

area of the baffle ring and support rods, as expressed in Equation (27).

$$A_b = [A_s - \pi/4(D_{bo}^2 - D_{bi}^2) - D_r L_r] \qquad (27)$$

Friction factors utilized with bare tube RODbaffle heat exchangers are conventional Fanning friction factors. For low-finned tubes, the Fanning friction factors are expressed as a function of fin pitch-to-height ratio ($p/\epsilon$) in Equations (28), (29), and (30), for laminar, transition, and turbulent flow, respectively.

$$f_F = 3.28 / (Re_p)^{0.696} \qquad (28)$$

$$f_F = 0.000339 (Re_p)^{0.42} \qquad (29)$$

$$f_F = 0.1086/(Re_p)^{0.22} \qquad (30)$$

Pressure loss results from configurations ARG, ARH, and ANA, with water and light oil used as test fluids, are seen in Figure 6 to be well represented by the baffle flow coefficient model defined in Equation (23), over the extended Reynolds number range from 300 to 300,000.

## CONCLUSIONS

In the present investigation, thermal hydraulic predictive models are presented which are applicable over a wider baffle spacing and a wider Reynolds number range than proposed in the earlier publications by Gentry, Young, and Small (3,4). Heat transfer correlations are developed in conventional dimensionless form, with geometric parameters being taken into account by geometric coefficient functions. These geometric coefficient functions have been extended in the present study to cover the baffle spacing range from 76 mm (3.00 in.) to 457 mm (18.0). The pressure loss predictive model is expressed as the sum of an unbaffled, longitudinal component and a baffle flow component, for both bare and low finned tube exchangers. Recent test data taken at high Reynolds numbers have confirmed the heat transfer and baffle flow coefficient pressure loss models over the expanded Reynolds number range from 300 to 150,000. Expected accuracy of the present heat transfer correlation is $\pm 15\%$ and of the pressure loss model is $\pm 25\%$.

## LITERATURE CITED

1. Small, W. M., and Young, R. K., <u>Heat Transfer Engineering</u>, Vol. 1, No. 2, pp 21-27, 1979.

2. Gentry, C. C., and Small, W. M. <u>Proc. 2nd Symposium on Shell-and-Tube Heat Exchangers</u>, Houston, pp 389-409, 1981.

3. Gentry, C. C., Young, R. K., and Small W. M., <u>Proc. 7th Int. Heat Transfer Conference</u>, Munich, Vol. 6, pp 197-202, 1982.

4. Gentry, C. C., Young, R. K., and Small W. M., <u>Proc. 22nd National Heat Transfer Conference</u>, Niagara Falls, pp 104-109, 1984.

5. Lawrence, A. E., and Sherwood, T. K., <u>Ind. and Eng. Chem.</u>, Vol. 23, pp 301-309, 1931.

6. Taborek, J., Personal Communications, 1983.

## NOMENCLATURE

$A_b$ = baffle net flow area, m$^2$
$A_\ell$ = leakage flow area, m$^2$
$A_s$ = shellside net flow area, m$^2$
$C_L$ = laminar heat transfer geometry function
$C_\ell$ = laminar heat transfer geometric parameter
$C_o$ = zero spacing geometric coefficient
$C_p$ = specific heat, J/kg K
$C_{s\ell}$ = laminar heat transfer surface coefficient
$C_{st}$ = turbulent heat transfer surface coefficient
$C_T$ = turbulent heat transfer geometry function
$C_t$ = turbulent heat transfer geometric parameter
$C_1$ = baffle pressure loss geometric coefficient
$C_2$ = baffle pressure loss geometric coefficient
$C_\infty$ = asymptotic geometric coefficient
$D_{bi}$ = exchanger baffle ring inner diameter, m
$D_{bo}$ = exchanger baffle ring outer diameter, m
$D_h$ = characteristic diameter for Nu and $Re_h$, m
$D_o$ = exchanger outer tube limit, m
$D_p$ = characteristic diameter for $Re_b$ & $Re_p$, m
$D_r$ = rod diameter, m
$D_s$ = shell inner diameter, m
$D_t$ = tube outer diameter, m
$f_F$ = Fanning friction factor
$k$ = thermal conductivity, W/m K
$K_b$ = baffle flow resistance coefficient
$L_b$ = baffle spacing, m m
$L_r$ = support rod length per baffle, m
$L_t$ = tube length, m
$n_b$ = number of baffles
$n_t$ = number of tubes
Nu = Nusselt number, $h_o D_h / k$
$p$ = fin pitch, m m
Pr = Prandtl number, $C_p \mu / k$
$P_t$ = tube pitch, m
$Re_b$ = baffle Reynolds number, $\rho D_p U_b / \mu$
$Re_h$ = heat transfer Reynolds number, $\rho D_h U_s / \mu$
$Re_p$ = parallel flow Reynolds number, $\rho D_p U_s / \mu$
$U_b$ = baffle velocity, m/s
$U_s$ = parallel-flow velocity, m/s
$\varepsilon$ = fin height, m m
$\mu_b$ = viscosity at bulk temperature, kg/m s
$\mu_w$ = viscosity at wall temperature, kg/m s
$\lambda$ = leakage-to-shell area parameter
$\xi_\ell$ = laminar heat transfer $L_t/D_{bo}$ correction
$\xi_t$ = turbulent heat transfer $L_t/D_{bo}$ correction
$\rho$ = density, kg/m$^3$
$\Phi$ = viscosity correstion, $(\mu_b/\mu_w)^{0.14}$
$\psi$ = baffle coefficient $L_t/D_{bo}$ correction

# HEAT AND MOMENTUM TRANSFER PROCESSES THROUGH BANKS OF FLEXIBLE TUBES IN AIR CROSS-FLOW

E.E. Michaelides and Y. Chang ■ Mechanical and Aerospace Engineering
University of Delaware, Newark, DE 19711

R.T. Bosworth ■ E.I. Du Pont De Nemours & Company
Finishes and Fabricated Products Department, Wilmington, DE 19898

This report summarizes the results of the initial phase of an ongoing study to examine the pressure drop and heat transfer characteristics on the air side (shellside) of banks of flexible tubes made of Teflon®.

The outside heat transfer film coefficient and the pressure losses of an air stream flowing perpendicular to the tube bank and being heated with 15 psig steam were determined. The tube bank contained from 1 to 20 rows of 0.175 OD tubes of Teflon® arranged in a 90° square array with a 2.0 pitch ratio. The air flow was varied to cover a Reynolds number range of between 400 and 10,000.

It was found that such flexible tube heat exchangers exhibit higher pressure drop and heat transfer than equivalent rigid tube exchangers. In many cases the enhancement of the heat transfer is as high as 30%.

## INTRODUCTION

Heat and momentum exchange between banks of rigid tubes and fluid streams have been the subject of several investigations in the past. Since the subject is of importance to engineers, all major studies resulted in engineering correlations, charts or design data banks. Examples of these forms of information are the studies by Grimison ([1]), Bergelin et al. ([2]), and Fraas and Ozisik ([3]). Several handbooks ([4,5]), compedia and texts ([6,7]) also present the results of other studies in succinct form.

It appears that all previous studies pertain to rigid (usually metal) heat exchangers and all design information applies to rigid tube banks. Studies on flexible tubes vibrating with high amplitudes are not found in the literature, and the problem of heat transfer through or around vibrating tubes continues to be the subject of scientific investigations ([8]).

Tubes of Teflon® are flexible and expand substantially more than metals when heated. The passage of air through banks of such tubes causes free vibrations of high amplitude (often 2 to 3 times the tube diameter) which substantially increase the air turbulance. Also, the outside diameter of these tubes is usually less than 0.25 inches to maximize the heat transfer area for a certain volume, whereas almost all past investigations dealt with tube diameters considerably larger. Because of the anticorrosive properties of Teflon® these heat exchangers are used in corrosive environments to cool hot gas streams containing condensible acids where typical operating temperatures can be up to 450°F with air velocities corresponding to shellside Reynolds numbers from 400 to 3000.

It was necessary to develop engineering correlations for predicting the thermal and hydraulic performance of such tube banks in order to adequately optimize the design of commercial exchangers. The parameters examined in this study are the air velocity (expressed as Reynolds Number) and the temperature and pressure change across the bundle. The number of tube rows varied from 1 to 20.

## EXPERIMENTAL FACILITY AND INSTRUMENTATION

The experiments were conducted at the Fluid Dynamics Laboratory of the University of Delaware under a grant by the Du Pont Co., F&FP Department, Wilmington, Delaware. A wind tunnel having an 18" x 18" square test section was used to supply air through the bundles of tubes. (Figure 1)

An 18" x 18" tube bank having 20 rows of tubes in the direction of flow was prepared by stringing tubes of Teflon® through two 1/4" thick Lucite® spacers containing holes arranged in a 90° layout with a 2.0 pitch ratio. Banks containing tubes having 0.125", 0.175", and 0.25" outside diameters have been prepared but only the 0.175 tube diameter module is discussed here. The module was inserted into the test section of the wind tunnel so that the spacers coincided with the top and bottom surfaces. The inlet and outlet ends of the tubes were each combined into an integrated "tubesheet".

Steam was used tubeside as the heating medium. The number of tube rows was reduced after each set of runs by cutting out the desired number of rows and plugging the remaining ends with fused plugs of Teflon®. Banks containing 20, 15, 10, 5, 2, and 1 rows were studied. The range of the geometric parameters of the tube banks examined is shown in Table 1.

### Table 1

#### Geometric Parameters of Tube Banks

Layout: In line

Tube external diameter: 0.175 inches

Tube columns: 46

Pitch ratios: Longitudinal 2
Transverse 2

Range of tube rows: 20, 15, 10, 5, 2, 1

Range of tube free lengths between spacers: 18", 9", 6"

Pressure measurements were taken with an inclined manometer filled with oil of specific gravity 0.829. Air velocity was measured wiht an industrial anemometer calibrated with a DISA constant temperature anemometer. The heat flow from the bank of tubes to the air was determined by measuring the temperature of the air stream with copper-constantan thermocouples. One temperature measurement was taken upstream of the heat exchanger module and five measurements were taken downstream at different locations of the cross-section to account for heating non-uniformities. These temperature measurements were averaged to determine a global heat transfer coefficient for the module.

## FLOW VARIABLES, PARAMETERS AND QUANTITIES OF INTEREST

Among the quantities measured and evaluated for this project were:

1) The velocity of air between the cylindrical tubes, $V_m$, for the geometries used is related to the free stream velocity, V, by:

$$V_m = V \left[ \frac{S_T}{S_T - 1} \right] \quad (1)$$

where $S_T$ is the transverse pitch ratio expressed in outside tube diameters.

2) The Reynolds number,

$$Re = \frac{V_m \rho d}{\mu} \quad (2)$$

3) The Euler number,

$$Eu = \frac{2 \Delta P}{\rho V_m^2} \quad (3)$$

and

4) The friction factor,

$$f' = \frac{2 \Delta P}{N \rho V_m^2} \quad (3)$$

where $\Delta P$ is the pressure drop through the entire module and N the number of rows.

The overall heat transfer coefficient based on the outside tube heat transfer area $U_o$, is defined by,

$$\dot{Q} = U_o A_o \, (LMTD) \quad (5)$$

where $A_o$ is known and LMTD and $\dot{Q}$ are calculated from the experimentally determined temperatures and flows.

$$\dot{Q} = \dot{m} c_p (\overline{T}_{in} - \overline{T}_{out}) \quad (5a)$$

$$LMTD = \frac{(T_s - \overline{T}_{in}) - (T_s - \overline{T}_{out})}{\ln \left( \frac{T_s - \overline{T}_{in}}{T_s - \overline{T}_{out}} \right)} \quad (5b)$$

For clean tubes, $U_o$ is composed of the two film conductances and the wall conductance

$$\frac{1}{U_o} = \frac{1}{h_o} + \left(\frac{d_o}{d_i}\right)\frac{1}{h_i} + \frac{(d_o)\ln(d_o/d_i)}{2K_T} \quad (6)$$

In this study, the wall conductance is known and the heat transfer coefficient for the condensing steam $h_i$ can be estimated from known engineering correlations. Hence, the outside film coefficient, $h_o$, can be calculated from equation 6. It must be noted that the $h_o$ is less than the wall conductance; which is much less than $h_i$, so the relative errors in $h_i$, $k_T$ and tube geometry result in relatively small errors for $h_o$.

The outside film heat transfer coefficient, $h_o$, may be presented in dimensionless forms as follows:

$$Nu = \frac{h_o d_o}{k} \quad (7a)$$

$$J_{colb} = \frac{h_o}{\rho c_p V_m}\left[\frac{c_p \mu}{k}\right]^{0.67}\left[\frac{\mu_w}{\mu}\right]^{0.14} \quad (7b)$$

and

$$J_B = \frac{h_o d_o}{k}\left[\frac{c_p \mu}{k}\right]^{-0.33}\left[\frac{\mu_w}{\mu}\right]^{0.14}$$

$$= J_{colb} \cdot Re. \quad (7c)$$

## MEASUREMENTS, OBSERVATIONS AND RESULTS

After steady state conditions were obtained in each run, the pressure drop through the heat exchanger module, $\Delta P$, was recorded, as well as the free stream velocity, V, and inlet and exit air temperatures. Runs were made using 7 and 15 psig steam as the heating fluid. The quantities of interest (f', $h_o$, $J_B$, Re, etc.) were calculated according to equations (1) to (7).

During the course of the experiments, it was observed that when the tubes were heated with the steam they became substantailly longer and more flexible. Consequently, the geometric configuration of the tube bank changed from a square arrangement to an almost random orientation. Also, the passage of air through the tube bank bowed the tubes in the direction of the flow.

Vibrations were observed at Reynolds numbers greater than 800 with amplitudes visibly large and often exceeding three tube diameters. Stroboscopic measurements were taken to determine the frequency of vibrations, but it was observed that the tubes vibrated at different frequencies and that both amplitudes and frequencies varied with time. The order of magnitude of the frequencies was 12 Hz corresponding to a Strouhal number of approximately 0.004. The corresponding niumbers for cold tubes (no steam condensing inside) was 18 Hz and 0.008. The Strouhal numbers observed are much lower than those observed in rigid-tube experiments. The difference is due to the extremely low rigidity of the tubes and the fact that, in this case, the shedding vortex influences the direction of motion of each tube. The variability in the frequencies of vibrations is due to the fact that the turbulence level generated in the module influences the vibrations of the tubes non-uniformly and neighboring tubes can influence the movement of each other. There were no audible vibrations observed during this study.

Since the tubes vibrated at different frequencies no resonance was observed. The relatively low frequency, high amplitude vibrations did not produce any observable mechanical wear and the only visible effect was the change in geometric arrangement. Long term vibrational effects are under observation at this time; no significant changes have been observed after 4 months.

Typical results for the module made of 0.175" diameter tubes are shown in Figures 2 and 3. The pressure drop is expressed in terms of f' versus Reynolds number for the module with a tube free length of approximately 18" and 20 rows of tubes. The solid line represents the prediction of the Grimison equation (1) (applicable for Reynolds numbers greater than 2,000). As in transition region flows, a minima in f' was noted in all experimental runs and always occured in the range 900 < Re < 1700. Figure 3 depicts the corresponding experimental data for the outside heat transfer coefficient, $h_o$, (expressed as $J_B$). The solid line represents predictions for the rigid tube data (1,2,7). It can be observed that the flexible tubes have a higher $h_o$ than the corresponding rigid tubes when the Reynolds number is more than 800. This is due to tube vibrations which increase the stream turbulence and thus enhance heat transfer.

As has been found for rigid tube heat exchangers, it was observed that for less than ten rows, the friciton factor per row was higher and the heat transfer coefficient lower at high Reynolds numbers. However, at low Reynolds numbers and before the flow transition occured the one- and two-row results fell very near to the other results of ten or more rows (Figure 4).

In the interest of space, only sample results are presented here. However, reference (9) contains the complete set of experimental data and design correlations and can be consulted for further details.

## HEAT TRANSFER CORRELATIONS

The experiments indicated that $J_B$ is a strong function of Reynolds number. The dependence of $J_B$ on the Prandtl number was not investigated since the experiments were only conducted with air as the outside fluid.

The data were correlated by the exponential equation,

$$J_B = C\, Re^m \tag{8}$$

Since the heat transfer and $\Delta P$ curves exhibit a change in slope at around $Re \sim 3000$ due to a flow transition, separate correlations were made for $Re < 3000$ and $Re > 3000$. Table 2 summarizes the values for C and m.

### Table 2

#### Correlation Coefficients for the Heat Transfer Data

|  | C | m |
|---|---|---|
| Re $\leq$ 3000 | 0.12 | 0.74 |
| No spacer (18" free length) Re > 3000 | 0.26 | 0.64 |
| Rigid Tubes (Reference 7) Re > 2000 | 0.25 | 0.62 |

The present experimental data and resulting correlations indicate that heat exchangers made with flexible tubes exhibit higher outside heat transfer coefficients than an equivalent array of rigid tubes due to tube vibrations which enhance the perturbations of the flow field and facilitate heat transfer. It is also apparent that heat and momentum exchange processes (pressure drop) follow the same trends here as in any other fluid dynamics processes.

The main source of error in the heat transfer data is the determinaiton of the flow velocity. Good insulation of the test section and use of copper-constantan thermocouples minimized experimental errors from heat losses or temperature measurements. The combined error on $J_B$ is expected to be less than 10% for all Reynolds numbers (9).

## CONCLUSIONS

Banks of flexible tubes exhibit higher heat transfer coefficients than an equivalent array of rigid tubes due to low frequency, high amplitude vibrations of the tubes.

Banks of flexible tubes also show higher pressure drop than an equivalent array of rigid tubes, for the same reason. Tube movement changes the layout of the tube bundle which is tantamount to instantaneously changing the geometric characteristics of the bundle.

All tubes do not vibrate at the same frequency and, therefore, vibration resonance or noise was never observed. Frequency measurements revealed that the Struohal number of the vibrations ranges between 0.004 and 0.008 compared to a value of 0.21 observed for rigid tubes.

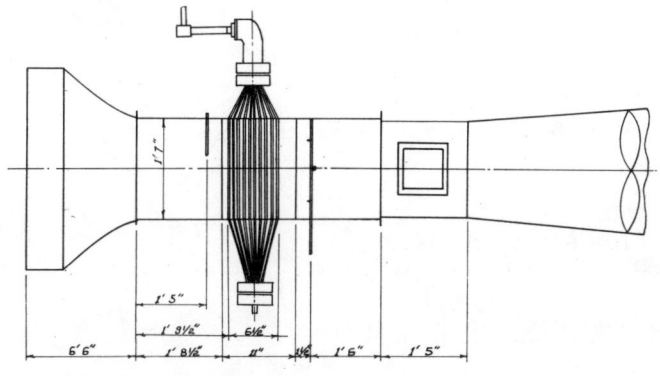

Figure 1. Schematic diagram of the experimental facility.

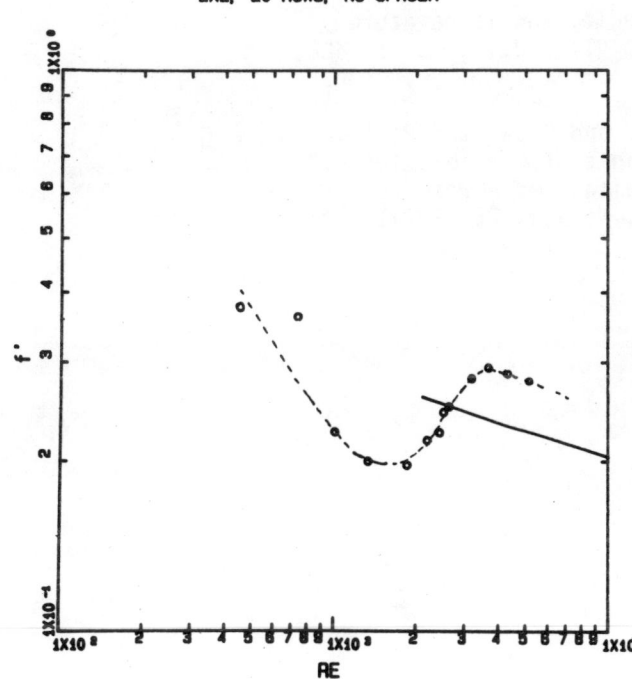

Figure 2. Pressure drop for a tube bundle of twenty rows: The solid line represents Grimison's correlation.

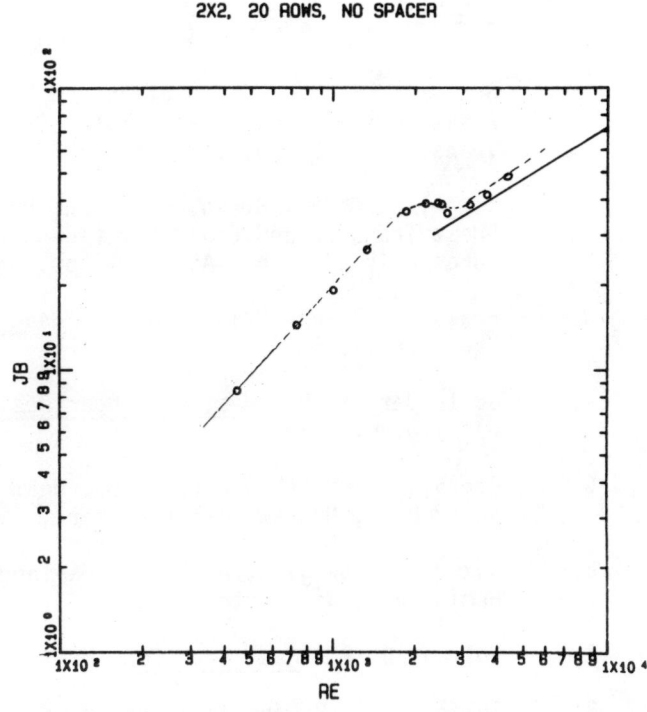

Figure 3. Outside heat transfer coefficient for flexible tubes; the solid line is for rigid tubes.

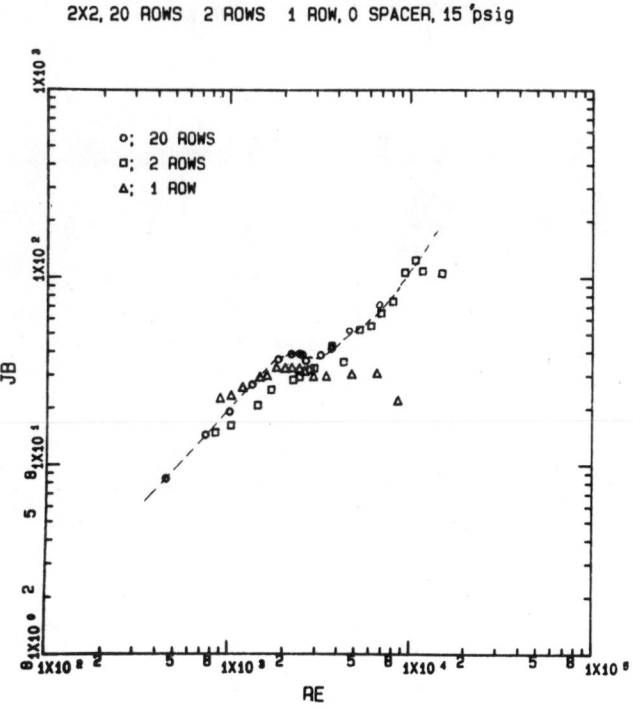

Figure 4. Heat transfer data for banks of twenty, two and one rows.

## LITERATURE CITED

1. Grimison, E. D., "Correlation and Utilization of New Data on Flow Resistance and Heat Transfer for Cross Flow of Gases Over Tube Banks," Trans. ASME 59, pp. 583-594, 1937.

2. Bergelin, O. P., Brown, G. A. and Doberstein, S. C., "Heat Transfer and Fluid Friction During Flow Across Banks of Tubes - IV.," Trans. ASME, 74, pp. 953-960, 1952.

3. Fraas, A. P. and Ozisik, M. N., Heat Exchanger Design, J. Wiley, New York, 1965.

4. Schlünder, E. U., Heat Exchanger Design Handbook, Hemisphere, Washington, 1983.

5. Afgan, N. and Schlünder, E. U., Heat Exchangers: Design and Theory Sourcebook, McGraw Hill, New York, 1974.

6. Kakac, S., Bergles, A. E. and Mayinnger, F., Heat Exchangers, Hemisphere, Washington, 1981.

7. Holman, J. P., Heat Transfer, McGraw Hill, New York, 1981.

8. Telionis, D. P. and Romaniuk, M. S., "Velocity and Temperature Streaming in Oscillating Boundary Layers," AIAA. J., 16 5, p. 488, 1977.

9. Michaelides, E. E., Chang, Y., Holt, T. T. and Bosworth, R. T., "Pressure Drop and Heat Transfer Across Banks of Flexible Tubes," Technical Report Number 262, 1984. Mechanical and Aerospace Engineering Dept., University of Delaware, Newark, DE 19711.

## NOTATION

| | |
|---|---|
| $A_o$ | Total outside tube area |
| $c_p$ | Air heat capacity per unit mass |
| $d_o$ | Outside tube diameter |
| $d_i$ | Inside tube diameter |
| $\Delta P$ | Pressure drop |
| Eu | Euler number |
| $h_o$ | Outside heat transfer coefficient |
| $h_i$ | Inside heat transfer coefficient |
| k | Air thermal conductivity |
| $k_T$ | Thermal conductivity of Teflon® |
| $J_B$ | Dimensionless number defined in equation (7c) |
| $J_{colb}$ | Colburn number |
| LMTD | Log-mean temperature difference |
| N | Number of tube rows |
| Nu | Nusselt number |
| Q | Rate of heat transfer |
| Re | Reynolds number |
| $S_T$ | Transverse pitch ratio |
| T | Temperature |
| $\bar{T}$ | Area-averaged temperature |
| $U_o$ | Overall heat transfer coefficient |
| V | Free stream velocity |
| $V_m$ | Air velocity between tubes |
| m | Air mass flow |
| $\mu$ | Air viscosity |
| $\mu_w$ | Air viscosity evaluated at wall temperature |
| $\rho$ | Air density |

# ANALYSIS OF HEAT TRANSFER IN PLATE HEAT EXCHANGERS

John J. Marano and John L. Jechura ■ Marathon Oil Company, Findlay, Ohio 45840

A model is developed to predict the steady-state performance of a plate heat exchanger. Previous models for heat transfer in plate heat exchangers have used Runge-Kutta methods to solve the set of first-order, ordinary differential equations (O.D.E.'s) describing the heat transfer between channels in the exchanger. The proposed method is superior to these Runge-Kutta methods in that initial guesses are not required for any of the outlet temperatures from the exchanger. In the method presented, the set of O.D.E.'s are expressed in matrix form, and their solution is expanded in terms of the eigenvalues and eigenvectors of the matrix. The QR-algorithm is used to determine the eigenvalues and vectors. The model described is quite general and can also be used to analyze multipass shell and tube exchangers and multiple fluid plate exchangers.

## INTRODUCTION

Over the years, a large number of mathematical models of compact heat exchangers have appeared in the literature. Many of these models are based on the analytical solution of the differential equations describing heat transfer in the exchanger and have been very specific to the configuration of the exchanger being modeled. Models have been presented to solve both the design problem and the performance problem (1). This paper deals exclusively with the prediction of the performance of plate-type heat exchangers, that is, the determination of the outlet temperatures when the inlet temperatures and area are known. These exchangers are sufficiently complex that past modeling efforts have relied on either numerical integration techniques (2, 3), such as Runge-Kutta, or dimensional analysis (4). The model proposed here is based on a generalized, analytical solution to the compact heat exchanger performance problem.

Heat transfer between streams in a heat exchanger can be described by a system of first-order, ordinary differential equations. The analytical solution of this type of system has been known for a long time, and involves an expansion in terms of the eigenvalues and eigenvectors of the system. The applicability of this solution technique to compact heat exchanger problems was originally pointed out by Wolf (5), and has recently been reviewed and expanded on by Zaleski (6). However, (5) does not apply the method, and (6) only considers relatively simple problems. More general problems requiring computer implementation to find the eigenvalues and vectors have not been addressed. The analysis here pays special attention to computer implementation. The model will be applied to plate heat exchangers, but is also applicable to other types of compact exchangers such as shell-and-tube exchangers.

The method proposed has several advantages over numerical integration techniques. Numerical integration requires the finite-difference heat transfer equations to be solved iteratively, since normally the boundary conditions are not all specified at one end of the exchanger. This introduces the problem of coming up with good guesses for the unknown temperatures at this end. Accuracy and stability problems can also occur.

## ANALYSIS

A plate heat exchanger consists of a number of thin plates which are clamped together and suspended from a frame. The streams flow through the channels left between the plates, entering and exiting through ports located in the corners. The plates are corregated to induce intense turbulence and, thus, high heat transfer rates. Gaskets, which run around the ports and the edges of the plates, are used to connect different channels so that the hot and cold streams flow through alternate channels. It is this basic design which gives the plate heat exchanger its very attractive properties: compact design, high thermal efficiency, flexibility and ease of maintenance. For a more complete description of a plate heat exchanger, see (7).

For the exchanger described above, the following conditions will be assumed:

1) The exchanger is at steady-state;

2) The flows of all streams are parallel to each other (i.e., no crossflow);

3) The temperature in any channel is uniform over any cross-section perpendicular to the direction of flow;

4) No phase changes occur and the heat capacities of all streams are constant;

5) The overall heat transfer coefficient for each plate is constant; and

6) Heat losses are negligible.

A differential heat balance for any one of the n channels yields:

$$i_{Di} W_i C_i \frac{dt_i}{dz} = (UA)_{i,i-1}(t_{i-1}-t_i)$$
$$+ (UA)_{i,i+1}(t_{i+1}-t_i) \quad i=1,2,\ldots,n \quad (1)$$

where $i_{Di}$ is $\pm 1$, and indicates the direction of flow. This can be written in matrix form as:

$$\frac{d\underline{t}}{dz} = \underline{I}_D \underline{A}_o \underline{t} \quad (2)$$

where $\underline{t}$ is a column vector with elements $t_i$, $\underline{I}_D$ is a diagonal matrix with elements $i_{Di}$, and $\underline{A}_o$ is a non-symmetric, tridiagonal matrix containing the coefficients from (1).

There are advantages to making a change of variables in (2) by defining:

$$\underline{h} = \underline{\Omega}\, \underline{t} \quad (3)$$

where $\underline{\Omega}$ is a diagonal matrix with elements $(W_i C_i)^{1/2}$. The elements of $\underline{h}$, the $h_i$, are the transformed temperatures. Then, upon substitution and rearrangement, (2) becomes:

$$\frac{d\underline{h}}{dz} = \underline{I}_D \underline{A}\, \underline{h} \quad (4)$$

The matrix product $\underline{I}_D \underline{A}$ will be referred to as the differential coefficient matrix. Note that $\underline{A}$ is similar to $\underline{A}_o$, since:

$$\underline{A} = \underline{\Omega}\, \underline{A}_o\, \underline{\Omega}^{-1} \quad (5)$$

The matrix $\underline{A}$ is tridiagonal and symmetric ($\underline{A}^T = \underline{A}$) with nonzero elements $a_{ij}$:

$$a_{ij} = \begin{cases} -[(UA)_{i,j-1}+(UA)_{i,j+1}]/W_i C_i, & j=i \\ (UA)_{ij}/(W_i C_i W_j C_j)^{1/2}, & j=i-1, i+1 \end{cases} \quad (6)$$

The $a_{ij}$ are dimensionless and resemble the quantity known as the number of heat transfer units (NTU's) of the exchanger (1). The motivation behind the transformation (3) was to make $\underline{A}$ symmetric and to scale the elements of $\underline{A}$, both of which improve the computational aspects of the problem.

The general solution to (4) is shown elsewhere (8) to be:

$$\underline{h}(z) = \underline{X} e^{\underline{\Lambda} z}\, \underline{k} \quad (7)$$

where the constant vector $\underline{k}$ can be expressed in terms of $\underline{h}(0)$:

$$\underline{k} = \underline{X}^{-1}\underline{h}(0) \quad (8)$$

The matrices $\underline{\Lambda}$ and $\underline{X}$ are determined from the solution of the eigenproblem:

$$(\underline{I}_D \underline{A})\underline{X} = \underline{X}\, \underline{\Lambda} \quad (9)$$

In the analysis presented here, $\underline{\Lambda}$ is a diagonal matrix with the eigenvalues, the $\lambda_i$, along the diagonal, and $\underline{X}$ is a matrix whose columns, the $\underline{x}_i$, are the eigenvectors of $\underline{I}_D \underline{A}$.

Premultiplying (9) by $\underline{X}^{-1}$ yields:

$$\underline{X}^{-1}\underline{I}_D \underline{A}\, \underline{X} = \underline{\Lambda} \quad (10)$$

Thus, the eigenvector matrix can be used as a similarity transformation to reduce $\underline{I}_D \underline{A}$ to diagonal form. This is the basis for many of the methods used for determining the eigenvalues and vectors of a matrix.

Important properties of the differential coefficient matrix which were used to select a specific procedure to solve the eigenproblem were noted in (6):

1) $\underline{I}_D \underline{A}$ is always singular;

2) When off-diagonal elements of $\underline{A}$, $a_{i,i+1}=0$, the matrix $\underline{I}_D\underline{A}$ can be split into block diagonal submatrices;

3) Only one eigenvalue per block will be zero as long as $\Sigma\, i_{D_i} W_i C_i \neq 0$ in each block diagonal submatrix of $\underline{I}_D \underline{A}$. However, when the summation equals zero, two zero eigenvalues result for that block;

4) All eigenvalues are real, hence, all eigenvectors are real; and

5) All eigenvectors corresponding to nonzero eigenvalues are distinct.

When multiple zero eigenvalues occur in a submatrix, $\underline{I}_D\underline{A}$ is said to be defective and this case will not be considered here. Note that while $\underline{A}$ is symmetric, $\underline{I}_D\,\underline{A}$ normally is not and a method must be used which is valid for a real, general matrix.

## Boundary Conditions

The flow patterns considered give rise to three types of boundary conditions (B.C.'s):

1) Inlet temperature specified at $z = 0$ ($h_i(0)$ known);

2) Inlet temperature specified at $z = 1$ ($h_i(1)$ known); and

3) For series flow in the exchanger, the inlet temperature of the ith channel equals the outlet temperature of the jth channel ($h_i(0\text{ or }1) = h_j(0\text{ or }1)$).

A problem must have at least two B.C.'s of Type 1 and/or 2, corresponding to the introduction of at least two feed streams.

The tranformed terminal temperatures can be related using Equations (7) and (8) evaluated at $z=1$:

$$\underline{h}(1) = \underline{X}e^{\underline{\Lambda}}\underline{X}^{-1}\underline{h}(0) = \underline{U}\underline{h}(0) \qquad (11)$$

where $\underline{U}$ is the matrix combination $\underline{X}e^{\underline{\Lambda}}\underline{X}^{-1}$ which will be called the integrated coefficient matrix.

The symmetry of $\underline{A}$ did not simplify the calculation of $\underline{\Lambda}$ and $\underline{X}$, but it does simplify the calculation of $\underline{X}^{-1}$ and, hence, $\underline{U}$. For any matrix, the eigenvalues of the matrix and its transpose are identical and the eigenvectors are orthogonal or can be made so (8). Then, in addition to (9):

$$(\underline{I}_D\underline{A})^T \underline{Y} = \underline{Y}\,\underline{\Lambda} \qquad (12)$$

and:

$$\underline{Y}^T \underline{X} = \underline{D} \qquad (13)$$

where $\underline{D}$ is a diagonal matrix as long as $\underline{I}_D\underline{A}$ is not defective. However, since $\underline{A}$ is symmetric, Equation (12) can be written as:

$$(\underline{I}_D\underline{A})(\underline{I}_D\underline{Y}) = (\underline{I}_D\underline{Y})\,\underline{\Lambda} \qquad (14)$$

which implies $\underline{X} = \underline{I}_D\underline{Y}$, and:

$$\underline{X}^T \underline{I}_D \underline{X} = \underline{D} \qquad (15)$$

Because $\underline{I}_D$ contains $+1$ elements along the diagonal, the diagonal elements of $\underline{D}$ can be positive or negative. Therefore, $\underline{X}$ will be normalized such that:

$$\underline{X}^T \underline{I}_D \underline{X} = \underline{I}_N \qquad (16)$$

where $\underline{I}_N$ is diagonal with elements of $\pm 1$. Then, the inverse of $\underline{X}$ can be given as:

$$\underline{X}^{-1} = \underline{I}_N \underline{X}^T \underline{I}_D \qquad (17)$$

and $\underline{U}$ as:

$$\underline{U} = \underline{X}e^{\underline{\Lambda}}\underline{I}_N\underline{X}^T\underline{I}_D \qquad (18)$$

Once $\underline{U}$ is calculated, the boundary conditions can be applied to (11) to solve for the unknown transformed terminal temperatures.

## COMPUTATION

A general computer program was written to demonstrate the feasibility of the above method to the analysis of plate exchangers. The program was written in FORTRAN-77 to execute on an IBM 370/3081 computer system. Double precision was used throughout to minimize rounding effects.

Below are the program steps followed to determine the unknown terminal temperatures and temperature profiles in an exchanger:

1) Assemble the differential coefficient matrix from the input;

2) Determine the eigenvalues and vectors of the differential coefficient matrix;

3) Form the integrated coefficient matrix and solve for all the unknown terminal temperatures; and

4) Calculate the constant vector for the particular solution corresponding to the boundary conditions and calculate the temperature profiles.

The procedure chosen to solve the eigenproblem was the QR algorithm (9, 10) as available from the EISPACK subroutine library (11). The routines needed are called from EISPACK's RG driver subroutine and are good for any general, real matrix. They perform the following operations:

1) Balance the matrix and isolate eigenvalues whenever possible;

2) Reduce matrix to upper Hessenberg form;

3) Perform QR method and accumulate similarity transforms to obtain eigenvectors of the balanced matrix; and

4) Form the eigenvectors of the original matrix by back transforming those calculated for the balanced matrix.

The second step is not really needed since $I_DA$ is tridiagonal and, hence, already in upper Hessenberg form. The QR algorithm is based on the transformation (10), however in this procedure, the differential coefficient matrix is only reduced to upper triangular form with the eigenvalues on the diagonal.

The determination of the terminal temperatures involves solving a system of linear equations. Equation (11) and the Type 3 B.C.'s can be combined to give:

$$\begin{bmatrix} \underline{U} & -\underline{I} \\ \underline{C}' & \underline{C}'' \end{bmatrix} \begin{bmatrix} \underline{h}(0) \\ \underline{h}(1) \end{bmatrix} = \underline{0} \qquad (19)$$

where $\underline{C}'$ and $\underline{C}''$ contain the information from the Type 3 B.C.'s. This partitioned matrix will be of dimension $(n+n_c) \times (2n)$. The Type 1 and 2 B.C.'s can be applied by shifting all of the known $h_i(0)$ and $h_i(1)$ values to the right-hand side, leaving the system as:

$$\underline{U}'\underline{u} = \underline{b} \qquad (20)$$

where $\underline{U}'$ is the $(n+n_c) \times (n+n_c)$ matrix containing the columns of the partitioned matrix corresponding to the unknown $h_i(0)$ and $h_i(1)$ values, $\underline{u}$ is the vector of unknown values, and $\underline{b}$ contains the contributions from the known values. This system of equations can be solved using LU decomposition employing partial pivoting and row equilibration (12). The terminal temperatures are then calculated using the inverse transformation of (3).

To determine temperature profiles, the constant vector $\underline{k}$ is calculated with Equation (8) after the boundary conditions have been applied and all of the transformed temperatures at z=0 have been found. Then, the internal temperature profiles are calculated using the general solution given by Equation (7) and the inverse of (3).

EXAMPLES

Many different heat exchanger configurations were studied to test the computer program including TEMA E, J and G type shell-and-tube exchangers. The results were verified with known solutions for these exchangers (1).

Four plate exchanger configurations are presented here. These are for looped, series, complex and three-fluid, looped flow, all of which are common configurations used to balance flowrates with allowable pressure drops and temperature approaches. The results are shown in Figures 1 through 4. Each figure contains a schematic of that configuration (for convenience, the plates have been shown as being horizontal) upon which the inlet temperatures, the predicted outlet temperatures from each channel and the average outlet temperatures are given. Where possible, these results were verified using the results presented in (3).

All of the example plate exchanger problems presented took less than 1.0 CPU second to run. The configurations in Figures 1 and 4 both have seven plates, and those in Figures 2 and 3 have five; however, exchangers with as many as 100 plates have also been successfully modeled. Also note that different heat transfer coefficients can be specified for different plates, as was done in Figure 4 for the two types of plates in the three-fluid exchanger. The details of these four examples, along with the predicted temperature profiles, are available from the authors upon request.

CONCLUSIONS

It has been shown here that it is feasible to model a certain class of compact heat exchangers using the generalized, analytical solution to the governing heat transfer equations. This was demonstrated for plate-type exchangers which can have quite complex configurations. A model such as this requires computer implementation to solve the associated eigenproblem and system of linear equations. However, it does have the advantage of being quite general in that many exchanger configurations, including multiple-fluid exchangers, can be studied with a single model. In addition, the output from the computer program can be used to generate performance curves (effectiveness vs. NTU), design curves (LMTD correction vs. effectiveness) or temperature profiles. Thus, design and synthesis problems can also be considered with the model.

Future plans include extending the model to handle defective cases, phase changes, non-tridiagonal systems and other boundary conditions. A comparison of computation times with those from numerical integration methods is also planned.

ACKNOWLEDGEMENT

The authors would like to express their appreciation to the Marathon Oil Company for permission to present this paper, and to Ms. Sandy May for preparing the manuscript.

NOTATION

| | |
|---|---|
| $A$ | area of a plate, $m^2$ |
| $a_{ij}$ | element of $\underline{A}$ |
| $C_i$ | heat capacity in channel i, J/kg-K |
| $h_i$ | element of $\underline{h}$, equal to $(W_i C_i)^{1/2} t_i$ |
| $i_{Di}$ | diagonal element of $\underline{I}_D$, equal to $\pm 1$ |
| $n$ | number of channels |
| $n_c$ | number of Type 3 boundary conditions |
| $t_i$ | temperature in channel i, °C |
| $U$ | plate heat transfer coefficient, $W/m^2$-K |
| $W_i$ | flowrate in channel i, kg/s |
| $z$ | fraction of length down a channel |
| $\lambda_i$ | ith eigenvalue, diagonal element of $\underline{\Lambda}$ |

Vectors and Matrices

| | |
|---|---|
| $\underline{A}$ | coefficient matrix in transformed system of equations |
| $\underline{A}_o$ | original coefficient matrix of system |
| $\underline{b}$ | vector used in matrix inversion |
| $\underline{C}'$ | matrix appearing in partition matrix |
| $\underline{C}''$ | matrix appearing in partition matrix |
| $\underline{D}$ | unscaled diagonal matrix |
| $\underline{h}$ | transformed temperature vector |
| $\underline{I}$ | identity matrix |
| $\underline{I}_D$ | diagonal directional matrix |
| $\underline{I}_N$ | scaled diagonal matrix with elements $\pm 1$ |
| $\underline{k}$ | vector with constant elements |
| $\underline{t}$ | temperature vector |
| $\underline{U}$ | integrated coefficient matrix of system |
| $\underline{U}'$ | matrix derived from partition matrix |
| $\underline{u}$ | vector of unknown $h_i(0)$ and $h_i(1)$ |
| $\underline{X}$ | eigenvector matrix for $\underline{I}_D \underline{A}$ |
| $\underline{x}_i$ | ith eigenvector, column i of $\underline{X}$ |
| $\underline{Y}$ | eigenvector matrix for $(\underline{I}_D \underline{A})^T$ |
| $\underline{\Lambda}$ | diagonal matrix of eigenvalues of $\underline{I}_D \underline{A}$ |
| $\underline{\Omega}$ | diagonal matrix for change of variables |

Subscripts/Superscripts

| | |
|---|---|
| i,j | refer to a channel or an element of a vector or matrix |
| T | transpose of a vector or matrix |
| -1 | inverse of a matrix |

LITERATURE CITED

1. Kays, W. M., A. L. London, Compact Heat Exchangers, 2nd Ed., pp. 13-24, McGraw-Hill, New York (1964).

2. McKillop, A. A., & W. L. Dunkley, Ind. Eng. Chem., 52 (9), pp. 733-744 (1960).

3. Jackson, B. W., & R. A. Troupe, Chem. Eng. Prog. Symp. Series, 62 (64), pp. 185-190 (1966).

4. Buonopane, R. A., R. A. Troupe & J. C. Morgan, Chem. Eng. Prog., 59 (7), pp 57-61 (1963).

5. Wolf, J., Int. J. Heat Mass Transfer, 7, pp. 901-919 (1964).

6. Zaleski, T., Chem. Eng. Sci., 39 (7/8), pp. 1251-1260 (1984).

7. Bell, K. J., Heat Exchangers Thermal-Hydraulic Fundamentals and Design, Edited by S. Kakac, A. E. Bergles & F. Mayinger, McGraw-Hill, New York, pp. 165-175 (1981).

8. Amundson, N. R., Mathematical Methods in Chemical Engineering, Chap. 5-8, Prentice-Hall, Inc., Englewood Cliffs, N. J. (1966).

9. Wilkinson, J. H., The Algebraic Eigenvalue Problem, Clarendon Press, Oxford, (1965).

10. Wilkinson, J. H., & C. Reinsch, Handbook of Automatic Computation, Vol. II: Linear Algebra, Springer, Berlin (1971).

11. EISPACK3 software package, NESC No. 534, National Energy Software Center, Argonne National Laboratory.

12. Dahlquist, G. and A. Bjorck, Numerical Methods, Chap. 5, Prentice-Hall, Inc., Englewood Cliffs, NJ, (1974).

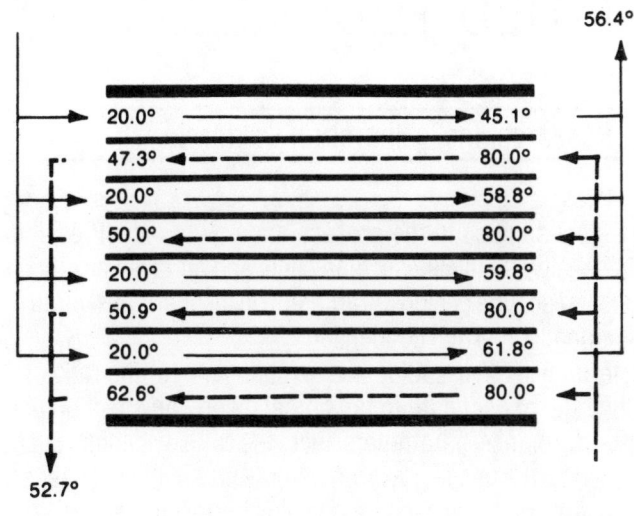

Figure 1. Seven plate, looped-flow exchanger.

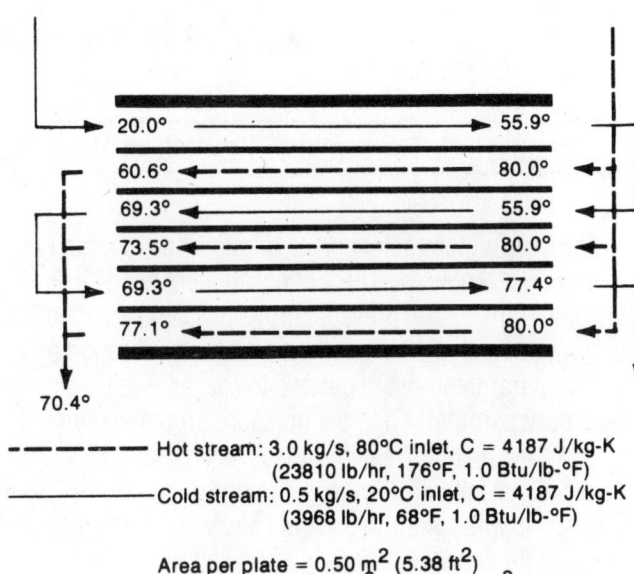

Figure 3. Five plate, complex-flow exchanger.

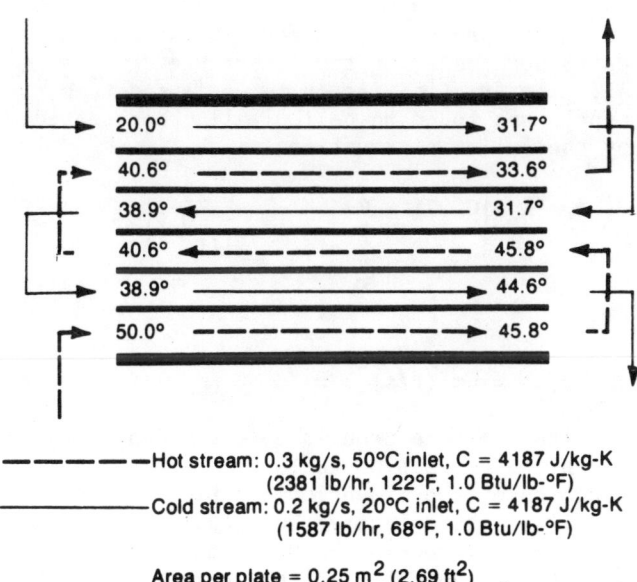

Figure 2. Five plate, series-flow exchanger.

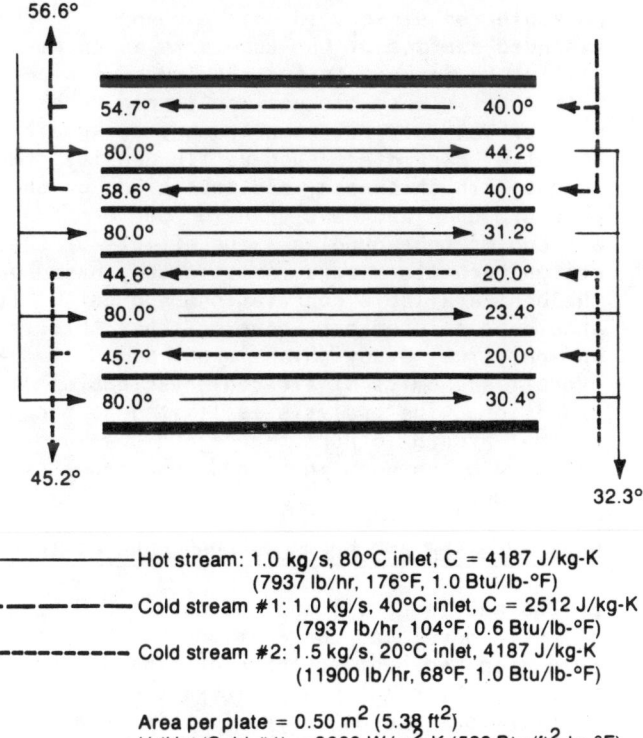

Figure 4. Seven plate, three-fluid, looped-flow exchanger.

# PARAMETRIC STUDY OF AIR-COOLED HEAT EXCHANGER FINNED TUBE GEOMETRY

A. Ganguli, S.S. Tung, and J. Taborek ■ Heat Transfer Research, Inc., Alhambra, California

For an optimum selection of tubes for air-cooled heat exchangers as used in the process and power industries, it is important to know the effect of parameters such as fin height, fin density, fin thickness, tube pitch, and air face velocity on the heat transfer and pressure drop. In this paper, effects of various finned tube design parameters on the thermal-hydraulic performance of a tube bundle were studied through an example of an isothermal condenser.

Two parameters were defined to characterize the thermal performance of air-cooled heat exchangers: (a) the ratio of a heat duty parameter and the pressure drop expended, and (b) the ratio of heat duty to fan horsepower. The effect of tube pitch is most important but also very complex. The thermal performance parameters increase with widening pitch, especially at lower air velocity, but this gain is partially compensated by the need for a larger overall size of the exchanger. Further, it was found that higher thermal effectiveness is possible at lower air face velocities, shorter fin heights, and medium fin densities. There was no appreciable effect of fin thickness variations.

While this paper shows some general trends, an optimum design can only be affected by a number of manufacturing and economic considerations which are outside the scope of the present study.

## INTRODUCTION

Due to the low airside heat transfer coefficient, finned tubes are practically indispensable for air-cooled heat exchangers. The extended surface of the tube acts as an amplifier to compensate for the lower airside film coefficient. To utilize finned tubes most effectively, it is necessary to know the effect of parameters such as fin density, fin height, fin thickness, and tube pitch on the pressure drop and heat transfer performance of air cooler tube bundles. The effects of these design parameters were investigated, based on the best available correlations and data. It should be pointed out, however, that this is an analytical study which is not aimed at the overall economics of air-cooler selection or operation. The analysis is limited to the thermal-hydraulic performance and how such is affected by changes in finned tube geometry parameters.

## HEAT TRANSFER AND PRESSURE DROP CORRELATIONS

Briggs and Young (1) and Robinson and Briggs (2) have developed correlations for both pressure drop and heat transfer of finned tube bundles. Webb (3) has reviewed the available correlations and data base in this area. Rozenman and Pundyk (4) have also published a compilation of all the data on finned tube banks available in the open literature. Based on these and other studies, it seems that the equation developed by Schmidt (5) and given below in a slightly modified form is most suitable for predicting the Colburn j-factor on the airside of finned tube bundles in staggered pitch.

$$j = 0.38 \, Re^{-0.4} \, (A_o/A_r)^{-0.15} \quad (1)$$

A new pressure drop correlation is presented here based on the literature data (3 and 4). The friction factor for a finned tube bundle was found to be logically represented by the following equation.

$$f = K_p \left[ 0.021 + \frac{27.2}{Re_{eff}} + \frac{0.29}{e_{eff}^{0.2}} \right] \quad (2)$$

$$K_p = 1 + \frac{2}{1+a} \exp(-a/4) \quad (3)$$

$$Re_{eff} = Re \, (\ell/s)^{-1} = (d_r \, G_{max}/\mu)(\ell/s)^{-1} \quad (4)$$

The pressure drop is severely dependent on the pitch correction factor, $K_p$, which is a function of the dimensionless fin tip clearance, $a$. The effective Reynolds number incorporates a correction based on the fin height to fin spacing ratio which can vary from 2.5 to 12.5. Note that the Reynolds number is based on the finned tube root diameter.

---

S.S. Tung is presently with Ralph M. Parsons Company

So called "optimization studies" have been attemped by various authors in the past (6, 7, 8, and 9). Most of these are not pertinent to the subject of this paper.

## DEFINITION OF $\epsilon$ AND $\eta$ FOR PARAMETRIC RESPONSE

For a better engineering design of air-cooled heat exchangers, it is essential to understand how the bundle performance is affected by changes in finned tube geometry and layout. The behavior of a finned tube bundle can be studied via the general pressure drop and heat transfer correlations as discussed above. Knowing the f- and j-factors, the pressure drop and heat transfer performance of a tube bundle can be estimated by

$$\Delta P = \frac{2 G_{max}^2 N_r}{\rho\, g_c} \quad (5)$$

$$h_a = h_c [1 - (1 - \Omega_a)(A_f)_o/A_o] \quad (6)$$

$$h_c = j\, C_p\, V_{max}\, \rho/Pr^{2/3} \quad (7)$$

To study the thermal effectiveness of a tube bundle, a parameter, $\epsilon$, is defined as

$$\epsilon = \frac{\text{heat duty/unit tube length/unit temp diff}}{\text{pressure drop per row of tubes}}$$

$$= \frac{h_a A_o}{\Delta P/N_r} \quad (8)$$

The bundle thermal conversion factor, $\epsilon$, is essentially a measure of how much heat flux may be expected per unit of pressure drop. However, as we know, the thermal performance of an air-cooled heat exchanger depends not only on the air velocity and the tube dimensions but also on the temperature driving force between the two process streams. Therefore, the performance of a tube bundle cannot be simply determined by the bundle thermal conversion factor, $\epsilon$, without referring to the temperature driving force. For this purpose, an additional parameter called the exchanger performance factor, $\eta$, was defined as

$$\eta = \frac{Q}{HP_{air}} \quad (9)$$

and $Q = UA\, \Delta T$ (10)

$$HP_{air} = C\,(CFM)(P_v + \Delta P_s) \quad (11)$$

Combining Equations (5) through (11) one finds that $\epsilon$ and $\eta$ are functions of Re, $A_o$, $\Delta T$, $P_t$, and $N_r$. The Reynolds number effect will be described in terms of the more familiar face velocity.

The above analysis is approximate and does not include secondary effects such as row effects on j and f, air maldistribution, backflow in lower tuberows etc. Both $\epsilon$ and $\eta$ appear to be a strong function of the air face velocity and the total outside surface area. The functional dependency is, however, too complex and cannot be described by a simple equation.

An example will be worked out next to illustrate the thermal-hydraulic performance variations of an air-cooled heat exchanger due to changes in finned tube design parameters.

For simplicity, an isothermal condenser was chosen for this study. The air and steam inlet temperature were 68 F (20 C) and 250 F (121.1 C), respectively. The tubeside heat transfer coefficient was assumed constant and equal to 2000 Btu/hr ft$^2$ F (11,350 W/m$^2$ K). For cases with fixed bundle geometry, the following dimensions were used:

```
Tube OD      : 1.0 in. (25.4 mm)
Tube ID      : 0.757 in. (19.2 mm)
Tube length  : 10.0 ft (3.05 m)
30 tubes per row, 4 rows, 1 pass
```

For cases with variable bundle geometry, the following range of dimensions was used:

Fin height     : 0.54, 0.64 and 0.75 in. (12.7, 15.2 and 19.1 mm)
Fin thickness  : 0.017 and 0.025 in. (0.43 and 0.64 mm)
Fin density    : 3 to 12 fins/in. (118.1 to 472.4 fins/m)
Tube pitch     : 2.2 to 3.2 in. (55.9 to 81.3 mm)

A 30° layout (equilateral, staggered) was assumed in all cases. The performance of several alternative finned tube bundle designs is discussed next, with reference to the parameters $\epsilon$ and $\eta$.

## Effect of Fin Density, Fin Height, and Fin Thickness

The effect of fin density and fin height on the previously defined factors $\epsilon$ and $\eta$ are shown in Figures 1 and 2. Figure 1 shows $\epsilon$ for a 1.0-in. (25.4-mm) diameter tube bundle arranged on a 2.50-in. (63.5-mm) pitch with a fin thickness of 0.017 in. (0.43 mm). Three fin heights of 0.5 in. (12.7 mm), 0.6 in. (15.2 mm) and 0.75 in. (19.1 mm) were considered, as shown parametrically in Figures 1 and 2. At a very low air face velocity of 200 ft/min. (1.02 m/s), the bundle thermal conversion factor, $\epsilon$, seems to show a maximum at around 7 to 9 fins/in. (276 to 354 fin/m). However, at air face velocities higher than 200 ft/min. (1.02 m/s), no optimum was observed. Instead, $\epsilon$ increased with the fin density ($N_f$) and approached an asymptotic value depending on the air face velocity. Note that considerably higher $\epsilon$ values were obtained for lower face velocities. Similar results were also observed, but not shown here, for a 2.0-in. (50.8-mm) diameter tube bundle on a 4.179-in. (106.2-mm) pitch.

Figures 1 and 2 show that there was little effect of fin height variations on $\epsilon$ and $\eta$. Generally speaking, $\epsilon$, increases with increasing fin density and with decreasing air velocity. The results for tubes with 0.54 and 0.64-in. (12.7 and 15.2-mm) high fins are practically the same. Tubes with 0.75-in. (19.1-mm) high fins, generally produced the lowest effectiveness. For the fixed pitch of 2.5 in. (63.5-mm), medium high fins with 0.6-in. (15.2-mm) fin height gave the highest effectiveness.

Figure 2 shows the effect of fin density and fin height on the exchanger performance factor, $\eta$, for a 1.0-in. (25.4-mm) diameter tube bundle. In contrast to $\epsilon$ shown previously, $\eta$ does not show any maximum, but decreases with increasing $N_f$, especially at lower values of $V_{face}$. At higher air velocities, the exchanger performance factor becomes insensitive to changes in fin density. It seems that air face velocity is the dominant factor as far as the thermal conversion factor and the exchanger performance factor are concerned. The major difference between the thermal conversion factor, $\epsilon$, and the exchanger performance factor, $\eta$, arises from the effect of the temperature driving force (or LMTD of the tube bundle). It can be shown that the LMTD of an exchanger decreases almost linearly with decreasing air face velocity and with increasing $N_f$. This effect can be logically explained by considering the exit air temperature for the various cases.

The exchanger performance factor, $\eta$, was found to decrease monotonically with increasing fin height. Highest values of $\eta$ were obtained with medium high fins. The difference, however, becomes smaller as the face velocity is increased, almost disappearing at 600 ft/min (3.05 m/sec) air face velocity. For further clarification of this effect, Figure 3 shows the heat duty per unit bundle face area as a function of the fin density, with face velocity and fin height as parameters. It can be seen that the heat duty of the exchanger always increases with increasing fin density, air face velocity, and fin height. However, it seems that the increase in heat duty with the fin height is more pronounced at higher air velocities. At air face velocities of 200 ft/min (1.02 m/s) or lower, the exchanger duty may approach a constant value irrespective of the fin height. To summarize, lower air velocities and shorter fins produce higher thermal efficiency, but higher face velocities and higher fins are required to maximize the heat duty.

Figures 1, 2, and 3 show that the bundles with the higher thermal efficiency will suffer from lower heat duty per unit area and vice versa. Therefore, a larger plot area and a bigger heat exchanger will be required for a fixed heat duty if the exchanger has to be operated with higher thermal efficiency (i.e., at lower air velocity and with fewer fins). Actually, a compromise based on the cost of plot area or exchanger size required (i.e., investment cost) and the benefits of higher thermal efficiency (i.e., lower operating cost) should be considered for each design case separately. Some practical drawbacks of using lower air velocities are discussed in the Conclusion.

The effect of fin thickness on the performance factors was found to be negligible. Two fin thickness of 0.017 in (0.43 mm) and 0.025 in. (0.63 mm) were considered with the given example. The effect was slight at 200 ft/min (1.02 m/s) and disappeared at 600 ft/min (3.05 m/s) for both $\epsilon$ and $\eta$.

### Effect of Tube Pitch

The effect of tube pitch on the thermal performance of finned tube bundles deserves a detailed study since this effect is rather complex. However, owing to limitations of scope and space, only a basic treatment is given here. It is understood that $P_t$ refers to the transverse pitch and only equilateral triangular layouts are covered. The fin density, fin height, and fin thickness were chosen to represent those commonly found for industrial 1.0-in. (25.4-mm) finned tubes. At a fixed fin density, both the bundle thermal conversion factor, $\epsilon$, and the exchanger performance factor, $\eta$, increased with the fin tip clearance ($P_t - d_f$) as illustrated in Figures 4 and 5 for a 1.0-in. (25.4-mm) finned tube bundle. As before, both $\epsilon$ and $\eta$ were also found to increase with decreasing air face velocity, indicating that a higher thermal efficiency can be obtained by operating at a lower air velocity. The rate of increase of both these factors as a function of increasing pitch was quite steep at 200 ft/min (1.02 m/s) face velocity and tapered off at 600 ft/min (3.05 m/s).

It may seem at first glance that the tube pitch should be as wide as possible. This needs to be evaluated in the light of several factors. First, due to the decreasing level of turbulence and tube-to-tube interactions, the heat transfer and pressure drop correlations need to be validated by data at wide pitches. Second, large pitches usually necessitate larger bundles and larger plot areas with an increase of associated construction costs. These and other practical considerations may override the performance criteria discussed here in terms of $\epsilon$ and $\eta$.

To further explain the improvements in $\eta$ with increasing pitch, Figures 6 and 7 indicate that at the same air face velocity the heat duty obtainable from a heat exchanger by relaxing the fin tip clearance $(P_t - d_f)$ is essentially the same, but the required air horsepower for the exchanger decreases dramatically with greater fin tip clearance. The slight increase in heat duty with $(P_t - d_f)$, as indicated in Figure 6, is primarily due to more air flow through the bundle resulting from increased width due to larger pitch. Figure 7 shows that a sharp drop in required air horsepower was observed by increasing tip clearances, especially at higher air face velocities. The pressure drop across a finned tube bundle, of course, diminishes with increasing fin tip clearance. Although air horsepower normally goes up with air flow rate, the lower pressure drop across the finned tube bundle seems to have greater effect in this case on the determination of air horsepower.

## CONCLUSION

Several conclusions based on the present idealized study are possible.

* The heat duty of an air-cooled exchanger remains essentially the same, under the same process conditions, if the tube pitch increases. However, due to a decrease in pressure drop across the bundle as a result of larger fin tip clearance, both the thermal parameters $\epsilon$ and $\eta$ increase with tube pitch. This study is limited to equilateral pitches only and based on generalized correlations which have not been tested at very wide pitches. In general, it appears that $\epsilon$ and $\eta$ will increase until such a fin tip clearance is reached that air bypassing would result.

* It is usually beneficial to operate an air-cooled heat exchanger at lower air face velocities strictly from a thermal-hydraulic point of view. A low air flowrate, however, may be impractical for reasons of capital cost and operational convenience. The poor heat transfer rate of the bundle has to be compensated by an increased bundle size, leading to higher capital expenditure. From an operating standpoint, low air face velocities lead to recirculation of the hot exit air and loss of performance due to wind effects. The air flow, in addition, cannot be controlled satisfactorily by either changing the fan pitch or using louvers.

* Under the same process conditions, the heat duty of an air-cooled exchanger increases with an increase of the fin density, fin height, and fin thickness of the attached fins. However, the increase in heat duty is offset by the increase in pressure drop across the tube bundle, resulting in some reduction of the bundle thermal conversion factor, $\epsilon$, and the exchanger performance factor, $\eta$.

The evaluation of economic and other factors which govern the final selection of an air-cooled heat exchanger is very complex and outside the scope of this paper. It is sufficient to note that a better thermal design is not necessarily an economic optimum choice.

* It should be made clear, however, that the effects of mean temperature difference (MTD) and the tubeside heat transfer coefficient have not been investigated. Therefore, somewhat different conclusions are possible for conditions such as gas cooling or viscous liquid cooling where the air film coefficient may not be dominating. However, effects of tube pitch and lower face velocities are expected to be the same.

## NOTATION

| | |
|---|---|
| A | total heat transfer area of the exchanger bundle, $m^2$, $ft^2$ |
| $A_{face}$ | face area of the tube bank, $m^2$, $ft^2$ |
| $(A_f)_o$ | fin outside area per unit length, $m^2/m$, $ft^2/ft$ |

|  |  |
|---|---|
| $A_o$ | total outside surface area per unit length, m²/m, ft²/ft |
| $A_r$ | root area or plain tube area per unit length, m²/m, ft²/ft |
| C | constant used for air HP calculation, 0.0009842, (SI) 0.0001573 (U.S.) |
| $C_p$ | specific heat of air at constant pressure, J/kg K, Btu/lb$_m$ F |
| CFM | volumetric flow rate of air, m³/s, ft³/hr |
| $d_f$ | fin OD, mm, in. |
| $d_r$ | root tube diameter, mm, in. |
| f | fanning friction factor |
| $G_{max}$ | maximum air mass velocity (= $\rho V_{max}$), kg/s m², lb$_m$/hr ft² |
| $g_c$ | conversion factor, 1, 4.17 (10)⁸ lb$_m$ ft/lb$_f$ hr² |
| $HP_{air}$ | air horsepower, kW, hp |
| $h_a$ | actual airside heat transfer coefficient, W/m² K, Btu/hr ft² F |
| $h_c$ | airside heat transfer coefficient based on 100 percent fin efficiency, W/m² K, Btu/hr ft² F |
| j | heat transfer j-factor |
| $K_p$ | tube and layout configuration factor |
| $\ell$ | fin height, mm, in. |
| $N_f$ | fin density, m⁻¹, in.⁻¹ |
| $N_r$ | number of tuberows |
| Pr | Prandtl number |
| $P_t$ | tube pitch, mm, in. |
| Q | heat duty of the exchanger bundle, W, Btu/hr |
| Re | Reynolds number based on root diameter |
| $Re_{eff}$ | effective Reynolds number |
| s | fin spacing, mm, in. |
| $t_f$ | fin thickness, mm, in. |
| U | overall heat transfer coefficient, W/m² K, Btu/hr ft² F |
| $V_{face}$ | air face velocity, m/s, ft/min |
| $V_{max}$ | maximum air velocity, m/s, ft/hr |
| $\alpha$ | dimensionless fin tip clearance, $(P_t - d_f)/d_r$ |
| $\Delta P$ | pressure drop across the tube bank, Pa in. Water |
| $\Delta T$ | LMTD of the exchanger bundle, C, F |
| $\epsilon$ | bundle thermal conversion factor (W/m K) (Pa), (Btu/hr ft F)/(in. Water |
| $\eta$ | exchanger thermal performance factor, (-), (Btu/hr)/hp |
| $\mu$ | air viscosity, Ns/m², lb$_m$/ft hr |
| $\rho$ | air density, kg/m³, lb$_m$/ft² |
| $\Omega_a$ | fin efficiency |

## LITERATURE CITED

1. Briggs, D. E. and Young, E. H. "Convection Heat Transfer and Pressure Drop of Air Flowing Across Triangular Pitch Banks of Finned Tubes," CEP Symp., Series No. 41, 59, 1 (1963)

2. Robinson, K. K. and Briggs, D. E., "Pressure Drop of Air Flowing Across Triangular Pitch Banks of Finned Tubes," AIChE, Eighth National Heat Transfer Conference (Aug. 1965)

3. Webb, R. L., "Air-Side Heat Transfer in Finned Tube Heat Exchangers," Heat Transfer Engineering, Vol. 1, No. 3, January-March 1980

4. Rozenman, T. and Pundyk, J. M., "Heat Transfer and Pressure Drop Characteristic of Dry Tower Extended Surfaces, Part I: Heat Transfer and Pressure Drop Data," BNWL-PFR 7-100, PFR Engineering System, Inc., March 1976

5. Schmidt, T. E., "Der Warmeubergang an Rippenrohre und die Berechnung von Rohrbundelnwarmeaustauschern." ("The Heat Transfer at Finned Tubes and Computations of Tube Bank Heat Exchangers"), Kaltetechnik, 15, No. 4, Part I, 98-102, Part II, 370-378, 1963

6. Kern, D. G., "Optmum Air-Fin Cooler Design," CEP Symp. Series, No. 59, Vol. 55, 1959

7. Lohrisch, F. W., "What are Optimum Conditions for Air-Cooled Exchangers," Hydro-

carbon Processing, Vol. 45, No. 6, June 1966

8. Baskerville, J. E., Robinson, T. R. and Rohsenow, W. M., "Performance Ranking of Crossflow Finned Tubular Heat Exchanger Surfaces," Proc. ASME/JSME Thermal Engineering Joint Conference, Honolulu, Hawaii, March 1983

9. Johnson, B. M., Kreid, Dennis K., and Hauser, Steven G. "A Method of Comparing Performance of Extended Surface Heat Exchangers," Heat Transfer Engineering, Vol. 4, No. 1, January-March 1983

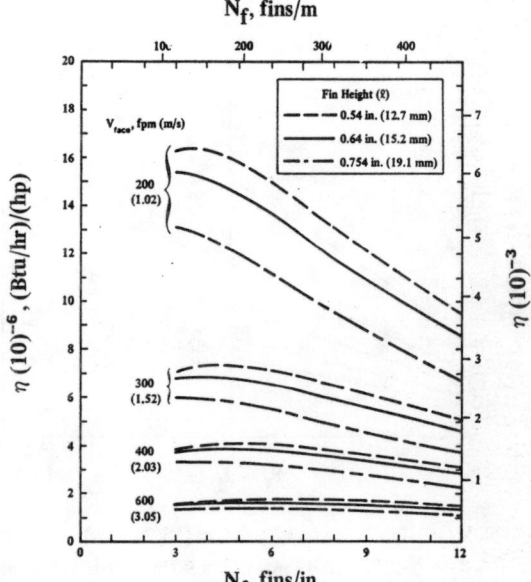

$d_r = 1$ in. (25.4 mm), $t_f = 0.017$ in. (0.43 mm)
$P_t = 2.5$ in. (63.5 mm), 30 tubes/row, 4 rows

**FIGURE 2** Effect of Fin Density and Fin Height on the Exchanger Performance factor, $\eta$

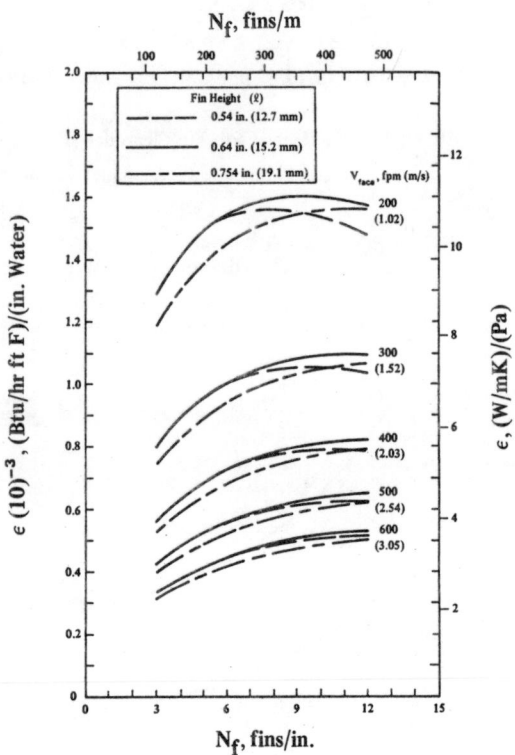

($d_r = 1$ in. (25.4 mm), $t_f = 0.017$ in. (0.43 mm), $P_t = 2.5$ in. (63.5 mm))

**FIGURE 1** Effect of Fin Density and Fin Height on the Bundle Thermal Conversion Factor, $\epsilon$

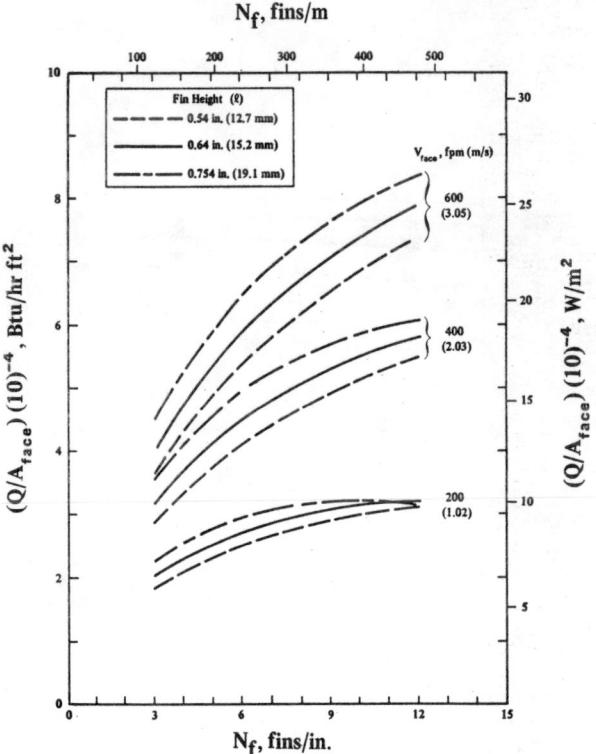

$d_r = 1$ in. (25.4 mm), $t_f = 0.017$ in. (0.43 mm), $P_t = 2.5$ in. (63.5 mm), 30 tubes/row, 4 rows

**FIGURE 3** Heat Duty per Unit Bundle Face Area as a Function of Fin Height

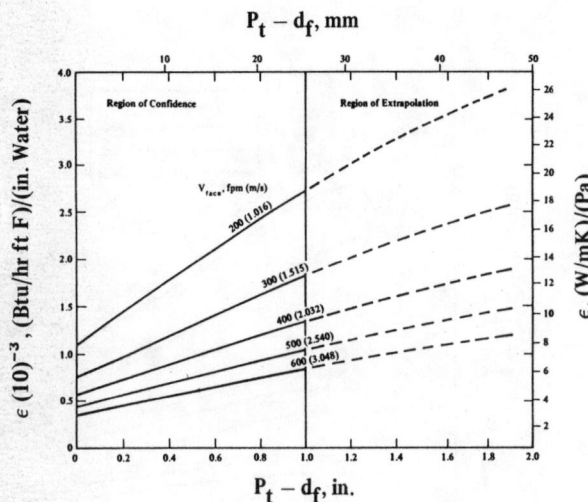

$d_r$ = 1-in (25.4 mm) ℓ = 0.6 in (15.24 mm)
$t_f$ = 0.017-in (0.43 mm) $N_f$ = 9 fins/in (354.3 fins/in)

**FIGURE 4** Effect of Fin Tip Clearance ($P_t - d_f$) to Bundle Thermal Conversion Factor, $\epsilon$

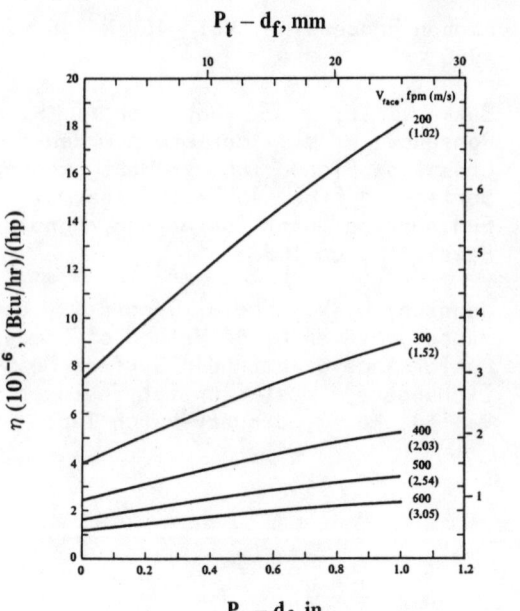

$d_r$ = 1.0 in. (25.4 mm), $N_f$ = 9 fins/in. (354.3 fins/m), ℓ = 0.6 in. (15.2 mm), $t_f$ = 0.017 in. (0.43 mm), 30 tubes/rows, 4 rows

**FIGURE 5** Effect of Fin Tip Clearance ($P_t - d_f$) on the Exchanger Performance factor, $\eta$

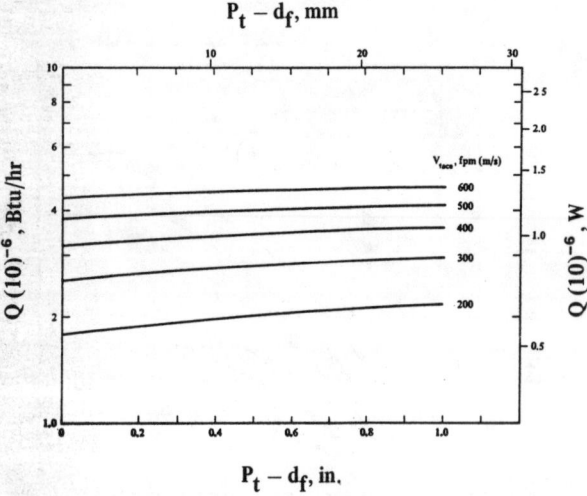

**FIGURE 6** Effect of Fin Tip Clearance ($P_t - d_f$) on the Heat Duty of the Example Condenser

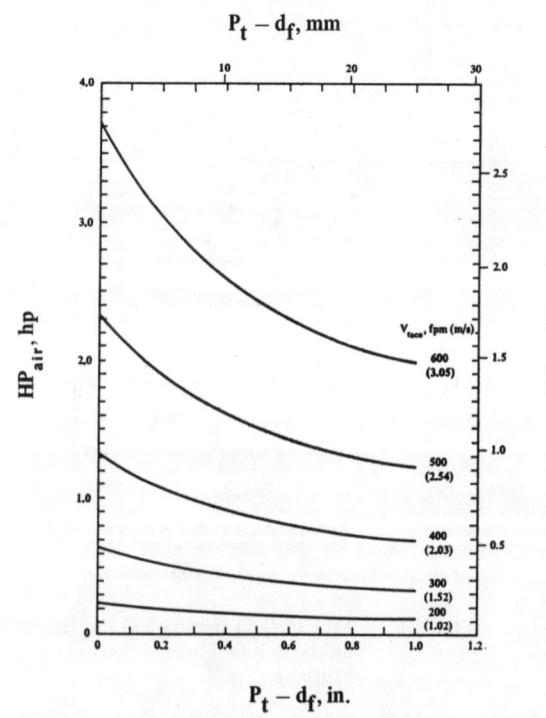

**FIGURE 7** Effect of Fin Tip Clearance ($P_t - d_f$) on the Air Horsepower of the Example Condenser

# THE GENERAL PREDICTION OF CONVECTIVE BOILING COEFFICIENTS IN PLATE-FIN HEAT EXCHANGER PASSAGES

J.M. Robertson and R.H. Clarke ∎ HTFS, AERE, Harwell, UK

The systematic study of the boiling heat transfer characteristics of the small finned passages of plate-fin heat exchangers using liquid nitrogen and R11 has already been reported (1-6). This study has now been extended to the low quality region where a liquid film two phase flow pattern cannot hold.

An analysis of the measured convective boiling coefficients, without nucleate boiling present, from tests with liquid nitrogen in a perforated-fin test section shows the separate effects of quality, mass flux and pressure. The influence of the Reynolds number (and flow regime) and also of the two phase flow pattern on test results at all qualities, is illustrated on a map. A model of boiling heat transfer with slug flow is advanced and used to analyze results at very low qualitites. The resolution of the anomalous boiling characteristics of serrated-fin passages at low qualities is also discussed.

## INTRODUCTION

Systematic laboratory investigations have been carried out by the Heat Transfer and Fluid Flow Service (HTFS) for several years into the boiling characteristics of various forms of finned passages used in industrial brazed-aluminium plate-fin heat exchangers. These studies into upflow boiling have used liquid nitrogen and Freon 11 (R11). The ranges of mass flux and heat flux used in the experiments have been representative of industrial applications in the chemical process industry.

The results of work using electrically-heated test sections 3.4 m long, constructed from perforated and serrated fins have already been reported [e.g. 1-6] together with details of the test rigs and test procedures. The dimensions of the finning used were: 6.35 mm high, 590 fins per metre, fins 0.2 mm thick, forming rectangular passages with an equivalent diameter of 2.4 mm. The serrated fins had serrations 3.125 mm long and the perforated fins had 5% perforation.

The general approach has been to measure local convective-boiling heat transfer coefficients, i.e. with little or no nucleate boiling present, together with local overall two-phase pressure gradients under a variety of test conditions. From these observations empirical correlation methods have been explored [1,3],
methods based on the characteristics of relevant two-phase flow patterns have been developed [4] and also taken into account the special nature of interrupted serrated finning [5]. The effects on the delay of boiling onset with very small temperature differences between streams, which are thermodynamically necessary for heat exchange at low temperature, have also been reported [2,6].

In a recent paper [4] a simple climbing liquid-film model of the two-phase flow pattern in perforated-fin channels, with constant shear, was used to interpret the results of tests with liquid nitrogen at higher qualities and where convective boiling was dominant. The experimental conditions ranged from sub-cooled inlet to high exit qualities, mass fluxes from 37 kg/m$^2$s to approximately 120 kg/m$^2$s, and pressures from 1.3 to 5.2 bar. Heat fluxes were restricted to less than 4000 W/m$^2$ on the primary surface of the finned passage.

In the present paper a study of the test results at lower qualities from these tests, where the liquid film model cannot hold, is now reported. For this study, a correlating scheme has been developed for (a) identifying the boundaries of the relevant two-phase flow patterns and (b) quantifying the separate effects of mass flux (in terms of Reynolds number), quality and pressure on convective boiling coefficients. Since the finned rectangular channels are so narrow it may be expected that single bubbles are able to span the gap (~ 1.5 mm) and thus form slug-flow readily. A simple model of boiling heat transfer with slug-flow has therefore been put forward. The resolution of the anomalous characteristics of serrated-fin passages at low qualities is also discussed.

## CORRELATING SCHEME FOR CONVECTIVE BOILING HEAT TRANSFER

**Main Features.** In order to illustrate the separate effects of quality and mass flux, the measured convective boiling coefficients at a nominal pressure of 1.4-1.8 bar and for various mixture qualities are plotted against $Re_{TL}$, the Reynolds number for the total flow regarded as a liquid (Fig.1). The corresponding heat transfer coefficients at zero quality are also plotted and these reflect the expected change in dependence on $Re_{TL}$ (or on mass flux since $d_e$ and $\mu_L$ are fixed) in the laminar, transition and turbulent regions.

This plot exhibits three major features. Firstly, $Re_{TL}$ has the same effect on heat transfer coefficients over the complete quality range: at $Re_{TL} < 2000$, convective boiling and single-phase coefficients tend to constant values; at higher values of $Re_{TL}$, coefficients all tend to be proportional to $Re_{TL}$ to the power of 0.8 approximately

Secondly, at qualities even as low as 0.05, the convective boiling coefficients are considerably greater than the equivalent liquid coefficient, for all values of $Re_{TL}$, i.e. small quantities of vapour have a great effect [9]. Thirdly, at high qualities, the effect of quality on boiling coefficients is very much greater at values of $Re_{TL}$ above 2000.

These three features are also apparent in results with nitrogen [1] and also Freon 11 (R11) [3] from other tests with a serrated-fin test section which has the same finned channel dimensions (Fig.6). Since $Re_{TL}$ has absolutely general application to flow in channels, it can therefore be reasoned that these three features are also absolutely general and they should be present in any channel with convective boiling where test conditions span the laminar and turbulent regions.

**Flow Patterns.** The two-phase flow patterns expected to be present in each region of Fig.1 are delineated in Fig.2. The lower

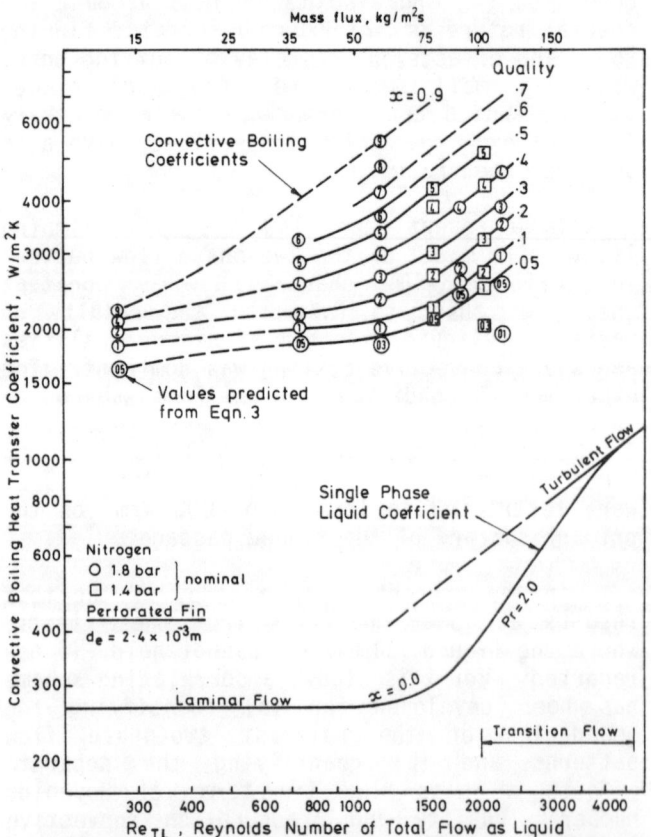

Fig.1. Plot of Measured Convective Boiling and Liquid Coefficients against Reynolds Number of Total Flow as a Liquid.

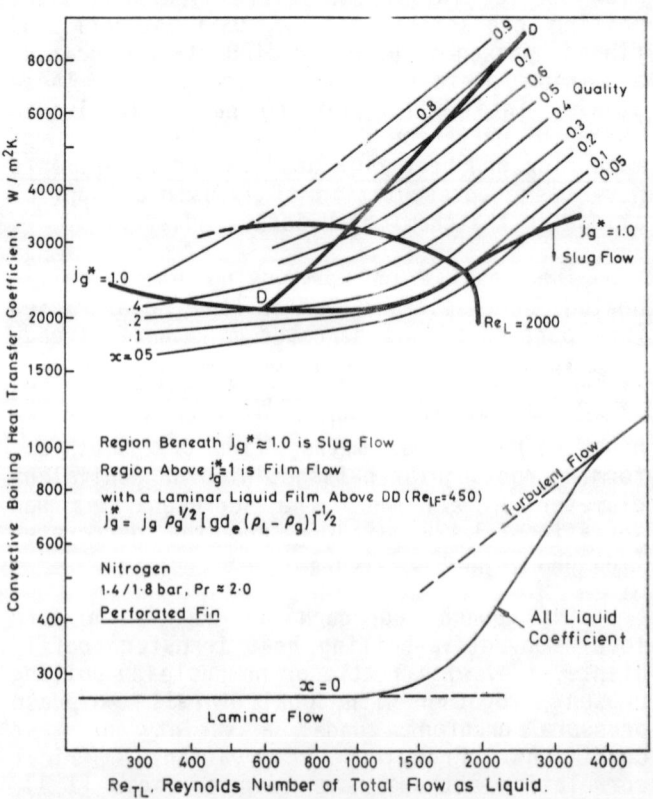

Fig.2. Regions of Two-Phase Flow Patterns on Plot of Convective Boiling Coefficients against $Re_{TL}$

boundary of film-flow with a turbulent film and upper boundary of slug-flow can, for present purposes, be taken as the locus of $j_g^* = 1$. [$j_g^*$ reflects the ratio of inertia to buoyancy forces in slug-flow [7].] Below $j_g^* = 1$, as plotted on Fig.2, slug-flow is shown to exist at higher qualities as $Re_{TL}$ decreases. As $Re_{TL}$ increases, for a fixed pressure, slug flow will exist over a decreasing region.

From previous analyses of the film-flow region, it has been shown [4] that the film becomes laminar at high qualities. By plotting results in terms of the Nusselt number of the film ($Nu_{LF}$) against its Reynolds number ($Re_{LF}$), for a fixed pressure and Prandtl number ($Pr_L$), it can be seen that the critical value of $Re_{LF}$ is approximately 450 for $Pr_L = 2.0$ (Fig.3). On this figure it can also be seen that at lower qualities, film-flow theory no longer holds as $Re_{LF}$ increases. This deviation from the film-flow prediction occurs for each mass flux close to $j_g^* = 1$ (or $\tau_w/\rho_L$ g m $< 4.0$) as film-flow breaks down to form slug or churn-flow. The locus of $Re_{LF} = 450$ (DD) is plotted on Fig.2: above DD film-flow exists with a laminar film; below DD the film is not laminar.

On Fig.1, the quite different characteristics of the regions above and below the critical Reynolds number for the perforated-fin channels can be separated by the locus of $Re_L = 2000$, where $Re_L = Re_{TL} (1 - x)$, and $Re_L \rightarrow Re_{TL}$ as $x \rightarrow 0$.

Effect of Pressure. With a test section of given dimensions a high Reynolds number can be obtained either by increasing mass flux or reducing viscosity by increasing the pressure. This latter results in an increase in vapour density which will reduce vapour velocities and the interfacial shear stress ($\tau_i$). This stress affects the thickness of the climbing liquid film and hence also the convective coefficients at high qualities.

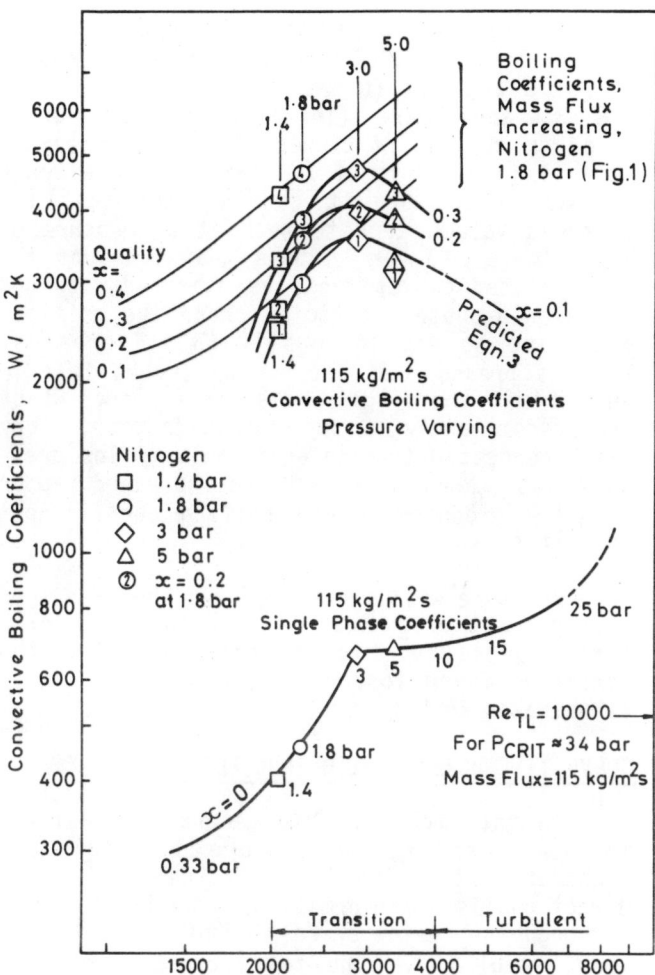

Fig.4. Plot of Measured Convective Boiling Coefficients against $Re_{TL}$ showing effect of Pressure (at a given Mass Flux)

The general effect as **pressure** is varied can now be examined by considering the range of convective boiling coefficients to be expected over the full range from near-critical to very low pressure. On Fig.4 are plotted the measured convective boiling and also the single-phase heat transfer coefficients for a chosen mass flux of ~ 115 kg/m²s over the pressure range 1.4 - 5 bar covered in the tests; corresponding pressures are noted on the plots. At this mass flux, coefficients at x = 0 can be seen to be very dependent on pressure between 0.33 bar and 3 bar (reduced pressures of ~ 0.01 and ~ 0.1) partially as a result of the effects of $Pr_L$ on the Graetz number (each length of finning is about 120 diameters). They then become relatively independent of pressure up to 25 bar where the proximity of the critical pressure begins to dominate.

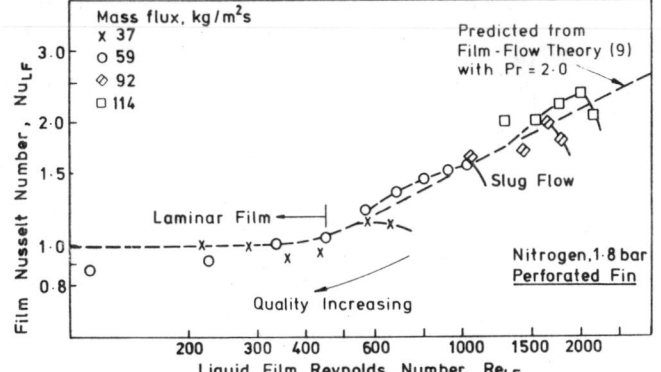

Fig.3. Plot of Liquid Film Nusselt Number against Film Reynolds Number.

On Fig.4 are also plotted measured convective boiling coefficients for qualities of 0.1, 0.2, 0.3 and 0.4 at nominal pressures of 1.4, 1.8, 3.0 and 5 bar. (These points are superimposed on the set of curves from Fig.1 obtained at a fixed pressure of 1.8 bar but for a range of mass fluxes.) It can be seen that the measured values tend to peak at a pressure of approximately 3 bar. Above that pressure the extrapolated curves tend, as expected, towards the liquid-phase coefficient near the critical pressure as vapour density ($\rho_g$) increases. Below a pressure of 3 bar, the curves fall in unison with the single-phase coefficients which are predominantly in the transition zone. They can be expected to rise again at very low pressures as $\rho_g$ decreases and the increased vapour velocity produces higher boiling coefficients in film flow.

Pressure will therefore have the effect of moving a set of curves, as e.g. in Fig.1 for a fixed pressure, in accordance with the variation found for an arbitrary mass flux as illustrated in Fig.4.

## CONVECTIVE BOILING WITH SLUG FLOW

In the light of this general analysis it is now possible to explore boiling heat transfer with slug-flow. However, little direct published information is available here, although the review by Collier of this topic is very useful [9]. The test results from this region can be analysed within our observations that (a) there is an enormous effect of vapour volume on coefficients at very low qualities and (b) the coefficients at low and zero qualities are equally influenced by Reynolds number.

In boiling heat transfer with slug-flow and with no nucleate boiling present there will be two simultaneous mechanisms occurring in parallel: evaporation of the falling liquid film surrounding each rising bubble and sensible heating of the climbing liquid slugs. The thickness of the falling liquid film will control its coefficient ($h_{\eta F}$) and the fraction of the length of passage occupied by bubbles will determine the relative importance of evaporation. The film will in general be laminar and its thickness can be estimated using a method by Wallis [7]. While it is not yet possible to estimate the length of individual bubbles, the void fraction can also be obtained from Wallis:

$$\alpha = \frac{j_g}{1.2 (j_g + j_L) + v_{b\infty}} \quad (1)$$

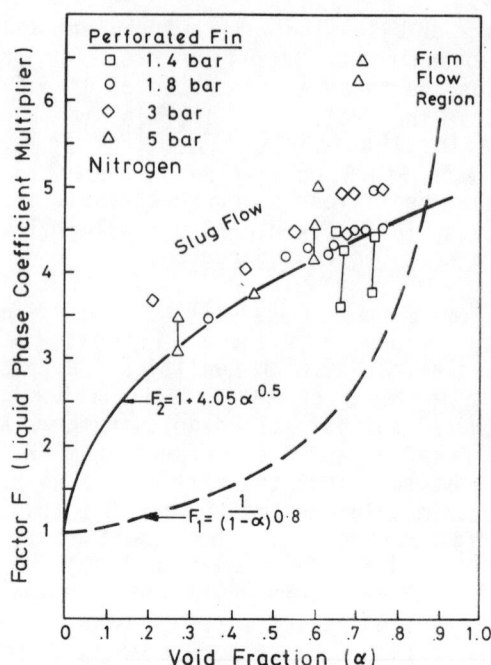

Fig.5. Plot of Factors $F_1$ and $F_2$ against calculated Void Fraction.

Thus the effective heat transfer coefficient associated with the falling laminar film, and taking into account the fin efficiency as shown before [1], is:

$$h_{\eta F} = k_L \alpha / m \quad (2)$$

It is presumed, for present purposes, that the individual bubbles are short enough that evaporation does not significantly thin the falling film and hence increase the mean value of $h_{\eta F}$ along the film.

The simple and apparently effective method for estimating the heat transfer to the liquid slugs put forward by Lunde [8], and recommended by Collier [9], assumes that the heat transfer coefficient of the liquid fraction is increased by a factor $F_1 = (1 - \alpha)^{-0.8}$ in the turbulent region, as void fraction and liquid-phase velocity increase with quality. This method has the advantage that the effect of Reynolds number ($Re_L$) should be the same as that found in our tests and discussed earlier.

Experimental results can now be compared with predictions from Lunde's method by assuming that the effective convective boiling coefficient ($h_{\eta SF}$) is obtained by adding the separate coefficients for the heating of the liquid slug ($h_{\eta S}$) multiplied by some unknown factor, and evaporation of the liquid film ($h_{\eta F}$). This implies that the temperature difference between wall and bulk (saturation) is the same for both processes and this can only hold if the liquid slugs are very short and any bulk superheating is immediately converted to vapour. Thus, from Equation (2):

$$h_{\eta SF} = k_L \alpha / m + F_2 h_{\eta S} \qquad (3)$$

Using the single-phase heat transfer characteristics of the perforated fins in the form of the Colburn J factor, the unknown factor $F_2$ can be obtained from measured values of $h_{\eta SF}$. The results for the available data spanning $Re_{TL}$ from 750 to 3000, and pressures from 1.4 to 5.0 bar are given in Fig.5. It can be seen that the values of $F_2$ are reasonably correlated with the calculated void fraction but are much greater than those for $F_1$. This implies that the simple increase of slug velocity with $\alpha$ as suggested by Lunde is insufficient explanation.

Lunde's model for heat transfer to liquid slugs does not take into account the liquid circulation within the slug [11] and the inevitable superheating of long slugs. The transient conduction mechanism put forward by Kusuda et al. [10] whereby this recirculation deposits a cold layer of liquid on the channel wall after each bubble, could possibly provide a better model.

Attempts to correlate the measured coefficients at low qualities using the velocity of the slug $(j_T)$ have not proved to be successful since the very high Reynolds number of the slug does not reproduce the observed effect of the critical value of $Re_{TL}$ as in Fig.1. The test results at all qualities have already been satisfactorily correlated by plotting F, the ratio of the boiling to single-phase coefficient, against the Martinelli parameter X [4].

The expression for $F_2$ given in Fig.5 can be used with Equation (3) to predict $h_{\eta SF}$ at low qualities at low values of $Re_{TL}$. Reasonable examples of predictions are given in Fig.1 at $Re_{TL}$ = 300 for low qualities (for $j_g^* < 1$), and Fig.4 for x = 0.1 at higher pressures.

## SERRATED FIN

As reported previously [1], the single-phase heat transfer coefficients of the perforated and serrated fins differ markedly (Fig.6), and at high qualities in film flow, the convective boiling coefficients of the serrated fin are also distinctly greater because of the continuous liquid entrainment produced by the interrupted fins [5]. However, at lower qualities, the coefficients from each fin surprisingly have the same magnitudes and exhibit the same features (Fig.1 and Fig.6). It can now be concluded that the reason for the similarity at low quality is the presence of the same two-phase flow pattern - probably slug-flow - and the totally different boiling characteristics of liquid-film flow.

## CONCLUSIONS

A map has been produced which identifies the regions where the various two-phase flow patterns may be expected to be present in convective boiling in perforated plate-fin heat exchanger passages, particularly at low Reynolds numbers. The Reynolds number of the total flow regarded as a liquid, $Re_{TL}$, has been used as a correlating parameter to clarify the separate effects of mass flux, quality, pressure and flow regime (laminar or turbulent) on the convective boiling coefficients. It is recommended that during design of equipment, the significance of $Re_{TL}$ of any boiling channel should be taken into account.

A simple treatment of boiling with slug-flow has been developed and it is concluded that a model, which takes into account the

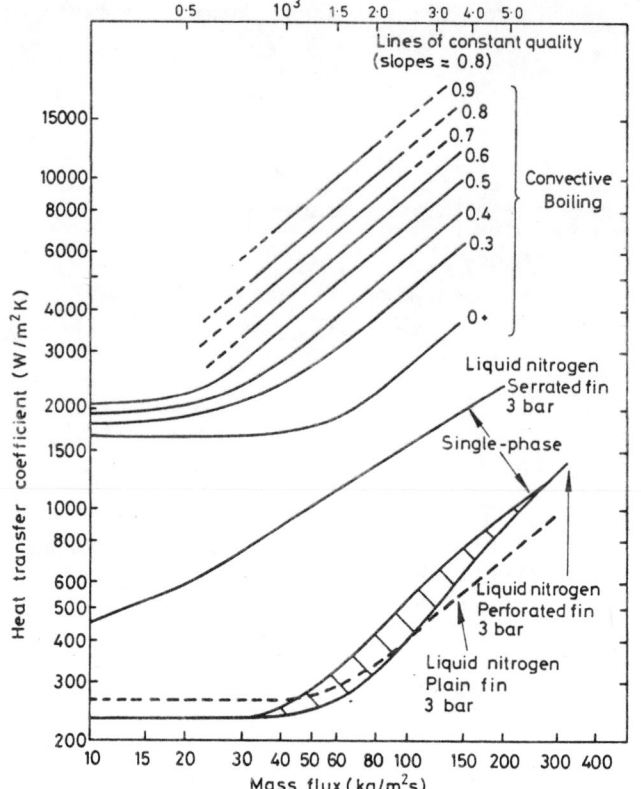

Fig. 6. Plot of Curves of Convective Boiling Coefficients for Serrated Fin against $Re_{TL}$ and Mass Flux (from (1))

fluid mechanics of the liquid-phase in slug-flow, should be more successful in predicting heat transfer to it.

It is suggested that the presence of slug-flow explains why convective boiling coefficients, at low quality with no nucleate boiling present, are similar in magnitude and nature with serrated and perforated fins.

ACKNOWLEDGEMENT

Much of the work described in this paper was supported by the UK Mechanical and Electrical Engineering Requirements Board.

NOTATION

| | |
|---|---|
| $d_e$ | Equivalent diameter of rectangular passage, m |
| $F_2$ | Factor (Equation 3) |
| $g$ | Gravitational acceleration, m/s² |
| $h_{\eta F}$ | Falling-film boiling coefficient, with fin efficiency, W/m²K |
| $h_{\eta S}$ | Liquid-phase heat transfer coefficient, with fin efficiency, W/m²K |
| $h_{\eta SF}$ | Slug-flow convective boiling coefficient, with fin efficiency, W/m²K |
| $j_g$, $j_L$ | Vapour, liquid volumetric fluxes, m/s |
| $j_T$ | Total volumetric flux ($j_g + j_L$), m/s |
| $j_g^*$ | Non-dimensional vapour velocity, Fig.2 |
| $J$ | Colburn J factor ($J = St_L Pr_L^{2/3}$) |
| $k_L$ | Liquid conductivity, W/mK |
| $m$ | Film thickness (calculated), m |
| $Nu_{LF}$ | Film Nusselt number of liquid film |
| $Pr_L$ | Prandtl number of liquid |
| $Re_L$ | Reynolds number of liquid fraction |
| $Re_{LF}$ | Reynolds number of liquid film |
| $Re_{TL}$ | Reynolds number of total flow regarded as liquid |
| $St_L$ | Stanton number of liquid |
| $v_{b\infty}$ | Bubble rise velocity in stagnant liquid (inertia controlled), m/s |
| $x$ | Mixture quality |
| $\alpha$ | Void fraction |
| $\rho_g$, $\rho_L$ | Vapour and liquid densities, kg/m³ |
| $\tau_i$, $\tau_w$ | Interface, wall shear stresses, N/m² |
| $\mu_L$ | Liquid viscosity, kg/m s |

LITERATURE CITED

[1] ROBERTSON, J.M. (1979) "Boiling Heat Transfer of Liquid Nitrogen in Brazed-Aluminium Plate-Fin Heat Exchangers". AIChE Symp. Ser., No.189, Vol.75, pp.151-164, National Heat Transfer Conference, San Diego.

[2] ROBERTSON, J.M. and CLARKE, R.H. (1981) "The Onset of Boiling of Liquid Nitrogen in Plate-Fin Heat Exchangers". AIChE Symp. Ser., No.208, Vol.77, pp.86-95, National Heat Transfer Conference, Milwaukee.

[3] ROBERTSON, J.M. and LOVEGROVE, P.C. (1983) "Boiling Heat Transfer with Freon 11 in Brazed Aluminium Plate-Fin Heat Exchangers". ASME J. of Heat Transfer, Vol.105, pp.605-610, August.

[4] ROBERTSON, J.M. (1983) "The Boiling Characteristics of Perforated Plate-Fin Channels with Liquid Nitrogen in Upflow". ASME, HTD-Vol.27, pp.35-41, National Heat Transfer Conference, Seattle.

[5] ROBERTSON, J.M. (1984) "The Prediction of Convective Boiling Coefficients in Serrated Plate-Fin Passages using an Interrupted Liquid-Film Flow Model". ASME, HTD-Vol.34, pp.163-173, National Heat Transfer Conference, Niagara Falls.

[6] CLARKE, R.H. and ROBERTSON, J.M. (1984) "Investigations into the Onset of Boiling with Liquid Nitrogen in Plate-Fin Heat Exchanger Passages under Constant Wall Temperature Boundary Conditions". AIChE Symp. Ser., No.236, Vol.80, pp.98-104, National Heat Transfer Conference, Niagara Falls.

[7] WALLIS, G.B. (1969) "One Dimensional Two-Phase Flow". McGraw Hill.

[8] LUNDE, K.E. (1961) "Heat Transfer and Pressure Drop in Two-Phase Flow". Chem. Eng. Prog. Symp. Series, Vol.57, 32, pp.104-110.

[9] COLLIER, J.G. (1972) "Convective Boiling and Condensation". McGraw Hill, 1st Edition.

[10] KUSUDA, H., MONDE, M., UEHARA, H. and OTUBO, K. (1980) "Bubble Influence in Boiling Heat Transfer in a Narrow Space (at Low Heat Fluxes)". Heat Transfer Jap. Res., Vol.9, No.2, pp.48-60.

[11] TAYLOR, G.I. (1961) Jnl. of Fluid Mechanics, 10, p.161.

# A STUDY OF GEOMETRIC SCALING OF CURVED-ARM PRIMARY STEAM/WATER SEPARATORS

A.M. Eaton and W.P. Preuter ■ The Babcock and Wilcox Company
Research and Development Divisions, Alliance, Ohio 44601
J.R. Wall ■ The Babcock and Wilcox Company, Nuclear Equipment Division, Barberton, Ohio 44203

The purpose of this study was to evaluate geometric scaling criteria for curved-arm primary (CAP) steam/water separators. These criteria were established with the intent of preserving desirable performance characteristics when a separator reference design is scaled to a larger or smaller size. The criteria were evaluated using air/water experiments to compare the performance of a larger and a smaller geometrically scaled separator with the reference design.

## INTRODUCTION

Over several decades, The Babcock & Wilcox Company (B&W) has acquired significant expertise in developing steam/water separation equipment and technology. The common approach used in this development has been to design a separator concept on an intuitive basis, test the concept to evaluate its performance, and then proceed through a series of experimental refinements/improvements until desired performance is achieved. This empirical approach has been -- and continues to be -- generally the only approach available because of the complex nature of two-phase flow mechanisms governing the separation process. No general theory of two-phase flow is available that provides a general descriptive mathematical model for the analysis of a multiphase flow process like separation.

Because the development of separation equipment is heavily dependent on experimental work that may be expensive and time consuming, the development of geometric scaling criteria that can be used to scale an existing separator design with desirable performance to a larger or smaller size is of interest to the separator designer. For example, significant cost savings can result if small-scale equipment is used in the development of a large design and full-scale evaluation is reserved for the final design prototype. Geometric scaling would also permit changes to be more easily made in application and system configurations that dictate larger or smaller equipment designs.

Such geometric scaling of equipment involving two-phase flows to larger or smaller sizes while preserving desirable performance characteristics is generally a difficult task. This is because the behavior of two-phase flow is strongly affected by geometry. A straightforward change in geometric size can sometimes lead to drastic and unexpected differences in behavior because of changes in flow regimes, even if the flow rates are scaled and the physical properties of the flow remain the same.

The attraction and difficulty of geometric scaling, and a desire to learn more about the effects of size on steam/water separator performance led to the study described in this paper. Beginning with an original design with good performance characteristics, scaling criteria were hypothesized that could be used to increase or decrease the size of the separator while preserving the performance characteristics when compared on the basis of separator flow rate per unit cross-sectional area. These scaling criteria were determined on the basis of present understanding of the fundamental physical mechanisms governing the performance characteristics of a separator.

---

A. M. Eaton is now with Hercules Aerospace, Magna, Utah.

The base design used in this study was a curved-arm primary (CAP) separator shown in Figure 1. This separator is contained within a 13-1/2-inch (34.3 cm) return cylinder (see Figure 2). Using the scaling criteria, one separator was designed for use in a 21-inch (53.3 cm) diameter return cylinder (2.4 times the original area), and another was designed for use in an 11-1/2-inch (29.2 cm) diameter return cylinder (0.73 times the original area). These sizes were the extremes in sizes that could be tested within the test facility capacity limits and still obtain useful data for performance comparison. Air/water testing was used to obtain performance data for the three separator designs.

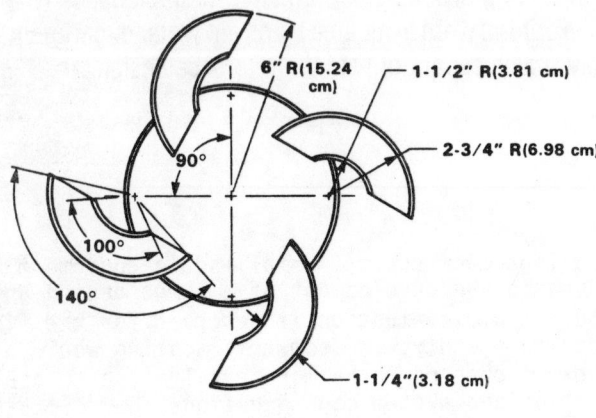

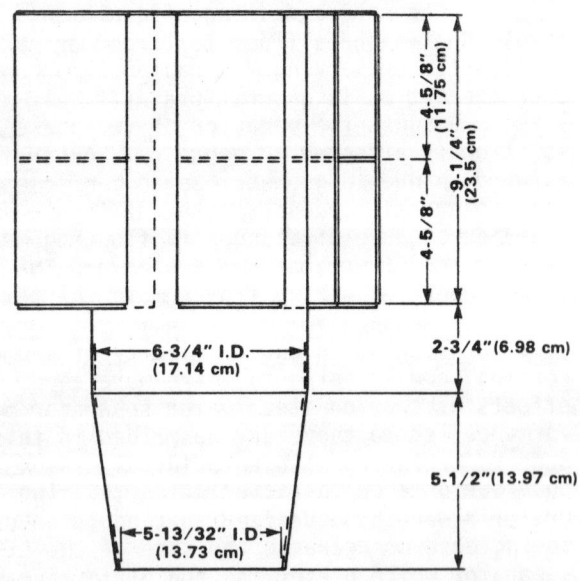

Figure 1. 13-1/2-inch CAP separator geometry (return cylinder not shown).

## DEFINITION OF SCALING CRITERIA

The method of geometric scaling established in this study is based on using an existing separator design with desirable performance characteristics as a scaling reference. The scaling criteria are then defined in such a way that the reference design can be scaled to a different size while preserving desirable performance features in terms of separator loading per unit cross-sectional area. The criteria take the form of non-dimensional ratios of the key separator dimensions shown in Figure 2. Scaling proceeds by matching the ratios in the scaled design with the ratios in the reference design. If the scaling criteria have been properly chosen, similar ratios should result in similar performance.

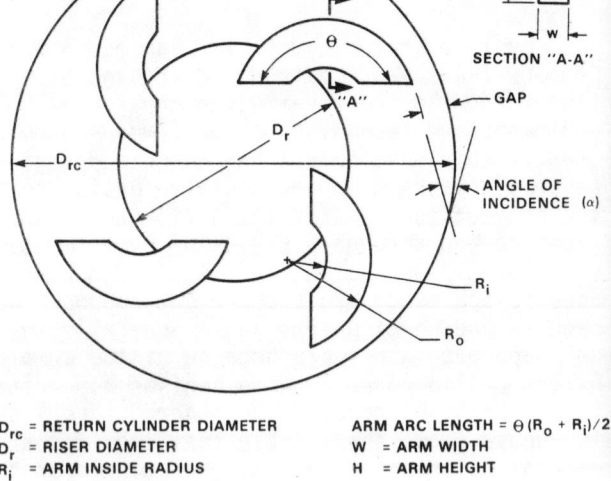

$D_{rc}$ = RETURN CYLINDER DIAMETER
$D_r$ = RISER DIAMETER
$R_i$ = ARM INSIDE RADIUS
$R_o$ = ARM OUTSIDE RADIUS
ARM ARC LENGTH = $\Theta(R_o + R_i)/2$
W = ARM WIDTH
H = ARM HEIGHT

Figure 2. Key dimensions in CAP separator.

The criteria are defined on the basis of extensive experience with the designing and testing of CAP separators. These criteria ratios can be divided into two groups: those based on geometric similarity and those based on performance similarity. The following criteria are those based on geometric similarity, and are to be maintained constant relative to the reference design:

- Ratio of riser to return cylinder diameter ($D_r/D_{rc}$)

- Ratio of arm width to arm height (aspect ratio, W/H)
- Ratio of gap to arm width (Gap/W)
- Ratio of gap to return cylinder diameter (Gap/$D_{rc}$)
- Angle of incidence ($\alpha$)

Performance similarity will be maintained in the scaled design if the following conditions are preserved:

- Separation efficiency in the arms of the separator
- Separator pressure drop
- Separation efficiency in the free drum
- Separated liquid height above the top of the separator arms

It is recognized that each of these processes is a function of complex, two-phase flow interactions. Relatively simple geometric ratios may therefore be inadequate for scaling all aspects of the separation fluid flow processes. However, it is postulated that certain physical mechanisms dominate the separation process and that certain global parameters can characterize these mechanisms. These parameters serve as a basis for developing scaling criteria and correspond to the following superficial velocities:

- Liquid and vapor in the riser
- Liquid and vapor in the arm
- Vapor in the free drum

The liquid and vapor superficial velocities in the riser and arm affect the separator pressure drop and separation efficiency in the arm. The liquid velocity in the arm will affect the pressure drop and liquid height formation, and the vapor velocity in the free drum will affect the separation efficiency in the return cylinder. Similar velocities will result in the scaled design if the following ratios are matched with the scaling reference design:

- Ratio of free drum area to return cylinder area
- Ratio of curved-arm area to return cylinder area (assumes that the same number of separator arms will be maintained)

The final criterion is based on maintaining separation efficiency in the arm. Assuming the velocities are the same, the separation efficiency in the arm should be preserved if the basic arm shape is the same. This will result if the ratio of the difference in the inside and outside arm radii and the arm arc length is maintained. All of the criteria are summarized in Table 1.

Table 1

CAP GEOMETRIC SCALING PARAMETERS AND SCALED GEOMETRY COMPARISONS

| Parameter | 13-1/2" CAP | 11-1/2" CAP | 21" CAP |
|---|---|---|---|
| **Scaling** | | | |
| Ratio of Riser to Return Cylinder Diameter | .5 | .5 | .5 |
| Arm Aspect Ratio | 7.2 | 7.2 | 7.2 |
| Ratio of Gap to Arm Width | .6 | .6 | .6 |
| Ratio of Gap to Return Cylinder | .056 | .056 | .057 |
| Angle of Incidence | 30° | 30° | 30° |
| Ratio of Free Drum Area to Return Cylinder Area | .6 | .6 | .6 |
| Ratio of Total Curved-Arm Outlet Area to Return Cylinder Area | .3 | .3 | .3 |
| Ratio of Arm Width to Average Arm Length | .265 | .265 | .265 |
| **Geometry** | | | |
| Return Cylinder Diameter, inches (cm) | 13.50 (34.3) | 11.50 (29.2) | 21.0 (53.3) |
| Riser Diameter, inches (cm) | 6.75 (17.1) | 5.75 (14.6) | 10.5 (26.7) |
| Lower Riser Diameter, inches (cm) | 5.41 (13.7) | 4.56 (11.6) | 8.44 (21.4) |
| Arm Height, inches (cm) | 9.00 (22.9) | 7.70 (19.6) | 14.0 (35.6) |
| Arm Width, inches (cm) | 1.25 (3.2) | 1.07 (2.7) | 1.94 (4.9) |
| Mean Arm Radius, inches (cm) | 2.125 (5.4) | 1.575 (4.0) | 3.11 (7.9) |
| Gap; Minimum Distance Between Arm and Return Cylinder, inches (cm) | 0.75 (1.91) | 0.64 (1.63) | 1.167 (3.0) |
| Included Angle Along Average Arm Radius | 127° | 147° | 135° |
| Free Drum Area, inches$^2$ (cm$^2$) | 82.6 (532.9) | 63 (406.4) | 210 (1354.8) |

Applying the scaling criteria begins with the definition of the return cylinder diameter. With the return cylinder diameter defined, all the other geometric parameters are obtained directly from the scaling criteria. The outside radius of the curved arm is assumed to be the distance between the riser and return cylinder. If necessary, the arc length of the arm intrudes into the riser to achieve the desired length. For this study, resulting dimensions of the scaled separators are listed in Table 1. Top views of the scaled CAP geometries are illustrated in Figure 3.

difference. The enclosure drum dimensions for each separator are also listed in Figure 4; all other dimensions shown in the figure were the same for each separator. The purpose of the enclosure drum was to simulate the typical area available to each separator when placed in a steam drum. The enclosure drum area was scaled in direct proportion to the return cylinder. The primary-secondary spacing corresponds to the distance previously tested with this type of primary-secondary combination -- 21 inches (53.3 cm) from the top of the curved arms of the primary separator to the bottom of the secondary separator.

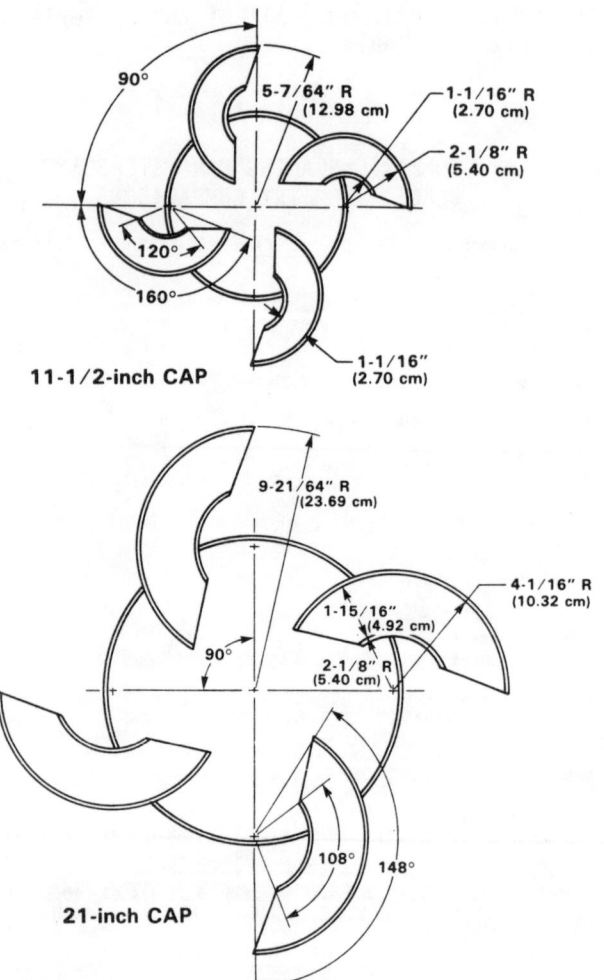

Figure 3. Geometrically scaled separator designs (return cylinders not shown).

## DESCRIPTION OF AIR/WATER TESTING

Each separator was installed for testing in the facility as shown in Figure 4. A single B&W secondary cyclone separator was used in series with the primary separator. For the 21-inch (53.3 cm) design, tests were conducted with and without a truncated cone (see Figure 4) extending from the enclosure drum atop the return cylinder to the inlet of the secondary separator. The cone's purpose was to direct flow toward the secondary separator, but tests with and without the cone showed no performance

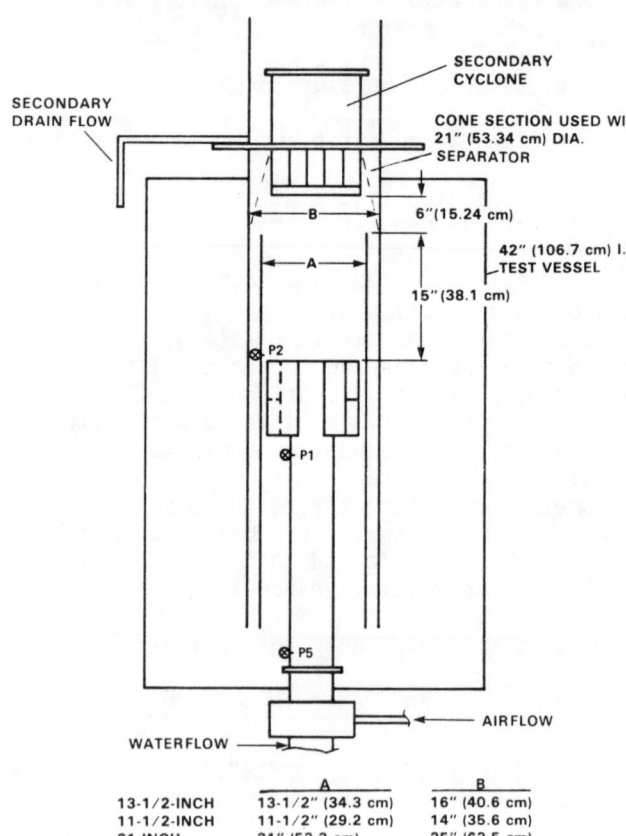

Figure 4. Installation of separators in air/water test facility.

Air and water were introduced into the test section through a mixer located at the bottom of the test vessel. The separators were tested at airflows ranging from 500 to 4500 scfm (0.24 to 2.12 $m^3$/s) and waterflows ranging from 110 to 600 gpm (6.94 to 37.85 l/s). Three repeat points were measured in each of the test sequences. The return cylinder area was used to convert the absolute airflow rates to rates based on unit separator area.

Instrumentation was included to measure the following quantities:

- Water carryover in the air from the primary separator (secondary separator drain flow)
- Primary separator pressure drop
- Riser pressure drop
- Liquid annulus height above the curved arms

Since carryover from the secondary separator discharge to atmosphere was negligible, it was assumed that the secondary drain flow was equal to the primary separator moisture carryover. This flow measurement was performed using an accumulating flow tank.

The liquid height was measured using marked graduations on the return and enclosure cylinders. The pressure drops were measured using pressure taps connected to manometers. The locations of the taps for the pressure drop measurements are shown in Figure 4.

DISCUSSION OF RESULTS

The performance of the separators was compared in three areas: water carryover, pressure drop, and height of the separated liquid annulus on the return cylinder above the separator arms. Comparisons of water carryover and pressure drop for the separators are shown for a waterflow rate of 150 gpm/ft$^2$ (102 l/s-m$^2$) in Figures 5 and 6, and for a waterflow rate of 250 gpm/ft$^2$ (170 l/s-m$^2$) in Figures 7 and 8. Comparisons are shown as a function of volumetric airflow per unit separator area (scfm/ft$^2$, m$^3$/s-m$^2$). For both waterflow rates, good agreement is observed when increasing size from 13-1/2 inches (34.3 cm) to 21 inches (53.3 cm). With the 11-1/2-inch (29.2 cm) separator, comparisons with the other two separators are in good agreement for some flow conditions, and in poor agreement at other conditions. The criteria were therefore successful in preserving consistent performance when scaling to a larger size, but not as successful in scaling to the smaller size.

Comparisons of separated liquid height above the top of the separator arms are shown for the two waterflow rates in Figures 9 and 10. Here, it is observed that when compared in terms of loading per unit area, the liquid height above the top of the arms increases as the separator size increases. It is hypothesized that the primary reason for this behavior is differences in liquid velocities between the geometries, which is related, in turn, to differences in void fraction. The void fraction and/or flow regimes in multiphase flows are known to be a strong function of the geometry as well as the flow conditions. Apparently the scaling criteria are not sufficient for preserving void fraction.

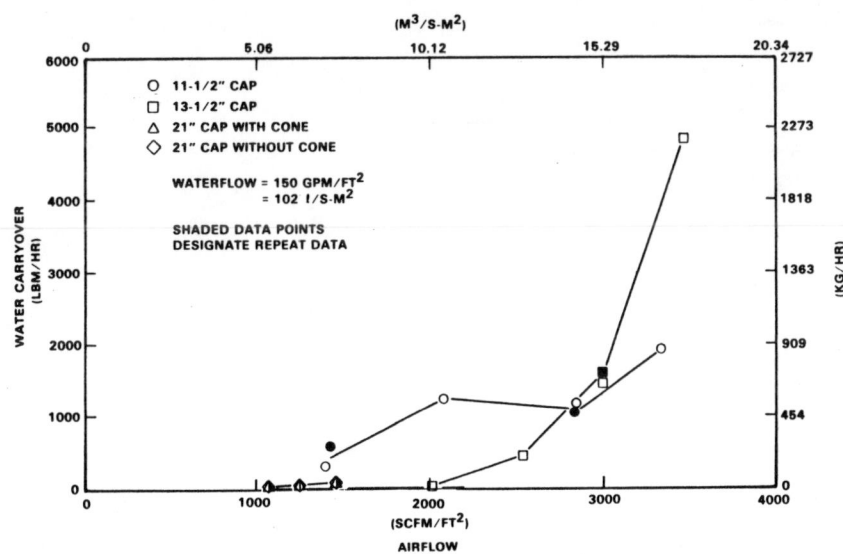

Figure 5. Water carryover comparisons.

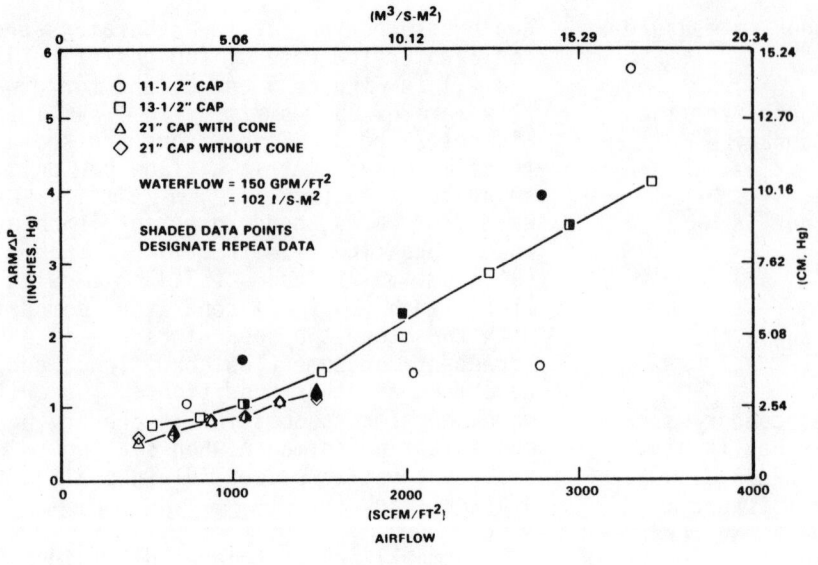

Figure 6. Pressure drop comparisons.

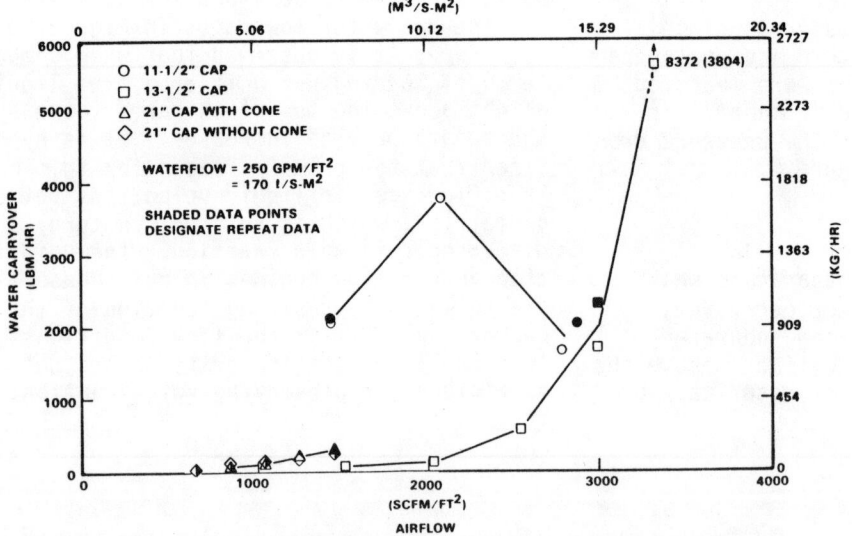

Figure 7. Water carryover comparisons.

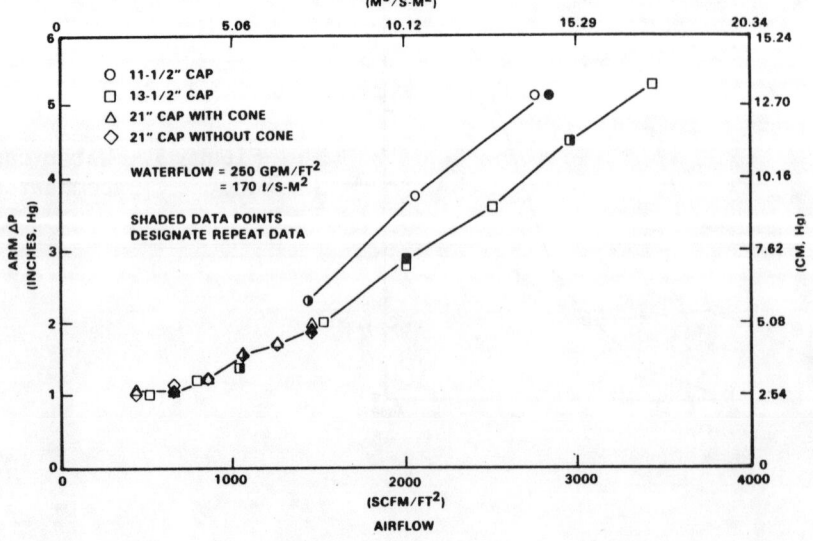

Figure 8. Pressure drop comparisons.

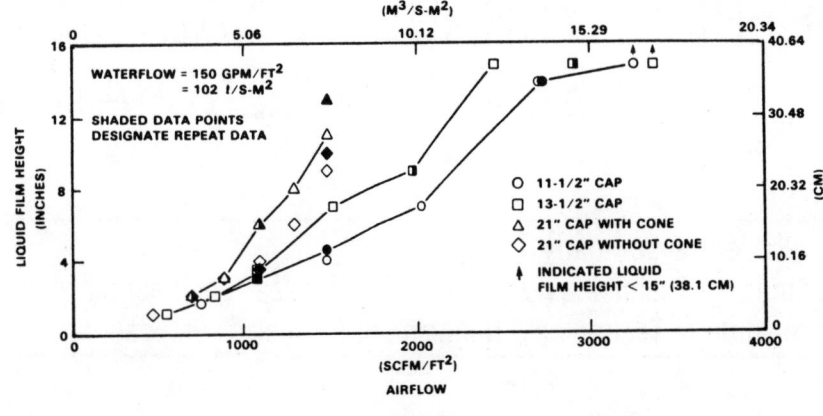

Figure 9. Liquid height comparisons.

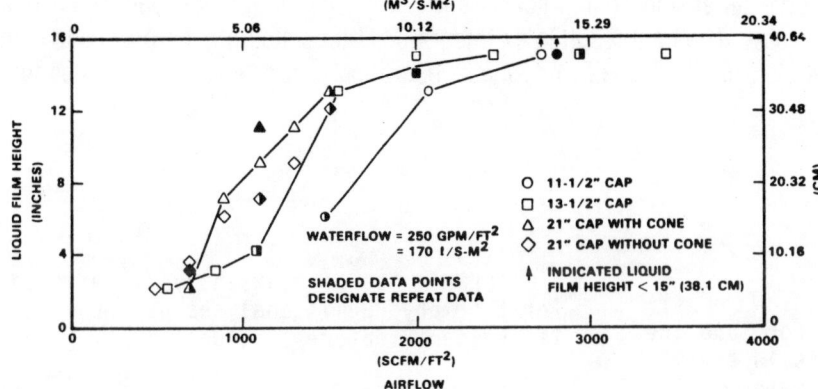

Figure 10. Liquid height comparisons.

## CONCLUSIONS

The criteria defined to arrive at the geometrically scaled separators were based on elementary assumptions about the dominate physical mechanisms governing the separation process. Despite the simplicity of the assumptions however, the good agreement between the large-scale separator and the original design leads to the conclusion that these assumptions were justified. These criteria have therefore been demonstrated as the best means presently available for geometric scaling of CAP separators if the goal of such scaling is to preserve the performance characteristics of the separator in terms of loading per unit of separator area. The success of these parameters suggests that similar criteria could also be developed for other types of separators where similar assumptions can be made about the mechanisms that dominate the separation process.

The success of the criteria represents only the first step in the development of a set of reliable geometric scaling criteria. The behavior of the liquid annulus above the separator arms and the inconsistent performance of the small-scale separator leads to the conclusion that there are still important mechanisms in the separation process that must be better understood before a fully adequate scaling method can be established. The results indicate that some important mechanisms were not adequately scaled with the defined criteria. An example of this is the hypothesis that the void fraction and corresponding fluid velocities were not adequately scaled, which led to the differences in liquid annulus height between the separators.

It is concluded that further studies of geometric scaling should be performed to better generalize the conclusions about geometric scaling beyond the specific results obtained in this study. It is also evident that further study of geometric scaling should include experiments with steam/water as well as air/water. While it is hypothesized that the scaling criteria are valid for steam/water as well as air/water, such a hypothesis must be demonstrated experimentally. This type of information is significant if separation equipment developed extensively with air/water testing is destined for use in steam/water systems.

# THERMIT-UTSG: A COMPUTER CODE FOR THERMOHYDRAULIC ANALYSIS OF U-TUBE STEAM GENERATORS

Hugo C. da Silva, Jr. ■ Yankee Atomic Electric Company
Neil E. Todreas ■ Massachusetts Institute of Technology, Department of Nuclear Engineering
David D. Lanning ■ Massachusetts Institute of Technology, Department of Nuclear Engineering

A benchmark computer code for U-tube steam generator thermohydraulic analysis has been developed. The starting point has been the two fluid, six equation, three dimensional computer code THERMIT, originally developed for the analysis of transients in light water reactor (LWR) cores. The final result is an integrated U-tube steam generator code with the capability of calculating (a) global parameters for both mild and severe operational transients, and (b) steady state parameter distributions.

## INTRODUCTION

The philosophy of, the need for, and the approaches to modeling transients in steam generator units for pressurized water reactor (PWR) systems have been previously reviewed (1). In particular, benchmark analytical representations of the steam generators can aid in applications of systems codes and in the development of fast codes (2) for operator assistance. At least four characteristics are desirable of such a benchmark model.

First, it should be capable of representing the entire unit for eventual interfacing with a system code. Second, it should be based on detailed physical models, supplemented by well-tested empirical correlations and utilize a reliable numerical method, while still allowing for the assessment of potentially simplifying assumptions. Third, it should be verified. Fourth, it should provide a basic framework for expansion to severe transient (accident) analyses. The present model exhibits these characteristics. The approach used in the present work has been to develop separate models for the steam dome, for the downcomer, evaporator and riser, and for the primary fluid and U-tubes, as well as to develop the procedures necessary to integrate the various regional models into the THERMIT (3) framework.

THERMIT has been extensively verified for thermohydraulic analyses of LWR cores. Its salient features include:

1. Thermal and mechanical non-equilibrium between phases described by a six equation model coupled by mechanistic constitutive relations.

2. Complete boiling curve at each cell with an elaborate heat transfer logic to select the appropriate heat transfer correlation.

3. Semi-implicit numerical technique particularly suitable for severe transients.

The following sections describe the developments required to integrate as well as to individually analyze the various U-tube steam generator (UTSG) regions. The effort to assess the final integrated model (THERMIT-UTSG) is also summarized.

## TWO-FLUID MODEL

The details of the two-fluid equations used in this work have been presented elsewhere (3).

The upper and lower slab-looking hatched areas in Figure 1 represent fictitious cells. These cells are used to define local boundary conditions for the two-fluid model. In the top fictitious cells, and in the steam dome the relevant quantities are calculated separately from the two-fluid model and updated prior to the initiation of each time step calculation. The separate procedure is called the recirculation model.

## RECIRCULATION MODEL

The regions of the steam generator for which the recirculation model is used are shown in Figure 2. They include the steam filled space and part of the liquid filled downcomer, down to slightly below the water level. The steam filled space excluding the separators (dotted region in Figure 2) is called the steam dome. The liquid filled space in the neighborhood of the downcomer water level ($V_2$ region in Figure 2) is called the upper downcomer. The position of the lower boundary of the upper downcomer will vary with changes in the water level.

The apparent complexity of utilizing two approaches, one for the evaporator, riser and downcomer (two-fluid model), and one for the steam dome and upper downcomer (recirculation model) is outweighed by two advantages.

First, unnecessary complex developments in the two-fluid equations in THERMIT can be circumvented. In order to model the steam dome in the two-fluid domain, mass sink and source terms would have to be added in the conservation equations to account for the liquid removed by and returning from the separators.

The second advantage is a reduction in computing. Since the velocity field in the steam dome is of no practical concern in the context of this work, the momentum equations in that region can be dropped. The steam separator momentum is accounted for in the form of a pressure drop given by a suitable empirical correlation ($\underline{4}$). Thus, a separate model for the steam dome saves considerable computing by eliminating THERMIT two-fluid model computational cells, while adding a set of equations which is smaller but sufficiently descriptive.

For each of the control volumes shown in Figure 2, $V_2$ and $V_1$, mass and energy conservation equations for liquid and vapor are written as follows:

$$\dot{M}_2 = w_{fd} + (1-x_r)w_r + w_{cond} - w_{fl} - w_d \quad (1)$$

$$\frac{d}{dt}(M_2 h_2) = w_{fd} h_{fd} + w_{cond} h_f - w_d h_2 - w_{fl} h_g$$
$$+ (1-x_r)w_r h_f + \dot{Q}_2 + (M_2/\rho_2)\dot{P} \quad (2)$$

$$\dot{M}_1 = w_r x_r - w_s - w_{cond} + w_{fl} \quad (3)$$

$$\frac{d}{dt}(M_1 h_1) = w_r x_r h_g + w_{fl} h_g - w_{cond} h_f$$
$$+ \dot{Q}_1 + (M_1/\rho_1)\dot{P} - w_s h_1 \quad (4)$$

Four state equations are written for enthalpy and density of liquid and vapor respectively as a function of pressure and temperature.

An additional equation is obtained by imposing the constraint that the total volume of the two regions is constant although they are separated by a moving liquid interface. Hence:

$$\frac{d}{dt}\left(\frac{M_1}{\rho_1} + \frac{M_2}{\rho_2}\right) = 0 \quad (5)$$

The basic model includes the four state equations plus Equations (1) to (5). The unknowns are thirteen: p (or $w_s$, if p is an unknown then $w_s$ must be prescribed and vice versa), $\dot{M}_1$, $\dot{M}_2$, $w_{cond}$, $w_{fl}$, $h_1$, $h_2$, $\rho_1$, $\rho_2$, $\dot{Q}_2$, $\dot{Q}_1$, $T_1$, and $T_2$.

The first step towards closure requires equations for the heat source/sink terms, $\dot{Q}_1$ and $\dot{Q}_2$. These are given by transient heat conduction relations formulated independently of the unknowns by assuming the wall boundary condition is the saturation temperature at the prevailing steam dome pressure.

The second step incorporates the assumption ($\underline{5}$) that neither phase can exist in non-stable form, i.e., the vapor can be saturated or superheated, but not subcooled; while the liquid can be saturated or subcooled, but not superheated. The assumption provides an extra set of constraints on four of the unknowns, viz.: $w_{fl}$, $w_{cond}$, $T_1$, and $T_2$.

Following closure, the nine recirculation model equations are solved iteratively for the remaining nine unknowns. The top two-fluid region boundary conditions must now be set to allow the next time step computation of the two-fluid model. The pressure, $p_d$, is prescribed in the topmost fictitious cell of the two-fluid model domain of each downcomer channel (Figure 1) such that the resulting pressure at the water level is the new time steam dome pressure, p.

The liquid temperature, $T_2$, resulting from the recirculation model is fed into all downcomer cells in the two-fluid domain located above the lower boundary of $V_2$ (Figure 2). The purpose of this is to obtain correct transport times for the liquid in $V_2$ down to the evaporator inlet.

## PRIMARY MODEL

The primary model domain includes primary coolant and tube metal, from inlet to outlet plena inclusive. The model computes the temperature of: primary fluid, primary-side wall, secondary-side wall and intermediate wall. These are calculated for each representative primary tube, in each secondary-side cell, in the tube bundle region. Each representative tube is called a tube bank. An example of an arrangement with three tube banks is given in Figure 3.

The energy equation for a tube bank is obtained by summing the energy equation for one tube over the total number of U-tubes in a tube bank. This allows tube-to-tube length differences to be accounted for in a global sense as described in (4). The primary- and secondary-side energy equations are linked through the heat flux at the outer wall of the U-tubes, which is taken explicitly in both equations. This allows primary- and secondary-side solution procedures to be carried out independently.

Commonly used resistance temperature detectors (RTD), can have non-negligible response times. Therefore, primary plena temperature predictions which are to be compared with measurements obtained from sensors with potentially large time constants, can be corrected in THERMIT-USG using a first order lag differential equation, where the time constant is an input.

## SYSTEM BOUNDARY CONDITIONS

Table I lists the set of system boundary conditions used as input in this work. These are the minimum number of input parameters needed for each type of calculation. Other available parameters are used for comparison with calculated results.

Although the power is chosen as input in steady-state, the feedwater flow rate could have been used equally as well since it is equal to the power divided by the saturated vapor to feedwater enthalpy difference.

For steady-state analysis at full power, it is desirable to match a given steam dome pressure. Thus, a fouling coefficient is calculated, since equating the primary enthalpy drop to the secondary heat output, for a fixed steam dome pressure leads to more equations than unknowns, unless the overall heat transfer coefficient is adjusted. At a power level other than full power, the steam dome pressure is calculated and the full power fouling coefficient is input.

## ASSESSMENT

In order to assess the integral model, both local parameter and global parameter perspectives are adopted.

### Local Parameter Assessment (Steady-State)

The local parameter assessment indicates that THERMIT-UTSG steady-state calculations at half and full power are in good agreement with measurements and the calculations of another computer model (6), as shown in Table II.

At full power, the fouling parameter was calculated to be one (Table II). Since the present model does not use an overall heat transfer coefficient, but, rather, the individual heat transfer mechanisms are represented at each computational cell, it is clear that the integrated effect of these calculations has implied an overall heat transfer coefficient consistent with the actual steam dome pressure, which was input.

At half power, the fouling parameter is the same as that calculated at 100 percent power. For this case, the steam dome pressure was calculated to be at its measured value. Since the full power fouling parameter is unity, this result validates the heat transfer package in steady-state at low powers.

Figure 4 shows primary fluid and tube wall temperature distributions at three elevations. Although the present model utilizes one channel for each hot- and cold-leg, respectively, there are three calculated primary temperatures in each channel by virtue of the tube bank method illustrated in Figure 3. Secondary-side axial temperature distributions are shown in Figure 5. Calculated values agree well with the measurements.

Global Parameters Assessment (Transient)

The global parameter perspective is adopted with the objective of assessing the overall steam generator model response to actual plant transients. Tests cover the range of input disturbances that could be expected during operational transients.

The measurements were performed during start up tests at the Arkansas Nuclear One power plant (7). There are two measured curves for the primary outlet temperature corresponding to data gathered for the two cold legs.

Comparisons of calculated and measured parameters are presented in Figure 6 for two tests: (a) Turbine Trip (TT) and (b) Loss of Primary Flow (LOF). Calculations reproduce well data trends and show good agreement in the magnitudes as well.

CONCLUSION

The assessment study shows that the present version of the THERMIT Code is capable of calculating (a) steady-state parameters with intermediate resolution and (b) global parameters for U-tube steam generator operational transients.

The present two-fluid model is also ideally suited for situations involving a higher degree of non-equilibrium, which are expected in severe non-operational transients, i.e., accident situations such as a steam line break or the rupture of (a) U-tube(s). However, assessment work in this area remains to be carried out.

ACKNOWLEDGEMENTS

The authors would like to thank Northeast Utilities Service Company and Yankee Atomic Electric Company for their financial contribution to this work. In addition, Hugo C. da Silva, Jr. gratefully acknowledges the fellowships from CAPES, an agency of the Brazilian Ministry of Education and from the Catholic University of Rio de Janeiro.

NOTATION

| | |
|---|---|
| h | Specific enthalpy |
| M | Mass |
| Ntb | Number of U-tubes in a tube bank |
| P | Pressure |
| Q | Heat |
| $\rho$ | Density |
| T | Temperature |
| t | Time |
| w | Mass flow rate |
| x | Flowing quality |

Subscripts

| | |
|---|---|
| 1 | Steam dome, dotted region in Figure 2 |
| 2 | Upper downcomer liquid region, Figure 2 |
| cond | Condensing |
| d | Downcomer |
| f | Saturated liquid |
| fd | Feedwater |
| fl | Flashing |
| g | Saturated vapor |
| r | Riser top and past the steam separators |
| s | Steam at the outlet nozzle |
| sep | Steam separators |

LITERATURE CITED

1. S. P. Karla, "Modeling Transients in PWR Steam Generator Units," Nuclear Safety, Vol. 25, No. 1, January-February 1984.

2. W. H. Strohmayer, "Dynamic Modeling of Vertical U-Tube Steam Generators For Operational Safety Systems," Ph.D. thesis, MIT, 1982.

3. W. H. Reed and H. B. Steward, "THERMIT: A Computer Program for Three-Dimensional Thermal-Hydraulic Analysis of Light Water Reactor Cores," MIT Report prepared for EPRI, 1978.

4. H. C. da Silva, "Thermohydraulic Analysis of U-Tube Steam Generators," Ph.D Thesis, MIT, 1984.

5. E. D. Moeck and H. W. Hinds, "A Mathematical Model of Steam-Drum Dynamics," AECL-5057, 1976.

6. G. W. Hopkins, A. Y. Lee, and O. J. Mender, "Thermal and Hydraulic Code Verification: ATHOS2 and Model Boiler No. 2 Data," EPRI NP-2887, 1983.

7. D. P. Siska, "NSSS Transient Tests at ANO-2," EPRI NP-1708, May 1981.

Table I. System boundary conditions.

| STEADY STATE | | TRANSIENT |
|---|---|---|
| 100% Power | Other Power Levels | |
| 1. Primary Inlet Temperature | 1. Primary Inlet Temperature | 1. Primary Inlet Temperature |
| 2. Average Primary Mass Flux | 2. Average Primary Mass Flux | 2. Average Primary Mass Flux |
| 3. Primary System Pressure | 3. Primary System Pressure | 3. Primary System Pressure |
| 4. Power Level | 4. Power Level | 4. --- |
| 5. Steam Dome Pressure | 5. --- | 5. Steam Dome Pressure (Outlet Nozzle Steam Flow) |
| 6. Downcomer Water Level | 6. Downcomer Water Level | 6. --- |
| 7. Feedwater Temperature | 7. Feedwater Temperature | 7. Feedwater Temperature |
| 8. --- | 8. --- | 8. Feedwater Flow Rate |
| 9. --- | 9. Fouling Coefficient | 9. Fouling Coefficient |

Table II. Comparison of calculated and measured parameters for MB-2.

| PARAMETER | EXPERIMENT (6) | ATHOS (6) | THERMIT-UTSG |
|---|---|---|---|
| Primary Outlet Temperature | 565.4 - 566.8 | NA | 566.0 |
| Circulation Ratio | 6.88 | 7.20 - 7.80 | 6.85 |
| Feedwater flow Rate (kg/s) | 1.69 | 1.69 | 1.69 |
| Riser Quality | 0.15 | 0.13 - 0.14 | 0.15 |
| Riser Void Fraction | 0.61 | 0.62 - 0.74 | 0.70 |
| Bundle Pressure Drop (KPa) | 38 | 37 - 39 | 38 |
| Separator Pressure Drop (KPa) | 19 - 20 | 17 - 22 | 19 |
| Overall Pressure Drop (KPa) | 71 | 64 - 67 | 70 |
| Steam Dome Pressure (MPa) | 7.24 | 7.24 | 7.24 |
| Primary Outlet Temperature | 566.2 - 566.8 | Not given | 566.6 |
| Circulation Ratio | 2.98 - 3.03 | 3.06 - 3.63 | 3.03 |
| Feedwater flow Rate (kg/s) | 3.7 | 3.7 | 3.7 |
| Riser Quality | 0.33 - 0.34 | 0.28 - 0.33 | 0.33 |
| Riser Void Fraction | 0.77 | 0.81 - 0.90 | 0.85 |
| Bundle Pressure Drop (KPa) | 32 - 33 | 28 - 36 | 28 |
| Separator Pressure Drop (KPa) | 33 - 34 | 25 - 35 | 34 |
| Overall Pressure Drop (KPa) | 72 - 74 | 63 - 73 | 69 |
| Fouling Parameter | --- | --- | 1.0 |

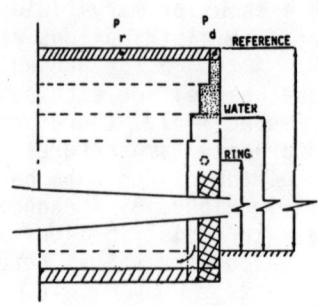

Figure 1. Two fluid model domain.

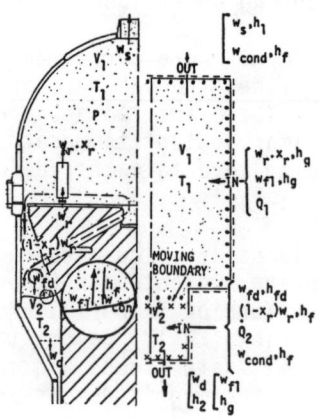

Figure 2. Recirculatioon model domain.

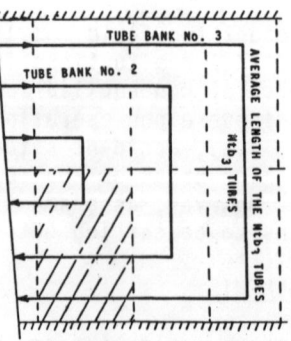

Figure 3. Primary size model domain showing three tube banks.

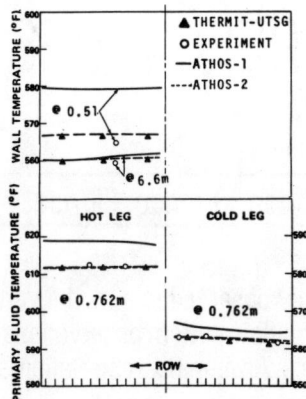

Figure 4. Half power tube wall temperatures at 0.51m and 6.6m above the tube sheet and full power primary side fluid temperatures at 0.762m.

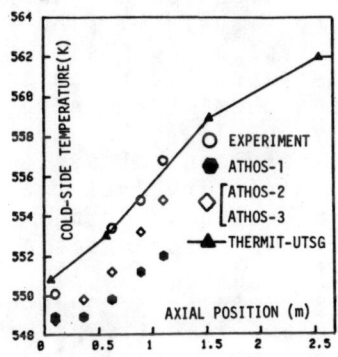

Figure 5. Secondary side axial temperature distribution in the cold leg at half power.

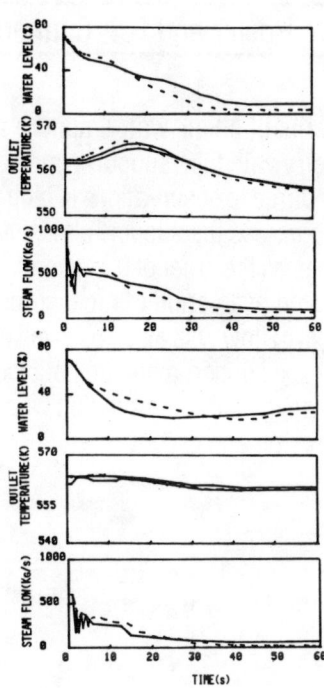

Figure 6. Comparison of calculated (—) and measured (——) parameters during (a) turbine trip (above) and (b) loss of primary flow (below) events at ANO-2.

# VISUAL OBSERVATIONS OF FLOW PATTERNS THROUGH TUBE SUPPORT PLATES OF CIRCULAR HOLE AND TREFOIL DESIGNS

G.W. Caille, E.R. Hosler and F.S. Gunnerson ■ University of Central Florida, Orlando, Florida

The flow patterns of an air-water mixture at atmospheric pressure were visually observed in a model vertical steam generator geometry with tube support plates of circular hole and trefoil designs. The flow pattern observations have been qualitatively compared to observations in high pressure steam-water heated systems. An empirical correlation is presented for pressure drop across the support plates.

Flow oscillations on the order of 0.4 seconds were observed in both support plates at similar mass flow combinations. The oscillation consisted of a period of rapid surging followed by a low flow or stagnation period on the order of 0.1 seconds. Reverse (downward) flow was also observed as part of the oscillation cycle in the trefoil support plate. The flow oscillations that exist in one tube support plate are coupled to the oscillation phase that exists in the preceding tube support plate. The operation of a steam generator in a region where this flow oscillation occurs could result in reduced steam generator life due to increased corrosion and dryout.

## INTRODUCTION

Current nuclear steam generator design requires the prediction of conventional engineering quantities for performance and material integrity evaluation. Historically little attention has been given to corrosion and the location where it may occur in the steam generator.

Recently this trend has been reversed due to corrosion causing potentially catastrophic steam generator failures which have resulted in costly repairs and decontamination. It is well documented that the flow patterns which result from the local geometry determine the pressure drop and heat transfer characteristics and these flow patterns also dictate the location and, to some extent, the type of corrosion that occurs.

In most commercial pressurized water reactor steam generators, the heat flux is less than the critical heat flux required to cause dryout or vapor blanketing (1). Vapor blanketing may still occur, however, as a result of internal structural components such as tube support plates, antivibration bars, and shroud supports causing vapor retention or prevention of rewetting of the tube surface. The mechanism or type of corrosion that occurs in the vapor blanketed area depends on the chemistry control system employed. There are two general chemistry control systems in us . The first involves the addition of a weak base or the salt of a weak acid which produces a basic pH enviroment. (The passivity of most metals is enhanced if the enviroment is basic provided it does not become excessively caustic.) In areas of vapor blanketing the pH agent concentrates and deposits on the tube surface resulting a very high pH and subsequent caustic stress corrosion (1). The second chemistry control involves the use of a volatile chemical which is boils off with the vapor. This results in a decrease in local pH near the vapor blanketed metal surface and will cause noticeable corrosion increases in less passive metals within the steam generator (usually structural components). It is only in the regions where vapor blanketing continually occurs which is a function of surface rewetting and local temperature rise that the corrosion effects are of significance, namely at some internal structural supports and possibly sludge deposits on heated surfaces.

The purpose of this study was to visually observe the flow patterns that exist through a trefoil (broached) shaped hole and a circular-hole tube support plate. The fluid is an air-water mixture at atmoshperic temperature and pressure with no heat additon. The air-water flow pattern was then compared with prototypical steam-water observations (2). The experimental apparatus is the same used by Yarizdeh (3) for which he developed an emprical pressure drop correlation across the circular tube support plate.

## EXPERIMENTAL APPARATUS

The test section consists of a large (57.15 mm inner diameter) transparent vertical Plexiglass tube with seven (12.7 mm diameter) internal solid vertical Plexiglass rods extending the length of the tube simulating the steam generator shell and tubes respectively. The rods are supported by 3 horizontal Plexiglass support plates that contain either the trefoil or circular hole orifices. Air and water enter the inlet mixing plenum at the bottom of the test section after being throttled and metered. The air-water mixture discharges to a drain system upon exiting the test section. The test section is fitted with eight individually valved pressure taps which are located 50.8, 152.4, 254, and 355.6 mm above and below the center support plate. The pressure taps also served as dye injection ports (Figure 1).

The support plate to be investigated was installed in the center of the test section. Each support plate contained seven equally spaced holes in a tri-angular pattern. The circular support plate holes were 13.46 mm in diameter with a hydraulic diameter of 0.76 mm with a total flow area of 0.76 mm$^2$ (3) (Figure 2). The trefoil support plate hole had a hydraulic diameter of 2.32 mm with a total flow area of 2.53 mm$^2$

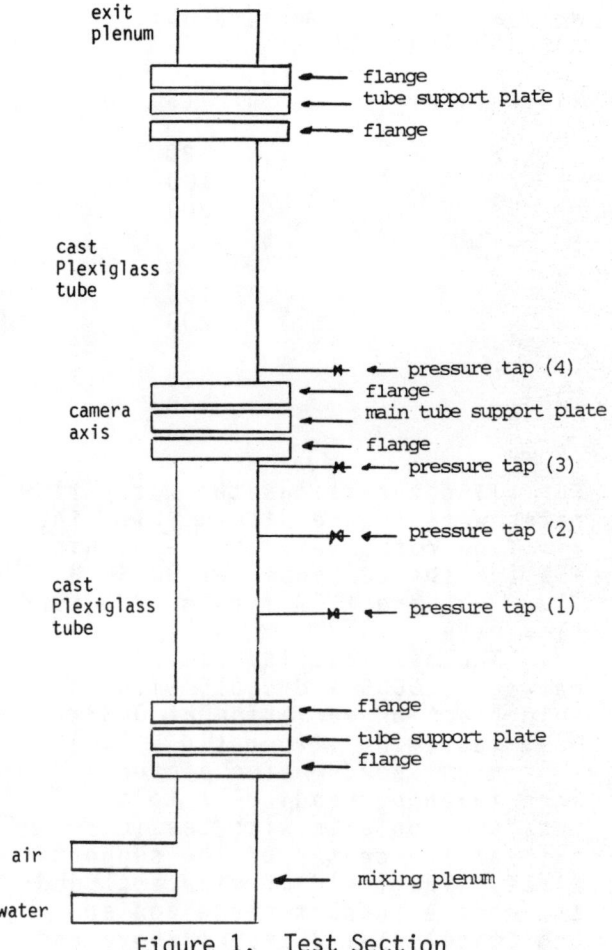

Figure 1. Test Section

(4) (Figure 3). Each support plate contained a flat vertical surface on the edge of the support plate which provided a normal surface for photography.

The flat vertical surface of the test support plate and internal holes were polished by hand to optical clarity using a powder toothpaste and water mixture. Small vertical and horizontal grooves which occurred in the holes from the fabrication process and were not removed. The grooves provided a surface irregularity but did not appear to affect the vapor bubbles.

## EXPERIMENTAL PROCEDURE

Moisture-free air and water at atmospheric pressure and ambient temperature were introduced into the inlet plenum of the flow tube where mixing occurred. Test runs were at the following flow combinations:

| Water (GPM) | Air (SCFH) |
|---|---|
| 2 | 20 |
|   | 100 |
|   | 200 |
| 3 | 20 |
|   | 100 |
|   | 200 |
| 5 | 20 |
|   | 100 |
|   | 200 |

For all observations the water flow rates were stable within ±5%. The air flow rates were stable within ±5% for the 20 and the 100 SCFH flow rate and ±10% for the 200 SCFH flow rate.

The mass quality varied between 0.0006 and 0.015 with a void fraction variation of 0.33 to 0.93 (assuming slip ratio = 1.0).

High speed motion pictures were taken perpendicular to the test section axis with the lens axis at the center of the support plate. The field of view included the entire support plate and approximately 1-1/2 inches above and below the support plate.

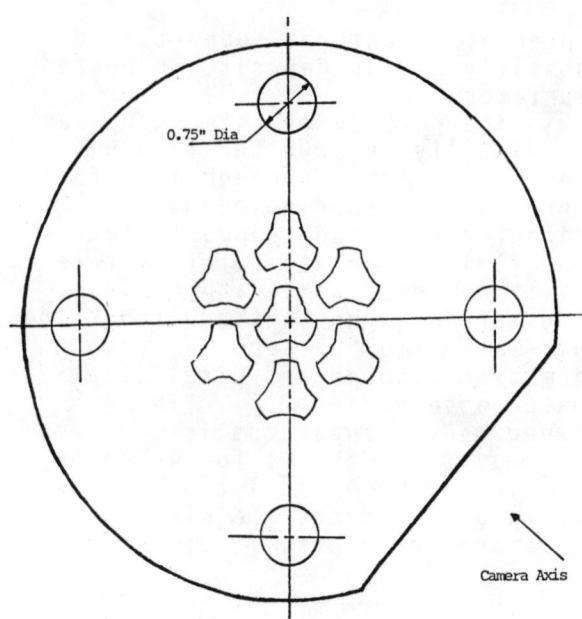

Figure 3. Side and Top Views of the Support Plate (Bashar 1983)

Additional pictures were taken scanning the distance between the lower and center support plates to determine the coupling of the oscillations. The framing rate was 1000 frames per second.

In determination of the empirical pressure drop correlation for the circular-hole support plate, differential pressure measurements were obtained using a diffe ential pressure transducer and digital voltmeter (3).

## FLOW PATTERN DESCRIPTIONS

The flow patterns were divided into two categories which are characterized by the presence or absence of oscillations in the main flow direction. The oscillations occurred at approximately the same mass flow rate combinations for the trefoil and circular hole support plates. There are no significant differences in the flow patterns between the trefoil and circular hole support plate prior to the occurrence of the oscillations. The primary difference in the oscillations is that reverse flow occurs as part of the oscillation cycle in the trefoil support plate. This reverse flow was not observed in the circular hole support plate.

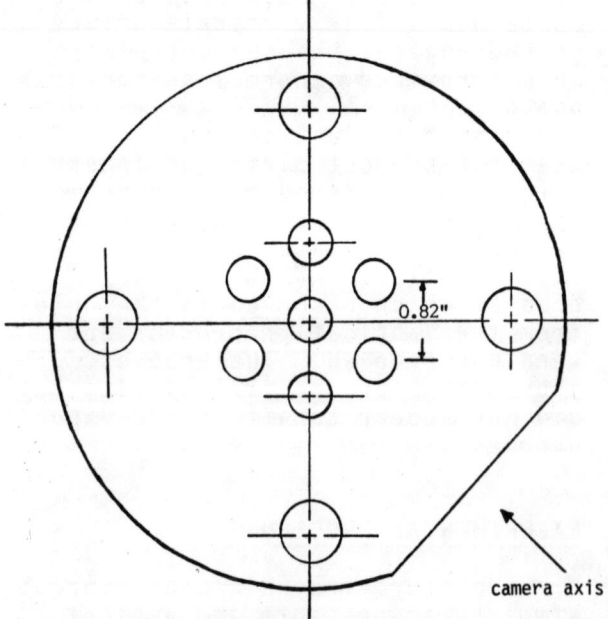

Figure 2. Top and Front View of the Circular-hole Tube Support Plate

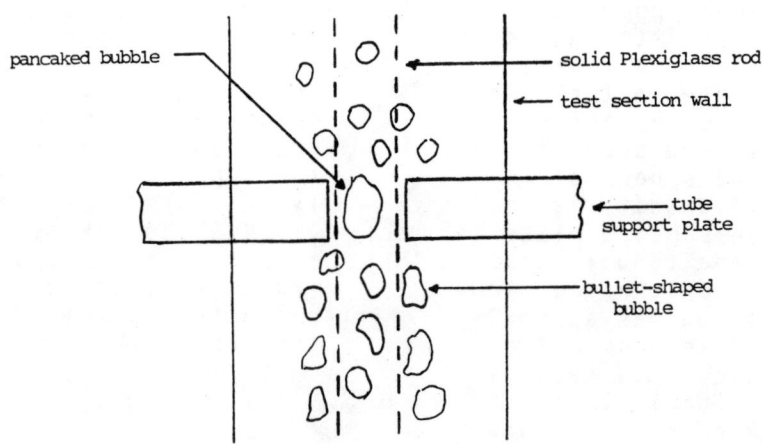

Figure 4. Bubbly Flow

The nonoscillatory flow patterns is characterized as bubbly flow (Figure 4). The bubbles are spherical and relatively uniform in size below the support plate. Bullet-shaped bubbles are interspersed at the low flow rate combination. The bubble velocity is approximately uniform and predominately upward. Some of the bubbles become smaller at the support plate due to collision with the support plate and other bubbles. As the bubble enters the support plate it becomes stretched and pancaked. The bullet-shaped bubbles become more circular as the pancaking occurs. The stretching effect is greater in the circular hole support plate resulting in the bubble covering a larger area. The velocity of the bubble through the plate was relatively uniform except near the top of the support plate where rapid acceleration occurs. Once the bubble exits the support plate, it was broken up by the chaotic flow. Above the support plate there was upward vertical flow along the tubes with slight downward flow as the radial distance from the tube increased. At the immediate top of the support plate, random horizontal flow was observed also.

The flow oscillations can be characterized as a surging effect. There exist two distinct flow patterns in the circular hole support plate. The first is a very rapid, violent upward flow of both water and air, called jetting. The second is an almost stagnant condition where the water velocity is assumed to be in the same direction and approximate magnitude as the bubble velocity. In the trefoil support plate, a third distinct pattern of reverse (downward) flow occurs immediately after stagnation.

In the circular hole support plate, the duration of the stagnation and jetting periods and the violence of the jetting is a function of the mass flow rate combination. In the 20SCFH/2GPM air-water combination, the jetting is not excessively rapid and lasts approximately 0.2 seconds. The fluid velocity then quickly decreases to a very small upward velocity (not complete stagnation). The duration of the low flow period is approximately 0.1 seconds. This periodicity appears to be constant within the accuracy of visual observation.

The surging propagates in a step-like fashion through each support plate in the flow tube. The surge of bubbles from the lower plate impact the center support plate which initiates the jetting through that support plate. the flow deaccelerates very rapidly to the stagnant condition when there is a substantial decrease in the bubble population. The bubble population below the plate starts to increase and slow upward flow begins. This slow upward flow continues until the surge from the lower plate causing a large increase in the bubble population below the plate and jetting occurs.

During the jetting phase, there is an increase in turbulence above and below the support plate. The bubbles above the support plate are very small and spherical in shape. The flow velocity is too rapid through the support plate to distinguish any individual bubbles.

The duration of the low flow or stagnation period was approximately 0.1 to 0.2 seconds. The jetting lasted about 0.3 seconds. As the air flow rate is increased (same water flow rate), the duration of the stagnant period decreases and the velocity, turbulence and duration of the jetting increases. In addition, the average bubble size decreases. As the water flow rate increases, the duration of the stagnant period is decreased and a higher air flow rate is required to obtain the same velocity of jetting that was observed under low water flow rates. In addition, the bubbles are slightly larger (for same air flow rates). For higher water flow rates, the sympatic effect of the surge from the bottom support plate is not as pronounced nor is there the large accumulation of bubbles below the plate prior to jetting. The absence of bubbles above and below the support plate following the jetting still occurs.

In the trefoil support plate, the oscillations are similar to those in the circular hole support plate with two additional phenomena: 1) there are velocity oscillations during the jetting flow period that were not present in circular hole jetting period; and 2) during the stagnation period of the oscillation there is reverse flow present.

The velocity oscillations during the jetting were random and the actual magnitudes were not measurable. They appeared to occur with greater frequency at the higher mass flow rates.

The reverse flow is initiated from below the support plate. Initially upward flow is still occuring in and above the support plate and downward flow below the support plate. Then a true stagnation of flow occurs where the bubble velocity, and presumably, water velocity are zero. The downward or reverse flow that is occurring below the support plate then cause reverse flow to occur in the support plate which subsequently initiates reverse flow above the plate. The reverse flow in the support plate stops when the surge from the lower support plate impacts the bottom of the center support plate (Figure 5). The

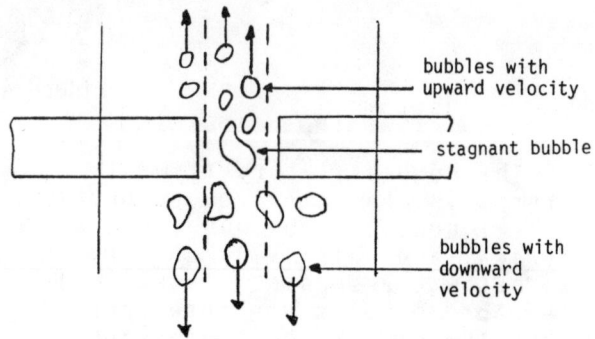

Reverse flow initiates from below support plate.

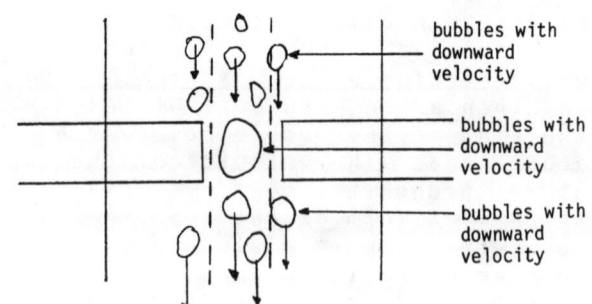

Reverse flow fully established.

Figure 5. Initiation of Reverse Flow

bubble surge was observed leaving the lower support plate and traveling upward as a group of bubbles. The group impacted the reverse flowing bubbles which were swept upward eventually through the center support plate (jetting). The concentration of bubbles above and below the group was markedly smaller.

The degree of backflow depends

primarily on the water flow rate. At 2 GPM water flow rate, the backflow lasts 0.1 to 0.2 seconds. As the water flow rate is increased to 5 GPM, the backflow lasts less than 0.1 seconds. During the stagnation and reverse flow phases, the bubble population above and below the support plate is very small.

In both support plate designs, the flow patterns and oscillations occurred uniformly across the flow tube. The randomness of the oscillations increased at higher mass flow rates.

Single-phase (water) observations were made using dye injected through pressure taps 1, 2, 3, 4 (Figure 1). The water flow rates were 2, 5 and 7 GPM. No surging or reverse flow was observed. Two-phase observations show increased mixing as expected and extensive reverse flow along the tube walls (taps 1, 2). Reduced reverse flow at taps 3 and 4 was also observed. No dye was observed to travel from above to below the support plate (dye injected at tap 4).

DISCUSSION

The test section simulates the tube support plate area that exists in many commericial steam generators. It is not intended to stimulate the entire steam generator since many of the complex feedback relations that exist in the steam generator, such as re-circulation, do not exist in the test section. The mass qualities and void fractions are similar to commercial steam generators. The mass flow rate of 360 pounds mass per hour per tube (100SCFH/5GPM) is comparable to commercial steam generators during normal operations.

Current research on two-phase flow oscillations can generally be divided into two catergories. The first deals with oscillations in heated tubes where fully developed flow occurs. The second deals with general flow observations in the steam generator where many complex oscillations modes and feedback reactions occur. The overriding concensus is that the actual system oscillations are usually complex combinations of fundamental oscillations described by Bergles (5). Additional research is required to close the gap between the two categories.

Since no heat is added to the test section, the oscillations attributed to the density and enthalpy changes can be eliminated. The density wave oscillation (instability) which has been observed in operating steam generators is also not applicable.

One possible explanation of the oscillations is a mechanism similar to the fundamental re-laxation instability. This instability is a result of the transition from bubbly flow with an associated large pressure drop to annular flow with a relatively smaller pressure drop. As the flow pattern shifts to annular flow, the flow rate increases rapidly. The flow rate is not sufficient to maintain the annular flow pattern and it reverts back to bubbly flow and a lower flow rate. This instability assumes a fully developed flow pattern. Bergles (6) showed that the fundamental relaxation instability occurs in a quality range of 0.0 to 0.002 for an air-water mixture flowing in a tube at low pressure (less than 100psia). He proposed that increased agitation caused by an increased heat flux or restrictions would slightly reduce the transition quality. Annular flow was not detectable from the photographs and the fundamental relaxation instability offers no explanation of the stagnant or reverse flow phases but a flow pattern transition during jetting remains a possibility.

Wallis and Heasley (7) describe how a traveling vapor plug can cause a decrease in system outlet pressure which initiates a surge. The surge is stopped when the exit quality is reduced and exit pressure increases. This surge (7), although derived for a natural circulation system, may offer a partial explanation to the reverse flow portion of the observed oscillation. Immediately after a jetting phase has occurred in the center support plate, there is a local quality increase above the bottom support plate. This may result in a locally lower pressure at the bottom support plate causing reverse flow. The vapor plug hypothesis is further supported by the observation that the vapor bubbles travel up the test section as a group.

Dryout or vapor blanketing may occur during the periods of stagnation and reverse flow. The dryout would have serious corrosion consequences since it would be continually occurring in the same location namely the support plate.

Smith and Armstrong's analysis (1) shows that the land area of a broached (trefoil or quartrafoil) support plate could become vapor blanketed over its entire width (analysis assumes constant flow with heat addition). In addition, heat circular hole support plate was demonstrated to have extensive vapor blanketing. Several methods of eliminating the degree of vapor blanketing were pointed out:

1. Limit the contact region between support plate and tubes.
2. Cause abrupt contour changes using flat lands vice curved lands in the broached design.
3. Use hourglass-shaped lands.
4. Vary the gap width in flow direction.

This analysis demonstrates inherent vapor blanketing caused by the hole design and when coupled to the revese flow and stagnation periods, corrosion could be substantially increased. High speed motion pictures edited by Hosler (2) of an operating steam genertor substantiate the extensive vapor blankeing in the tube support plate region for the trefoil design support plate.

Yarizadeh (3) using standard curve fitting techniques for pressure measurements taken along the test section developed the following empirical relation for the pressure drop across a circular hole support plate:

$$\Delta P = \frac{(K_{TP})_{TSP} \, G^2}{2 \, g_c \, \bar{\rho}}$$

where

$(K_{TP})_{TSP} = 1.779 + 2.754 \exp(B*X)$

$B = [-2.607 - 0.424 \, (GPM)^{-1} *$ water flow rate $(GPM)] * 100$

The correlation is only applicable to this apparatus and an air-water mixture. In a report on actual steam generator operating conditions (8), the differential pressure behavior showed similar trends to the air-water differential pressure. For the same quality, the air-water differential pressure is predicted to be higher than the steam-water mixture. This suggests the oscillations may occur at higher mass flow rates than the air-water mixture if the oscillations are related to the pressure drop. If the oscillations are related only to the mass flow combination, the steam-water oscillations should occur at approximately the same flow rates as the air-water combination.

As heat is added along the tube length, the quality increases vertcally which should result in increased oscillations. Due to the higher temperatures of an operating steam generator, the corrosion effects will be significantly accelerated implying less vapor blanketing required for damage to occur.

## CONCLUSIONS AND RECOMMENDATIONS

In summary, the observed oscillations do not appear to be described entirely by any of the single classical flow oscillation mechanisms. They are related to the void fraction and are present in both tube support plate geometries. As suggested by Smith and Armstrong (1) and Hosler (2), potential accelerated corrosion can occur in the support plate area due to vapor blanketing and less passive metal, and is in fact found in operating steam generators.

This investigation has raised many questions concerning these oscillations that need to be addressed in future research:

1. How the water velocity corresponds to the bubble velocity.
2. Real time pressure measurements during the oscillation.
3. Heat addition effects on the oscillations.
4. Are the oscillations present in the same form in steam-water mixtures and what is the pressure-temperature dependence.
5. Oscillation dependence on hole size/hydraulic diameter.

With these questions answered, it may be possible to design a steam generator that avoids the oscillation region and a prototype model tested at low temperature and pressure with air-water at reduced cost.

## NOTATION

$G$     Mass flux through support plate crevices, lbm/s-ft$^2$

$g_c$    Conversion factor in Newton's law, 32.2 lbm-ft/lbf-s$^2$

$\Delta P$   Pressure drop, PSI

$X$     Mass quality (mass flow rate of air/total mass flow rate), dimensionless

$\bar{\rho}$     Mean density of the two-phase mixture, lbm/ft$^3$

## Subscripts

TP    Two-phase

TSP   Tube support plate

## LITERATURE CITED

1. Smith, A.C. and Armstrong, R.C. "Effects of Geometry on Thermal and Hydraulic Conditions in Steam Generator Tube Support Crevices." August 1983. Manuscript submitted for publication.

2. Hosler, E.R. "Crevicle Boiling Studies in Model Vertical Tube Steam Generators" motion picture, University of Central Florida, 1982.

3. Yarizadeh, F. "Pressure Drop Across A Restriction of Annular Geometry." M.S. Thesis, University of Central Florida, 1982.

4. Bashar, R.H. "Pressure Drop Across A Tube Support Plate of Trefoil Geometry Used in Steam Generator." M.S. Research Report, University of Central Florida, 1983.

5. Bergles, A.E.; Collioer, J.G.; Delhaye, J.M.; Hewitt, G.F.; and Mayinger, F. <u>Two-Phase Flow and Heat Transfer in the Power and Process Industry</u>. Washington: Hemisphere, 1981.

6. Bergles, A.E.; Lopina, R.F.; and Fiori, M.P. "Critical-Heat-Flux and Flow Pattern Observations for Low Pressure Water Flowing in Tubes." <u>ASME Transactions - Journal of Heat Transfer</u> 89 (February 1967): 69-74.

7. Wallis, G.B. and Heasley, J.H. "Oscillations in Two-Phase Flow Systems." <u>ASME Transactions - Journal of Heat Transfer</u> 83 (August 1961): 363-369.

8. Cassell, D.S. and Vroom, D.W. "Thermal-Hydraulic Tests of Steam Generator Tube Support Plate Crevices." EPRI-NP-2838, January 1983.

# MIXED LAMINAR CONVECTION IN TROMBE WALL CHANNEL

G.C. Huang and S.K. Chaturvedi ■ Mechanical Engineering and Mechanics
Old Dominion University, Norfolk, VA 23508

The two-dimensional, steady, combined forced and free convection in a vertical channel is investigated for the laminar regime. The vertical walls of the channel are maintained at constant but different temperatures while the horizontal walls are insulated. A finite difference method using up-wind differencing for the non-linear convective terms, and central differencing for the second order derivatives is employed to solve the governing differential equations for the mass, momentum and energy balances. The solution is obtained for stream function, vorticity and temperature as the dependent variables by an iterative technique known as successive substitution with over relaxation.

The flow and temperature patterns in the channel are obtained for air for Reynolds number and Grashof number ranging from 15 to 200 and 10,000 to 1,000,000, respectively. Both local and overall heat transfer coefficients are computed for the channel aspect ratio varying from 5 to 15. For a given value of Grashof number, as Reynolds number is increased, the flow patterns in the vertical channel exhibit a change from natural convection like flow patterns in which a large recirculating region is formed in the vertical part of the channel, to a forced flow type pattern. At low Reynolds number, the stream function and isotherms are qualitatively similar to those reported for the natural convection in rectangular slots.

## INTRODUCTION AND PROBLEM STATEMENT

Rapidly growing acceptance of solar energy as means of heating and cooling has stimulated research in the area of thermo-gravitational flows in open-ended cavities and parallel wall channel configurations. A particular configuration analyzed extensively in recent years is the Trombe wall channel (1) shown in Figure 1. This system is used for collecting and transporting incident solar energy to the conditioned space. The present work analyzes the fluid-dynamical and heat transfer characteristics of the mixed convective flow in the passage shown in Figure 1.

Convection caused by the heating of a vertical wall in a parallel channel configuration has been the subject of several previous investigations (2-9). A literature search identifying research work concerning the convective flow in a vertical channel with particular reference to the Trombe wall converged to the following papers.

Akbari and Borgers (10) have considered the laminar free convection flow in a parallel channel Trombe wall configuration, and have solved the fully elliptic equations by finite difference technique. Chung and Thompson (11) and Ramakrishnan (12) have solved a similar problem by using the finite element technique. The above-mentioned studies have neglected the coupling between the solid wall and the channel flow. Furthermore, the neglect of sharp corners due to reentrant geometry also precludes the formation of separation bubbles at the sharp corners.

From the review of existing literature, it is evident that little information exists regarding the fluid dynamical and heat transfer characteristics of Trombe wall channel operating in the mixed convection mode. To supplement the existing literature, it is proposed here to investigate the steady, two-dimensional, combined forced and free convection in a vertical channel. To simulate the Trombe wall channel geometry properly, the horizontal inlet and exit sections have been added to the vertical channel. The main thrust of this study is to analyze the fundamental nature of the fluid dynamical and heat transfer mechanisms. The analysis of the problem involves the solution of partial differential equations governing the transport of momentum, energy and mass. A finite difference method using the up-wind differencing for the non-linear convective terms and the central differencing for the second order derivatives is employed. The details of problem formulation and numerical methodology are presented in the next section.

# MATHEMATICAL FORMULATION AND NUMERICAL SOLUTION PROCEDURE

## Governing Equations

The physical model considered in this study is illustrated in Figure 1. The interior vertical wall, FG, is maintained at a constant temperature $T_H$, whereas the exterior wall, BC, is kept at a lower temperature $T_o$. The fluid entering from the lower horizontal channel is motivated by the heated wall, and an external mechnanical device, thus resulting in mixed convection mode. The flow is assumed to be steady and two-dimensional. The non-dimensional governing equations, subject to the Boussinesq approximation, can be written in the conservative form as

$$\frac{\partial}{\partial x}\left(\frac{1}{Pr\,Re}\frac{\partial \theta}{\partial x} - u\theta\right)$$
$$+ \frac{\partial}{\partial y}\left(\frac{1}{Pr\,Re}\frac{\partial \theta}{\partial y} - v\theta\right) = 0 \qquad (1)$$

$$\frac{\partial}{\partial x}\left(\frac{1}{Re}\frac{\partial \omega}{\partial x} - u\omega\right) + \frac{\partial}{\partial y}\left(\frac{1}{Re}\frac{\partial \omega}{\partial y} - v\omega\right)$$
$$= -\frac{Gr}{Re^2}\frac{\partial \theta}{\partial x} \qquad (2)$$

$$\frac{\partial^2 \psi}{\partial x^2} + \frac{\partial^2 \psi}{\partial y^2} = -\omega \qquad (3)$$

where $\psi$ and $\omega$ are non-dimensional stream function and vorticity respectively. The other symbols in above equations are defined as follows.

$Re$ = Reynolds number = $\frac{\rho V d}{\mu}$

$Pr$ = Prandtl number = $\frac{\mu c_p}{k}$

$Gr$ = Grashof number = $\frac{g\beta d^3 (T_H - T_o)}{\nu^2}$

$\theta = \frac{T - T_o}{T_H - T_o}$

where d is the channel width and V is the characteristic velocity.

## Finite Difference Scheme

The governing equations, Equations (1)-(3) are solved by a finite difference scheme which utilizes central differencing for the second order derivative and upwind or one-sided differencing for the non-linear first order convective terms. The role of upwind differencing procedure in stabilizing the numerical scheme has been well documented. The application of this scheme to free convection flow at a high Grashof number and forced convection flow at a high Reynolds number are discussed in references (13) and (14) respectively.

Following the procedure in reference (13), the governing finite difference equation for $\omega$, $\theta$ and $\psi$ can be written in the standard five point formula form. These finite difference equations subject to appropriate boundary conditions, are solved by an iterative method known as successive substitution. If $\theta^s$, $\omega^s$ and $\psi^s$ denote functional values at the end of $s^{th}$ iteration, the values of $\theta$, $\omega$ and $\psi$ at $(s+1)^{th}$ iteration level are calculated from the following expressions,

$$\theta^{s+1}_{i,j} = (1-F_\theta)\,\theta^s_{i,j} + \frac{F_\theta}{A_\theta}(a_\theta \theta^s_{i+1,j} + b_\theta \theta^{s+1}_{i-1,j}$$
$$+ c_\theta \theta^s_{i,j+1} + d_\theta \theta^{s+1}_{i,j-1}) \qquad (4)$$

$$\omega^{s+1}_{i,j} = (1-F_\omega)\,\omega^s_{i,j} + \frac{F_\omega}{A_\omega}[a_\omega \omega^s_{i+1,j}$$
$$+ b_\omega \omega^{s+1}_{i-1,j} + c_\omega \omega^s_{i,j} + d_\omega \omega^{s+1}_{i,j-1}$$
$$+ \frac{0.5 Gr}{Re^2}(\theta^{s+1}_{i+1,j} - \theta^{s+1}_{i-1,j})] \qquad (5)$$

$$\psi^{s+1}_{i,j} = (1-F_\psi)\,\psi^s_{i,j} + \frac{F_\psi}{4}(\psi^s_{i+1,j} + \psi^{s+1}_{i-1,j}$$
$$+ \psi^s_{i,j+1} + \psi^{s+1}_{i,j-1} + h^2 \omega^{s+1}_{i,j}) \qquad (6)$$

where h is the step size and $F_\theta$, $F_\omega$ and $F_\psi$ are over relaxation parameters which depend on the mesh size and fluid mechanical parameters (13). Expressions for $A_\theta$, $a_\theta$

etc. are given in reference (15).

A computer code was written to determine the flow pattern in the channel. The computational sequence is arranged such that, starting from some arbitrarily assigned initial values, first $\theta_{i,j}^{s+1}$ is computed for the $(s+1)^{th}$ iteration level. The known temperatures at points $(i+1,j)$ and $(i-1,j)$ are then substituted in the finite difference analog of vorticity equation to estimate $\omega_{i,j}^{s+1}$. Finally, the stream function $\psi_{i,j}^{s+1}$ is computed from Equation (6).

### Boundary Conditions

The boundary conditions for temperature, stream function and vorticity are illustrated in Figure 1. The value of stream function on the inner wall in forced or mixed convective flow depends on the velocity distribution specified at the inlet of the channel. In the present study, a parabolic entrance velocity profile is chosen arbitrarily. From this velocity distribution, the stream function and vorticity are obtained in terms of location in the y-direction at the entrance. At the upper exit section of the channel, the zero first derivative boundary conditions of $\theta$ and $\omega$, and vanishing second derivative boundary conditions of $\psi$ are imposed. The dimensionless temperature boundary values of 0 and 1.0 are imposed on the outer and inner vertical walls respectively. All the horizontal walls are kept insulated, and a zero value of nondimensional fluid temperature is also specified at the inlet of the channel. The boundary values of vorticity on the wall have been evaluated by using the Wood's method (13).

### DISCUSSION OF RESULTS

The stability and convergence of solution, and the conservation of energy and mass on a global basis are taken as indications of the success of numerical scheme used in the present study. For most cases, setting the values of $F_\theta$, $F_\psi$ and $F_\omega$ equal to 1.0, 1.6 and 0.6 respectively yielded converged solutions in about 250 iterations. The selection of step size, in general, requires a compromise between the requirement of higher solution accuracy and the low computational cost. A comparison of streamline pattern of air in the channel as determined by the present numerical scheme for the grid sizes of 0.0125 and 0.0083 (corresponding to 25x81 and 37x121 mesh respectively) is shown in Figure 2. The streamline contours indicate only a minor difference between the two step size cases. For most cases, the deviation between the mass flow rate at the entrance and those computed at various sections in the channel is within 3%. Similarly, the deviation from energy balance for the selected global control volume is within 7% for the 25x81 grid.

### Streamline Contour Maps

Figure 3 compares the streamline patterns of air in a channel with aspect ratio of 5.0 (where aspect ratio is defined as the ratio of vertical channel height to its width) for Grashof number of $3.34 \times 10^4$, and Reynolds number varying in the range of 15 to 150. For the lowest Reynolds number case, for which the effect of free convection prevails over the effect of forced convection, the streamline pattern looks more like that encountered in pure free convection flow in rectangular slot (3). It is important to note that due to strong buoyancy effects at the inner wall, there is no flow separation at the lower inner corner. Forced convection effects increase with increasing Reynolds number, and the recirculation flow pattern in the middle of the channel disappears gradually. As the Reynolds number is further increased, the closed eddies are finally "swept away" by the rapidly moving stream. Furthermore, due to increased inertia effect, the flow in the lower horizontal channel is unable to negotiate the bend and it separates at the lower inner corner forming a separation bubble on the hot vertical surface. The reattachment length of this separation bubble is found to increase with increasing Reynolds number. Also, it is noted in Figure 3 that a dividing streamline ($\psi=0$) exists in the vertical channel which divides the recirculating region from the forced flow coming from the inlet. At low Reynolds number, the flow separation becomes stronger at the top inner corner due to strong fluid lift off near the hot surface as a result of stronger buoyancy force.

Figure 4 shows the streamline patterns of air for an aspect ratio of 15.0, Re = 100 and Gr ranging from $10^4$ to $10^6$. One can observe that as Gr is increased, the effect is similar to keeping Gr fixed and reducing Reynolds number (see Figure 3 for Ar = 5.0). For the highest Gr value, the flow exhibits multi-cellular structure, similar to the one reported in reference (3).

### Isotherm Contour Maps

Figure 5 indicates the effects of variation of Re on the temperature contour maps of air. The Grashof number for these cases is $3.34 \times 10^4$. Temperature contour lines are always perpendicular to the horizontal walls due to the specification of an insulated wall boundary condition. At low Reynolds number, the isotherms are found to be inclined towards the cooler left wall. As Re is increased, the inclination shifts towards the right side of the upper region in the vertical channel. This illustrates the tendency of forced convection to dominate the buoyancy force as the inlet fluid velocity increases. Also, note that at low Reynolds number, there is some upstream influence, indicated by the warming of fluid in the lower horizontal channel even before it enters the vertical channel. As Re increases, this upstream influence decays since convection dominates the thermal diffusion of heat in the upstream direction.

Figure 6 shows the isotherms for Ar=15, Re=100 and Gr ranging from $10^4$ to $10^6$. At Gr=$10^4$ and $10^5$, the isotherm patterns are qualitatively similar to those seen in Figure 5. At moderately high Gr, e.g. $10^5$, the contour lines show nearly uniform temperature in the core region of the vertical channel and a steep temperature gradient near walls. The temperature contour maps for low Re and moderately high Gr are also qualitative similar to the contour maps for free convection in vertical slots (3). At Gr=$10^6$, the isotherm patterns show a multicellular structure.

### Heat Transfer Coefficient

In this section, the dependence of Nusselt number on Reynolds number and Grashof number is discussed. For the sake of clarity, the average Nusselt numbers on the hot face and the cold face are referred by the symbols $Nu_h (\equiv h_h d/k)$ and $Nu_c (\equiv h_c d/k)$ respectively. Note that d is the channel width. Figure 7 shows the variation of $Nu_h$ and $Nu_c$ for aspect ratio of 5 and 10 and Gr = 10,000. As Reynolds number is increased, $Nu_h$ increases somewhat gradually while $Nu_c$ decreases sharply due change in flow regime. Since $Nu_c$ is calculated from the heat transferred from the fluid to the opposite wall, it represents heat loss from the system. Thus, increasing the value of Re implies that larger net energy is delivered to the conditioned space. Although the effect of aspect ratio is relatively small for Gr=$10^4$, its effect is somewhat more pronounced at Gr=$10^5$ (15).

The curve marked A in Figure 7 gives the Nusselt number ($Nu_h$) as calculated from the boundary layer approximation for forced flow on the hot face for an aspect ratio of 10. Since difference between the curves marked A (boundary layer approximation in forced flow) and B (the present case) is small, one may conclude that the forced flow limit has been reached at Re=100 Numerical experiments in the current study suggests that this occurs at $Gr/Re^2 \approx 1.0$.

### CONCLUSIONS

The prominent conclusions of this study can be summarized as follows.

1. As shown in previous studies, the ratio $Gr/Re^2$ is an important parameter for estimating the effect of buoyancy force for mixed convection problems in Trombe wall channels. For $Gr/Re^2 > 1.0$, the buoyancy force effects can be neglected. For these limiting cases, the Nusselt number on the hot surface as calculated from forced flow boundary layer approximation is in good accord with the results obtained in this study by the numerical technique.

2. For a given value of Grashof number, as Reynolds number is increased, the flow patterns in the vertical channel exhibit a change from natural convection like flow patterns with large recirculating region, to a forced flow type pattern. The net energy picked up from the channel increases rapidly due to decreased heat losses at the cold face.

3. At low Reynolds number, the stream function and isotherms are qualitatively similar to those reported for the natural convection in rectangular slots.

## LITERATURE CITED

1. Robert, J. F., Peube, J. L., and F. Trombe, *Energy Conservation in Heating and Cooling and Ventilating Buildings*, Hemisphere Publishing Corp., Washington (1978).

2. Sparrow, E. M. and J. L. Gregg, *Journal of Applied Mechanics*, 26, 133 (1959).

3. Elder, J. W., *Journal of Fluid Mechanics*, 72 (1), 87 (1965).

4. de Vahl Davis, G. and G. D. Mallinson, *Journal of Fluid Mechanics*, 72 (1), 87 (1965).

5. Miyatake, O. and T. Fujii, *Heat Transfer Japanese Research*, (Tokyo), 1, 30 (1972).

6. Miyatake, O., Fuji, T. Fuji, M. and H. Tanaka, "*Heat Transfer Japanese Research*, (Tokyo), 2, 25 (1973).

7. Bodoia, J. R. and J. F. Osterle, *Journal of Heat Transfer*, 84, 40 (1962).

8. Szewczyk, A. A., *Journal of Heat Transfer*, 86, 501 (1964).

9. Zeldin, B. and F. W. Schmidt, *Journal of Heat Transfer*, 94, 211 (1972).

10. Akbari, H. and T. R. Borgers, "Free Convection Laminar Flow Within the Trombe Wall," Technical Report Submitted to the U. S. Department of Energy (1978).

11. Chung, K. S. and D. H. Thompson, "A Calculational Method for Combined Natural Convection and Forced Circulational Flow in a Channel," ASME Heat Transfer Conference, Orlando (1980).

12. Ramakrishnan R., "Finite Element Analysis of the Trombe Wall Channel," M.S. Thesis, Florida Institute of Technology, (1982).

13. Nogotov. E. F. *Application of Numerical Heat Transfer*, Hemisphere Publishing Corp., Washington (1977).

14. Gosman, D. *Heat and Mass Transfer in Recirculating Flows*, Academic Press, New York (1969).

15. Huang, G. C., "Combined Forced and Free Convection in the Trombe Wall Channel," M.S. Thesis, Old Dominion University (1985).

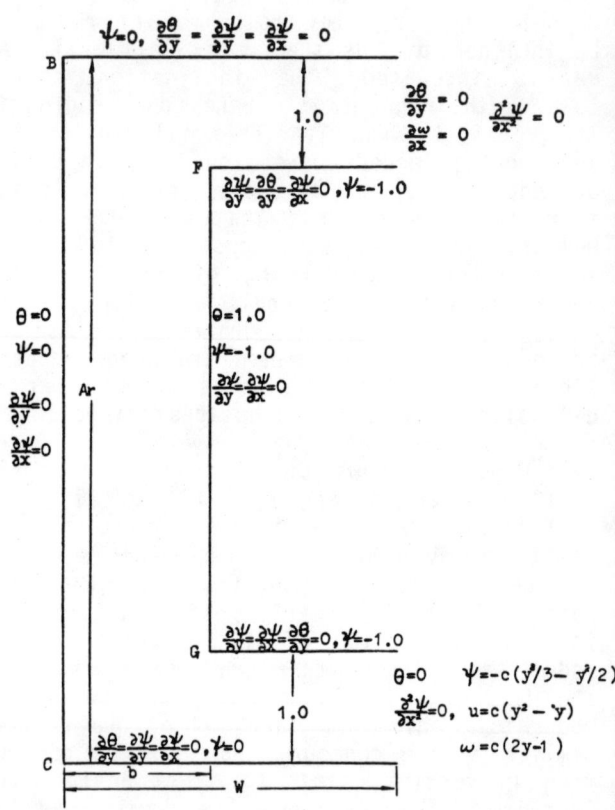

Figure 1. Analytical system and boundary conditions.

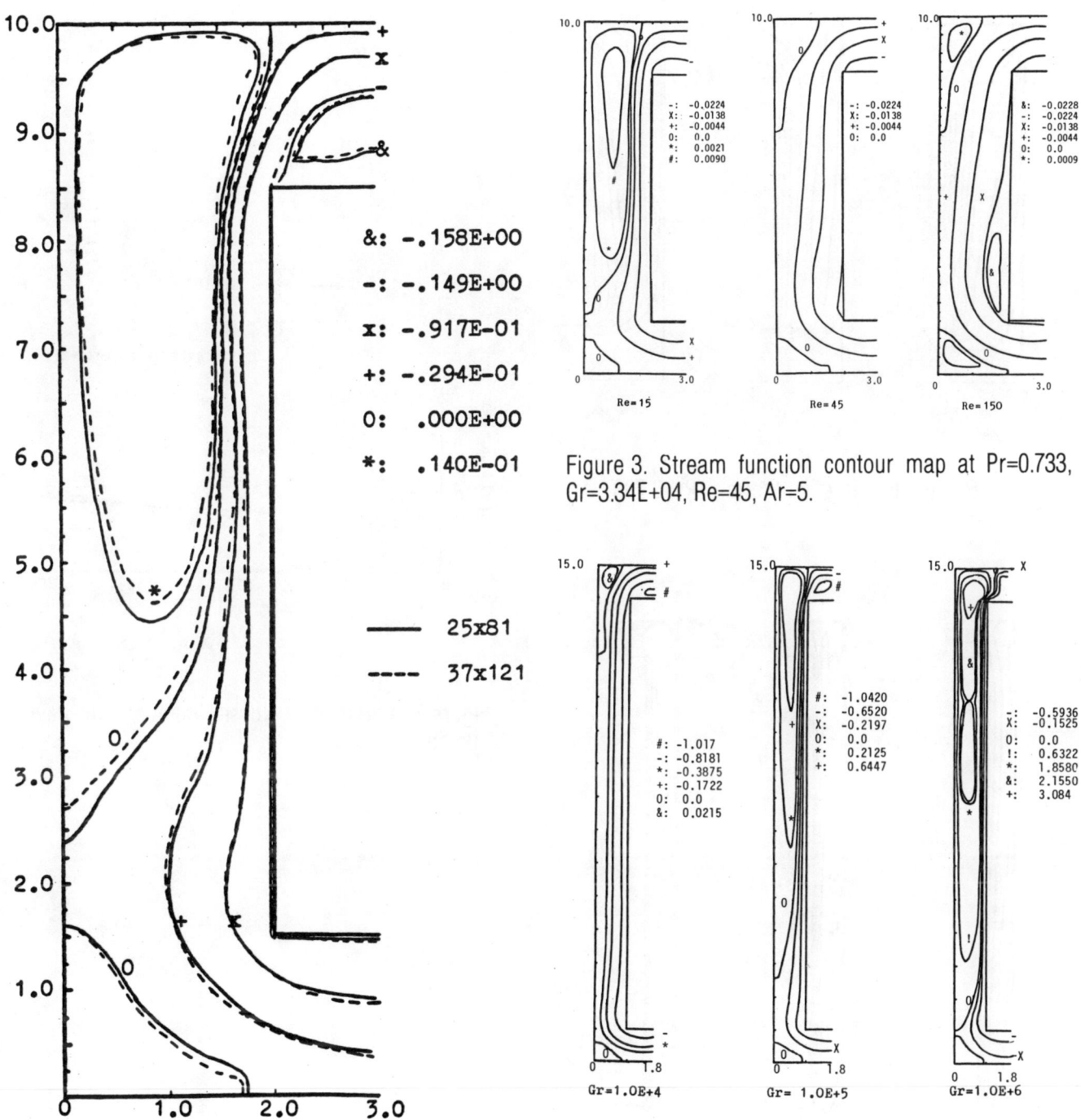

Figure 2. Steady state streamlines, Pre=0.733, Ar=5, Gr=3.34E+04.

Figure 3. Stream function contour map at Pr=0.733, Gr=3.34E+04, Re=45, Ar=5.

Figure 4. Steady state streamlines, Pr=0.733, Ar=15, Re=100.

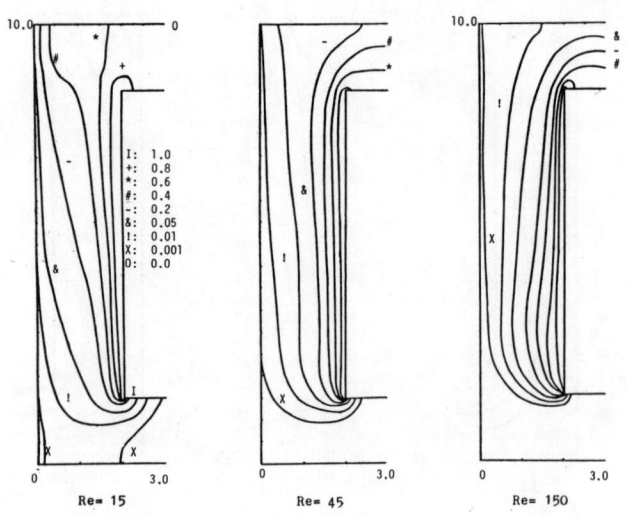

Figure 5. Steady state isotherms, Pr=0.733, Ar=5, Gr=3.34E+04.

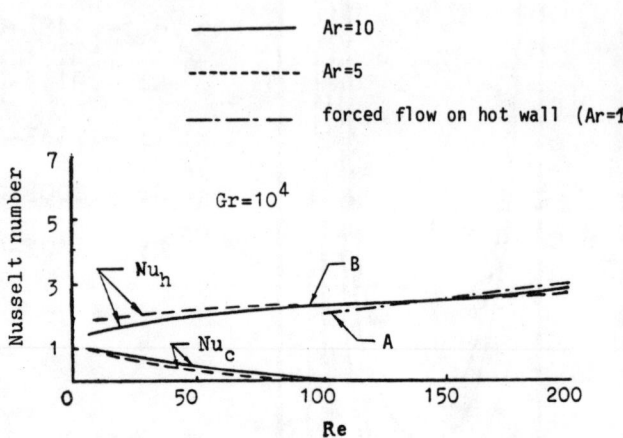

Figure 7. Variation of Nusselt number with Reynolds number.

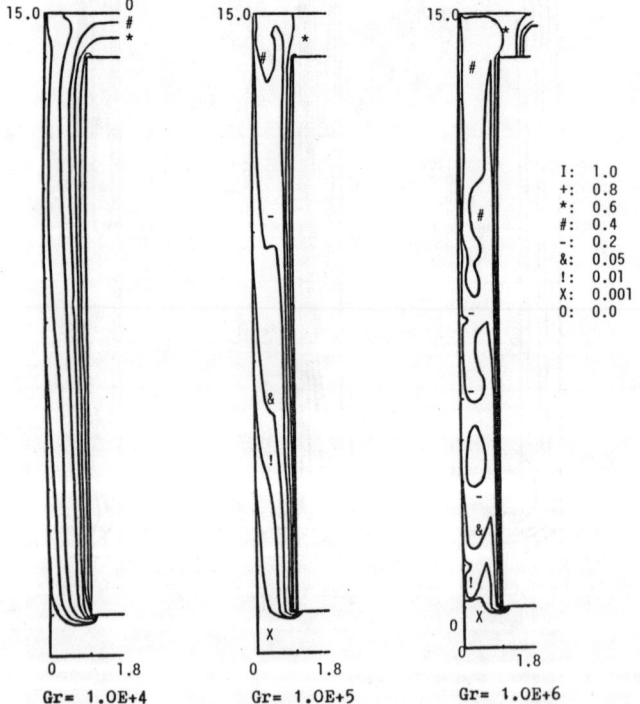

Figure 6. Steady state isotherms, Pr=0.733, Ar=15, Re=100.

# PERFORMANCE OF A FIXED MIRROR-DISTRIBUTED FOCUS SOLAR THERMAL SYSTEM: THE CROSBYTON SOLAR POWER PROJECT

L. Davis Clements ■ Dept. of Chemical Engineering, University of Nebraska, Lincoln, NE 68588-0126
Karan L. Watson ■ Dept. of Electrical Engineering, Texas A&M University, College Station, TX 77843
John D. Reichert ■ Radtech, Inc., Albuquerque, NM 87110

The Fixed mirror-distributed focus (FMDF) concept offers solar thermal power conversion at high concentration ratios with a fixed, hemispherical mirror, and a single, tracking receiver. This paper describes the optics and gives simplified performance equations for such a system. The actual performance of the Crosbyton (Texas) ADVS unit in water-steam operation is compared with the predictions.

## INTRODUCTION TO THE CONCEPT

Since January 23, 1980, the Analog Design Verification System (ADVS) of the Crosbyton Solar Power Project (CSPP) has been in nearly continuous operation. The solar-thermal concept used in the Crosbyton Solar Power Project is a Fixed Mirror-Distributed Focus (FMDF) system. The FMDF concept uses a hemispherical segment mirror to concentrate incident solar radiation and a once-through, high-pressure steam boiler to capture the concentrated light energy for use in process heating or for electrical power generation.

In the FMDF concept the collector is a fixed, spherical segment mirror, as indicated in Figure 1. Advantages of the spherical shape include:

1. a large area concentrating device possible with a single, two-axis tracking system for the receiver,

2. the accuracy of the mirror optics very high because the mirror support structure and the individual mirrors are very rigid, without sacrificing surface survivability,

3. the basic shape and dimensions of the focal region constant throughout the day, and

4. the focal region well shaped for energy capture with a simple receiver design.

The focal region of a spherical segment, when illuminated by a point source infinitely far away, is a line segment. This line segment extends from the collector surface toward the center of curvature of the sphere for a length equal to half the radius of the sphere, as indicated in Figure 2. The focal line segment lies on the line through the center of curvature of the sphere and the point source. As the point source moves, the focal region continues to be a line segment, but moves within the collector to always lie on the line containing the center of curvature and the source.

The actual focal region of a solar collector will deviate from a line segment because the sun is not a point source. The finite disk of the source causes the line segment focal region to expand to be the frustum of a cone whose vertex angle is the same as the angular diameter of the source and whose length is the same as the line segment described above. Deviations in the focal region occur if the collector deviates

---

Work performed at Texas Tech University, Lubbock, TX.

in any way from a perfect sphere of the correct curvature. In order to intercept all of the energy the collector concentrates, the receiver must be enlarged slightly to account for deviations from the spherical shape.

The traditional approach to predicting concentration profiles in concentrating solar collectors has been to use ray tracing techniques (Meinel and Meinel (1) and Wood (2). The Ratio of Solid Angles (ROSA) technique developed by Reichert and Brock (Reichert and Brock (3); Brock (4); and Reichert (5)) was developed specifically to compute the optical concentration profiles for the FMDF system. Although highly accurate, the ROSA technique is extremely time consuming.

In order to reduce the computation time required for describing FMDF optics, the Approximate Azimuthal Average Approach (AAAA) was developed by (Leung (6); Reichert and Leung (7)). In the AAAA technique, essential optical characteristics are retained, but computation time is reduced (Clements and Reichert (8)). Plots of optical concentration profiles obtained from AAAA are shown in Figures 3a to 3c for inclination angles of $0°$, $30°$, and $60°$, respectively. In these plots the abscissa indicates the position along the reciever tube in normalized arc lengths. Normalization is done by dividing the arc length along the tube by the sphere radius for the collector. Notice that the basic pattern of the plots is always the same.

The AAAA procedure which generated the optical profiles accounts for the finite size of the sun and mirror inaccuracies by using $0.5°$ as the solar radius, rather than $0.26°$. The $0.24°$ allowance for mirror surface defects has been shown to be adequate, based upon measurements with new mirrors from the ADVS. The optical profiles are also based upon a perfectly aligned receiver. While truly perfect alignment is not possible in practice, the actual misalignment (tracking) error for the receiver is typically less than $0.5°$. Because the actual receiver is somewhat oversized, this amount of misalignment is negligible in terms of net heat capture.

However, at various inclination angles the details of the optical profile do vary significantly. These variations occur because of the loss of symmetry of a hemisphere (as viewed from the receiver) if the receiver is located anywhere except the very center of the bowl ($\alpha=0°$). Except for this special case, the reflective collector surface on one side of the receiver is less than the amount of surface available in other directions. This means there are varying amounts of energy being reflected into the focal region from the various directions. This creates a "shadow" or loss of optical power on one side of the receiver. Because the arc length spirals down and around the cone, the rapid azimuthal variations of optical power can be seen in the figures. Visual inspection of the receiver in operation confirms this "shadowing" effect.

The special optics of the FMDF concept give rise to an equally specialized receiver design. The receiver design chosen for the ADVS was a cylindrical support with two parallel tubes wrapped around it. Two tubes were chosen because of the size of the ADVS. In practice, the number of tubes in parallel depends upon the size of the mirror dish. The trade-off between optics and heat transfer performance which leads to the spiral tube geometry is outlined by Clements (9). Preliminary performance predictions for a full scale ADVS system have been presented by Clements and Shankar (10). This paper will focus upon the comparison between actual and predicted performance for the ADVS operation.

## ANALOG DESIGN VERIFICATION SYSTEM

The Analog Design Verification System is a fully instrumental FMDF unit which was brought to operational status on January 22, 1980, in Crosbyton, Texas. The largest single solar collector ever built is part of this system. The collector is a quarter-sphere composed of 430 mirror panels of 12 different shapes laid out to form a segment of a 75 foot (22.9 meters) diameter sphere. The rim angle of the collector is $60°$, so the actual aperture diameter is 65 feet (19.8 meters).

The collector is tilted $15°$ to the south. The annual energy capture would be maximized by tilting the bowl to an angle equal to the latitude of the collector, which is $33-5/8°$ for the ADVS. However, 85% of the maximum benefit available from tilting the bowl is obtained with a $15°$ tilt. The influence of the costs involved has led to some compromise in performance parameters.

Economics as well as performance technology also played a significant role in determining the shape of the receiver for the

ADVS. The shape of the focal region is conical, suggesting a conical receiver. The actual receiver is cylindrical, not conical. The receiver has a 6 inch (15.24 cm) diameter right cylindrical shape. As a result, some energy near the foot of the focal region is not captured by the receiver. The receiver is made of Inconel 617, painted black with Pyromark paint to provide an average absorptivity of 90%.

The mirror panels for the ADVS are made of 3/32 inch (2.4 mm) thick standard grade (high iron content) float glass with a silver backing. The mirror reflectivity is 0.88. The mirrors were shaped by pressing the glass over molds with the proper curvature and then gluing the glass to a paper honeycomb support structure. There is a "rollover" in the shape near the edges of the panels. This "rollover" region results from a band about 3 to 4 cm wide around the edge of the panel where there is no shaping to the radius of curvature. This "rollover" band is required to bond the glass to the honeycomb. The "rollover" is important because the shape near the edges deviates enough from the desired curvature that light reflected from that region is not guaranteed to intersect the $1^\circ$ focal region.

## PREDICTION OF FMDF PERFORMANCE

The performance of the receiver for the ADVS was modelled using the Thermal-Fluid Analysis Program (TFAP) developed by Clements and Shankar (10) and Shankar (11). The objective of this model is to be able to trace the path of the fluid and to predict the state of the fluid (water/steam in the case under consideration) as it traverses through the tubes around the receiver. This is achieved by a differential heat balance on the receiver after dividing it into a number of small elements. The model predicts the total heat lost to the surroundings by convective and radiative mechanisms.

The TFAP model used to describe the receiver for the Fixed-Mirror Distributed Focus System is a detailed analysis of the heat transfer and fluid flow behavior for flow in a long pipe with an external heat source. It handles different receiver geometries, sizes and flow paths. The model handles cylindrical and conical receivers with any size of tubes and any number of tube channels running in parallel. As the product of tube diameter and number of tubes in parallel becomes large, the width of an element becomes large and certain assumptions cannot be approximated because of changes in the optical profile. The model is capable of handling different flow paths in that the inlet and outlet of the fluid could be at any point along the receiver and the flow direction is a provision for bypassing any element(s).

The interrelation between tube diameter, number of tubes, and overall concentrator diameter is important, but not immediately obvious. The operating philosophy adopted for the ADVS is to supply the working fluid, steam, for example, to the system load at constant conditions of pressure, temperature and quality. The total energy concentrated upon the receiver varies instantaneously with time of day, wind speed, ambient temperature, etc., but the total energy potentially available from the system is set by the concentrator diameter. Since the outlet and inlet conditions for the working fluid are fixed, the variable used to respond to the instantaneous changes in energy input is the fluid mass flow rate.

The variability in mass flow rate creates a constrained optimization problem in receiver design. Low resistance to heat transfer from the tube wall to the working fluid and, hence, low net heat losses to the ambient resulting from low tube wall temperatures, dictate high velocities in the tubes. High velocities suggest either many, small tubes or one somewhat larger tube. The pressure drops associated with the high velocities suggest a few, larger tubes. As outlined by Clements (9) the final design is a compromise between heat transfer integrity (reasonably high velocity over a range of mass flow rates) and pressure drop.

The TFAP code computes the partitioning of the concentrated energy between the working fluid, convective losses, and radiative losses, once structural parameters (sizes, shapes, materials, etc.) for the collector and receiver have been specified. Several optical and mirror design factors combine into a scaling factor, called the attendance, for the power, $P_W$, thermalized into the receiver walls. The name "attendance" was chosen for this parameter because it represents the fraction of the total optical power actually reaching and absorbed by the receiver, i.e. the fraction of the input energy actually attending the receiver.

The attendance is defined as

$$\alpha = R \alpha_R \eta_C \eta_R \eta_A \eta_B \quad (1)$$

where

- $R$ = reflectivity of the mirrors (0.88 for the ADVS)
- $\alpha_R$ = absorptivity of the receiver surface (0.9 for the ADVS)
- $\eta_C$ = concentrator collection factor = $f_1 f_2 f_3$
- $f_1$ = panel quality factor. This is a measure of the average accuracy of the radius of curvature across a panel. For a focal region with the vertex angle less than or equal to $1°$, $f_1 = 1$. This factor is less than 1 for focal region vertex angles greater than $1°$. The measured focal region vertex angles for the mirrors used in the ADVS were all $\leq 1°$, so $f_1 = 1$ for the ADVS.
- $f_2$ = interception factor. This factor represents the fraction of the energy concentrated into the focal region which actually is optically intercepted by the receiver. It depends upon the receiver size and shape. In the case of the ADVS, the 15.2 cm OD, cylindrical receiver does not optically intercept 2 percent of the focal region, so $f_2 = 0.98$ for the ADVS.
- $f_3$ = tracking misalignment factor. This factor accounts for systematic errors in receiver tracking greater than the $1°$ focal region vertex angle, in terms of the fraction of the power in the focal region intercepted. Measured performance for the ADVS shows $f_3 = 1$.
- $\eta_R$ = mirror alignment factor. This parameter is a measure of the fraction of the input power actually directed into the focal region as a result of properly aligned mirrors. In practice $\eta_R = 1$ because it is easy to align the mirrors and the alignment is maintained.
- $\eta_A$ = aperture effectiveness factor. This factor is the ratio of the net (mirror surfaced) aperture area to the gross aperture area. This correction arises because there are gaps between the mirror panels and because the edge rollover area on each mirror is not counted as active mirror surface. In the case of the ADVS, $\eta_A = 0.969$
- $\eta_B$ = blocking and shading factor. This factor accounts for the fraction of the effective aperture area remaining after allowing for shadowing from the receiver support structure. The blocking and shading factor for the ADVS is computed to be 0.912.

The details of calculating each of the factors are given in Watson (1981) for the actual ADVS mirror layout shown in Figure 4. Using the values for the factors noted above, the maximum theoretical attendance for the ADVS is

$$\alpha_{th} = (0.88)(0.9)(0.98)(0.969)(0.912)$$
$$= 0.686 \quad (2)$$

The expression "maximum theoretical attendance" is used deliberately. With the exception of the mirror reflectivity, all of the factors entering into the attendance are fixed by material or geometry. However, the reflectivity of the mirrors is not simply the reflectivity of light shining directly onto the mirrors, ($R = 0.88$). It also includes a factor arising from the fact that some light is reflected more than once from the mirror surface before reaching the focal region. Power is lost with each reflection; i.e., light which is reflected N times is weighted with the Nth power of the reflectivity, $R^N$. The fraction of the solar power directed into each order of reflection is (for a fixed rim angle) a function only of the inclination angle, $\alpha$, and may readily be computed.

Once the fraction, $F_N$, of power going into each order is computed as a function of and weighted with the appropriate power of the direct reflectivity, R, then the effective reflectivity, $R_{eff}$, can be described:

$$R_{eff}(\alpha) = R\, B_{in}(\alpha) \quad (3a)$$

where

$$B_{in}(\alpha) = \frac{1}{R} \sum_N R^N F_N(\alpha) \quad (3b)$$

For a bowl with $60°$ rim angle:

$$B_{in}(\alpha) = 1 - \frac{B(\alpha/75°)}{1 + (0.972)(\alpha/75°)^5} \quad (3c)$$

where

$$B = 0.375(1 - R) \quad (3d)$$

Thus, when R is 0.88, the value of $B_{in}$ is 0.45. At $\alpha = 0°$, $R_{eff}$ has its maximum value of 0.88.

The actual attendance is measured daily at solar noon by comparing actual power capture with the theoretical power capture for the instantaneous conditions measured. Once the daily attendance factor is determined, the actual power captured may be easily compared with the predicted power capture. The ratio of measured to theoretical attendance varies as shown in Figure 5. There is some indication of a slow downward trend in performance. This trend is due to some deterioration of the mirror surface, dirt accumulation and the breakage of a few mirrors. The drop in performance at about day 110 was due to heavy rains putting a dirt film on the mirrors. The mirrors were washed and realigned and performance returned to about 95 percent of theoretical. Subsequently, the performance has degraded very slowly due primarily to deterioration of the radius of curvature of the mirrors because of debonding.

The daily attendance can be used as a measure of the optical condition of the system, for example, to determine when to wash, realign, or replace mirrors. Also, the daily attendance can be used as a measure of optical condition of the system in computing mass flow rates in the thermal process control algorithms.

## PREDICTED THERMAL POWER OF THE SYSTEM

The actual thermal energy delivered to the receiver at any time depends upon the attendance, the gross aperture area, and the direct normal insolation. The inclination angle between the sun and the symmetry axis of the fixed aperture, $\alpha$, changes during the day, resulting in a cosine loss of direct normal insolation and the reflectivity loss described by Equation 3c. When all of these attenuation factors are included, the thermal energy delivered to the receiver is given by

$$P_W = A_{ng} B_{in}(\alpha) I_{DN} \cos\alpha \quad (4)$$

where

$I_{DN}$ = direct normal insolation (KW/m$^2$)

$\alpha$ = the inclination angle between the optical axis (sun) and the normal to the fixed aperture symmetry axis

$A_{NG}$ = the nominal gross aperture area (ft$^2$) [For the ADVS, $A_{NG}$ = 308.3m$^2$ (3318.3 ft$^2$)]

$\alpha$ = the attendance, defined in Equation 1.

$B_{IN}(\alpha)$ = the multiple bounce input factor, defined in Equation 3c.

The estimated power thermalized into the receiver wall (Equation 4) may be found for any $I_{DN}$ and $\alpha$ by using $\alpha_{th}$. This power, $P_W$, is redistributed (in steady state) into three modes:

$$P_W = P_F + P_{CL} + P_{RL} \quad (5)$$

where

$P_{CL}$ = power lost to convection
$P_{RL}$ = power lost to radiation

The "fluid power capture factor" is defined as

$$\sigma_F = P_F/P_W = 1 - (P_{CL} + P_{RL})/P_W \quad (6)$$

The fluid power capture represents the fraction of the total available energy available to the receiver that actually enters the working fluid.

Unlike $\alpha$, the factor $\sigma_F$ changes from moment to moment in a complicated manner which depends upon the instantaneous values of the insolation $I_{DN}$, the mass flow rate $\dot{m}$ the wind speed $V_W$, as well as the value of the inclination angle

Two options are available for computing the fluid power capture factor. The TFAP program may be used to generate a large data base of computed fluid power capture factors over a range of conditions, and the values in this data base can then be represented in the form of empirical performance equations.

The TFAP program was used to generate fluid power capture factors for high pressure steam (50 to 70 atm) delivered at temperatures ranging from 280 to 540 C. The results of empirical fitting of this data base are given as a set of performance equations. These performance expressions require input of the inclination angle and the windspeed. The ambient temperature and the inlet temperature of the water were found to have minimal effect on the performance.

Typical patterns of power delivered by an FMDF system, using constant outlet conditions are shown in Figures 6 and 7. The basic form of the curves is a cosine, based on the solar inclination angle. The short term variations in power reflect changes in insolation, windspeed, and wind direction throughout the operating day.

In order to test the performance equations, 46 instantaneous operating points were chosen at random. The actual energy absorbed by the water/steam working fluid was compared with the predicted values. The theoretical attendance, $_{th}$ = 0.686 was used in the comparison. The average error was 9.5 percent for all the points. When three points which were not at steady state were removed, the average error was -3.0 percent.

SUMMARY

The optics and the energy capture of an FMDF system are quite complex. The ROSA and AAAA approaches to computing the optical concentration profiles for an FMDF concentrator offer detailed information concerning both concentration ratio and spatial structure of the energy within the focal region. Similarly, the detailed thermal analysis for an FMDF receiver is laborious and somewhat complicated.

This paper describes a direct approach for coupling optical calculations and detailed heat capture predictions through the use of simplified performance equations and the use of the attendance parameter. The attendance consolidates a number of optical and structural attenuation factors into a single term. The theoretical attendance is useful in predicting performance of an FMDF system. The measured daily attendance is a figure of merit which is useful in quantifying the effects of dirt, misalignment, and mirror or receiver degradation on the long term performance of an FMDF device.

LITERATURE CITED

1. Meinel, A. B. and M. P. Meinel, *Applied Solar Energy, An Introduction*, Ch 6 -7, Addison-Wesley, Reading, PA (1976).

2. Wood, R. W., *Optical Physics*, Ch II, Dover Publications, New York (1967).

3. Reichert, J. D. and B. C. Brock, "Crosbyton Solar Power Project Phase I Interim Technical Report," Vol. II, Appendix C, Texas Tech University, Lubbock, TX (ERDA Contract No # (29-2)-3737) (1977)

4. Brock, B. C., "Optical Analysis of Spherical Segment Solar Collectors," Ph.D. Dissertation, Texas Tech University, Lubbock, TX (1977).

5. Reichert, J. D., "A Strategy for Calculations of Optical Concentration Distributions in Fixed Mirror Systems," *Proc. of ERDA Solar Workshop on Methods for Optical Analysis of Central Receiver Systems*, 155-174, Houston, TX (1977).

6. Leung, Hipsun, "Optical Power Concentrations on Aligned and Misaligned Receivers in Solar Gridiron Power Systems," M. S. Thesis, Texas Tech University, Lubbock, TX (1978).

7. Reichert, J. D. and Hipsun Leung, "Optical Concentrations on Aligned and Misaligned Conical Receivers for Gridiron Power Systems," Crosbyton Solar Power Project Report CSP-OPC-2, DOE Contract EY-76-C-04-3737, Mod A003 (1978).

8. Clements, L. D. and J. D. Reichert, "Optical-Thermal Performance Analysis for a Fixed Mirror-Distributed Focus Solar-Thermal-Electric Power System," Proc. 14th Intersociety Energy Conversion Engineering Conf., Vol. 1, 11-14 (1979).

9. Clements, L. D., "Design Considerations for the Energy Receiver in a Fixed Mirror-Distributed Focus (FMDF) Solar Energy System," Vol 1, 295-311, *Alternative Energy Sources: An International Compendium*, N. T. Veziroglu, ed., Hemisphere Publ. Corp., Washington (1978).

10. Clements, L. D. and Hariharan Shankar, "Thermal Performance Analysis of a Fixed Mirror-Distributed Focus (FMDF) Steam Receiver," paper presented at 86th AIChE National Meeting, Houston, TX (1979).

11. Shankar, Hariharan, "Simulation of the Receiver in a Fixed Mirror-Distributed Focus Solar Power System," M.S. Thesis, Texas Tech University, Lubbock, TX (1981).

12. Watson, K. L., "Performance Analysis of a Solar Gridiron Design Verification System," M. S. Thesis, Texas Tech University, Lubbock, TX (1981).

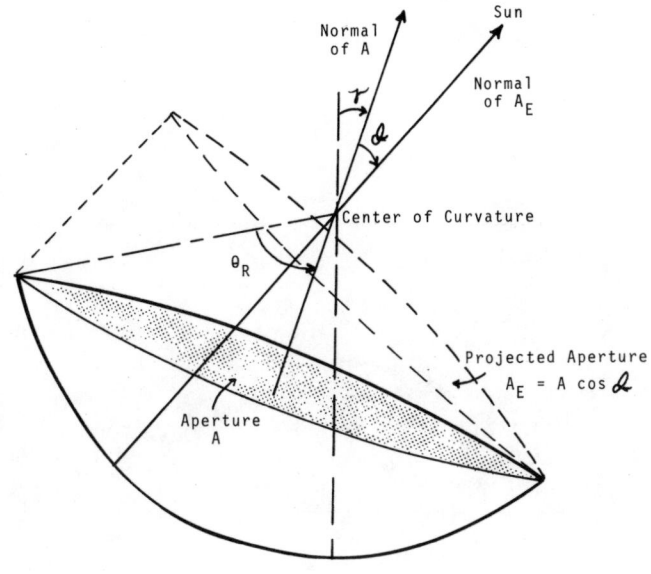

Figure 1. Schematic diagram and geometry of an FMDF concentrator.

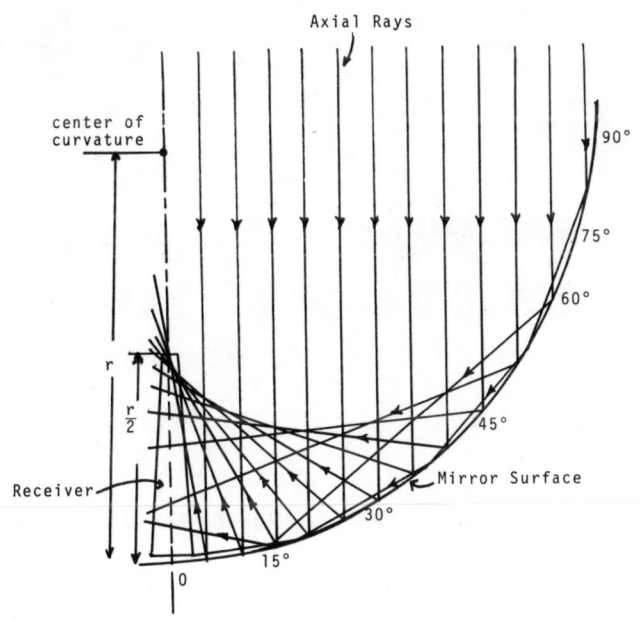

Figure 2. Optical concentration patterns for an FMDF concentrator.

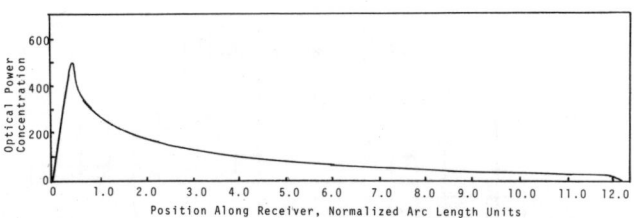

Figure 3(a). Optical power concentration profile for an aligned receiver, inclination angle = 0°.

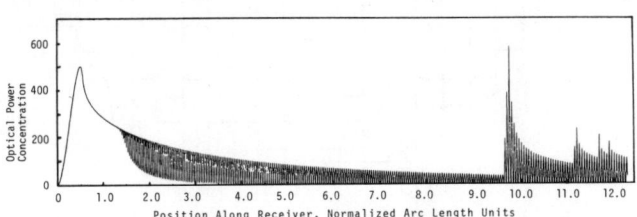

Figure 3(b). Optical power concentration profile for an aligned receiver, inclination angle = 30°.

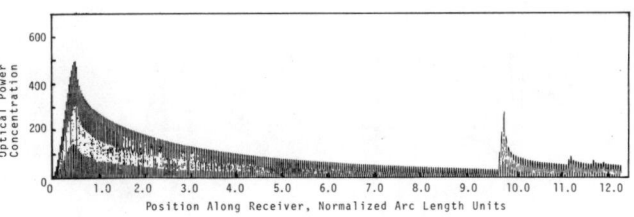

Figure 3(c). Optical power concentration profile for an aligned receiver, inclination angle = 60°.

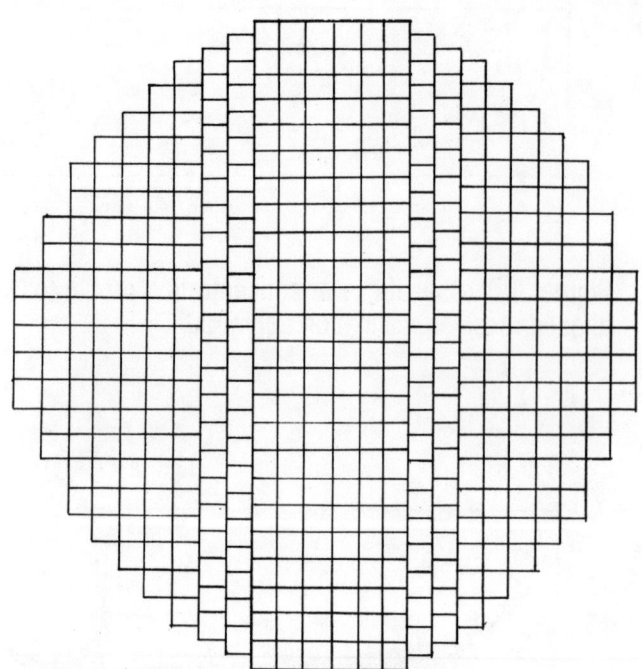

Figure 4. Mirror panel layout for the Crosbyton ADVS.

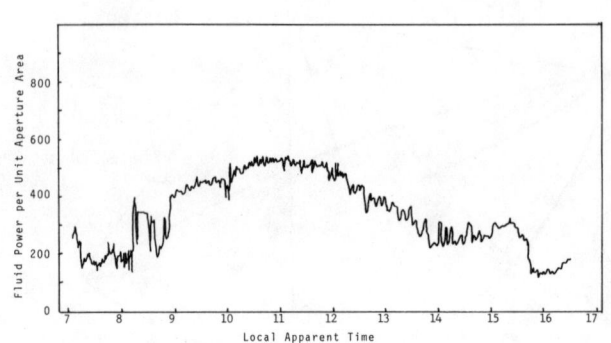

Figure 6. Daily thermal performance — May 24, 1980.

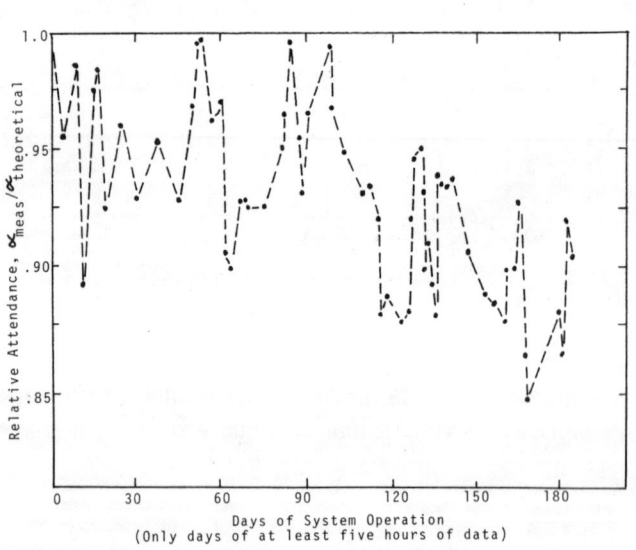

Figure 5. Measured daily attendance compared with theoretical attendance during operation from March 20, 1980 to September 22, 1980.

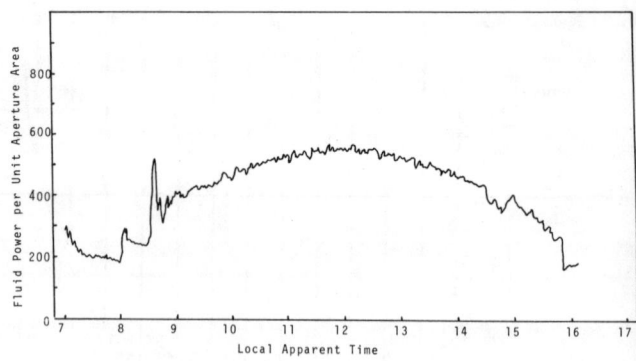

Figure 7. Daily thermal performance — June 14, 1980.

# TIME DEPENDENT THREE DIMENSIONAL THERMAL ANALYSIS OF A SOLAR HEATING SYSTEM

A.A. Arafa ■ Cleveland State University, Cleveland, Ohio 44115

A transient simulation model for a solar heating system which is used for heating both domestic water and the space in a building has been developed. This model considers a three dimensional thermal analysis in the flat plate collector and the stratification in the storage tank. The hourly average meteorological data measured in two different sites in West Germany, 1973, were used as input for the solar heating system. The hourly heating load of the building for the corresponding site as calculated from a transient simulation model (1) was used as input load for the solar heating system. Two different methods have been used in the numerical solution of the coupled nonlinear differential equations resulting from the model. One uses the Runge-Kutta method and the other uses a self-devised method. The self-devised method gives very small deviations in the results compared to the Runge-Kutta method and reduces the computing time considerably. The effect of the collector area, the normalized pipe spacing in the absorber plate, the ratio of the storage volume to the collector area and the collector arrangement on the system performance is investigated for each site. Qualitative comparisons between data and predictions are presented for a domestic solar heating system installed in two different houses in the northern and southern parts of West Germany. These comparisons show fair agreement.

## INTRODUCTION

The shortage in the fossil fuel resources and the inevitable increase in fuel costs enhance the attraction of solar energy as an alternative source. The full utilization of solar energy can be achieved through optimization procedures. The large number of parameters involved in a solar heating system, the time dependency of the meteorological data as well as the different user habits make the experimental optimization of such a system impractical. Therefore, the use of validated simulation models as optimization tools is a practical approach. Most of the available simulation models use the Hottel (2) equation to describe the collector performance which is based on steady state working conditions. Also, the black box approach for describing the performance of different components in a solar heating system is commonly used. An exact investigation of the dynamic thermal performance of the solar collector leads to a complicated set of differential equations, the solution of which requires relatively long computing time. The main purpose of this work is to investigate the dynamic thermal performance of a solar heating system precisely while keeping the computing time as minimal as possible.

## THE PHYSICAL MODEL AND ITS INPUT DATA

This model consists of a flat plate solar collector, storage tank, auxiliary heater, pumps, valves and controllers. Figure 1 shows a schematic illustration of the system.

The collector loop uses a mixture of 50% water and 50% glycol as a heat transfer medium. The collector loop's pump operates when the fluid temperature at the collector outlet is higher than the water temperature in the storage tank. The storage tank uses water as a heating medium and is connected to the house through a water heating system. The water has a minimum temperature of 45°C at the inlet to the house heating system and its mass flow rate is controlled to provide the required heating load and to maintain a constant temperature of the return water to the storage tank. The return water can either enter the storage tank at three different levels or bypass it. Starting at the bottom of the storage tank, if the water temperature at this level is higher than the temperature of the return water then the return water will go through the storage tank. Otherwise it will bypass this level to the next level. The water passage is controlled by three 3-way valves and two non-return valves. The working function of these valves is illustrated in the table at the bottom on figure 1.

The domestic hot water also has a minimum temperature of 45°C. If the solar radiation is not enough to maintain the 45°C for either domestic hot water or space heating, the auxiliary heater will provide the energy needed to maintain this temperature.

The meteorological data of two sites in

West Germany (Hamburg and Freiburg) were chosen as input data for the solar heating system to investigate the sensitivity of the system parameters to small differences in the meteorological conditions. The meteorological data of Freiburg reflects a slightly higher solar radiation and a lower ambient temperature than the meteorological data of Hamburg (1). These meteorological data were measured in 1973 and were recorded as hourly average values. The solar radiations were measured on a horizontal surface with the total of 982 and 1198 kWh/m$^2$.a for Hamburg and Freiburg respectively. The formula given by Liu and Jordan (3) is used here to calculate the global solar radiation on a tilted surface from the measured values on a horizontal surface. For the qualitative parameter investigation, the meteorological data were processed according to the standard day averaging method (4). In this method the hourly meteorological data for each hour of a day are averaged over a period of one month. This produces data for the so called standard days. Thereby the data for the whole year can be represented by the data of twelve standard days. These standard days are used in all the simulation presented in this paper.

## HEAT DEMAND FOR HOT WATER

The daily water consumption per person is between 60 and 100 liters. Based on the percentage distribution of the daily water consumption given by (5) and for three persons with a daily consumption of 75 liters each, the annual heat demand for hot water is 3146 kWh. This is calculated for a temperature increase of (45 - 12) = 33 K and 365 days.

## HEAT DEMAND FOR SPACE HEATING

In another study (1) a transient simulation model has been developed for investigating the dynamic thermal performance of buildings. This model provides the hourly as well as the annual heat demand "$Q_{SH}$" for a given house design. For a family house with an area of 100m$^2$ (other parameters are to be taken from (1)), the hourly heat demands were calculated for the whole year using the hourly meteorological data of both Hamburg and Freiburg. The calculated values were then processed according to the standard days averaging method (1). The annual heat demands were calculated to be 14780 kWh for Hamburg and 16365 kWh for Freiburg.

## COLLECTOR MODEL

The flat plate collector consists of a fin - pipe - fin system as an absorber plate, a working fluid, insulation on the back of the collector and one or more transparent cover plates. Figure 2 shows a symmetrical element of such collector. The variation of the thermal capacity across the absorber plate and in the vertical direction of the collector as well as the different amount of energy received by each component in the collector implies the consideration of a three dimensional problem to achieve precise thermal analysis. The mathematical analysis of such collector is presented in (6).

## STORAGE MODEL

A closed cylindrical tank with a diameter of 0.75 m and a height of 1.5 m is used as a storage. The mathematical model for the storage tank is based on the following assumptions:
a) The storage tank is filled with water all the time.
b) The heat capacities of the tank walls and the heat exchangers are negligible.
c) Convection does not take place in the storage tank.
d) Three isothermal layers with a height $h = h_S/3$ and a diameter D are considered.
e) The temperature of the fluid inside the heat exchanger in each layer is equal to the water temperature of the corresponding layer in the tank.

The accuracy of the storage model and the computing time increases by increasing the number of layers. Three layers, however, have been chosen from previous work (7) as a compromise between accuracy and computing time.

The energy balance equations for the three nodal points are presented in (8). The solar energy transferred from the collector to the storage tank $Q_{sol}$, the useful energy extracted from the storage for either space heating and/or domestic hot water $Q_u$ and the stored energy $Q_s$ over a period of time $\Delta t$ are determined as follows:

$$Q_{sol} = \sum_{t=0}^{t} F \cdot (\dot{M}c_F \cdot (\theta_0 - \theta_{S,3}) \cdot \Delta t) \quad (1)$$

where: $F = 0$ for $\theta_0 < \theta_{S,1}$,
$F = 1$ for $\theta_0 > \theta_{S,1}$

$$Q_u = \sum_{t=0}^{t} [(\dot{m}_{SH} \cdot c_F \cdot (\theta_{S,1} - \theta_{o,H})) \cdot F \quad (2)$$
$$+ \dot{m}_{HW} \cdot c_F \cdot (\theta_{S,1} - \theta_{in})] \cdot \Delta t$$

$F = 0$ for $\theta_{S,1} < \theta_{o,H}$, $F = 1$ for $\theta_{S,1} > \theta_{o,H}$

$$Q_S = V_S \cdot \rho_F \cdot c_F \cdot \frac{1}{3} \sum_{i=1}^{3} [\theta_{S,t}(t) - \theta_{S,i}(t=0)] \quad (3)$$

The auxiliary energy $Q_{aux}$ will be needed whenever $\theta_{S,1} < \theta_{SH}$,

$$Q_{aux} = \sum_{t=0}^{t} [\dot{m}_{SH} \cdot c_F \cdot (\theta_{SH} - \theta) + \dot{m}_{HW} \cdot c_F \cdot (\theta_{HW} - \theta_{S,1})] \cdot \Delta t \quad (4)$$

$\theta = \theta_{S,1}$ when $\theta_{o,H} < \theta_{S,1}$;
$\theta = \theta_{o,H}$ when $\theta_{o,H} > \theta_{S,1}$

$\dot{M}$, $\dot{m}_{HW}$, $\dot{m}_{SH}$, $\theta$ and $\theta_S$ are time dependent.

## SOLUTION METHODS

The system performance can be determined by solving the simple algebraic Equations (1) through (4). These equations, however are coupled with the time dependent temperatures of the collector, and the storage tank. The solution of the system equations is essential for predicting the time dependent temperatures. The system equations also include the coupled non-linear partial differential equations for the collector (6).

The partial differential equations can be converted into difference differential equations with the use of a nodal model (6). There are subroutine programs that can solve such system equations numerically e.g. Dascru (4). In the present work in addition to using such a subroutine, a self-devised method is used which will be referred to as "e- Function". Both solution methods are described in (1) and (8). A package of computer programs has been developed in order to include both solution methods and simulate the transient thermal performance of the solar heating system.

## CRITERIA FOR THE DESIGN OPTIMIZATION

The annual average thermal effieciency "$\bar{\eta}$" and the solar fraction "f" were used as criteria for obtaining the optimal design of the solar heating system. The annual average thermal efficiency (system efficiency) is the ratio of the total useful energy collected and the incident solar radiation and is calculated by summing the following equation over a period of one year:

$$\eta = (Q_S + \sum_{t=0}^{t} \dot{Q}_u \cdot \Delta t) / (\sum_{t=0}^{t} I_S \cdot A \cdot \Delta t),$$
$$(\Delta t = 6 \text{ min.}) \quad (5)$$

The stored energy $Q_S$ is considered in Equation (5) because it may be used directly after the simulation interval. If it is not used directly, it will cause an increase in the initial storage temperature in the following year. The solar fraction is the ratio of the annually used collected solar energy to the annual heating demand and is given by:

$$f = (\sum_{t=0}^{t} \dot{Q}_u \cdot \Delta t) / (\sum_{t=0}^{t} \dot{Q}_u \cdot \Delta t + Q_{aux}) \quad (6)$$

## THE NUMERICAL AND THE e- FUNCTION RESULTS

An initial design for the solar heating system is proposed here for further investigations. This design accounts for a collector area of 14m$^2$ and a storage volume of 0.662m$^3$, whereas the collector has a non-selective absorber plate 1 m long with a pipe spacing of 5 cm and double glass covers. Other constant parameters are given in the appendix.

Both the numerical and the e- function solution methods were applied in calculating the performance of such a system using the standard day meteorological data and the heat demand of Hamburg. For both solution methods, the dominant variables for determining the system performance according to Equations (5) and (6) with the corresponding computing time are given in Table 1.

Table 1: Comparison between the results of the numerical and the e- function solution methods.

| parameter method | $\dot{Q}_u$ kWh | $\dot{Q}_u + Q_s$ kWh | $Q_{aux}$ kWh | $\bar{\eta}$ % | f % | Computing time (sec) |
|---|---|---|---|---|---|---|
| numerical | 2174.2 | 2260.9 | 15821.8 | 16.76 | 12.1 | 461.6 |
| e-function | 2205.5 | 2299.0 | 15793.8 | 17.05 | 12.2 | 60.0 |
| Deviation in % from num. method | +1.44 | 1.68 | -0.17 | 1.7 | 0.8 | |

This table shows that the largest deviation in

the results obtained by the e- function is 1.7% of that obtained by the numerical solution. On the other hand the computing time of the numerical solution is 7.7 times larger than that of the e- function solution method. Based on the very short computing time and the small deviations, the e- function solution method is chosen to carry out the following calculations.

## PARAMETER VARIATIONS

To investigate the effect of the different parameters on the system performance, only one system or component parameter (e.g. collector area or collector length) is varied while the others are kept constant. Only the variable parameters will be referred to in the figures. The values for the other parameters are to be taken from the initial design mentioned previously and from the appendix. The standard days data for both meteorological and heat demand are used in all subsequent simulations. Results for both Hamburg and Freiburg will be presented only if they show significant differences, otherwise the results for Hamburg will be presented.

### Collector Area

The variation of the system efficiency with respect to the collector area is shown in Figure 3 for two different ratios of normalized pipe spacing (L/d = 5 and 7) for Hamburg. This figure shows that the system efficiency decreases by increasing the collector area. From the definition of the system efficiency given by Equation (5), it is obvious that the value of the denominator increases linearly by increasing the collector area. The value of the nominator, which represents the useful energy collected, does not increase linearly by increasing the collector area. For a given storage volume, a larger collector area will cause a faster rate of increase in the storage temperature than a smaller collector area. Consequently the collector operating temperature increases; which will cause an increase in the heat losses, i.e. decrease in the useful energy. Therefore, the rate of increase in the nominator of Equation (5) is smaller than the rate of increase in the denominator when the collector area is increased. On the other hand, a small collector area which provides a high system efficiency needs more auxiliary energy since the storage temperature will be low. This contradictory trend is not considered in Equation (5). Contrary to the system efficiency, the solar fraction increases by increasing the collector area. Figure 4 indicates that the solar fraction shows a slight curvature when the collector area is increased. The maximum progress in the solar fraction occurs by increasing the collector area from 6 $m^2$ to 14 $m^2$. This increase in the collector area yields an increase in the solar fraction of 6%. Both criteria show no optimal value for the collector area. Moreover, the collector area should by optimized in combination with the storage volume.

### Pipe Spacing

The system efficiency shows optimal values for the normalized pipe spacing L/d. The results are shown in Figure 5. For each collector area there is an optimal value of L/d by which the system efficiency is maximum (the dashed line in the figure). For values of L/d smaller than the optimal values, the absorber mass (including the water content) is large and thereby the time needed for warming up the collector in the early morning from the ambient temperature to the storage temperature is large (6). For values of L/d larger than the optimal values, the absorber temperature is higher and consequently the heat losses are higher. An optimal ratio of L/d = 7 is obtained for a collector area of 14 $m^2$. The system efficiency for L/d = 7 is about 4% higher than that of L/d = 1 (i.e. pipe-pipe absorber). The solar fraction shows the same trend with reversed order of the collector areas, where the largest area has the highest solar fraction (Figure 6). It should be pointed out that a collector area of 14 $m^2$ and L/d = 7 has a higher solar fraction than a collector area of 22 $m^2$ and L/d = 1. In the subsequent investigations a ratio of L/d = 7 is used.

### Storage Volume

The storage volume is varied from 0.331 $m^3$ to 1.324 $m^3$ (i.e. 0.5 $V_S$ to 2 $V_S$). The initial storage dimensions are $H_S$ = D = 0.75m. The small storage volume is achieved by reducing the initial height to one half of its value. The large volume considers two identical storage tanks with the same initial dimensions and connected in parallel. This connection method assures that the layer height and the ratio of volume and surface area of the storage tank stays constant (except for 0.5 $V_S$).

The relationship between the system efficiency and the storage volume for different collector areas is shown in Figure 7 for Hamburg. For a given collector area, an increase in the storage volume yields an increase in the system efficiency. Contrary to this, the solar fraction shows optimal values of the storage volume as illustrated in Figure 8. This figure shows that for each collector area there is an optimal storage volume which provides the highest solar fraction. For values of storage volumes smaller than the optimal values, the temperature in the storage tank increases at the earlier stages with faster rate. Consequently the solar radiation would not be enough to increase the temperature in the collector considerably so that more thermal energy could be stored. For storage volume values larger than the optimal values, the storage temperature is lower which causes a decrease in the solar fraction. This Figure shows that the optimal ratio of storage volume to collector area is $V_S/A = 0.047$ m for both 14 $m^2$ and 18 $m^2$ collector areas. Löf (10) has shown that a ratio of $V_S/A = 0.0487$ m is economically optimal for Albuquerque (U.S.A.). Practically, this optimal ratio gives the optimal collector area for a given storage volume.

## Collector Length

In the previous investigations a collector length of 1 m was considered. For a collector area of 14 $m^2$ this leads to a collector arrangement of 14 collectors, each 1 m long, connected in parallel. This arrangement has a total mass flow rate of $\dot{M}$. In this section, another two collector arrangements are considered (Figure 9). The product of the total mass flow rate and the collector length must stay constant, in order to maintain a constant flow rate in each pipe in the absorber and a constant pumping power for all collector arrangements. The relationship between the solar fraction and collector length is illustrated in Figure 10 for Hamburg and Freiburg. This figure shows that the optimal collector length is 2 m for Hamburg. For Freiburg, a collector length of 1 m has the highest solar fraction. The optimal collector length depends on the intensity of the solar radiation and the operating temperature (6). Freiburg has a relatively higher solar radiation and lower ambient temperature than Hamburg (1). The higher solar radiation leads to higher operating temperature and the lower ambient temperature increases the heat losses. A longer collector provides a higher operating temperature and thereby the heat losses increases. The lower operating temperature for Hamburg can be concluded from the same Figure. The solar fraction curve for Freiburg is between 4% and 5% higher than that of Hamburg, although the heat demand for Freiburg is higher (1). The system efficiency showed also the same trend (8).

## COMPARISON BETWEEN MEASURED AN PREDICTED RESULTS

The measured data of two identical solar heating systems for hot water located in two different unoccupied houses, one in Wahlstedt north of Hamburg, the other in Rottal near Freiburg, was used to examine the accuracy of the e- Function solution method. The solar heating system is described in (11). The measured data were available as weekly average values for a three year period (11). Because hourly data was not available, a qualitative comparison, in which the meteorological data of Hamburg was used for the system in Wahlstedt and that of Freiburg for the system in Rottal. The total solar radiation for Wahlstedt is close to that of Hamburg and that of Rottal is close to that of Freiburg.

The comparison between measured and predicted data for the system efficiency and the solar fraction is shown in Figure 11. In reference to the measured data, the predicted results show a relative deviation of 21.4% and 8.4% for the system efficiency for Wahlstedt and Rottal respectively. A small absolute deviation between measured and predicted system efficiencies yield a large relative deviation because of the small values of the system efficiency (12.5% for Wahlstedt and 13.1% for Rottal).

The comparison of the solar fraction shows better results. The relative deviation in the solar fraction is only - 1% for Wahlstedt and 3.3% for Rottal.

## CONCLUSIONS

The motivation for developing the present simulation model was to include the detailed collector parameters (see Figure 2) explicity in the model and investigate their effect on the system performance for two different weather conditions in West Germany. The results show that the optimal value of the normalized pipe spacing in the absorber (L/d) depends on the system parameters, e.g. collector area and the storage volume. For both weather conditions, an optimal ratio of L/d = 7 was obtained for a ratio of storage volume to collector area of $V_S/A = 0.047$ m. This ratio shows an optimal value when the solar fraction was used as design criterion. The system efficiency failed to provide the optimal ratio of $V_S/A$. An optimal collector length of 2 m was obtained for the northern part of West Germany (Hamburg) while for the southern part (Freiburg), the shortest collector length considered in this work proved to be more effective than longer ones. The use of the e- function solution method showed a considerable decrease in the computing time compared to the Runge Kutta method.

The practical applications of this simulation program are:
a) To determine the optimal flat plate solar collector parameters for a (known) weather and working condition (e.g. desire temperature, hot water consumption) which may be helpful for the solar collector manufacturers.
b) To determine the optimal working condition as well as the system design for a known weather condition and detailed parameters of a solar heating system.

## ACKNOWLEDGEMENTS

The author wishes to express his appreciation to the West German Ministry for research and technology for funding this work. The advice and encouragement of Professor Eric Hahne, Director of the Institute of Thermodynamics and Heat Transfer, University of Stuttgart, West Germany are also greatly appreciated.

## NOTATION

$A$ = collector area, $m^2$ (10.76 $ft^2$)
$c$ = specific heat, J/kgK (2.388 · E - 4 Btu/lbm F)
$D$ = storage tank inside diameter, m (3.28 ft)
$d$ = tube outer diameter, m
$f$ = solar fraction
$I_s$ = solar radiation, $W/m^2$ (0.317 $Btu/ft^2$ hr)
$L$ = pipe spacing, m
$\dot{m}$ = mass flow rate, kg/sec (2.204 lbm/sec)
$M$ = total mass flow rate to the storage tank, kg/sec
$Q$ = heat energy, kWh (3.4144 E + 3 Btu)
$\dot{Q}$ = heat flow rate, W (3.412 Btu/hr)
$t$ = time, sec
$V$ = volume, $m^3$ (35.32 $ft^3$)
$\alpha$ = thermal conductanc, $W/m^2K$ (0.176 $Btu/hr\ ft^2\ °R$)
$\alpha^*$ = absorptivity
$\delta$ = thickness, m
$\varepsilon^*$ = emissivity
$\bar{\eta}$ = annual system efficiency
$\theta$ = temperature, °C ( °F = 1.8 θ + 32)
$\lambda$ = thermal conductivity, W/mK (6.938 $Btu/hr\ ft^2\ °R$)
$\rho$ = density, $kg/m^3$ (0.0624 $lbm/ft^3$)

### Subscripts

| | | | |
|---|---|---|---|
| A | Absorber | o | outlet |
| aux | auxiliary | p | pipe |
| F | fluid | r | thermal radiation |
| f | fin | S | Storage tank |
| G | glass cover | s | sotred |
| H | house | sol | solar |
| HW | hot water | SH | space heating |
| I | insulation | ∞ | ambient |
| in | inlet, inner | | |

## LITERATURE CITED

1. Arafa, A. A., The dynamic Thermal Performance of Buildings, considered for publication in the 23rd ASME/AIChE National Heat Transfer Conference.

2. Hottel, H. C. and Whillier, A., Evaluation of Flat-Plate Solar Collector Performance Proc. of the Conference on the Use of Solar energy, Univ. of Arizona, Vol. II, p. 74, (1957).

3. Liu, B. Y. H. and Jordan, R. C., The Long-Term Average Performance of Flat-Plate, Solar energy Collectors, Solar Energy 7 (2), 53 (1963).

4. Siebers, D. L., Viskanta, R., Analytical Study of Flat-Plate Solar Collector Performance: "Realistic Models and Design Concepts", Technical Report No. ME-HTL-752, Purdue University, December 1975.

5. Beckmann, W. A., Klein, S. A. and Duffie J. A.; Solar Heating Design by the f-Chart Method, John Wiley, New York, p. 5 (1977).

6. Arafa, A., Fisch, N., and Hahne, E., Transient Behavior of Solar Flat Plate Collectors, Second Int. Solar Forum, July 1978, Hamburg, (West Germany), Vol. I, pp. 549-566, (1978).

7. Hahne, E. and Arafa, A., Optimale and wirtschaftliche Solare Heizungssysteme, Warme, West Germany, 87, pp. 63-68, (1981).

8. Arafa, A., "Ein Beitragzur Optimale Auslegung Eines Solaren Heizungs-Systems Fuer Ein Wohngebaeude", PhD thesis, Stuttgart University, Stuttgart, West Germany (1982).

9. DASCRU, Automatic Step change Merson Differential Equation Solver, International Mathematical and Statistical Librarys, Inc., Houston, Texas.

10. Pogany, D., Ward, D. S. and Löf, G. O. G., The Economics of Solar Heating and Cooling Systems, Proceedings of the ISES Conference, Los Angeles, July 1975.

11. Reiss, H., A Long Term Data Accumulation Program Using Six Differently Located Solar Energy Heating Installations for the Supply of Warm Water in the Federal Republic of Germany, Solar Energy Vol. 30, No. 2, pp. 133-146, 1983.

## APPENDIX

$c_A$ = 500.00 J/kgK  
$c_G$ = 774.00 J/kgK  
$c_I$ = 800.00 J/kgK  
$d$ = 0.01 m  
$d_{in}$ = 0.009 m  
$\alpha_A^*$ = 0.95 -  
$\delta_A$ = 0.001 m  
$\delta_f$ = 0.003 m  
$\delta_G$ = 0.003 m  
$\delta_I$ = 0.1 m  

$\lambda_A$ = 21.00 W/mK  
$\lambda_G$ = 0.76 W/mK  
$\lambda_I$ = 0.032 W/mk  
$\varepsilon_A^*$ = 0.95 -  
$\varepsilon_G^*$ = 0.876  
$\rho_A$ = 7880.00 kg/m³  
$\rho_G$ = 2500.00 kg/m³  
$\rho_I$ = 50.00 kg/m³  

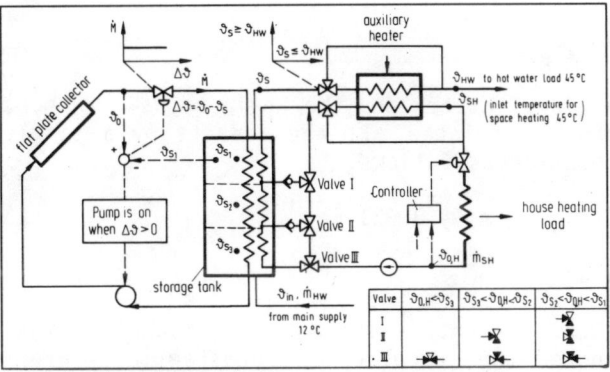

Figure 1. Schematic of a solar heating system (collector facing south, tilted 45°).

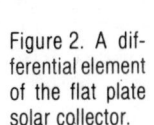

Figure 2. A differential element of the flat plate solar collector.

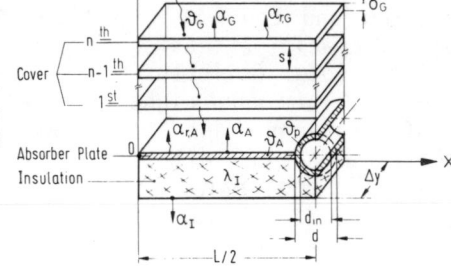

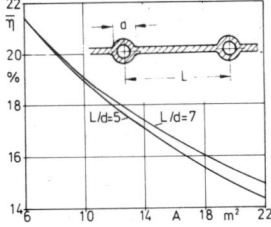

Figure 3. Effect of the collector area on the system efficiency.

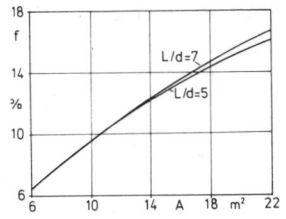

Figure 4. Effect of the collector area on the solar fraction.

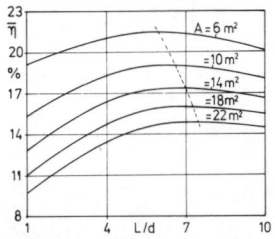

Figure 5. Effect of L/d on the system efficiency.

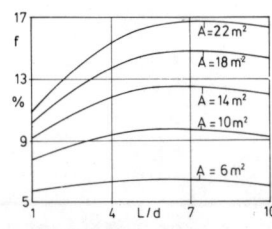

Figure 6. Effect of L/d on the solar fraction.

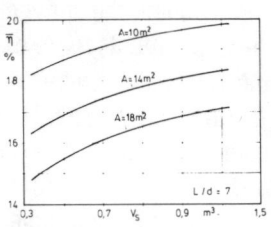

Figure 7. Effect of the storage volume on the system efficiency.

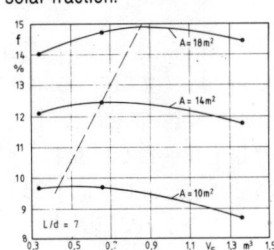

Figure 8. Effect of the storage volume on the solar fraction.

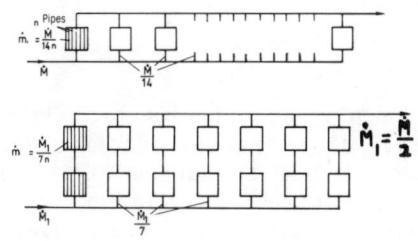

Figure 9. Different collector arrangements with a total area of 14 m² (Reynolds = 160.00 in each pipe).

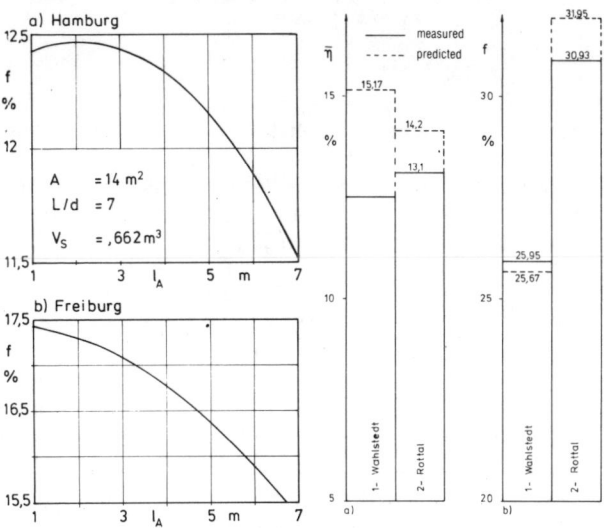

Figure 10. Effect of the collector length on the solar fraction.

Figure 11. Measured and predicted: (a) system efficiency, (b) solar fraction.

# NEW CORRELATIONS FOR CONVECTIVE HEAT TRANSFER COEFFICIENTS FOR FLAT PLATE SOLAR COLLECTORS

Said Shakerin ■ Mechanical Engineering Department, Colorado State University, Fort Collins, CO 80523

Experiments were performed to evaluate the convective heat transfer coefficient for a flat plate mounted in a wooden model of a roof of a building. The experiments were carried out in a closed-circuit wind tunnel and included parametric adjustments of the roof pitch and Reynolds number. The roof pitch was varied from zero to 90 degrees and Reynolds number, based on the plate length, ranged from 58000 to 250000. A transient, one lump, thermal approach was used for heat transfer calculations. Due to a separation bubble at the leading edge of the model, i.e. the roof, at the angles of attack of less than 40 degrees, the flow became turbulent after reattachment. This resulted in a higher heat transfer coefficient than previously reported in the literature. At the higher angles of attack, the flow was not separated at the leading edge and the heat transfer coefficient was found to be approximately independent of the angle of attack.

## INTRODUCTION

The first published experimental study of the convective heat transfer coefficient for flat plates is due to Jurges (1). He investigated the forced convection heat transfer between an isothermal, square plate and air flowing parallel to the plate. The outcome of that experiment was a linear equation for estimating the convective heat transfer coefficient as a function of airstream velocity as shown here,

$$h_w = a + bV \qquad (1)$$

where a, b are contants and documented by McAdams (2). Equation (1) has been used to predict the convective heat transfer coefficient due to the wind for the flat plate solar collectors for many years.

Sparrow and Tien (3) reported the limitations of Equation (1) and, hence, performed a series of wind tunnel experiments to obtain the convective heat transfer coefficient for a flat plate. Through a sublimation technique, and via mass transfer calculation, they evaluated the convective heat transfer coefficient for a flat plate. Their experiment encompassed a wide range of angles of attack and angles of yaw and extended over a Reynolds number range from about 20000 to 100000. Due to three dimensional flow effects, the heat transfer coefficients were insensitive to both the angle of attack and the angle of yaw. Hence, they correlated all the results by a single equation as follows,

$$j = 0.931 \, Re^{-0.5} \qquad (2)$$

where $j = (\frac{h}{\rho c \, U_\infty}) Pr^{2/3}$ with an accuracy of ±2.5 percent. Through an example, they showed that Equation (1) substantially overestimated the heat transfer coefficient due to wind for the flat plate solar collectors. In another sublimation experiment, Sparrow and Lau (4) investigated the effect of adiabatic extension surfaces on wind-related solar collector heat transfer coefficients. They used eight different test plate configurations and reported a "global" correlation of all the results in the following equation,

$$j = 0.86 \, Re^{-0.5} \qquad (3)$$

with a deviation of ±10 percent.

Attention will now be turned to the limitations of the above studies. First, in order to avoid the blockage, three of the four edges of the cassette, which hold the sublimed material, was beveled as shown schematically in diagram A of Figure 1. However, in reality, either the solar collector's thickness or the roof's overhang thickness might block the wind. Second, as Sam et al. (5) reported, there exists a separation bubble at the leading edge of a model of rectangular cross section with an aspect ratio of 6 for angles of attack from 0 degree to about 40 degrees as shown in diagram B of Figure 1. The flow after reattachment was found to be turbulent. In view of these phenomenons, it was proposed to perform experiments with a more realistic model of a solar collector which is usually mounted on the roof of a building. This realization was the initial motivation for the present research.

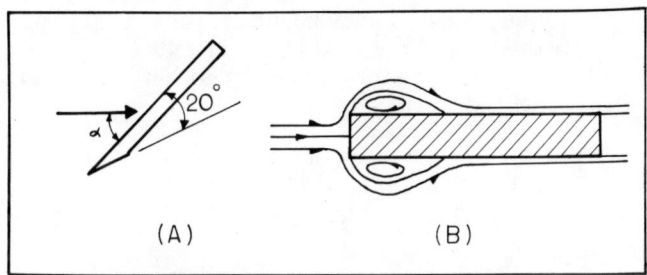

Figure 1. Illustrations of the geometries used in (3) and (5), diagrams (A) and (B), respectively.

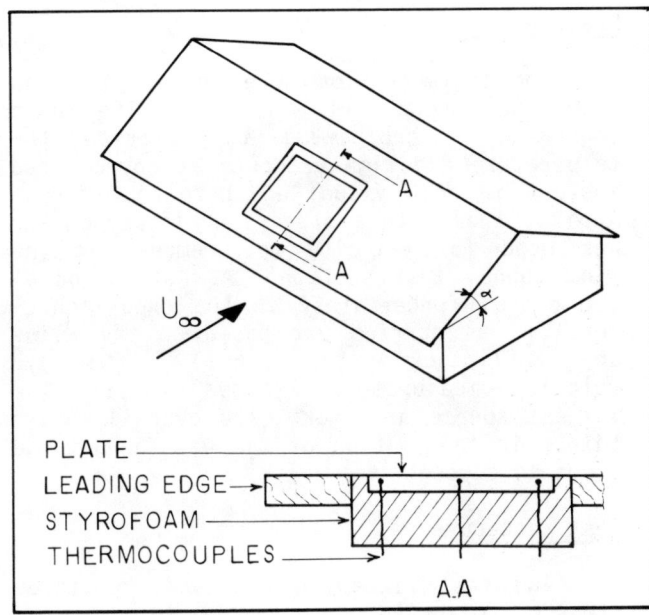

Figure 2. Schematic of the model

## THE EXPERIMENTS

### Apparatus

A rectangular aluminum plate with dimensions of 30.17 x 29.21 x 0.97 cm (11.88 x 11.50 x 0.38 in) was mounted on a roof of a model of a building. A 6.35 cm (2.5 in) thick styrofoam block was glued to the plate's back-face to minimize the unwanted heat losses. Through proper machining, three Copper-Constantan thermocouples were located within a few millimeters from the surface at the centerline of the plate to measure the plate temperature. Due to the high conductivity of the aluminum plate, the thermocouple's reading is expected to be very close to the actual surface temperature. The model was constructed from plywood as shown schematically in Figure 2. Through a careful design, the roof pitch could be set to any angle from 0 degree to 90 degrees at 5 degrees increments. The model was fitted exactly into the width of a closed circuit wind tunnel with a test section of 1.5 m (4.92 ft) wide by 1.2 m (3.9 ft) high and 9.2 m (30.2 ft) long.

### Experimental Method

A transient, one lump, thermal method was used in this study. It consists of heating the plate to a high temperature and then letting the plate cool to the ambient temperature under a fixed air flow rate. Measurements of the rate of the temperature decay can be analyzed to calculate the convective heat transfer coefficient. Although this method is the easiest to apply; but a meaningful result could be only achieved if the entire plate undergoing a thermal decay is at a uniform or nearly uniform temperature. A nearly uniform temperature distribution throughout the plate could be obtained if the conduction resistance in the plate is much less than the convection resistance at the plate surface. As shown by Shakerin (6), an error of about 4% in uniformity of temperature was never exceeded in this research. An experiment was carried out to evaluate the conduction losses through the styrofoam block as a function of the plate temperature. The conductance was found to be approximately a constant over a plate temperature range from 49°C (120°F) to 60°C (140°F).

## Measurements

The thermocouples were connected to an Esterling Angus data logger. The data logger was programmed to provide the temperature and time readouts at any desired instant in both visual and hard copy forms. A pitot tube with a precision micromanometer was used for velocity measurements at the wind tunnel test section. An estimation of the overall uncertainty in the experimental results was carried out based on the Kline and McClintock method (6,7). The air velocity measurements showed to be the largest source of error. The overall uncertainty in calculation of j was found to be about ±4 percent.

## Test Procedure

The test procedure consisted of three steps as following,

1. A temperature controlled hot plate with cast aluminum top was inverted on the aluminum plate for the heating purpose. To reduce the heat losses and allow for a uniform heating, the hot plate was moved around the aluminum plate. The portions of the aluminum plate which were not covered with the hot plate were insulated by a fiberglass blanket. Once the temperature at the center of the plate reached about 76.6°C (170°F), the hot plate was turned off and removed from the model.

2. A thick insulation blanket was set on the entire roof for about 15 minutes in order to ensure a uniform temperature distribution in the aluminum plate. A maximum deviation of 1.7°C (3°F) among thermocouple readings was allowed but this seldom exceeded 1.1°C (2°F).

3. The insulating blanket was removed from the wind tunnel and the air flow was allowed to pass over the model, at a predetermined velocity, until the temperature at the plate's center dropped to 37.8°C (100°F). Depending on the rate of the temperature decay, the scan time for temperature recording was typically between 10 to 30 seconds. Data in the temperature range of 49°C (120°F) to 60°C (140°F) were used for the heat transfer calculations. The above steps were repeated for five different angles of attack and six different air velocities.

## MATHEMATICAL MODEL

The unsteady energy equation for the aluminum plate with a uniform temperature is shown here

$$\rho v c \frac{dT}{dt} = -[H+K(T)](T-T_\infty) \quad (4)$$

where $K(T)$ represents the conduction losses through the styrofoam block and H represents the convective losses from the aluminum plate which, for simplicity, is assumed to be temperature independent. Note that, as described earlier, $K(T) \approx K$ if 49°C < T < 60°C. After rearranging Equation (4) and integrating both sides, we obtain,

$$\bar{h} = -\frac{\rho v c}{tA} \ln\left(\frac{T-T_\infty}{T_0-T_\infty}\right) - \frac{K}{A} \quad (5)$$

where $\bar{h} = \frac{H}{A}$.

By substituting the average Nusselt number into Equation (5), the Colburn factor, j, can be written as,

$$j = \frac{Pr^{2/3}}{\rho'c'U_\infty}\left[-\frac{mc}{tA} \ln\left(\frac{T-T_\infty}{T_0-T_\infty}\right) - \frac{K}{A}\right] \quad (6)$$

where $j = \frac{\overline{Nu}}{Re_L Pr^{1/3}}$

Equation (6) was used to calculate j from the experimental data.

## RESULTS

Figure 3 shows the results of the present study along with the results previously published in the literature. It is evident that at the angles of attack of 0 degree and 30 degrees, the heat transfer coefficients are higher for given Reynolds numbers. This is due to the formation of a separation bubble at the leading edge of the roof at these angles of attack. In reality, the solar collector thickness would cause a similar separation in the flow. The flow after reattachment became turbulent and hence, resulted in a higher heat transfer coefficient. Based on the previous published results (5), the critical angle below which the flow could become turbulent after reattachment is about 40 degrees. At higher angles of attack (i.e. above 40 degrees), however, the flow was not separated and remained laminar over the length of the plate. The results of the present study for laminar flows are in a good agreement with those given in (3,4). The result for the zero degree angle of attack is in a reasonable agreement with Jurges linear equation. The overall results of the present study are correlated in the following equations,

$$j = 1.23 \, Re_L^{-0.5} \quad \alpha < 40° \quad (7)$$

$$j = 0.90 \, Re_L^{-0.5} \quad \alpha \geq 40° \quad (8)$$

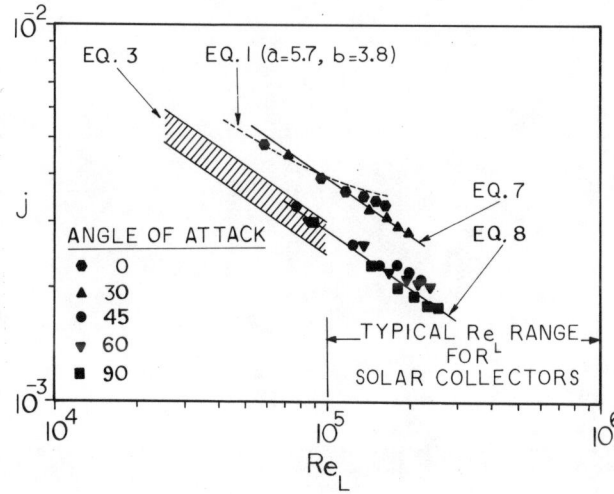

Figure 3. Plots of Colburn factor vs. Reynolds number of present and previous studies.

## CONCLUSIONS

Correlations for the convective heat transfer coefficient for the flat plate solar collectors as a function of Reynolds number are given for the two different ranges of the angle of attack. The flow over the flat plate solar collector is turbulent for the angles of attack of less than about 40 degrees due to a separation bubble at the leading edge. However, at the higher angles of attack the flow is laminar.

## ACKNOWLEDGMENT

The author would like to thank Professor M. B. Larson at Oregon State University for his many valuable suggestions through the completion of this research.

## NOTATION

a,b    constants

A    surface area of aluminum plate

c    specific heat of aluminum plate

c'    specific heat of air

$h_w$    convection coefficient due to wind

$\bar{h}$    average heat transfer coefficient per unit area

H    total heat transfer coefficient

j    Colburn factor

K    conductance loss through insulation material

L    plate length

m    mass of aluminum plate

Nu    average Nusselt number

Pr    Prandtl number

Re    Reynolds number

$Re_L$    Reynolds number based on the plate length

t    time

$T_0$    initial plate temperature

T    plate temperature

$T_\infty$    ambient temperature

v    volume of the plate

V    average wind velocity

$U_\infty$    free stream velocity in the wind tunnel test section

$\alpha$    angle of attack

ρ     density of aluminum

ρ'     density of air

## LITERATURE CITED

1. Jurges, I. W., Gesundh.-Ing., 19 (1), 1 (1924).

2. McAdams, W. A., <u>Heat Transmission</u>, Third Edition, p. 249, McGraw-Hill, New York (1954).

3. Sparrow, E. M., and Tien, K. K., Journal of Heat Transfer, 99, pp. 507-512 (1977).

4. Sparrow, E. M., and Lau, S. C., Journal of Heat Transfer, 103, pp. 268-271 (1981).

5. Sam, R. G., Lessmann, R. C., and Test, F. L., Journal of Fluids Engineering, 101, pp. 443-448 (1979).

6. Shakerin, S., "The Correlation Between Convective Heat Transfer Coefficient, Wind Velocity and Angle of Attack for Flat Plates -(An Application to Flat Plate Solar Collectors)," Master's Project, submitted to Mechanical Engineering Department, Oregon State University (1981).

7. Holman, J. R., <u>Experimental Methods for Engineers</u>, Third Edition, pp. 43-48, McGraw-Hill, New York (1978).

# PERFORMANCE ANALYSIS OF HVAC SYSTEM FUELED BY SOLAR ENERGY OR A COMBINATION OF SOLAR, WOOD OR NATURAL GAS

O.J. Hahn, M.K. Richardson, R.C. Birkebak ■ Department of Mechanical Engineering
University of Kentucky, Lexington, KY 40506

W.R. Curtis ■ Berea Forestry Science Laboratory, U.S. Department of Agriculture
Forest Service, Berea, KY 40403

The heating ventilation and air conditioning (HVAC) system for a 2100 m² laboratory and office complex of the Berea Forestry Sciences Laboratory of the United States Forest Service was evaluated. The system consists of flat plate water cooled solar collectors, a storage tank, two small wood or gas backup boilers and an absorption cooler. Initial startup and shakedown problems were encountered ranging from premature deterioration of collector panels, system layout limitations to corrosion and misfunction of check valves. Efficiencies of all components were measured and combined into an overall computer model of the building and HVAC system to arrive at a predictable analog for aid in system operation.

As a result of this study system modifications were made to reduce operator time, maintenance, and ease of preheating the building in the morning.

## INTRODUCTION

Rising costs of conventional fuels have resulted in increasing interest in the application of solar energy for the heating and cooling of buildings, both residential and commercial/industrial. The present study deals with the analysis of an HVAC system capable of being fueled by solar energy or a combination of solar and wood or natural gas. The Berea Forestry Sciences Laboratory of the United States Forest Service, located in Berea, Kentucky has a 2100 m² (22,500 ft²) building comparable to other small industrial, commercial and institutional structures with energy requirements on the order of 0.26 - 0.51 GJ (0.25-0.5 x $10^6$ Btu). The energy system consisting of flat plate collectors, boilers, an absorption chiller and storage tank was designed by Kaiser-Taulbee Associates, Lexington, Kentucky.

In this investigation, the efficiency of the solar collector circuit (the primary energy system) and the boiler circuit (the auxiliary energy system) were determined and compared with the efficiencies specified by the manufacturers. A number of problems were discovered and rectified to decelerate the deterioration of the system and improve the overall efficiency to best utilize the energy it consumes.

## DESCRIPTION OF THE FACILITY

The Berea Forestry Sciences Laboratory is a 2100 m² (22,500) ft² facility located one mile east of Berea, Kentucky. It serves as an office facility for up to thirty employees and offers nine laboratories, a small computer room, a library, a conference room and various work areas.

In the design of the building, an effort was made to increase the energy efficiency of the building. Insulation was increased to prevent excessive energy loss to the environment. Roughly 48 per cent of the building is below grade (i.e., below the immediate ground level) (1).

If the external wall is maintained as closely as possible to the temperature of the environmental air, the convective energy loss is a minimum. The radiative loss component is assumed to be much less significant. The cooling load of the passive solar energy system when the solar rays are incident upon the building has to be considered. To protect the east and west walls from direct insolation, and thereby keep the surface temperature of the wall closer to the environmental air temperature, a long wooden baffle was placed parallel to each of these walls. These baffles are composed of fixed vertical louvers. The baffles not only block the direct solar gain which would normally be expected during the early morning and late afternoon hours during the summer, but they also reduce the air flow over the wall which reduces the effective heat transfer coefficient. The baffles are set at a distance from the wall sufficient to allow diffuse sunlight to enter the office windows facing these directions.

The north wall is almost entirely below grade and therefore has no appreciable surface area from which it can transfer heat via convection. The primary mode of energy exchange on the north wall is heat conduction into the soil. Due to the effective insulation of soil next to a wall, little can be done to reduce the energy losses on this side. It should be noted that the laboratories which require air conditioning systems independent of the main HVAC system to balance the heat produced by laboratory equipment are all situated with their external walls along the north side.

The south side of the building is exposed to the greatest possible direct solar gain. The manner in which this problem was addressed was to slope the soil against the wall to a point just below the office windows facing this side. An overhang was constructed above the windows long enough to block the direct solar gain during the summer, but short enough to permit the favorable use of the solar radiation during the winter. Between the grade line immediately adjacent to the wall and the lower edge of the office windows, a line of shrubbery was planted for not only the obvious esthetic purposes, but to help break up the air flow over the wall so as to reduce the convective heat transfer coefficient.

## DESCRIPTION OF THE HVAC SYSTEM

The solar collector circuit was designed to provide the majority of the energy required by the HVAC system. Figure 1 shows the general layout of the HVAC system. There are 120 independent flat plate collectors (copper tubes, on a copper plate with black chrome paint, Grumman model 432A) solar collectors, each one 1.2 m by 2.4 m (2). They are connected in parallel banks of ten, with four banks comprising a row. The collectors are mounted atop the building where they are sloped at an angle of 36.4 degrees to the horizontal, facing the south with an azimuth of 13 degrees. To allow for an enhancement of the solar radiation impingement on the collectors, the roof was corrugated in such a way that during periods of intense insolation the radiation would be reflected off the roof surface onto the collectors. To promote this, the roof was coated with CRM 100, a reflective roofing surface (3). A study showed that this coating increased the collected solar energy by less than 2%. This energy gain is not sufficient to justify the coating.

The water storage tank, located immediately underground on the north side of the building is 8.2 m (27 ft) long, 2.4 m (8 ft) in diameter, and has a storage capacity of 45.5 $m^3$. In addition to the natural insulation provided for the storage tank by the so it is surrounded by 0.1 m of styrofoam insul tion and gravel fill. At its nearest point the building, the storage tank is a mere 0.3 from the outside wall. Water from the stora tank passes through the solar collectors wit a flowrate of 4.5 x $10^{-3}$ $m^3$/s (72 gpm), unde the force of a base mounted 1 1/2 horsepower circulating pump, called the solar pump, to return to the storage tank.

Auxiliary energy is added to the system by one of two water-wall boilers, models LB-30 and LB-70, manufactured by Riteway Manufacturing Company. These two boilers are capable of being fueled by either firewood o natural gas. Water is circulated through th boilers in one of two ways. Water is circulated by circulator-1, a 5.7 x $10^{-3}$ $m^3$/s pum with 15 m of head from the storage tank, through the boilers and back to the storage tank to charge the tank, or circulator-3, a 3.8 x $10^{-3} m^3$/s pump with 5.9 m of head, can circulate water through the boilers and into the heat exchanger to transfer energy to the water entering the building. The heat exchanger is a double-pass model with copper tubing and a heating surface of 1.4 $m^2$.

For cooling, a Arkla model WFB 300 absorption chiller with a design delivered capacity of 90 KW (306,000 Btu/hr) is used. H water is supplied to the chiller from the storage tank through a 6.3 x $10^{-3}$ $m^3$/s, 15 m head circulating pump (circulator-2). The chilled water is circulated through the building by a 5.7 x $10^{-3}$ $m^3$/s, 18.3 m head pump (circulator-4).

## DATA ACQUISITION SYSTEM

The data acquisition system consists of Digital PDP-114/34 host computer, an Omega OM-104 multiplexer, a smaller multiplexer co structed by Raymond Willis (an employee of t Forest Service), an Eppley pyranometer mounte horizontally, twelve duel element copper-constantan (T-type) thermocouples in the HVAC system, a thermocouple to measure the enviro mental air temperature, a thermocouple to me sure the indoor space temperature, and other input from gas flow and fuel input measurements. The dual element thermocouples place in the wells of the piping of the HVAC syste allow the temperatures to be read by the system operator on an Omega 2166A digital

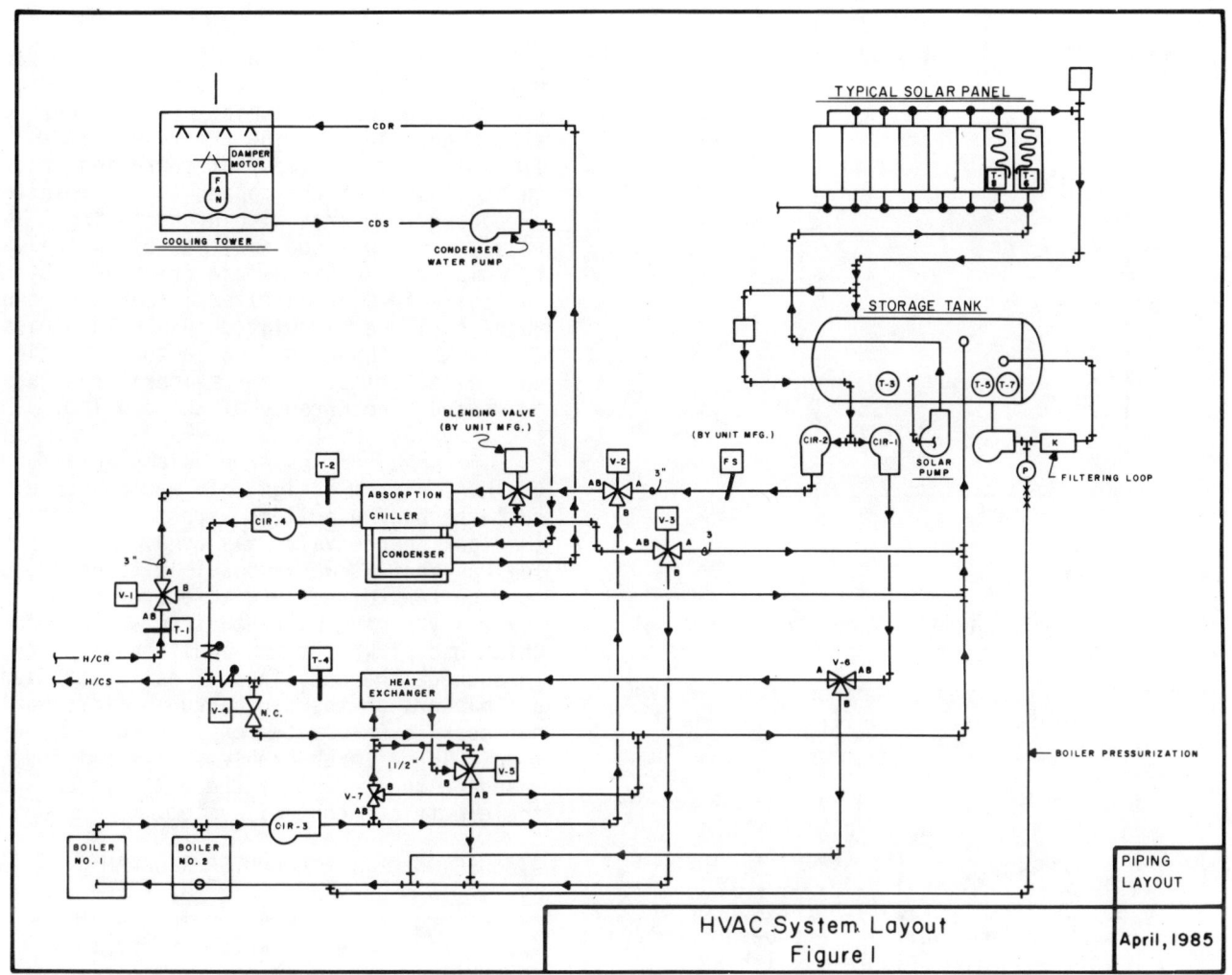

Figure 1. HVAC system layout

thermometer located in the boiler room for the purpose of system monitoring.

The thermocouple and other voltages were read from the multiplexer once every fifteen seconds and averaged over one minute. Once sixty such averages were recorded, each voltage was averaged over one hour and stored in one of two disc files: HVAC temperatures and the solar flow, wind, outside air temperature, and other data.

## ANALYSIS AND MODIFICATIONS

To determine how well the HVAC system was utilizing the energy made available to it, efficiency tests were performed on the collector system and the backup boilers. The efficiency was defined as the energy absorbed by the cooling water divided by the energy available to that system component.

$$\eta = \frac{\text{Energy Absorbed}}{\text{Energy Available}}$$

The boiler efficiency was found to be 56 percent, compared to the manufacturer's rated value of 75 percent. This efficiency was determined by an energy balance of combustion products of natural gas and enthalpy increase of coolant. Combustion with 30% excess air resulted in 8.2% $CO_2$, 0.6% $CO$, 5.6% $O_2$ and 85.6% $N_2$. A graph of the solar collector efficiency as a function of solar time is shown in Figure 2 along with the manufacturer's efficiencies for the same parameters. The average efficiency was determined as 35 percent compared to 52 percent from the manufacturer's data.

The cause of this reduction in the efficiencies was determined to be corrosion inside the collector tubes, outside the tubes and on the collector surfaces. The internal corrosion was electrogalvanic due to the various

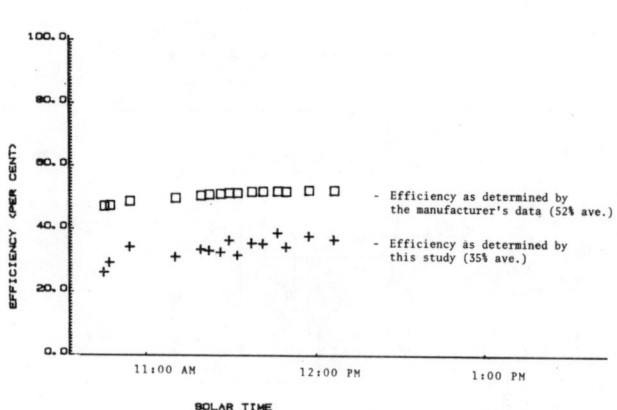

Figure 2. Collector efficiency versus solar time.

HVAC components. Water samples from the HVAC system were analyzed to confirm the presence of copper.

A corrosion protection has been added in January 1985 consisting of a KB 609 Closed System Corrosion Inhibitor (a solution of sodium borate, sodium nitrate, tolytriazole, and a polymeric dispersant). The corrosion products which have built up are being removed in the filter system (Figure 1) with an acid solution.

Corrosion was detected on the black chrome-painted copper absorber plates in 95 percent of the solar collectors. This was due to the formation of water droplets on the inside of the inner cover plates. These droplets subsequently fall onto the absorber plate. The plate becomes green in the areas which corrode, causing a reduction in the absorbed solar energy as mentioned above. It was not determined how much of the collector efficiency reduction was due to this type of corrosion. To alleviate this problem, dry air can be periodically circulated through each collector through small holes drilled into either end of the collectors.

The most severe problem encountered with the solar collectors was the total failure of several collectors on two separate occasions. Each incident occurred in the later afternoon of a very cold day with a clear sky. The solar circuit is a flow-back system permitting the water in the solar collectors to flow directly back into the storage tank when the control system measures an air temperature within the collectors of less than 125 F. On the two days in questions, however, the water did not evacuate the copper tubing before freezing, expanding, and bursting the tube. It is believed that the buildup of corrosion products on the inside of the copper tubing resulted in a reduced flow area within the tubing. This reduced area was too small to allow the water to completely drain before freezing. Since the collectors are arrayed in banks of ten, water could be circulated in a path bypassing the banks with damaged collectors. Draining the collectors at higher temperatures has prevented a recurrence of the problem.

A number of lesser problems were found in the HVAC system design. It was possible to fuel the backup boilers with firewood only if the tank charge valve was open. Another problem was that it was impossible to charge the storage tank directly with energy from the auxiliary system without circulating the water which had picked up the additional energy throughout the building. Finally, the ashpit blowers on the two backup boilers would not operate unless the chiller was in operation. These three problems were eliminated by changing the piping of the HVAC system and in the electrical control system.

## DISCUSSION OF THE COMPUTER PROGRAM

A program was written for an HP-1000 Series F computer to model the performance of the HVAC system using the results of the efficiency tests performed on the solar collectors and boilers (4). The program is capable of performing a 24-hour energy balance on the system given the data recorded in the disc files. It is also capable of predicting the insolation and energy load for any day of the year through the use of typical days for each month (5). These two capabilities are referred to as the Balance Model and the Predictive Model.

In the Balance Model, the program requests hourly values of the temperature of the water entering and exiting the solar collectors, the boilers, and the building. It also requires hourly values for indoor space temperature, outside air temperature, and solar radiation in addition to the total weight of firewood or the total volume of natural gas consumed by the boilers for the day. From these data, the program calculates the solar energy available, the solar energy collected, the working efficiency of the solar collector system, the energy available to the auxiliary energy system, the auxiliary energy absorbed by the system, and the working efficiency of the boilers. The load placed on the system is determined by performing a mass/energy balance on the water circulating through the building.

The predictive model contains files of average hourly data for each of the twelve months for the indoor space temperature, the solar radiation, and the environmental air temperature. These averages were determined from hourly data collected from November 1981 to June 1984, for each hour of the day. The program requests the day of the year for the model and uses that to translate the radiation measured horizontally into the radiation striking the tilted collector surface (6). The temperature difference across the building skin is used with one of the two building U-factors to calculate the load placed on the system for each hour. The sum of the insolation and the load are then used to predict the auxiliary energy which would be required to meet the demands of the HVAC system. This energy value is then translated into the weight of firewood or the volume of natural gas necessary for the auxiliary system.

Figure 3 shows the normal maximum solar

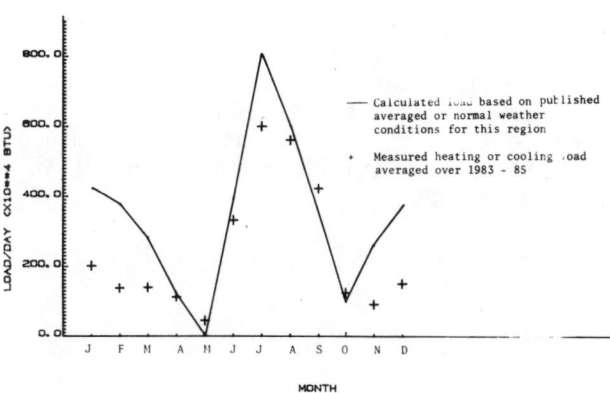

Figure 4. Energy load on the HVAC system.

temperatures for this region) placed on the HVAC system (using an indoor air temperature of 65 F for heating and 70 F for cooling) and the average monthly load as determined in this study. The computer model with given solar and weather information will predict the actual load to within ± 8%. This data has limited value except in estimating variances in predicted loads based on published yearly averages and actual demands. Figure 5 shows the actual load demand of the HVAC system during 1983-85 and the available solar energy as

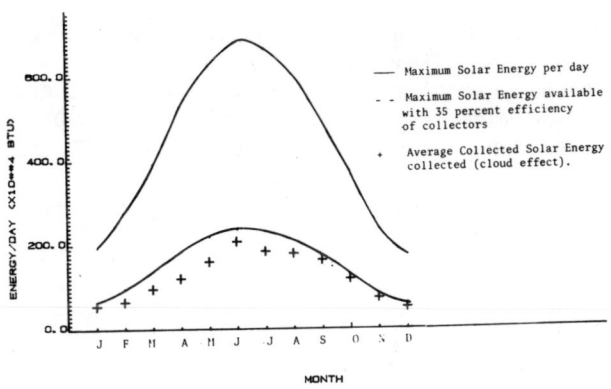

Figure 3. Solar radiation data.

energy available for each month (7), the maximum solar energy which could be collected (using a collector efficiency of 35%), and the average monthly solar energy collected as determined in this study. The scatter of the collected solar energy is the result of cloud cover. Figure 4 shows the calculated normal load (based on published yearly average

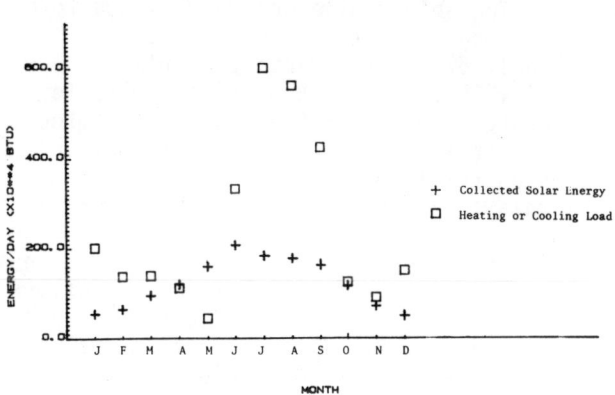

Figure 5. Energy for the HVAC system solar supply and system load.

obtained in this study. With increased solar collector efficiency (to 40%) no additional energy source would be needed for 5 months. This work can be used as a predictive tool for the operator either by running the computer model or using the degree-day method. Figure 6 shows the load on the HVAC system (heating and cooling) as a function of the number of degree-days, as determined from the data in this study. The energy requirements for cooling are higher per cooling degree day than for heating due to the 1.55 coefficient of performance of the absorption chiller and the electrical building input.

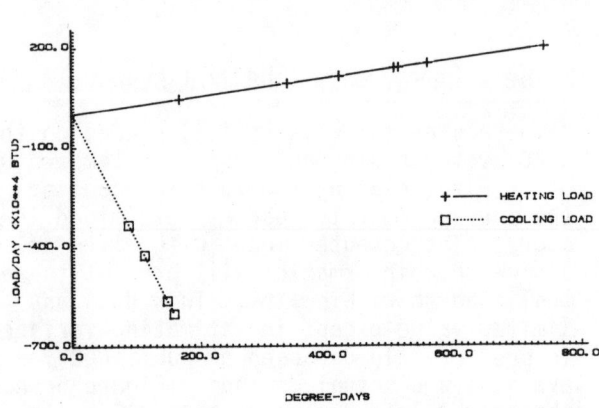

Figure 6. HVAC system load versus degree days.

## CONCLUSIONS

The solar collectors were found to be operating at an efficiency of 35% ($\pm$ 3%) which is only 67% of the manufacturer's value of 52% ($\pm$ 2%). The backup boilers were found to operate at 56.5% ($\pm$ 5%), only 75% of the manufacturer's value of 75%. Electrogalvanic corrosion resulting from the presence of different metals in the HVAC system with no method of corrosion inhibition was determined as the cause of efficiency loss. This corrosion is also responsible for the premature failure of one valve, to date, and is considered a prime factor in the total failure of several solar collectors. Corrosion inhibitors and filters were added. Corrosion products were collected. Future measurements will indicate if system deterioration has been halted.

Using the information from the data files of the computer program Predictive Model, the annual load on the HVAC system is $6.73 \times 10^{11}$ J ($0.64 \times 10^9$ Btu) and the annual solar energy available is $4.15 \times 10^{11}$ J ($0.4 \times 10^9$ Btu). The difference between them gives an auxiliary energy requirement of $2.6 \times 10^{11}$ J ($0.24 \times 10^9$ Btu). This would require $16.5 \times 10^3$ Kg of wood at a cost of $600.00 or $6.9 \times 10^3$ m$^3$ of natural gas at a cost of $1500.00 (using factors of $0.0162/pound of firewood and $6.06/$10^6$ Btu). A complete economic analysis would include life cycle cost, tax credits (if privately owned), etc. This evaluation has been deferred until maintainance and fuel costs have been established after the recent system modifications.

The modification in the operating system have reduced the operators time and the requirement to preheat the building in the morning. The fuel costs of wood and gas suggest that gas is the optimum backup fuel if the operators time is considered.

Had the HVAC been equipped with corrosion protection, the system would certainly have functioned closer to the prediction of the system designer who believed that the system would supply 75% of the building energy needs for heating and cooling. In the absence of corrosion, the results of this study conclude that the HVAC system would receive 63% of its fuel requirements from the sun. With the onset of the corrosion in the system at the present level only 55% of the energy needs could be met by the sun. The recent but belated installation of corrosion protections has stabilized the system. Corrosion product have been collected in filter. System performance seems to have stabilized (or improved as determined with only one measurement).

Final judgements on the economic worth of the system have to be based on at least one year of relatively trouble free operation. The modifications in the HVAC operation system and corrosion protection look promising.

## ACKNOWLEDGEMENT

The support of the U. S. Forest Service of the U. S. Department of Agriculture for this project under cooperative agreement #23625 is gratefully acknowledged. The University of Kentucky Research Foundation provided matching funds to make this project possible. Raymond Willis spent many hours in assisting in the data acquisition and establishment of data files. The assistance of Jim Clark, the HVAC system supervisor and David Thomas, a University of Kentucky senior contributed greatly to the successful completion of the project.

## LITERATURE CITED

1. Jorgensen, S.A., Editor, International Society for Ecological Modelling, vol. 6, "Handbook of Environmental Data and Ecological Parameters", Pergamon Press, New York, New York, 1979.

2. Grumman Sunstream Solar Products Information Sheet, Grumman Energy Systems Company, Melville, New York, p. 3.

3. "Berea Research Laboratory and District Ranger Office", U.S. Department of Agriculture, Specifications, Webb Dillehay Design Group, p. 7A-1.

4. Richardson, M.K., Master's Thesis, University of Kentucky (1985), "Performance Analysis of HVAC System Fueled by Solar Energy or a Combination of Solar and Wood".

5. "Solar Engineering of Thermal Processes", Duffie, J.A., Beckman, W.A., John Wiley and Sons, Inc., New York, New York, 1980.

6. Klein, S.A., Solar Energy, 19, 325 (1977).

# TEMPERATURE DISTRIBUTIONS AND HOT SPOTS IN A LAMINAR FLOW PHOTOCHEMICAL REACTOR

Falin Chen and Arne J. Pearlstein ■ Aerospace and Mechanical Engineering Department
University of Arizona, Tucson, Arizona 85721

The determination of temperature distributions, maximum temperatures, and hot spot locations in photochemical reactors is of considerable importance in their design and operation. Typically, unwanted thermal reactions will compete with the desired photochemical reaction(s), resulting in reduced yield and selectivity, and increased byproduct formation. The kinetics of the thermal reactions typically have an Arrhenius type temperature dependence, and so good thermal control is essential. In addition, the formation of wall deposits (due to the physical or chemical adsorption at a solid boundary of free radicals or other reactive intermediates formed in the primary photochemical event) depends strongly on the temperature at the wall where the deposition occurs. Wall deposit formation has been a significant problem in the scale-up of various photochemical processes, including the photonitrosation of cyclohexane to give an intermediate of the nylon-6 monomer caprolactam.

In the present work, we compute the temperature field in a laminar flow circular tube photoreactor subjected to an azimuthally uniform, radially incident monochromatic light source. The mass transfer and kinetics problem for this photoreactor has been treated by a number of authors. We have solved a steady energy equation of elliptic type, in which we have, for generality, retained the axial conduction term. We consider a photosensitized reaction and take the photosensitizer concentration to be a constant.

The most interesting result concerns the movement of the hot spot away from the centerline and towards the wall as the optical density (OD) increases. This is due to the fact that, as the OD increases, the light absorption (and internal heating resulting from molecular energy transfer processes) shifts closer and closer to the wall. The effect is to move the hot spot out towards the wall which is held at a fixed temperature by, say, forced convection cooling. As a result, the maximum temperature in the reactor decreases, but the temperature near the wall increases. This will probably be advantageous for product selectivity and yield in the central core of the reactor, but disadvantageous with respect to the formation of light absorbing deposits on the interior wall.

Axial heat conduction has significant effects for Peclet numbers less than 50, especially in the entry and exit regions of the illuminated section. One effect is a considerable upstream displacement of the hot spot at all optical densities, along with a reduction in the maximum temperature.

The results show that there is a critical value of the optical density, depending on the Peclet number below which the hot spot is located on the center-line, and above which the hot spot begins to move toward the wall.

# SOLAR RADIANT PROCESSING OF GAS-PARTICLE SYSTEMS FOR PRODUCING USEFUL FUELS AND CHEMICALS

A.J. Hunt, J. Ayer, P. Hull, F. Miller, R. Russo, and W. Yuen ■ Lawrence Berkeley Laboratory
University of California, Berkeley, CA 94720

This paper reviews the research program at LBL studying direct radiant heating for the production of fuels and chemicals. The research is investigating the use of gas-particle suspensions to absorb concentrated sunlight to supply energy for endothermic chemical reactions. The goal of the research is to understand the optical, thermodynamic, and chemical processes in solar heated particle suspensions through a balanced program of analytical and experimental investigations. The use of small particles as solar absorbers is discussed and spectral extinction efficiencies are calculated from Mie theory for various particle sizes. An equation for calculating the heat transfer from a particle to the surrounding gas at arbitrary Knudsen number is developed. The experimental section outlines the current laboratory studies of direct radiant heating that utilizes a high intensity radiant source to simulate concentrated sunlight. Some preliminary experimental results are presented.

## INTRODUCTION

LBL has been involved for the past seven years in the development of solar thermal receivers that utilize suspended particles as solar absorbers and heat exchangers to heat gases for power and industrial process heat applications ( 1 ). This program resulted in the design, construction ( 2 ), and successful solar demonstration ( 3,4 ) of the Mark I, Small Particle Heat Exchange Receiver (SPHER) at the solar test facility at Georgia Institute of Technology in 1982.

Our present research extends the earlier work to new methods of initiating chemical reactions. The unique combination of high direct solar flux density and high temperatures in a gas-particle suspension offers a new and unexplored environment for chemical processes. Direct radiant heating of gas-particle mixtures may be used for heating a working gas, processing chemical feedstocks or inducing chemical reactions in the suspending gas. However, before efficient and effective receiver/reactor designs can be developed it is imperative to understand the underlying physical processes. The particles must be absorbing in order to convert the radiant solar energy to thermal or chemical energy. The complex index of refraction, size, and shape of the particles have profound effects on the optical and thermal properties of gas-particle mixtures. The choice of particle size and mass loading (mass of particles per unit volume) have important effects on the size and shape of the receiver, reaction time and process conditions. Particle entrainment methods and the particle suspension flow path are also critical to the successful operation of a direct absorption receiver.

## ANALYTICAL STUDIES

To initiate the study of chemistry in radiantly heated gas-particle suspensions, a survey of chemical reactions in or between gases and absorbing solids was performed. A number of suitable reactions were identified and optical, physical, and chemical data were obtained the for the solids involved. Using this data, the optical properties of the particle suspensions were calculated to determine the energy absorbed by the particles. Having determined the heat input to the particles, the heat transfer between the particles and the surrounding gas was calculated to determine the temperatures and heating rates of the particles. To do this, a heat

---

*This Work was supported by the Solar Thermal Fuels and Chemicals Program Managed by the San Francisco Operations Office under the Assistant Secretary for Conservation and Renewable Energy and by the Director's Discretionary Fund through the U.S. Department of Energy under Contract No. DE-AC03-76SF00098.

transfer model was developed that was suitable for all particle sizes treated in this work.

The phenomena involved in the direct radiant heating of small particle suspensions are basically similar for a wide range of particle sizes. However, the importance of some effects can vary widely with particle size and mass loading. This study is concerned with particles small enough to be entrained in flowing gases, e.g., having diameters of a few hundred micrometers down to submicron sizes. The optimum particle loading per volume of gas depends on the application being considered. Very low particle densities are sufficient for gas phase reactions in which the particles act as the radiant heat exchanger or catalyst for gas phase reactions. On the other hand, if thermal processing of a particulate feed stock is desired, wherein the gas basically provides the transport system, much higher loading densities are appropriate. A case involving intermediate particle loading densities is encountered when it is desired to initiate reactions between the gas and particles. One of the goals of this research is to define the range of mass loading for various applications and its effect on receiver design.

## Optical Calculations

The absorption and scattering of sunlight by spherical particles can be calculated once the wavelength-dependent complex refractive index and the particle size distribution are determined. Tabulations of the refractive index were obtained from a literature search, and incorporated in a computer data base for several materials identified in the chemical survey. Mie calculations were used to determine the absorption and scattering efficiencies (cross sectional area for absorption or scattering divided by the geometric cross section of the particle). A computer program using a Mie subroutine was written to calculate the attenuation of monochromatic radiation by a suspension of particles with a known size distribution. The calculation was limited to single scattering. A second computer program was written to calculate the attenuation of energy from radiant sources possessing a broad spectral distribution. In this program the spectral distribution of an arc lamp or sunlight was convolved with the spectral attenuation of the particle distribution from the Mie calculation.

In Figure 1a, the calculated extinction efficiency (sum of the absorption and scattering efficiencies) is plotted versus wavelength for various particle sizes. Because of multiple scattering in particle suspensions, the effective attenuation in a reactor falls between values predicted by extinction and absorption. In figure 1b the attenuation of polychromatic radiation (in this case the solar simulator) is plotted versus the product of particle mass loading and path length for three particle size distributions. This figure permits the rough determination of the size of the receiver required to absorb a desired fraction of light within the particle suspension when the particle mass loading is specified or vice versa. The information in this figure must be combined with the spatial dependence of the radiant flux pattern to determine the heat input to the particles. All graphs were calculated using the optical properties of magnetite.

## Heat Transfer Calculations

As the particles absorb radiant energy they begin to heat and transfer energy to their surroundings by conduction, convection, radiation, and possibly through chemical reactions. Both the initial particle heating rate and the final steady state temperature will be determined by the heat transfer rates. If the particle is small, natural convection is negligible in most cases because the Grashof number is also small (Gr ~ $d^3$,

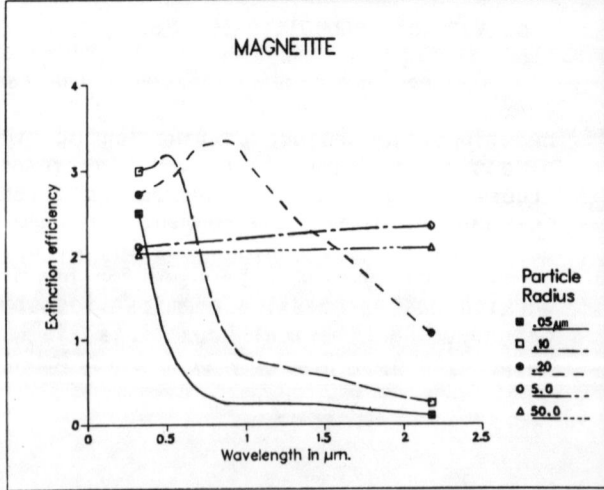

**Figure 1a.** Graphs of the extinction efficiency of a magnetite particle as a function of the wavelength of radiant flux. Graphs for several different particle sizes are shown.

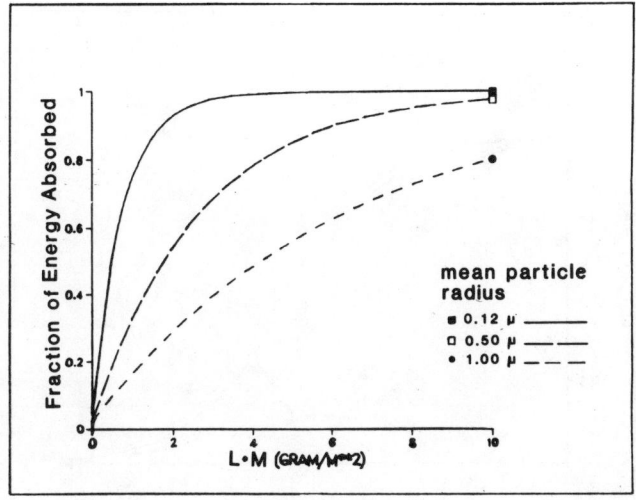

**Figure 1b.** Fraction of the energy absorbed from a Xenon arc-lamp as it passes through a suspension of magnetite particles as a function of the optical depth (product of the mass of particles per unit volume and the path length of the light). Graphs for three different gaussian distributions of particle sizes are shown to illustrate the effects of particle size on the ability of the gas-particle suspension to absorb light.

where d is the particle diameter). Forced convection is not important as long as there is no relative motion between the particle and gas, which is usually the case for micron sized particles. Assuming the particle is not reacting, conduction is thus the only mechanism by which energy is transferred to the gas. (The gas is taken to be transparent; only the particles interact radiantly.) To complete an energy balance on the particle it is necessary to know this conductive term; we propose a model for it that is applicable regardless of the Knudsen number (5).

The characteristics of conductive heat transfer from a particle to the surrounding gas may be broken into three regimes depending on the value of the Knudsen number (Kn), defined as $\lambda/d$, where $\lambda$ is the gas molecule mean free path and d is the characteristic dimension of the particle. For $Kn < 10^{-3}$ the continuum approximation for heat transfer applies. For $Kn > 10$ free molecular flow conditions prevail and expressions for the heat transfer based on molecular collisions apply. In the transition region, $10^{-3} < Kn < 10$, analytical modeling of heat transfer is difficult because neither the continuum nor the kinetic theory approach is strictly correct.

The model we used for the conductive heat transfer is the following: a spherical particle with radius a is stationary in an infinite gaseous medium with temperature $T_\infty$ as $r \to \infty$ (r is the radial coordinate with origin at the particle center). The region outside the particle is divided into two zones. Beyond a spherical boundary of radius $\lambda + a$, continuum conduction is assumed to hold, where $\lambda$ is the mean free path of the gas molecules. Within one mean free path of the surface it is assumed that the gas molecules collide only with the particle and not with one another. The molecules striking the particle are assumed to have a Maxwellian velocity distribution at temperature $T_B$, the zone boundary temperature. The particle is maintained at a fixed temperature $T_p$; the required energy is supplied or removed by radiation or chemical reaction.

To obtain the steady state heat flow, the Laplace equation is solved in spherical coordinates for the temperature field outside the boundary of radius $a+\lambda$. The solution, $T = A/r + T_\infty$ contains a constant A, determined by the zone boundary condition. Next, the energy carried to and from the particle is calculated from kinetic theory assuming a boundary temperature $T_B$. A is then determined by equating energy flows at the boundary. With A known, the continuum temperature distribution may be used to solve for $T_B$ and the resulting expression for the heat transfer Q is:

$$Q = \frac{4\alpha a k \Phi (T_p - T_\infty)}{Kn + \frac{\Phi\alpha}{(2Kn+1)\pi}} \quad (1)$$

where $\alpha$ is the accommodation coefficient (a measure of how well the molecules thermally accommodate to particle temperature), k is the thermal conductivity of the gas, and $\Phi$ is a numerical constant which depends of the internal energy of the gas molecule ($\Phi = 34/75$ for monatomic gas and $48/95$ for a diatomic gas).

A nondimensionalized boundary temperature may be defined as:

$$\frac{T_B - T_\infty}{T_p - T_\infty} = \theta; \quad (2)$$

which results in a quantity that varies between 0 and 1. Θ vs. Kn is plotted in Fig. 2. As Kn approaches 0, $T_B$ approaches $T_p$ and the continuum temperature gradient (with no temperature jump) results. For air at STP this corresponds to a particle diameter of ≥ 15 μm. As Kn increases toward infinity $T_B$ goes to $T_\infty$ and a temperature jump at the surface appears. This happens for particles of diameter less than .075 μm in air at STP. By calculating Kn for a particle of interest, reference to this plot reveals to what extent there is an effective temperature gradient around it.

The Nusselt number as a function of Kn can also be calculated for the general case from the heat transfer equation and the result is plotted in Fig. 3. Again by finding the appropriate Kn, Nu can easily be determined from the graph. It is important to note that Nu decreases as Kn increases, and the heat flux per area from a particle is $q = (Nuk/2a)(T_p - T_\infty)$. Therefore, as the particle becomes smaller, q increases since Nu does not decrease faster than a. This means that it is increasingly difficult for a small particle to be at a temperature which is different from the surrounding gas as the particle size decreases.

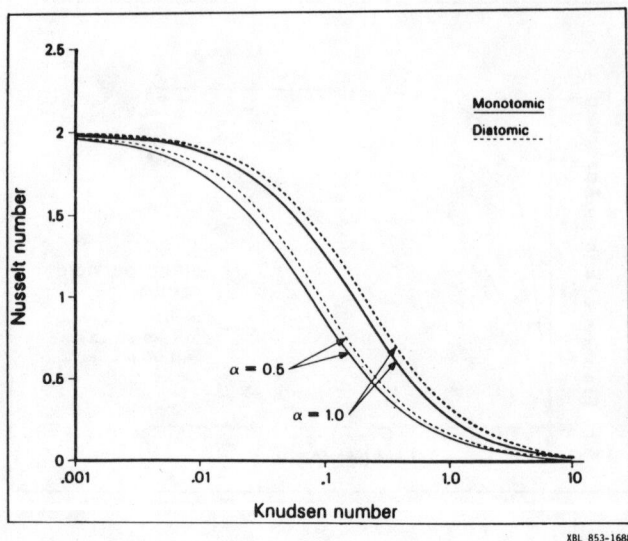

**Figure 3.** The Nusselt number as a function of the Knudsen number for a monatomic and a diatomic gas, respectively. Graph for values of the accommodation coefficient of 0.5 and 1.0 are shown for each gas.

Shown in Fig. 4 is a comparison of our result with those of two other workers ( 6 , 7 ). The experimental points were taken by Takao using a sphere in a rarefied gas to achieve a mean free path on the order of sphere radius (Kn ~ 1). (In order for a particle to have an equivalent Knudsen number at STP, its diameter would be about 0.07 μm.) Our treatment matches his data at least as well as his analytic expression does, and ours is far simpler to apply outside regions not shown on the graph. Also plotted for comparison is Sherman's empirical formula.

Because Equation 1 can be used in an energy balance to calculate the particle temperature for arbitrary Knudsen numbers, it is now possible to calculate particle temperatures for any radiant heating condition ( 8 ). The equation is limited to cases in which the particle size is not large enough to cause gravitational settling which would induce an additional forced convection term. However, the present treatment provides an upper bound to radiatively heated particle temperatures for falling particles because the forced convection will tend to reduce the temperature difference between particles and gas. The primary value of the equation lies in its ability to predict the heat transfer at values of Kn ~ 1 where a useful expression with an analytic basis was not previously in

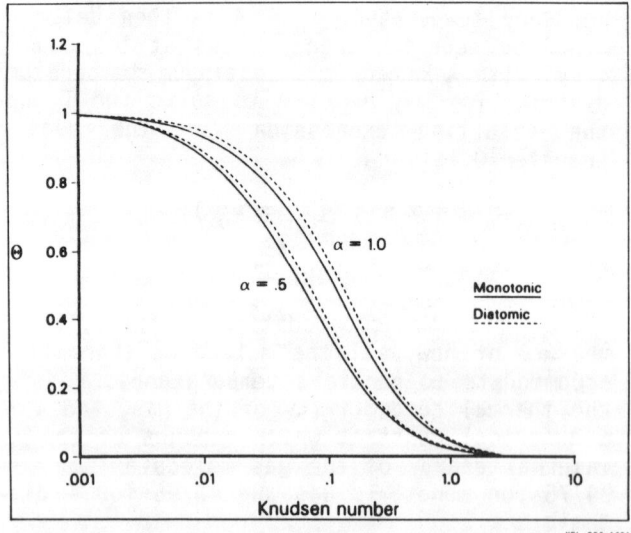

**Figure 2.** The non-dimensionalized boundary temperature as a function of the Knudsen number for both a monatomic and a diatomic gas. Graphs for values of the accommodation coefficient of 0.5 and 1.0 are shown for each gas.

use. The expression should find use in other areas where particle temperature is important, such as combustion or particle-gas heat transfer in the atmosphere.

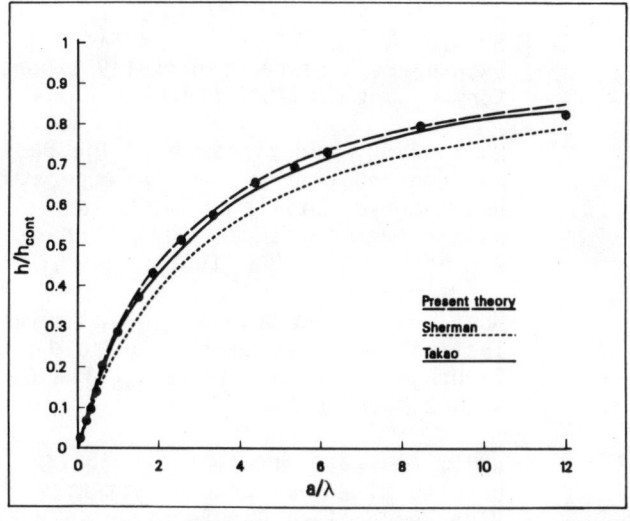

**Figure 4.** The non-dimensionalized heat transfer coefficient as a function of the ratio of particle radius to mean free path of a gas molecule. The points shown are experimental data from Takao. ( 7 )

## EXPERIMENTAL STUDIES

In order to simulate concentrated solar radiation in the laboratory, an arc-image furnace was constructed. A high intensity xenon arc lamp was mounted at one focus of a deep ellipsoid that reflects light coming from the arc to the other focus of the ellipsoid. A two-dimensional translational stage was constructed to provide stable mounting for the ellipsoidal reflector insuring reproducible flux profiles. A similar, but three-dimensional translation stage was constructed to allow precise positioning of the reactor vessel at the focus of the reflector. It also serves as the mount for a scanning calorimeter used to make spatial measurements of the radiant flux. A spatial integration of these measurements was made to determine total power available in the reactor zone. These measurements indicate a peak flux density on the order of 4000 kW/m$^2$ and a total input power to the reactor of 490 Watts, assuming an 8% reflection due to the vessel window.

The detailed three-dimensional map of the flux density was a critical factor in the design of the receiver/reactor. In the reactor in current use, the gas-particle mixture enters the reactor vessel at the top, is swirled in a cyclone fashion toward the bottom and is exhausted through a central tube. Light from the solar simulator passes through the bottom of a flat quartz window and is focussed just below the exhaust tube to ensure that all particles pass through the high intensity portion of the beam (see Figure 5). A number of experiments were conducted to observe the heating of particle-gas mixtures in the reactor. When the reactor was wrapped with insulation, temperatures in excess of 1200°K were reached with particle mass loadings less than 10 grams per cubic meter of carrier gas. Temperatures were measured with a Chromel-Alumel thermocouple placed in the exhaust tube of the reactor chamber.

Work was completed on a mechanical shaker and cyclone chamber for entraining small particles in a gas stream. Commercially available powders of carbon, hematite, and magnetite were successfully entrained in a gas stream and optical extinction measurements were performed on the particle suspensions. These particle materials were chosen

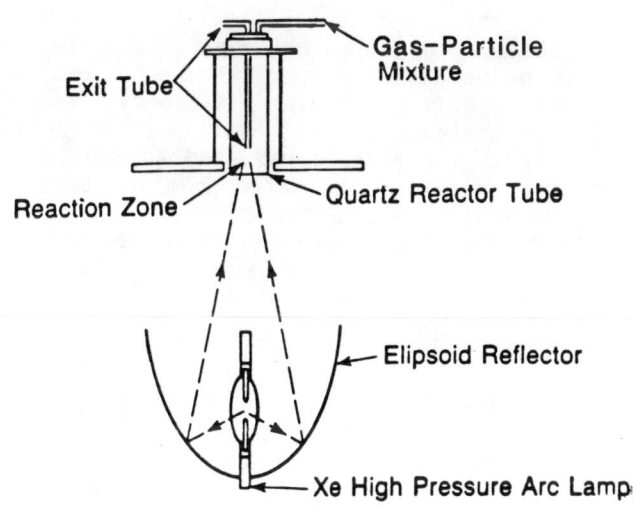

**Figure 5.** Diagram of the reactor vessel placed at the focus of the arc-image solar simulator.

for initial experimentation because of their possible roles in various reactions of interest for the production of useful fuels and chemicals The mass loading of the particles in the suspension was determined by drawing a known volume of the gas-particle mixture through a filter and weighing the filter. Samples were cut from the filter and scanning electron micrographs were made to determine particle size distributions.

As another application of absorbing suspensions we have undertaken preliminary studies of solar detoxification of hazardous chemical waste using the solar simulator. The toxic wastes most resistant to destruction by incineration are the polychlorobiphenyls (PCB's). Dichlorobenzene was chosen as a surrogate for PCB because it is considerably less toxic but has chemically similar properties. An injection system for introducing the hazardous waste material into the particle-gas stream was installed and tested. In tests using carbon particles as the absorbing material and air as the carrier gas, temperatures near $1200^{\circ}K$ were reached and the carbon particles were oxidized during passage through the reactor. Although the carbon particle suspension was sufficiently dense to absorb enough of the radiation passing through the quartz window to reach this temperature, the mass of carbon oxidized was a small fraction of the fuel that would have been needed to achieve the detoxification by direct combustion. After leaving the reactor, the gases were passed through a water bubbler for chemical analysis. Equipment with the accuracy to determine if 99.999% destruction or removal efficiency (DRE) had been achieved was not available, so an ion selective electrode was used to monitor the HCl content of the water from the bubbler. The measurement indicated that 108% of the dichlorobenzene has been destroyed with an instrument error of $\pm 10\%$.

CONCLUSION

The use of a gas-particle mixture as a solar absorption medium for heating gas has already been demonstrated. In this paper we have discussed research directed towards extending the small particle absorption concept to other applications, including the solar production of fuels and chemicals, and the destruction of toxic wastes. The field is entirely new and many issues still need to be addressed. We have broken ground by identifying and gathering data on possible reactions, calculating absorption characteristics of particle suspensions, developing a satisfactory particle to gas heat transfer model, and constructing a laboratory system that will allow these reactions to be studied experimentally.

LITERATURE CITED

1. Hunt, A.J., "Small Particle Heat Exchangers," Lawrence Berkeley Laboratory report LBL-7841 (1978).

2. Hunt, A. J. and Evans, D., "The Design and Construction of a High Temperature Gas Receiver Utilizing Small Particles as the Heat Exchanger (SPHER) Phase I Report ", LBL-13755 (1981).

3. Hunt, A. J. and Brown, C. T., "Solar Testing of the Small Particle Heat Exchange Receiver (SPHER), Phase II Report," LBL-15756 (1983).

4. Hunt, A.J. and Brown, C.T., "Solar Test Results of an Advanced Direct Absorption High Temperature Gas Receiver (SPHER)", presented at the 1983 Solar World Congress, International Solar Energy Society, Perth, Australia, August 15-19, LBL-16497.

5. Yuen, W.W., Miller, F.J., and Hunt, A.J., "A Two-Zone Model for Conduction Heat Transfer from a Particle to a Surrounding Gas at Arbitrary Knudsen Number", Submitted to the International Journal of Heat and Mass Transfer, LBL-18449 (1984).

6. Sherman, F.S., "A Survey of Experimental Results and Methods for the Transition Regime of Rarefied Gas Dynamics," Rarefied Gas Dynamics 3rd Symposium Vol II, Academic Press, New York (1963).

7. Takao, K., "Heat Transfer from a Sphere in a Rarefied Gas," Rarefied Gas Dynamics 3rd Symposium Vol II, Academic Press, New York (1963).

8. Yuen, W. and Hunt, A.J., "On the Heat Transfer Characteristics of Gas-Particle Mixture Under Direct Radiant Heating," submitted to International Communications of Heat and Mass Transfer.

# MOMENTUM AND ENERGY EXCHANGE IN A SOLID PARTICLE SOLAR CENTRAL RECEIVER

J.M. Hruby and P.K. Falcone ■ Sandia National Laboratories, Livermore, CA 94550

The use of free falling, sand sized solid particles as a direct heat transfer medium in a solar central receiver is under investigation at Sandia National Laboratories, Livermore. A series of models has been developed to describe the particle-air momentum and energy exchange in such a receiver. The earliest models were developed to assess the technical feasibility of the concept, and the later models were developed to more accurately describe the exchange. Early models were one dimensional and single sphere cases. More advanced models are two dimensional, and use a control volume consisting of particles and intervening air. Momentum and energy exchange models are combined at each level of complexity to describe the thermal processes in a cavity receiver.

## INTRODUCTION

The use of free-falling, sand-sized solid particles as a direct heat transfer medium in a solar central receiver is under investigation at Sandia National Laboratories, Livermore (1,2). The solid particle central receiver concept is being considered for high temperature applications of solar energy. Solid particles as a heat transfer medium have several advantages over gas or liquid working fluids: high volumetric heat capacity compared to gases, potential to be heated to temperatures in excess of 850 K, ability to absorb direct irradiation, and fewer hazards than molten salts or metals. The disadvantages of solid particles as a working medium include their abrasive characteristics and the difficulty associated with transporting solids. A conceptual design of a solid particle receiver is shown in Figure 1.

This paper focuses on the approach taken in modeling the momentum and energy exchange in a solid particle central receiver. The momentum and energy models are coupled and the results used to predict final particle temperature and receiver efficiency.

## PARTICLE-AIR MOMENTUM EXCHANGE

Particle-air momentum coupling is an important aspect of solid particle receiver modeling. Particles in disperse suspensions do not behave as single particles, and the particle velocity and volume fraction are important parameters in determining the final particle temperature.

### One-Dimensional Momentum Exchange Model

A one-dimensional, steady flow momentum exchange analysis was performed to establish particle and air velocities and volume fractions. An energy exchange analysis employing the resultant particle velocity and volume fraction can then be used to determine if particles can be expected to reach temperatures in excess of 850 K in reasonable fall heights (10 to 20 m) for a solar central receiver.

The approach used in modeling the particle-air suspension was to write the point equations of continuity and momentum for each phase, and then volume average the point equations to obtain averaged equations of motion. Assumptions employed were: constant area particle curtain, negligible particle-particle interaction, constant mass flow rate of gas, and negligible stress coefficient at the edge of the particle curtain. The governing equations are:

Particle Continuity:
$$\frac{d(\rho_s f_s v_s)}{dz} = 0 \qquad (1)$$

Particle Momentum:
$$\rho_s f_s v_s \frac{dv_s}{dz} = f_s(\rho_s - \rho_g)g - F_{gs} \qquad (2)$$

Air Continuity:
$$\frac{d(\rho_g f_g v_g)}{dz} = 0 \qquad (3)$$

Air Momentum:
$$\rho_g f_g v_g \frac{dv_g}{dz} = -f_g \frac{dP_g}{dz} + \rho_g f_g g + F_{gs} \qquad (4)$$

where the drag is an empirical value described by: $F_{gs} = 1 + 0.15 Re^{2/3}$

Figure 2 illustrates the resultant trends. The particle velocity increases monotonically and reaches a terminal velocity in a short distance. The gas velocity remains nearly constant throughout the fall distance because the gas mass flow rate is assumed to be constant, and the gas volume fraction is always near one. The particle volume fraction decreases as the velocity increases and attains a steady value after a short distance of fall. The terminal particle velocity increases with increasing mass flow rate.

### Experimental Particle Velocity Results

Laser Doppler velocimetry was employed to measure particle velocity as a function of distance of fall (3). In these tests, room temperature particles were released in quiescent air and allowed to free-fall. The results of these experiments along with the two limiting cases for one-dimensional particle-air flow are shown in Figure 3. The upper velocity limit is free-fall with no drag (in a vacuum), and the lower limit is the case described in the previous section where constant mass flow rate of air (no influx of air) is assumed. Predicted single sphere velocity is near the no influx case but slightly higher. Particle velocity measurements illustrate that the particles behave somewhere in between these two bounds. However, the gas velocity predicted by the model is too low since air entrainment is not included. A two-dimensional aerodynamic model which allows gas influx is required.

### Two-Dimensional Momentum Exchange Model

The approach taken for two-dimensional modeling was different than the one-dimensional approach. For this case, the particles were modeled with the Langrangian approach, while the fluid was modeled with the Eulerian approach. An established computer code using this trajectory technique called Particle-Source-In Cell (4) was employed. In this code the gas flow is obtained with no particles present, and then the particle trajectories are calculated. The particles are assumed to act as sources of momentum and energy in a gas cell and the gas field is recalculated. This procedure is repeated until the solution converges. Predictions with this model accurately describe the experimental measurements when a small, lateral initial particle velocity is introduced. The small lateral velocity allows the curtain cross-sectional area to increase, thereby simulating the actual flow.

## PARTICLE-AIR ENERGY EXCHANGE

Energy exchange analysis has focused on understanding the particle-particle radiative exchange mechanisms. The energy balance was performed in two different ways -- on a single representative particle, and on a volumetric group of particles and intervening air. The models developed and their associated results are discussed below.

### Single Particle Models

Several models in which a single representative particle is irradiated were developed. These models included direct irradiation of a representative particle as well as attenuated irradiation of a representative particle (5). The energy balance which describes these two cases is:

$$0.5\alpha AGI + 0.5\alpha AG \epsilon_w \sigma T_w^4 - A\epsilon_s \sigma T^4 \qquad (5)$$
$$= \rho_s c_p V \frac{dT}{dt} = \rho_s c_p V v_s \frac{dT}{dz}$$

where $G = 1$ for the direct case and $G = \frac{1 - e^{-\tau}}{\tau}$ for the attenuated case.

It was shown (5) using these single particle models that particles in attenuated flux excluding convection reached an equilibrium temperature of 1260 K in 3.7 s with a 1.0 MW/m² incident flux.

## Two-Flux Model

A limitation of the single particle calculations just described is that interparticle effects such as scattering are not included. An energy balance on the entire curtain is required. However, inclusion of the scattered energy terms results in a complex and difficult-to-solve energy equation.

A simplified model of scattering, in which the directionality of the scattered energy is limited to two directions (two-flux model) has been developed (6). A one-dimensional solution where the flux is assumed to be divided into two hemispherical components aligned along the x-axis can be solved with this method. The energy equation is

$$k_{eff}\frac{\partial^2 T}{\partial x^2} + a(I + J) - 2a\sigma T^4 = \rho_s c_p v_s f_s \frac{\partial T}{\partial z} \quad (6)$$

The value of $k_{eff}$ can be defined to include conductive and convective effects.

The major difficulty in the two-flux model is determining the absorption and scattering coefficients. For this analysis, absorption and scattering coefficients from Reference (7) were used. In addition, the back wall was assumed to emit and reflect radiation.

The temperatures predicted by the two-flux model were higher than the single particle model with attenuation. The predicted temperatures have uncertainties resulting from the choice of absorption and scattering coefficients, and effective thermal conductivity. The results of altering the scattering coefficient are not discussed in this section because the discrete ordinates results will be discussed and the trends for the two cases are in good agreement. Detailed results of the two-flux model can be found in Reference (5).

## Discrete Ordinates Model

A model which accounts for the directional nature of the propagating radiation field, interparticle scattering, and wavelength dependence of the particle optical properties has recently been developed (8). This model uses the discrete ordinates method of solution (9). The resultant energy equation is

$$k_{eff}\frac{\partial^2 T}{\partial x^2} - \frac{\partial q_r}{\partial x} - hb(T-T_g) = \rho_s c_p v_s f_s \frac{\partial T}{\partial z} \quad (7)$$

where:

$$q_r = \int_{all\,\lambda} \int_0^{2\pi} \int_0^{\pi} I_\lambda(x,\theta,\phi)\sin\theta\,d\lambda\,d\theta\,d\phi \quad (8)$$

and the equation defining $I_\lambda$ is

$$\cos\theta \frac{dI_\lambda(x,\theta,\phi)}{dx} = -(\sigma_\lambda + \kappa_\lambda)I_\lambda(x,\theta,\phi)$$

$$+ \kappa_\lambda I_{b,\lambda}(T)$$

$$+ \frac{\sigma_\lambda}{4\pi}\int_0^{2\pi}\int_0^{\pi} I_\lambda(x,\theta',\phi')p_\lambda(\theta',\phi'\to\theta,\phi)\sin\theta'\,d\theta'\,d\phi' \quad (9)$$

Optical property measurements, for integration with the discrete ordinates model, have been performed on candidate particle materials in free-falling curtains (10).

A parametric study was conducted using the discrete ordinates model (8). Decreasing the optical depth of the particle curtain increases the uniformity of particle heating through the curtain but reduces the overall capture of solar energy by the curtain. This same result is seen when the albedo, $\omega_\lambda$, is increased. Increasing the wall reflectivity increases the rate of absorption at the back of the curtain and provides more uniform and increased overall absorption. However, increasing the wall reflectivity may also increase radiative losses from the cavity.

The particle temperature after three seconds in a 0.1 m deep particle curtain for a flux level of 0.3 MW/m$^2$ was 1300 K near the front of the curtain and 615 K at the rear of the curtain. The parameters used in this case were: particle volume fraction of 0.005, particle diameter of 500 $\mu$m, scattering albedo of 0.10 (representative of silicon carbide), wall reflectivity of 0.9, and Nusselt number for convection of 2. The parameters used in this case are realistic, but the effects of the cavity on the particle heating have not been included.

## RECEIVER MODELING

Momentum and energy calculations have been combined to calculate cavity receiver performance estimates in two ways. These results are described in this section.

### Preliminary Receiver Modeling

As a first-attempt to determine receiver efficiency and trends, a model was developed which employed single particle velocity calculations and the attenuated single particle radiation model. Convective losses in this model were calculated as losses from cavity walls (11) and the back wall was assumed to be at the particle temperature. Energy exchange in the cavity was computed using the computer code RADSOLVER (12) which is based on the radiosity method of analysis. The back wall of the cavity was assumed to have effective properties that represented the particle curtain-back wall combination. The incoming solar flux was calculated with the computer code HELIOS (13) for the heliostat field at the Central Receiver Test Facility (CRTF). A cavity energy balance was performed until convergence between the particle curtain absorption and cavity energy absorption occurred.

The receiver used for analysis was designed to minimize spillage, radiative, and convective losses for the CRTF heliostat field configuration. Particle mass flow rates in the range from 225 to 425 kg/min were studied. The calculated receiver efficiency (excluding spillage) ranged from 65 to 79 percent. The final particle temperature decreased with increasing mass flow rate and varied from 1350 K to 1075 K respectively for 225 to 425 kg/min. Other trends were established and are discussed in detail in (14).

The convective losses calculated for this cavity were on the order of 10 percent of the incoming solar energy and are believed to be underestimated in this analysis. (In an actual cavity, the hot air produced by particle-to-air heat transfer will be used to preheat the particles and therefore will not be a system loss.) By modeling a receiver for the CRTF ($\sim$5 MW$_t$), an optimized high temperature receiver could not be designed. An ideal solid particle receiver would receive a high incident solar flux over the shortest possible distance for ample heating. At the CRTF, the heliostat field determines the size of the receiver and aperture, as well as the incident solar flux levels (peak flux 0.7 MW/m$^2$).

### Advanced Receiver Modeling

Receiver modeling employing the two-dimensional aerodynamic model and the discrete ordinates radiation model has been initiated. The solution is two-dimensional in height and depth of the particle curtain. Convective losses from the particles are calculated with a single sphere correlation. Convective losses from the walls are calculated by Reynold's analogy and agree well with the cavity correlations described in (11).

The efficiency of the CRTF cavity design discussed previously was predicted by this model to be approximately 50 percent (15). This low efficency is primarily due to particle-to-air convective losses of 20 percent. In order to increase efficiencies, higher incident flux levels and mass flow rates are required. A receiver efficency of over 70 percent has been predicted for the same cavity geometry with a 1.0 MW/m$^2$ incident flux level and an increase in mass flow rate by a factor of 5. In this cavity, the radiative losses are 15 precent, and the particle-to-air convective losses are about 10 percent. As mentioned earlier, use of the convectively generated hot air in a particle preheating process is likely.

Studies to determine the optimum cavity size and incident flux level which tradeoff the aperture size, receiver size and incident flux for larger heliostat fields are currently being initiated.

## CONCLUSION AND FUTURE WORK

A series of models has been developed to describe the particle-air momentum and energy exchange in a solid particle central receiver. The earliest models were developed to assess the technical feasibility of the concept, and the later models were developed to more accurately describe the exchange. Early models were one-dimensional and single sphere cases. More advanced models are two-dimensional, and use a control volume consisting of particles and intervening air. Momentum and energy exchange models are combined at each level of complexity to describe the thermal processes in a cavity. The modeling indicates that particle-air convective losses will play a major role in the heat transfer in a solid particle solar central receiver. Future modeling efforts will concentrate on the evaluation of small, high flux solid particle cavity receivers.

The analysis and testing of the solid particle receiver concept has led to an understanding of the receiver design considerations and demonstrated many aspects of the technical feasibility. Studies concerning components other than the receiver and the integration of these components into a working system are now being initiated. Conceptual designs of a solar plant which employs a solid particle receiver to produce high temperature energy for a specific application are planned. These conceptual designs will provide cost and performance predictions for use in evaluating this concept with others in the Department of Energy Solar Thermal Technology Program.

## ACKNOWLEDGEMENT

This work was supported by U.S. Department of Energy under contract DE-AC04-76DP00789.

## NOTATION

$A$ = particle surface area, $m^2$ (10.76 $ft^2$)
$F$ = drag force, N (0.22 lbf)
$G$ = attenuation factor = $(1 - e^{-\tau})/\tau$
$I$ = radiative intensity in +x direction, $W/m^2$ (0.317 Btu/(h $ft^2$))
$I_\lambda$ = monochromatic radiative intensity, $W/(m^2$ sr $\mu m)$ (0.317 Btu/(h $ft^2$ sr$\mu m$))
$I_b$ = Planck's function, $W/(m^2$ sr) (0.317 Btu/(h $ft^2$ sr ))
$J$ = radiative intensity in -x direction, $W/m^2$ (0.317 Btu/(h $ft^2$))
$P$ = pressure, $N/m^2$ (0.02 $lbf/ft^2$)
$Re$ = Reynold's number = $\rho_g(v_s-v_g)d/\mu$
$T$ = particle temperature unless subscripted, K ((T($^\circ$F)+460)/1.8)
$V$ = particle volume, $m^3$ (35.2 $ft^3$)

$a$ = volumetric absorption cross section for two-flux model, $m^3/m^2$ (3.28 $ft^3/ft^2$)
$b$ = ratio of area of particle to volume of mixture = $6f/d$, $m^{-1}$ (0.30 $ft^{-1}$)
$c_p$ = specific heat, kJ/(kg K) (0.239 Btu/(1bm $^\circ$F))
$d$ = particle diameter, m (3.28 ft)
$f$ = volume fraction
$g$ = acceleration due to gravity, 9.8 $m/s^2$ (32.17 $ft/s^2$)
$h$ = convective heat transfer coefficient, $W/m^2K$ (0.18 Btu/h $ft^2$ $^\circ$F)
$k_{eff}$ = effective thermal conductivity of particle-air volume, W/(m K) (0.58 Btu/(h ft $^\circ$F))
$p_\lambda$ = scattering phase function
$q_r$ = radiative heat flux, $W/m^2$ (0.317 Btu/hr $ft^2$)
$s$ = volumetric scattering cross section for two-flux model, $m^3/m^2$ (3.28 $ft^3/ft^2$)
$t$ = time, s
$v$ = velocity, m/s (3.28 ft/s)
$x$ = direction in depth of curtain, m (3.28 ft)
$z$ = distance in direction of particle fall, m (3.28 ft)

$\alpha$ = particle absorptivity
$\epsilon$ = emissivity
$\kappa_\lambda$ = wavelength dependent absorption coefficient for discrete ordinates model, $m^{-1}$ (0.30 $ft^{-1}$)
$\omega_\lambda$ = single scattering albedo = $\sigma_\lambda/(\sigma_\lambda + \kappa_\lambda)$
$\mu$ = kinematic viscosity, N $s/m^2$ (2.42 x $10^3$ lbm/(h ft))
$\rho$ = density, $kg/m^3$ (0.062 $lbm/ft^3$)
$\theta$ = polar angle of radiation coordinate system, radians
$\sigma$ = Stefan-Boltzmann constant, 5.67 x $10^{-8}$ $W/m^2K^4$ (0.171 x $10^{-8}$ Btu/(h $ft^2$ $^\circ R^4$)
$\sigma_\lambda$ = wavelength dependent scattering coefficient for discrete ordinates model, $m^{-1}$ (0.30 $ft^{-1}$)
$\tau$ = optical depth of particle curtain
$\phi$ = azimuthal angle of coordinate system, radians

subscripts
g = gas
s = solid
w = wall
$\lambda$ = wavelength

## LITERATURE CITED

1. Falcone, P. K., J. E. Noring and C. E. Hackett, "Evaluation and Application of Solid Thermal Energy Carriers in a High Temperature Solar Central Receiver System," 17th IECEC, Los Angeles (1982).

2. Martin, J. and J. Vitko, Jr., "ASCUAS: A Solar Central Receiver Utilizing a Solid Thermal Carrier," Sandia National Laboratories, SAND82-8203 (1982).

3. Hruby, J. M. and V. P. Burolla, "Solid Particle Receiver Experiments: Velocity Measurements," Sandia National Laboratories, SAND84-8238 (1984).

4. Crowe, C. T., M.P. Sharma and D. E. Stock, "The Particle-Source-In Cell (PSI-Cell) Model for Gas-Droplet Flows," Journal of Fluids Engineering, pp. 325-332 (June 1977).

5. Falcone, P. K., J. E. Noring and J. M. Hruby, "Assessment of a Solid Particle Receiver for a High Temperature Solar Central Receiver System," Sandia National Laboratories, SAND85-8208, (1985).

6. Hamaker, H. C., Phillips Res. Rep. 2, pp. 55-67, 103-15, 420-5 (1947).

7. Chen, J. C., "Radiant Heat Transfer in Packed Media," PhD Thesis, Univ. of MI (1961).

8. Houf, W. G. and R. Greif, "Radiative Transfer in a Solar Absorbing Particle Laden Flow," presented at 23rd National Heat Transfer Conference, ASME/AIChE, Denver (1985).

9. Chandrasekar, S., Radiative Transfer, Dover Publications, New York (1960).

10. Falcone, P. K., "Technical Review of the Solid Particle Receiver Program January 25-26, 1984," Sandia National Laboratories, SAND84-8229 (1984).

11. Siebers, D. L. and J. S. Kraabel, "Estimating Convective Energy Losses from Solar Central Receivers," Sandia National Laboratories, SAND84-8717 (1984).

12. Abrams, M., "RADSOLVER - A Computer Program for Calculating Spectrally-Dependent Radiative Heat Transfer in Solar Cavity Receivers," Sandia National Laboratories, SAND81-8248 (1981).

13. Vittitoe, C. N. and F. Biggs, "A User Guide to HELIOS: A Computer Program for Modeling the Optical Behavior of Reflecting Solar Concentrators Part I. Introduction and Code Input," Sandia National Laboratories, SAND81-1180 (1981).

14. LaJeunesse, C. A., "Thermal Performance and Design of a Solid Particle Cavity Receiver," Sandia National Laboratories, SAND85-8206, (1985).

15. Evans, G. H., et al, "Gas-Particle Flow within a High Temperature Cavity, including the Effects of Thermal Radiation," presented at 23rd National Heat Transfer Conference, ASME/AIChE, Denver (1985).

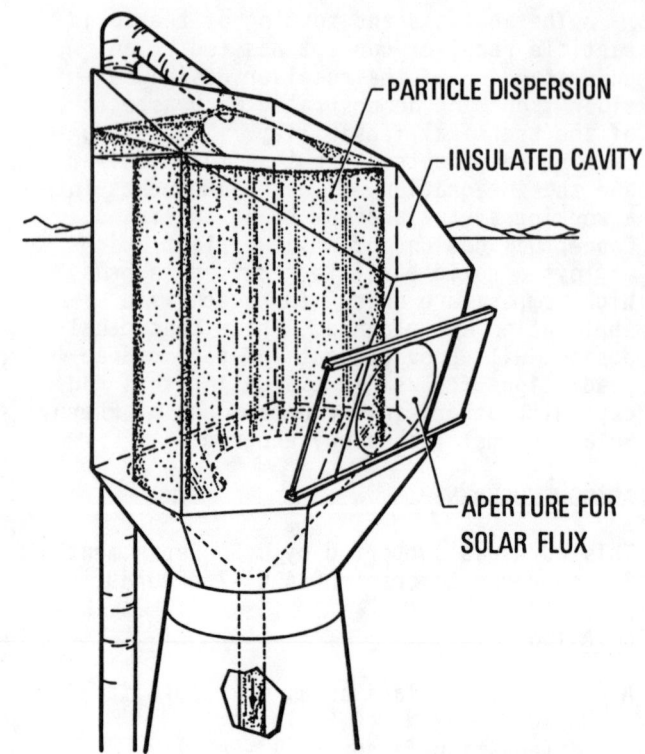

Figure 1. Conceptual design of a solid particle receiver.

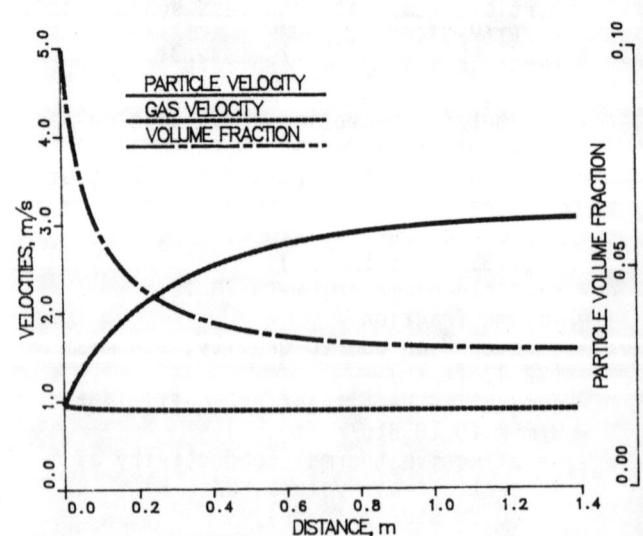

Figure 2. Particle velocity and volume fraction as a function of distance of fall.

Table 1. Transition and decomposition temperatures of zinc sulfate.

| | This Study | Tagawa (1984) | Ducarroir et al. (1982) | Mu and Perlmutter (1981) | Hosmer and Krikorian (1980) |
|---|---|---|---|---|---|
| Onset of α → β Phase Transformation, °C | 752 to 757 | - | - | - | 727-757 |
| Initial Decomposition Temperature, °C | 580 to 615 | 615 to 630 | 665 to 675 | 590 | - |
| Final Decomposition Temperature, °C | 757 to 847 | - | - | 712 | - |
| Sample Weight, mg | 16.5 to 23.4 | ~120 | ~300 | 10 to 20 | - |
| Sample Form | fine powder | Ground to 200 mesh | Powder Compacts | Powdered, 60-80 μm particle size | - |
| Sample Preparation | dehydrated at <380 °C | $ZnSO_4, 7H_2O$ | Dehydrated for 12 hrs. at 450°C | None | - |
| Sample Boat | Alumina | Platinum | Platinum | ? | - |
| Carrier Gas | Helium | Air and $N_2$ | Nitrogen | Nitrogen | - |
| Flow Rate, ml/min | 55 | 100 | 80-100 | 80 | - |
| Heating Rate, °C/min | 1.2 to 17.1 | 2 and 5 | 6.15 | successively at 1, 5 and 1.5 | - |
| Location of Sample Thermocouple | below the sample boat | between boat and tube wall | ? | outside sample boat | - |
| Instrument Used | Setaram TGA/DTA | Cahn | - | Dupont | Drop Calorimeter |

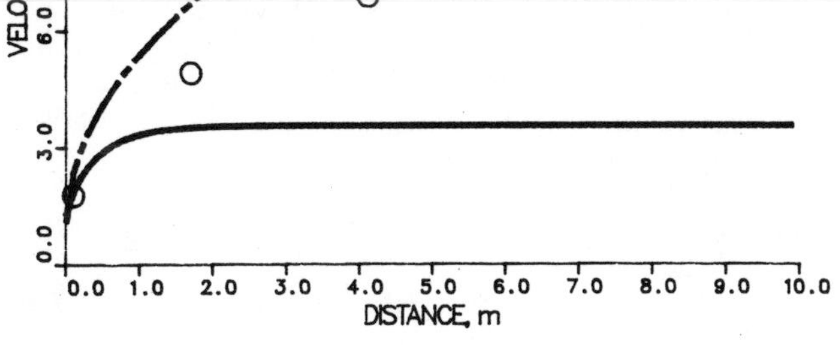

Table 2. Comparison between temperatures recorded by embedded and Setaram's sample thermocouples.

| Temperature of Embedded Sample Thermocouple (°C) | Recorded Sample Temperature by Setaram's Thermocouple | |
|---|---|---|
| | for 5°C/min & 12.15 mg | for 17.5°C/min & 52 mg |
| 198 | 195.5 (-2.5*) | 184 (14*) |
| 479 | 477.5 (-1.5) | 472 (-7) |
| 762 | 759 (-3) | 759 (-3) |
| 915.8 | 919 (+3) | 917.5 (+1.7) |
| 981 | 981 (-) | 983 (+2) |

*$T_{setaram} - T_{embedded}$

Table 3. Experimental conditions, DTA and DTG results for low heating rate decomposition of anhydrous zinc sulfate.

| Carrier Gas, Helium ml/min | | 54 | 54 | 57 | 54 | 55 | 55 |
|---|---|---|---|---|---|---|---|
| Sample weight, mg | | 16.51 | 23.41 | 19.65 | 18.75 | 18.65 | 18.85 |
| Residue Weight, mg | | 8.32 | 11.80 | 9.90 | 9.45 | 9.40 | 9.51 |
| Total Weight lost, mg | | 8.20 | 11.61 | 9.74 | 9.30 | 9.25 | 9.35 |
| Measured heating rate, °C/min | | 1.2 | 2.4 | 5.05 | 12.75 | 15 | 17.1 |
| @ region; oxy → ZnO* | | 1.2 | 2.4 | 5.05 | 10.13 | 12 | 11 |
| Measured temperature for α to β phase transition, °C | start | - | - | 757 | 757 | 755 | 752 |
| | peak | - | ~760 | ~to | 766 | 760 | 760 |
| | end | - | - | 765 | 777.5 | 779 | 771 |
| Zinc sulfate decomposition temps., °C | start | ~610 | ~610 | ~610** | ~580 | ~595 | ~615 |
| | peak | ~710 | ~753 | ~769 | 794 | 807 | 809 |
| | end | 747 | 790 | 812 | 838 | 841 | 847 |
| Final decomposition temperatures to ZnO, °C | start | 747 | 790 | 812 | 838 | 841 | 847 |
| | peak*** | 826 | 878 | 894 | 922 | 931 | 931 |
| | end | 846 | 890 | 911 | 941 | 946 | 948 |

*Drop in sample heating rate observed only within oxy → ZnO decomposition region.
**From DTG curves constructed by computer from TGA data.
***Temperature at maximum decomposition from constructed DTG curves.

Figure 3. One-dimensional particle velocity limits.

# EFFECTS OF HEATING RATE ON SOLAR THERMAL DECOMPOSITION OF ZINC SULFATE

A.T. Raissi and M.J. Antal, Jr. ■ Renewable Resources Research Laboratory
University of Hawaii, Honolulu, Hawaii 96822

Studies of the solid phase decomposition of zinc sulfate as a function of solids' temperature, heating rate and the incident radiative heat flux are presented. The experimental results are obtained in a novel thermogravimetric analysis system. This apparatus employs an intense beam of radiant energy, derived from an arc image furnace, to rapidly heat the solid sample. The interpretation of new data (including heating rates in excess of 1000 K/min), and comparison to more conventional low heaing rate results, reveal the role and effects of radiative heat transfer and rapid heating rates on the decomposition behavior of $ZnSO_4$.

## INTRODUCTION

One of the most promising applications of concentrated solar energy is direct solid phase decomposition of materials for the production of fuels and chemicals. A fundamental understanding of high temperature, high heating rate solid phase decomposition chemistry is essential in developing technologies needed to solve the future fuel crisis. Since solar furnaces are expensive to build and operate, new conversion techniques are needed which enjoy process economy, high efficiency and product selectivity.

The recent interest in the solar community to explore new means of utilizing concentrated solar energy has stimulated new research which emphasizes tne use of solar unique processes at elevated temperatures, as high as 1700 K, and radiant flux densities in excess of 100 $W/cm^2$ (1). These processes seek to take advantage of the unique aspects of solar radiation by effecting the process chemistry.

In this paper, we present the recent results of our research concerned with fundamental studies of the high temperature solid phase decomposition of zinc sulfate at low and high heating rates. The low heating rate studies were carried out using a conventional thermogravimetric system (Setaram Simultaneous TGA/DTA unit). The high heating rate results were obtained in a novel, Fast-TGA system capable of rapidly weighing materials exposed to concentrated radiant energy derived from an arc image furnace.

The goal of this work is to identify unique effects of radiant heating on the decomposition behavior of zinc sulfate. This paper presents a set of low heating rate results and compares them to those of other investigators. While the fundamental low heating rate decomposition studies of solid materials using conventional thermal analysis systems is now routine, no such instrumentation is currently available to study these processes under high temperature, high radiative flux environments similar to that of a solar furnace. Finally, some typical decomposition results obtained from intense radiative heating of zinc sulfate will be presented and discussed in conjunction with the design of a novel thermogravimetric system.

## PRIOR RESEARCH ON ZINC SULFATE DECOMPOSITION

The thermal decomposition of zinc sulfate is one step in many thermochemical hydrogen production cycles (2). In sulfuric acid-based thermochemical cycles, it is used to replace the reactions involving the boiling of azeotropic sulfuric acid and the decomposition of sulfuric acid vapor at high temperatures (3).

Low heating rate, thermogravimetric studies of the controlled dehydration and decomposition of several inorganic sulfates, including zinc sulfate, was reported recently by Mu and Perlmutter (4). They carried out their experiments at heating rates between 1 and 10 K/min and noted that the results were very sensitive to heating rates. The effects of parameters such as sample size and the type of carrier gas used to drive off the gas product of the reaction was also studied and found not to influence their experimental results. Kinetic parameters were also given for zinc sulfate dehydration and decomposition. More recently, Ducarroir et al. (5) presented a set of kinetic results for thermal decomposition of zinc sulfate by isothermal and nonisothermal TGA techniques. Relatively large sample sizes (~300 mg) were used in the form of loose and compacted dehydrated powder which were then heated in their thermogravimetric system. The results of their isothermal experiments revealed that the carrier gas flow rate through the reaction tube influenced the rate of decomposition of all the metal sulfates except zinc sulfate. They found that for decomposition of zinc sulfate pellets, the chemically controlled shrinking core model, gave a good representation of their results. The initial decomposition temperatures were also determined, which were generally much higher than those reported by Mu and Perlmutter (4). These inconsistencies in reported experimental results are not limited to findings of these investigators alone.

Clearly, the decomposition temperatures, kinetics and possibly reaction mechanisms and formation of intermediate compounds generally depend on many parameters. These factors include sample size and physical properties (particle size, shape and surface to volume ratio), shape and size of the sample holder, the heating rate of the sample and the flow rate of the carrier gas, which effects the partial pressure of the gaseous product evolved (6). These parameters are often ill defined in most thermogravimetric studies. Inconsistancies in reported kinetic values, decomposition temperatures and even experimental conclusions have been pointed out and tabulated by many investigators (4,5,7,8).

Tagawa (6) studied the thermal decomposition of some sixteen metal sulfates including zinc sulfate by thermogravimetry at heating rates of two and five K/min in flowing air and high-purity nitrogen. The initial decomposition temperatures in air for all the metal sulfates studied in his work showed a shift towards higher temperatures relative to those made in high-purity nitrogen. The magnitude of this shift for anhydrous zinc sulfate was found to be five and ten K at heating rates of two and five K/min respectively. His conclusion in itself is in disagreement with that of Mu and Perlmutter (4) who did not observe any effects due to the gaseous environment of the sample. According to Tagawa (6), the influence of the partial pressure of the gaseous product over the sample might be an important factor in effecting the initial decomposition of metal sulfates.

In the case of zinc sulfate, the formation of an intermediate oxysulfate has been reported by many investigators. Initially zinc sulfate is believed to decompose quickly into an oxysulfate with a chemical composition widely accepted as $ZnO.2ZnSO_4$ (9,10,11). A different composition for the oxysulfate intermediate has also been reported as $3ZnO.ZnSO_4$ (12) and $2ZnO.3ZnSO_4$ (13). The subsequent decomposition of this oxysulfate into the final product, zinc oxide, takes place at higher temperatures. To alleviate the problems associated with the formation of an intermediate oxysulfate, Krikorian and Shell (3) used radiant heating of zinc sulfate in an arc image furnace using four 1 KW tungsten lamps. Although they could not explain with confidence the behavior of zinc sulfate under high radiant flux densities, their qualititative results appear to indicate that the rapid heating of fine zinc sulfate particles enhanced their decomposition rate appreciably. Later Shell et al. (14) carried out similar experiments at a much larger scale in the focus of a 30 KW solar furnace at White Sands Solar Facility, New Mexico, but the results of their effort provided only limited, qualitative information and raised more questions than it answered. For example, they found that zinc sulfate decomposed at high heat fluxes at temperatures much lower than that required to decompose zinc sulfate using conventional heating techniques. This result seems to be contrary to the well established fact that the decomposition temperature of the reacting solid in a single reaction pathway, shifts toward higher temperatures as the heating rate increases (15).

The foregoing attributes of solid phase

decompositions at high heating rates requires an experimental approach that must be selected with the utmost care. The quantitative data relating the decomposition rate of the sample, its temperature and incident solar radiation must be obtained under well controlled conditions. The main thrust of the research effort presented in this paper has relied on the design and fabrication of a novel, Fast Thermogravimetric system capable of operating under high radiative environments similar to that of solar furnaces.

## APPARATUS AND PROCEDURES

### Low heating rate experiments

The thermal analysis instrument used for these tests consisted of a Setaram simultaneous TGA/DTA system. The unit was capable of operating at temperatures up to 1273 K and heating rates to about 20 K/min. This unit enjoys specially designed, fabricated and housed thermocouples (below the sample and reference boats) to minimize errors in sample temperature measurements, often present in most commercial TGA instruments.

All experiments were conducted on free flowing, dehydrated zinc sulfate powder samples. A drying technique suggested by Shell et al. (16) was used which consisted of first removing the first water of hydration of zinc sulfate heptahydrate (Baker Analyzed Reagent grade) at about 370 K for several days. The sample vial was connected to a vacuum pump while the zinc sulfate dried. The temperature in the vacuum oven was then increased to about 533 K and the solid was allowed to dry at this temperature for several more days. Finally the zinc sulfate sample was heated in air at a temperature of about 653 K continuously until no further weight loss could be detected. A total of about 120g of anhydrous zinc sulfate was prepared using this technique.

In our low heating rate experiments, we used high purity helium at a rate of approximately 55 ml/min. Table 1 presents the experimental conditions and the range of the main parameters investigated. Calibration of all balances used in this TG study was performed using Rice Lake Bearings, Inc. precision, class 1 (2 g) and 1.1 (1, 10, 100 mg) standard weights.

To evaluate the possible heat transfer effects on the sample temperature measurements under actual reaction conditions, several experiments were conducted using anhydrous zinc sulfate samples in a Setaram TGA/DTA. To measure sample temperatures, a type K thermocouple was carefully placed inside the sample boat and embedded beneath the sample. The thermocouple leads were then insulated by placing them inside a high purity alumina tube and brought out of the heated zone (TGA oven) and the instrument. During these tests no weight loss measurements could be obtained, however the DTA trace provided a means to identify the times for initiation and duration of reactions within the sample. In one experiment a 52 mg sample of anhydrous zinc sulfate was heated to 1253 K at a nominal rate of 17.5 K/min. In the second test the sample size and its nominal heating rate was reduced to 12.15 mg and 5 K/min respectively. The results of these experiments are summerized in Table 2.

From these results, it can be concluded that: 1) the difference between measured and actual sample temperature is a function of sample size, heating rate, and temperature as well as endothermicity of the chemical reactions; 2) a pronounced heat transfer effect seems to be present in the case of large sample size and high heating rates. Values of sample size and heating rates below which no significant heat transfer effects are present in the temperature range of interest, perhaps should be determined for every sample and heating condition. In our case, however, the limiting values of heating rate and sample size appeared to be about 5 K/min and 10 mg respectively. The discrepancy between actual and reported values of temperature decreased monotonically as the sample temperature incresed. This appeared to be due to enhanced heat transfer by radiation at higher temperatures.

Several experiments were conducted in order to calibrate sample thermocouples at both low (Setaram TGA/DTA) and high (MARK II Fast-TGA) heating rate instruments. These tests were carried out using the National Bureau of Standards new Curie Point Reference Materials (SRM GM-761) covering a temperature range of 520-1020 K. The results of these experiments did verify (less than 5K temperature difference between sample and sample thermocouple) performance of the Setaram's thermocouples but have not yet provided definitive measures of the performance of the Fast-TGA system.

### High heating rate instrumentation

A schematic diagram of the MARK II Fast-TGA system is shown in Figure 1. The apparatus consists of an electro-microbalance (Mettler model AE163) used as a thermobalance, a 30 $KW_e$/2$KW_{th}$ downward facing beam arc image furnace (17), 35 mm ID and 410 mm in length fused silica reaction tube, a nitrogen gas supply system and a vacuum pump. The Mettler balance was utilized in a top loading configuration in order to avoid unnecessary exposure of the hang down tube to high flux densities of incoming radiation at the top of the reflective cavity. A thin alumina sample boat holder (9.5 mm OD and 3 mm in height) was connected to a long (~500 mm), thin alumina tube (1.6 mm OD) which rests upon the balance arm. Each sample boat is fabricated out of high purity alumina (9.5 mm OD, 3 mm in height and 1.6 mm depth) and contains a S type thermocouple junction on the bottom of the boat in direct contact with the sample material. The sample boat thermocouple leads extend out from the bottom of the boat and make proper connections upon insertion into the sample boat holder on the top of the alumina tube. A specially designed heat exchanger serves to keep the heated length of the sample holder tube to a minimum.

A 60 mm OD and 380 mm long water cooled, silver coated reflective cavity serves as a photon bucket surrounding the reaction tube. A fairly uniform flux density is obtained within this volume due to multiple reflections. The radiant flux density incident upon the sample was varied by vertical displacement of the reflective cavity away from the focal plane. Continuous irradiant fluxes in the range of ten to several hundred watts/$cm^2$ were easily obtained at the sample surface.

The measurement of irradiant flux at the sample location was conducted using a Medtherm Calorimeter (a Gardon gauge type calorimeter) and found to be linearly proportional to the lamp's operating current. The instantaneous radiant flux densities at the sample location was therefore continuously monitored through the recording of the lamp's operating current by means of a strip chart recorder.

The collection and reduction of the data is carried out simultanously by an IBM-PC unit interfaced with the Fast-TGA system. The Mettler Micro-Macro balance is capable of taking at least five distinct weight change readings per second. These readings are than transmitted to the PC unit and stored for later analysis.

### RESULTS AND DISCUSSION

This section outlines results of ongoing research within the Renewable Resources Research Laboratory ($R^3L$), on low and high heating rate zinc sulfate decomposition. It was important to learn if data derived from conventional low heating rate studies can be correlated safely to infer solid phase decomposition chemistry effected by concentrated solar radiation. In addition the underlying assumption that high heating rates engendered in solar furnaces effect desirable chemistry needed to be verified. The heat and mass transfer effects are also important and must be addressed.

### Low heating rate thermal analysis results

Figures 2, 3, and 4 present reduced data for low heating rate TGA studies of anhydrous zinc sulfate. The TG (Figure 2) and DTA (Figure 3) curves evidence a two step weight loss mechanism corresponding to (a) the formation of a zinc oxysulfate intermediate which undergoes (b) subsequent, higher temperature decomposition to form zinc oxide. Figure 4 depicts reduced data for the rate of weight loss (with respect to temperature) versus weight fraction for heating rates of 1.2, 2.4 and 5.05 K/min. The "plateaus" at these heating rates are deduced from Figure 4, to be at weight fractions of about 85.7%, 84.2% and 83.3% for heating rates of 1.2, 2.4 and 5.05 K/min respectively.

A "plateau" at weight fraction of about 83.5% will be consistent with the widely accepted oxysulfate composition of $ZnO.2ZnSO_4$ reported in the literature. The discrepancy between the values of weight fraction at the plateau associated with initial decomposition of zinc sulfate derived from these low heating rate experiments with that of $ZnO.2ZnSO_4$ (83.5% weight fraction) seems to indicate the possibility of alternative pathways in the formation of the oxysulfate(s). The Friedman signiture (18-20 (Figure 5) provides additional evidence for the role of concurrent reactions in the formation of the oxysulfate(s).

The subsequent decomposition of the oxysulfate(s) apparently occurs via a simple, single step mechanism with an apparent activation energy of 57 Kcal/gmol. In the same Figure, values of apparent activation energies given by Mu and Perlmutter (4) and Ducarroir et al. (5) has been shown for the sake of comparison. Further work is required to identify the product(s) of the two competing reactions (possibly by utilizing advanced analytical techniques such as the use of an X-ray diffraction system) which consume the zinc sulfate. These considerations have direct bearing on the results of high heating rate experiments.

Figure 6 presents reduced DTA data from earlier experiments using anhydrous zinc sulfate at a nominal heating rate of 12.5 K/min (sample heating rate of 12.75 K/min). Examination of this DTA trace reveals three endotherms associated with $\alpha$ to $\beta$ phase transition, formation and decomposition of an intermediate zinc oxysulfate. This transition which is accompanied by a sharp endotherm appears to initiate at about 1030 K and occurs on top of a much broader endotherm associated with the formation of oxysulfate. Table 3 summarizes the experimental conditions and significant decomposition results obtained from some of our low heating rate experiments. The onset of the $\alpha$ to $\beta$ phase transformation, initial and final decomposition temperatures of anhydrous zinc sulfate obtained in this study are compared to that reported from several more recent investigations in Table 1. It is important to note that due to very gradual weight loss during the early stages of decomposition of zinc sulfate, the uncertainty regarding its initial decomposition temperatures can be appreciable. This, in turn, might explain in part the large discrepancies in the reported values of the initial decomposition temperatures. These temperatures are effected by the sample heating rates and might be influenced by the reaction chemistry as well as heat and mass transfer effects as discussed before. Higher values of initial decomposition temperatures given by Ducarroir et al. (5) compared to those reported in this study could be in part due to the effect of much larger sample sizes that were used by these investigators. Lower values of the final decomposition temperature of zinc sulfate given by Mu and Perlmutter (4) compared to that reported is believed to be due to prevalent heat transfer effects present within the Dupont TGA system and the location of its sample thermocouple.

Finally, as seen from the results of Table 3, a strong heat transfer effect is evident at higher heating rates which result in a significant drop in sample heating rates during the decomposition of the oxysulfate. This effect seems to have been enhanced by the strong endothermicity of the oxysulfate decomposition reaction. In that regard, the quantity of the decomposing sample used plays a pivotal rule in reducing such effects as discussed previously.

High heating rate TGA results
----

Figure 7 presents a set of preliminary thermogravimetric results obtained in the Fast-TGA system at two levels of radiant flux densities. The most striking aspect of these results is the apparent location of the "plateau" at 92.35% weight remaining on the TGA curve 2. Unfortunately, part of the data at the lower flux density (curve 1) near this plateau was lost during the experiment, and no such clear plateau is evident from the first thermogram. Although the reliability and reproducibility of a high heating rate sample temperature measurement needs further evaluation, nevertheless, the weight measurements are quite dependable. In that regard, it will be extremely interesting to see if such an effect is in fact real. A back calculation based on the value of the weight loss recorded at this "plateau" provides an oxysulfate composition of $2ZnO.11ZnSO_4$. The location of this plateau is far removed from that of $ZnO.2ZnSO_4$ (83.5%) and strongly suggests the possibility of unknown alternative high heating rate pathways active during early decomposition of zinc sulfate. This supports the evidence from our lower heating rate experiments, as discussed previously (Friedman signiture, Figure 5). Of course much more research needs to be done before a definitive conclusion can be drawn regarding the effects of radiant heating on solid phase chemistry of zinc sulfate decomposition.

Proposed reaction mechanisms at low and high heating rates
----

The evidence accumulated thus far, from both low and high heating rate zinc sulfate decomposition results suggest an initial pair of competitive reactions followed by a single step reaction to zinc oxide. A possible reaction mechanism is given below:

$$ZnSO_4 \xrightarrow{E_1} \text{intermediate} \xrightarrow{E_2 \geq E_1} ZnO$$
$$ZnSO_4 \xrightarrow{E_3} ZnO$$

It is also important to mention the typical values of the sample heating rates for a representative experiment. In the case of a higher flux experiment (25 W/cm$^2$), the initial heating rates measured were in excess of 2200 K/min, but dropped sharply to an average value of about 100 K/min over the entire range of the sample weight loss. Although these preliminary experiments were conducted at only 20 to 25 W/cm$^2$ capabilities within the furnace exist to achieve fluxes well in excess of several hundred watts/cm$^2$. Finally, it is interesting to note the initial decomposition temperatures obtained from these tests.

Initial decomposition temperatures in the range of 998-1023K are determined from Figure 7 for these flux levels and compare favorably with that of low heating rate results from this study (Table 3) and of the others (Table 1).

## CONCLUSIONS AND THE DIRECTION OF THE FUTURE RESEARCH

The most striking outcome of the research reported here is the accumulating evidence which suggests the possibility of the existence of alternative pathways during decomposition of zinc sulfate. The results of high heating rate, radiative solid phase decomposition of zinc sulfate justifies further research toward a better understanding of this important class of solid phase reaction under conditions of high temperature and radiant flux environments similar to that of a solar furnace. This will, of course, only be possible through the development of advanced instrumentation similar to the one presented in this work. The possible existance of alternative decomposition routes under radiant heating, immediately poses the question of whether these solar unique, decomposition pathways could be further enhanced to engender more desired chemistry. Identification of these possible intermediates, and a better understanding of relevant reaction mechanisms comprises the main thrust of future research. Our immediate research goal is to compile more high heating rate decomposition results using the Fast-TGA system, define and resolve questions regarding sample temperature measurements under these conditions and possible heat transfer effects.

The development and implementation of sophisticated mathematical models to interpret both reaction mechanisms and effects of heat and mass transfer to and from the reacting sample is also being pursued.

## ACKNOWLEDGEMENTS

This research was supported by the U. S. Department of Energy under contract No. DE-AC03-84SF12200. The authors wish to thank F. Wilkins (U.S. DOE) and Dr. G. Nix (SERI) for their interest and support of this work. The authors also thank W. S-L Mok, R. Narayan, B. Respicio and N. Castle, Jr. (all with the University of Hawaii) for their assistance in conducting this research.

## LITERATURE CITED

1. Manvi, R. Technical Summary. Presenting Jet Propulsion laboratory's view at Solar Thermal Research Workshop, Georgia Institute of Technology, Atlanta (Sept. 7-8, 1983).

2. Bowman, M. G., "The utilization of solar thermal sources for thermochemical production,"Proc. Ann. Meeting Am. Sect. Int. Solar Energy Soc., 3 (Sec.1) (1980).

3. Krikorian, O. H. and P. K. Shell, "The utilization of $ZnSO_4$ decomposition in thermochemical hydrogen cycles," Int. J. Hydrogen Energy,7(6), 463 (1982).

4. Mu, J. and D. D. Perlmutter,Ind. Eng.Chem. Process Des. Dev., 20, (4), 641-646 (1981).

5. Ducarroir, M., et al. "On the kinetics of the thermal decomposition of sulfates related with Hydrogen water splitting cycles," Proceedings of the 4th World Hydrogen Energy Conference, 2, 451 (June 1982).

6. Tagawa, H.,Thermochim. Acta, 80, 23-33 (1984).

7. Kolta, G. A. and M. H. Askar, Thermochim. Acta, 11, 305 (1975).

8. Ostroff, A. G. and R. T. Sanderson,J. Inorg. Nucl. Chem. 9, 45 (1959).

9. Hosmer, P.K. and O. H. Krikorian, "The High-Temperature enthalpies of zinc sulfate and zinc oyxsulfate," Presented at the 7th European Thermophysical Properties Conf., Antwerpen (June 30 - July 4, 1980).

10. Lau, K. H. et al. "High temperature chemistry of hydrogen production cycles," technical status report published by SRI International PYU-6788, (October 1979).

11. Ingraham, T. R. and P. Marier, Can. Met. Quart., 6(3), 249- 261 (1967).

12. Wantanabe, M. and T. Yoshida, Sci. Rep. Res. Inst. Tohoku Univ. Ser. A, 11(1), 66-72 (1959).

13. Hoschek, G. Monatsh. Chem. 93, 826-840 (1962).

14. Shell, P. K. et al. "Zinc sulfate decomposition in a solar rotary kiln," presented at the Annual Business Meeting of the Solar Thermal Test Facilities User's Association, (April 6-7, 1982), Houston, (Lawrence Livermore National Laboratory preprint, report UCRL-83634, March 10, 1982.)

15. Zakharov, V. Y., Russian Journal of Physical Chemistry,, 55(10), 1538-1539 (1982).

16. Shell, P. K. etal. "Solar thermal decomposition of zinc sulfate," preprint published in Lawrence Livermore National Laboratory report UCRL-53370, (January 1983).

17. Tabatabataie-Raissi, A. and M. J. Antal, Jr., "Design and operation of a 30 $KW_e/2KW_{th}$ downward facing beam arc image furnace," to appear in the J. Solar Energy.

18. Friedman, H. L., J. Polym. Sci., C6, 185 (1964).

19. Antal, Jr., M. J. et al., Comb. Sci. and Tech., 21, 141 (1980).

20. Antal, Jr., M. J. "Thermogravimetric signitures of complex solid phase pyrolysis mechanisms and kinetics" in Thermal Analysis Miller, B.(Ed.), Wiley, N.Y.(1983).

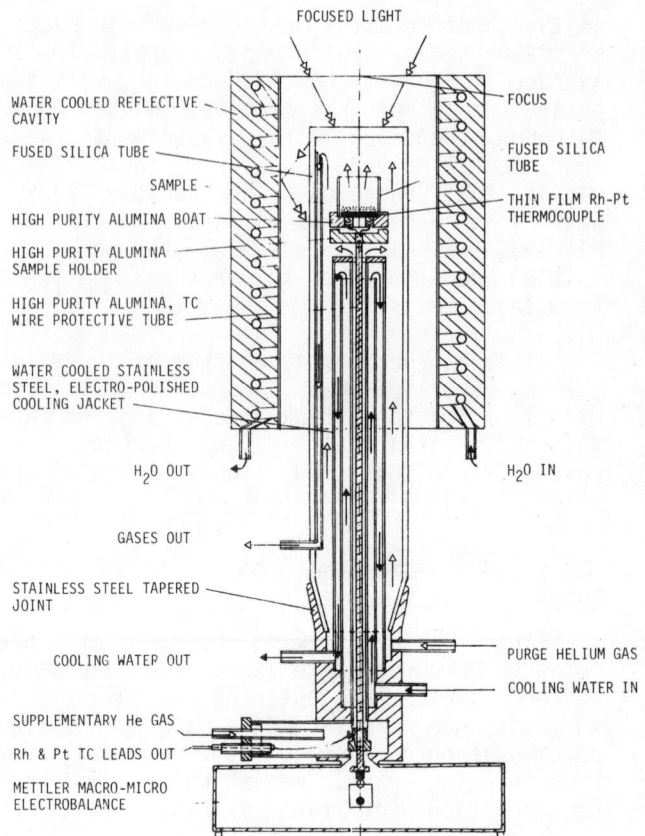

Figure 1. MARK II fast thermogravimetric analyzer system.

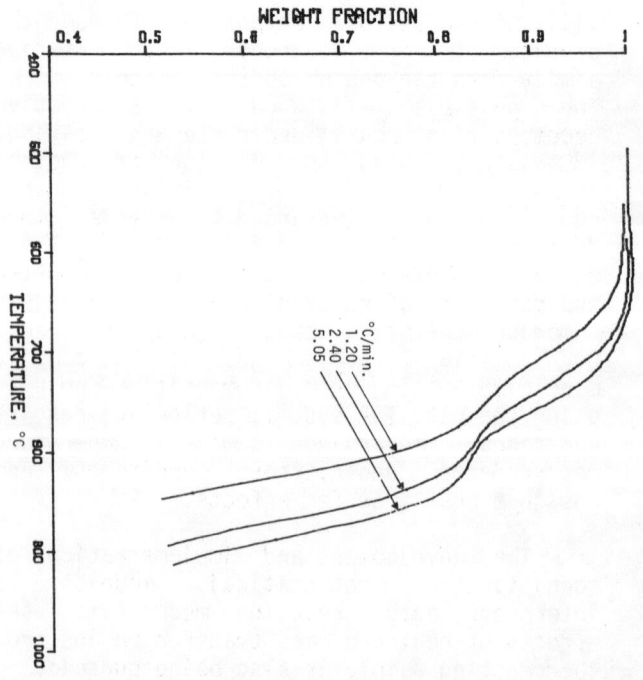

Figure 2. TG curves for anhydrous zinc sulfate in flowing pure helium at low (1-5 °C/min.) heating rates.

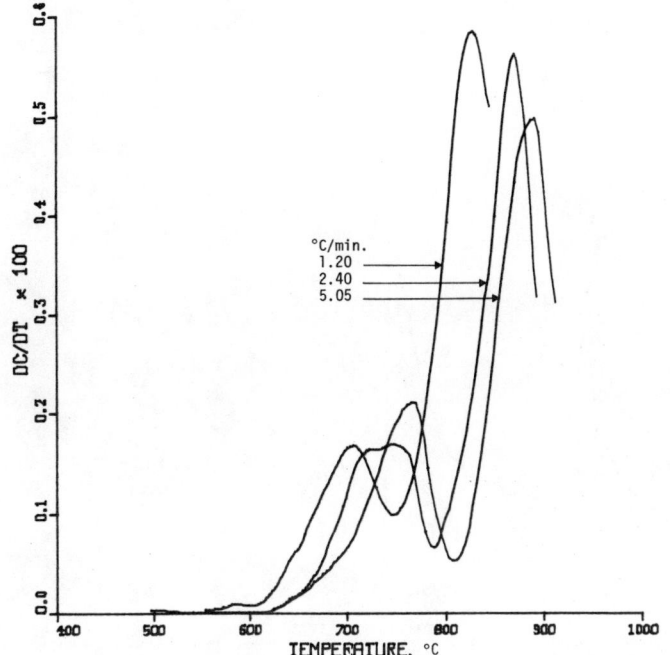

Figure 3. DTG curves for decomposition of anhydrous $ZnSO_4$ at low (1-5 °C/min.) heating rates.

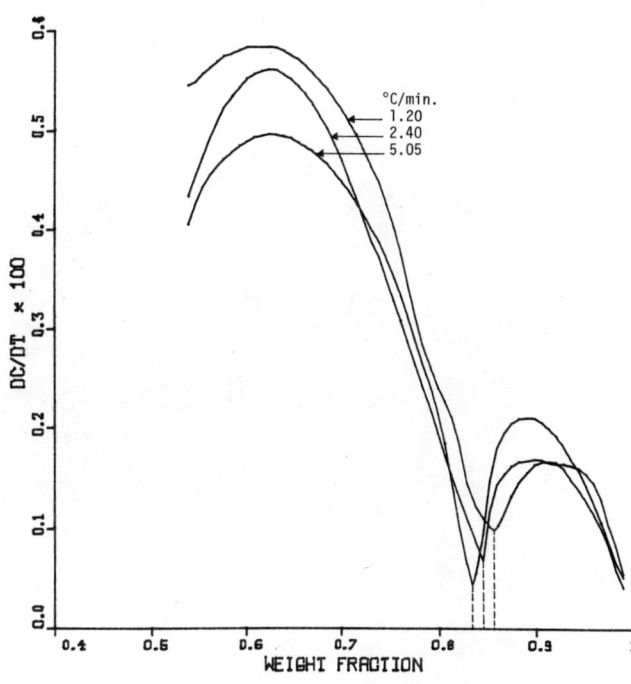

Figure 4. Rate of weight loss versus fraction sample weight remaining for anhydrous $ZnSO_4$ at low (1-5 °C/min.) heating rates.

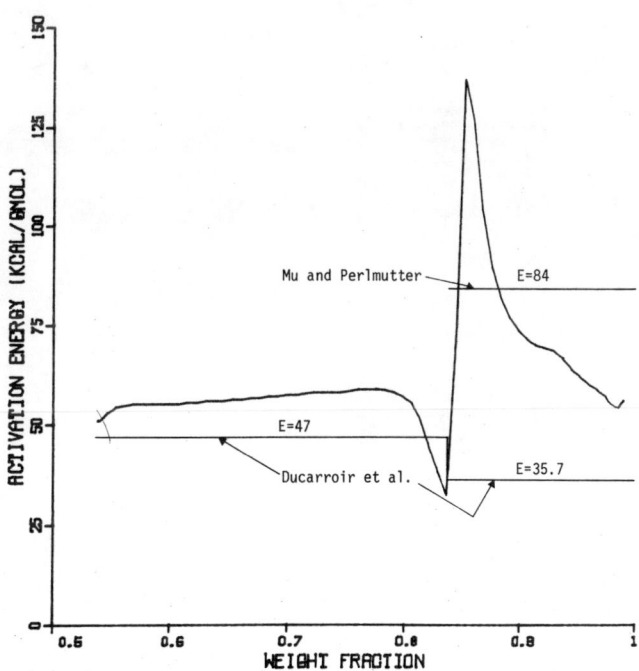

Figure 5. Friedman Signiture for decomposition of anhydrous zinc sulfate at low (1-5 °C/min.) heating rates.

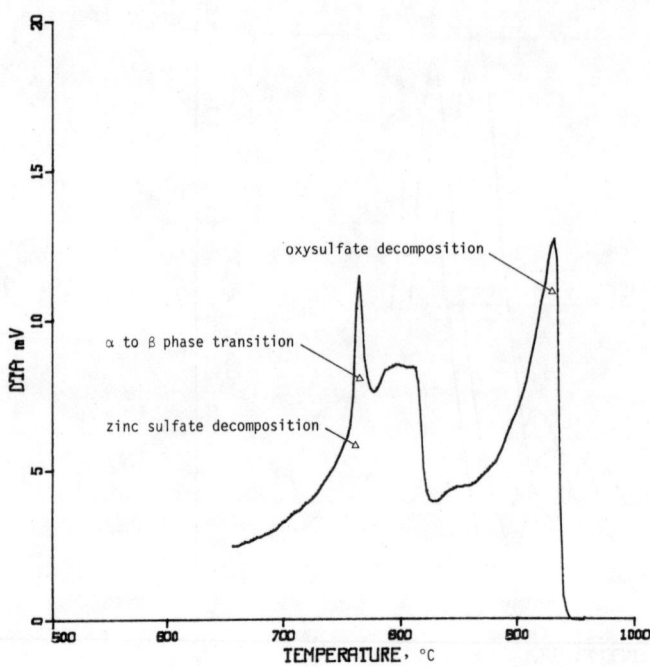

Figure 6. DTA curve for decomposition of anhydous ZnSO₄, 10 mg sample and 12.75 °C/min. heating rate.

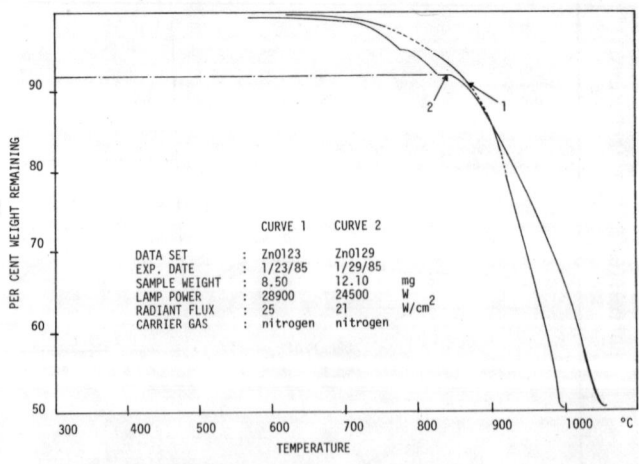

Figure 7. DTA curves for radiative decomposition of anhydrous ZnSO₄ at high heating rates.

# GAS-PARTICLE FLOW WITHIN A HIGH TEMPERATURE SOLAR CAVITY RECEIVER INCLUDING RADIATION HEAT TRANSFER

G. Evans and W. Houf ■ Sandia National Laboratories, Livermore, CA 94550
R. Greif ■ Mechanical Engineering Dept., University of California, Berkeley, CA 94720
C. Crowe ■ Mechanical Engineering Dept., Washington State University, Pullman, WA 99164

The flow of air and particles and the heat transfer inside a heated, open cavity containing a falling cloud of 100-1000 micron solid particles has been modeled numerically. The model includes radiative transport within the particle cloud and two-way momentum and thermal coupling between the particles and the air. The flow field is assumed to be two dimensional with steady mean quantities. The PSI-Cell (particle source in cell) computer code is used to model the gas-particle interaction. The radiative transfer within the cloud is modeled using the method of discrete ordinates.

An analysis has been made which includes the incident solar flux and the radiation transport among the cavity surfaces. Velocity and temperature profiles of the particles and the air have been predicted. The model has been used to examine the thermal performance of the solid particle solar receiver as a function of particle size, particle mass flow rate, particle scattering albedo, and incident solar flux.

## INTRODUCTION

Solid particles are being considered as the working medium for a solar central receiver at Sandia National Laboratories, Livermore ([1], [2]). Several advantages of using a solid particle receiver over systems using conventional fluids are: (1) the particles absorb the concentrated solar flux directly, eliminating the need for fluid conduit, (2) higher temperatures are theoretically possible, and (3) the particles can also serve as the storage medium. Proposed receiver designs consist of a cavity 5-10m tall mounted on a tower with a side facing aperture. Particles falling from a hole in the top of the cavity would be irradiated by the incident solar flux passing through the aperture.

### Initial Experiment Studies

Hruby et al.[3,4] have shown that the flow of particles from a hopper into ambient air is dilute (absence of particle-particle collisions), and that air entrainment must be accounted for in determining particle velocities. When a radiant heat source was applied to particles smaller than 500 microns, buoyancy effects in the air reduced the downward acceleration of the particles.

### Receiver Modeling

Radiative transport within the particle cloud and between the cavity walls must be accounted for in determining particle heating rates and cavity wall temperatures. Since Grashof numbers based on cavity height and cavity wall temperature (1000K) are of order $10^{13}$ the air flow field is turbulent with temperature dependent properties. The experimental studies have shown that two-way momentum and thermal coupling (each phase provides significant source terms for other phase) is important.

## TWO PHASE MODELING

A model (PSI-Cell, i.e., Particle Source in Cell) of dilute gas particle flows with heat and mass transfer has been developed by Crowe et al.[5]. Briefly, the model consists of a steady, two dimensional, elliptic, Eulerian description of the gas flow field coupled with a Lagrangian description of the particle flow field. The gas flow field is determined using TEACHT [6], which solves the conservation equations on staggered control volumes with the pressure, density, and temperature evaluated at control volume centers and the velocities evaluated at the control volume faces. A two equation $(\kappa - \epsilon)$ model of turbulence is included with constants established for a forced flow [7]. This turbulence model has been used [8] to predict flows with buoyancy effects; fair agreement with experimental results was obtained for mean velocity and temperature profiles.

### Gas Equations and Relations

The integrated form of the conservation equations for mass, momentum, and energy (neglecting viscous dissipation and compressibility effects) are given by:

$$(\rho u)_w^e \, \Delta y + (\rho v)_s^n \, \Delta x = 0 \qquad (1)$$

$$\left(\rho u u - \mu^* \frac{\partial u}{\partial x}\right)_w^e \Delta y + \left(\rho v u - \mu^* \frac{\partial u}{\partial y}\right)_s^n \Delta x = (P_w - $$

$$P_e)\Delta y - \rho g \Delta x \Delta y + \mu^* \frac{\partial u}{\partial x}\bigg|_w^e \Delta y + \mu^* \frac{\partial v}{\partial x}\bigg|_s^n \Delta x + S_p^x \quad (2)$$

$$\left(\rho uv - \mu^* \frac{\partial v}{\partial x}\right)_w^e \Delta y + \left(\rho vv - \mu^* \frac{\partial v}{\partial y}\right)_s^n \Delta x = (P_s -$$

$$P_n)\Delta x + \mu^* \frac{\partial u}{\partial y}\bigg|_w^e \Delta y + \mu^* \frac{\partial v}{\partial y}\bigg|_s^n \Delta x + S_p^y \quad (3)$$

$$\left(\rho uT - \frac{k^*}{c_p}\frac{\partial T}{\partial x}\right)_w^e \Delta y + \left(\rho vT - \frac{k^*}{c_p}\frac{\partial T}{\partial y}\right)_s^n \Delta x = S_p^T \quad (4)$$

where $e, w, n, s$ indicate that the corresponding terms are to be evaluated at the east, west, north, and south faces of the control volume, and $\Delta x$ and $\Delta y$ are the control volume dimensions in the vertical and horizontal directions, respectively. The convective-diffusive flux terms in equations (2-4) are evaluated using hybrid differencing (9). The source terms, $S_p^{x,y,T}$, refer to the momentum and energy added to the gaseous phase by the particles. A pressure correction equation using SIMPLE (10) is solved to insure local continuity, and effective conductivity, $k^*/c_p$, is set equal to effective viscosity, $\mu^*$. Variation of air properties with temperature is included in the model. No attempt has been made to account for the effects of buoyancy or particles on the structure of the turbulent field.

Particle Model and Fluid Source Terms

The particle momentum and energy equations are given by:

$$\rho_p \frac{\pi d_p^3}{6}\frac{Du_p}{Dt} = 3\pi d_p \mu \lambda (u - u_p) - \rho_p \frac{\pi d_p^3}{6} g \quad (5)$$

$$\rho_p \frac{\pi d_p^3}{6}\frac{Dv_p}{Dt} = 3\pi d_p \mu \lambda (v - v_p) \quad (6)$$

$$\rho_p c_p \frac{\pi d_p^3}{6}\frac{DT_p}{Dt} = \text{Nu} k \pi d_p (T - T_p) + Q_{rad} \quad (7)$$

The determination and implementation of $Q_{rad}$, the particle radiative heating rate, are discussed in subsequent sections.

The force in the $x$-direction on the gas in the computational cell due to the particles is given by:

$$S_p^x = F_p^x \sum_i \dot{N}_i \Delta t_i \quad (8)$$

where $F_p^x$ is the aerodynamic force on the gas due to a particle, $\dot{N}_i$ is the particle number flow rate along the $i^{th}$ trajectory, and $\Delta t_i$ is the particle transit time across the cell for the $i^{th}$ trajectory. The sum is applied over all trajectories passing through the computational cell for which the source term is being evaluated. A similar expression holds for the $y$-direction source term. The gas energy equation source term is given by:

$$S_p^T = (Q_p/c_p) \sum_i \dot{N}_i \Delta t_i \quad (9)$$

where $Q_p$ is the convection heat transfer rate from a particle to the air. Single particle drag coefficients and Nusselt number correlations have been used (11, 12).

RADIATION MODEL

The radiation model used in this study considers the interaction of an incident radiation field with the falling particle cloud and receiver rear wall. The model accounts for the directional nature of the radiation field, particle scattering, thermal emission and the wavelength dependence of the particle optical properties. The equation which governs the radiation field within the particle cloud at any elevation is (13)

$$\cos\theta \frac{dI_\lambda}{dy'}(y',\theta,\phi) = -(\sigma_\lambda + \kappa_\lambda)I_\lambda(y',\theta,\phi) + \kappa_\lambda I_{b,\lambda}(T_p)$$
$$+\frac{\sigma_\lambda}{4\pi}\iint_\Omega I_\lambda(y',\theta',\phi')p_\lambda(\theta',\phi' \to \theta,\phi)d\Omega' \quad (10)$$

where the quantity $I_\lambda(y',\theta,\phi)$ is the monochromatic intensity at $y'$ in the direction $(\theta,\phi)$, and the subscript $\lambda$ denotes wavelength. The depth, $y'$, into the particle cloud at any elevation is measured from the edge of the cloud receiving incident solar flux.

The intensity is the fundamental quantity that governs the radiation field within the particle cloud. The absorption coefficient, $\kappa_\lambda$, scattering coefficient, $\sigma_\lambda$, and phase function, $p_\lambda$, are monochromatic optical properties which depend on size, complex refractive index, and concentration of the particles in the cloud. The absorption and scattering coefficients characterize the attenuation per unit path-length along a traversing beam of radiation due to the respective effects of scattering and absorption. The phase function represents the probability that a beam moving in the direction $(\theta',\phi')$ and confined to a solid angle $d\Omega'$, will be scattered into a solid angle $d\Omega$, about the direction $(\theta,\phi)$.

To simplify the analysis, the dimensionless parameters $\omega_\lambda = \sigma_\lambda/\beta_\lambda$ and $d\tau_\lambda = \beta_\lambda dy'$, were introduced, where the extinction coefficient, $\beta_\lambda$, is the sum $(\sigma_\lambda + \kappa_\lambda)$. The quantity $\tau_\lambda$ is termed the optical depth, and the total optical thickness of the cloud is defined as

$$\tau_{\lambda,d} = \int_0^d \beta_\lambda dy' \qquad (11)$$

where $d$ is the thickness of the particle cloud at a particular elevation. The total optical thickness $\tau_{\lambda,d}$, is a measure of the ability of the particle cloud to attenuate radiation of a given wavelength. The quantity $\omega_\lambda$ is the single scattering albedo and is a measure of the relative importance of scattering to absorption in the interaction of radiation with a single particle. The albedo ranges from a value of zero for purely absorbing particles, to a value of one for particles that only scatter radiation. If the optical properties and lateral temperature profile for the solid carriers are known, then the radiative transfer equation can be solved subject to boundary conditions at the front and rear of the particle cloud to determine the radiation field $I_\lambda(y',\theta,\phi)$. The total radiative flux, which is used in determining the source term, $Q_{rad}$, in the particle energy equation, is then determined from the expression:

$$F(y') = \int_{all\,\lambda} \iint_\Omega I_\lambda(y',\theta',\phi')d\Omega'\,d\lambda \qquad (12)$$

The method of discrete ordinates is an accurate solution technique which has been applied to radiative transfer problems; it is the approach used in this study. The details of the method are given elsewhere (14) and will not be repeated here. For the purposes of this study, the scattering distribution, $p_\lambda$, was represented by a function of the form

$$p(\xi) = 1 + \sum_{k=1}^{100}(2k+1)g^k P_k(cos\xi) \qquad (13)$$

where the $P_k$ are Legendre polynomials of order k, and $\xi$ is the angle between the incident $(\phi',\theta)$ and scattered beams $(\theta,\phi)$. The value $g$ is an asymmetry factor which may be varied from -1 (strictly backward scattering) to 1 (strictly forward scattering) in order to alter the shape of the scattering function. A value of g equal to 0, which corresponds to an isotropic scattering distribution, is used in this study. Calculations over the range g = -0.5 to g = 0.5 show a variation of 15 percent in the absorbed solar radiation for a typical set of conditions (14).

## SOLUTION METHOD

A numerical solution is iterative and consists of the following steps:

1. The air flow field is first determined in the abscence of particles.
2. Particles are then introduced with an initial velocity and temperature from a point source at the top of the receiver. Particle velocity, position (trajectory), and temperature are determined by integrating equations (5-7). The thickness, $d$, of the particle cloud at each elevation corresponding to a computational cell boundary for a fluid scalar quantity is determined from the two extreme trajectories. Average particle number density $n$, and an average particle temperature are also determined for these elevations.
3. The radiative transport equation is solved for each horizontal slice (with vertical extent $\Delta x$) of the particle cloud; specification of the incident irradiation at the front edge of the cloud and the temperature and reflectivity of the back wall are required.
4. Particle temperatures are then recalculated with radiation source terms obtained from step 3, where $Q_{rad}$ for a particle between $y'$ and $y' + \Delta y'$ is determined by:

$$Q_{rad} = \frac{[F(y') - F(y' + \Delta y')]}{n \cdot \Delta y'} \qquad (14)$$

Iteration between this step and the preceding one is continued until there is little change in particle temperature. Fluid source terms are then accumulated.
5. The conservation equations for the fluid are now solved including the source terms. Wall element energy balances include radiation, convection, and a specified amount of conduction through the wall. These wall energy balances are embedded within a radiation enclosure calculation.

Steps 2 to 5 are repeated until convergence criteria based on the total residuals of the fluid mass (0.01), momentum (1.0), and energy (4.0) equations are obtained. Negligible changes occurred when the above criteria were satisfied simultaneously.

## RESULTS

### Geometry and Boundary Conditions

The receiver consists of a two dimensional, rectangular cavity (cf. Figure 1). The particles are introduced at a point (x= 6.0m, y = 0.965m) with the mass flow rate divided into ten equal parts. An initial spray cone of 0.6 radian is applied. Particles exit at the bottom of the receiver. Zero gradient conditions for the dependent variables are set at the top outflow opening. A hydrostatic pressure gradient is imposed between the aperture and the top opening $(\Delta P = \rho_\infty g L)$. A provision is made for applying a back pressure at top opening by modifying the hydrostatic pressure at this position with the constant, $K$ $(\Delta P = K\rho_\infty g L)$. Zero pressure correction and a direct application of the continuity equation are used across the fluid inlet and outlet areas.

Results were computed on a uniform 16 by 20 mesh ($\Delta x = 0.33$ m, $\Delta y = 0.21$ m) with wall functions used to apply boundary conditions to the first interior control volume. No attempt has been made to resolve the thin boundary layers which would result from these high Grashof number flows. Reynolds analogy was used to relate the convective heat transfer at the wall to wall shear stress.

## Nominal Parameter Values

The following nominal parameter values were used in the calculations: 0.25 mw/m$^2$ incident solar energy (uniformly applied to front of cloud), 1.0 kg/s particle mass flow rate, 650 micron diameter spherical particles, particle density and specific heat of 3.13X10$^3$ kg/m$^3$ and 10$^3$ J/kg-K, initial particle temperature and downward velocity of 293K and 0.3 m/s, a particle scattering albedo of 0.2, cavity wall reflectivity of 0.2, and a back pressure constant, $K$, of 0.95.

## Air and Particle Fields

Figure 2 shows the air flow field with and without particles. The air flow streams in the aperture, turns upward at the hot back wall of the cavity, and flows out the top opening. Additional upward movement of air occurs along the upper front wall of the cavity. A significant increase in upward air movement occurs as a result of the buoyancy generated by convective heat transfer from the particles. Isotherms are shown in Figure 3. The convection from the particles to the air is evident in the lower portion of the cavity where the cold inflowing air meets the hot falling particles.

Partical temperature and vertical velocity for the extreme (front and back) trajectories are shown in Figure 4. The particles have essentially reached an equilibrium condition between radiative gain and convective loss, after 6m of fall, with temperatures approaching 850 K. Particle vertical velocity is still increasing after 6m of fall.

## Results of Varying Parameters

Parameters were varied independently to determine their effect on cavity efficiency and particle exit temperature. Figure 5 shows the effect of varying particle mass flow rate on the cavity efficiency, defined as the ratio of sensible energy gain of the particles to the incident solar energy, and on the average exit temperature of the particles. Increasing the particle mass flow results in an increase in cavity efficiency but a decrease in particle temperatures. Clearly, for a fixed incident flux, an energy balance shows that an increase in the mass flow rate leads to a decrease in the maximum temperature difference. Figure 6 shows the variation of the average exit particle temperature with incident solar flux. Cavity efficiency remains fairly constant at 35 percent over the flux range studied.

Reducing the particle size might be expected to increase the cavity efficiency and particle temperature, since smaller particles result in a greater optical thickness and also remain in the flux field longer due to their smaller fall velocities. However, as Figure 7 shows, the convective loss fraction, defined as the ratio of the total convective heat transfer between the particles and the air to the incident solar energy, increases dramatically for the smaller particles. The result is that cavity efficiency and particle exit temperature are fairly insensitive to variations in particle size, over the size range considered.

The effect of using selectively absorbing particles was investigated by assigning one value of the particle scattering albedo, defined as the ratio of scattering to the sum of absorption and scattering, for the incident solar energy and another value for the reradiated energy from the cavity walls. A wavelength of 0.5 micron was used to demarcate the solar and infrared regions. A value of 0.2 was used for the albedo for the incident solar energy. By using a larger value of the scattering albedo for the infrared energy, the particles emit less energy and might be expected to retain a larger amount of the absorbed solar energy. However, as shown in Figure 8, cavity efficiency decreases as a function of increased infrared scattering albedo. This is a result of the decreased absorption of the (large amount of) infrared energy emitted from the cavity walls which is incident on the particle cloud.

As noted earlier, higher particle exit temperatures occur as a result of larger incident solar fluxes; in addition, higher cavity efficiencies result when the particle mass flow rate is increased. By combining larger fluxes with higher mass flow rates, both particle exit temperature and cavity efficiency increase. Specifically, an incident solar flux of 0.5 mw/m² combined with a particle mass flow rate of 1.5 kg/s results in particle exit temperatures of 1140 K and a cavity efficiency of 42 percent. Further increases in solar flux and mass flow rate (1.0 mw/m² and 2.5 kg/s), coupled with a reduction in cavity size to 2m deep by 3m high with a 1.5m aperture to reduce radiative and convective losses, resulted in cavity efficiency of 72 percent and particle exit temperatures of 1165 K.

SUMMARY

A model has been developed to predict gas particle flow and heat transfer in a two dimensional, steady, solid particle solar central receiver. The model includes the elliptic equations for conservation of mass, momentum, and energy of air, allowing for the temperature dependence of the physical properties. The PSI-Cell model provides a Lagrangian description of the particle flow and allows for two-way coupling of momentum and heat transfer between the particles and the air. Radiation transport within the particle cloud is determined by solving the radiative transport equation on a monochromatic basis, including the effects of anisotropic scattering. Radiation transport with the cavity walls and the front surface of the particle cloud is also included. A parameter study has resulted in the following conclusions:

1. increases in cavity efficiency are accompanied by lower particle exit temperature when mass flow rate is increased,
2. smaller particles (in conjunction with a fixed mass flow rate) result in greater optical thickness, longer residence time, and higher convective loss fraction,
3. larger incident fluxes result in relatively constant cavity efficiency and higher particle exit temperature,
4. Larger infrared scattering albedo (reduced particle emission) results in lower particle exit temperature and lower cavity efficiency. This is due to reduced absorption of the infrared radiation from the high temperature walls,
5. convective losses from the particles represent a significant fraction of the incident solar energy, and
6. the optical thickness of the particle cloud decreases with distance from the top of the cavity due to dilution resulting from particle acceleration in the vertical direction.

NOTATION

$I_{b,\lambda}$ = Planck's function, w/(m² -micron -sr)

$Nu$ = Nusselt number = $h d_p / k$
$c_p$ = specific heat, J/(kg K)
$d_p$ = particle diameter, m or microns
$g$ = acceleration due to gravity, m/s²
$\epsilon$ = turbulent dissipation, m²/s³
$\theta, \theta'$ = polar angle of coordinate system
$\phi, \phi'$ = azimuthal angle of coordinate system
$\kappa$ = turbulent kinetic energy, m²/s²
$\rho$ = density, kg/m³
$\lambda$ = dimensionless single particle drag coefficient
$\mu$ = viscosity, kg/(m s)
$\Omega$ = solid angle, sr

subscripts

$p$ = particle phase
$\infty$ = evaluated at ambient conditions

LITERATURE CITED

1. Martin, J. and J. Vitko, "ASCUAS: A Solar Central Receiver Utilizing a Solid Thermal Carrier," Sandia National Laboratories, SAND82-8203 (1982).

2. Falcone, P.K., J.E. Noring, and C.E. Hackett, "Evaluation and Application of Solid Thermal Energy Carriers in a High Temperature Solar Central Receiver System," presented at 17th IECEC, Los Angeles (1982).

3. Hruby, J.M. and V.P. Burolla, "Solid Particle Receiver Experiments: Velocity Measurements," Sandia National Laboratories, SAND84-8238 (1984).

4. Hruby, J.M., B.R. Steele, and V.P. Burolla, "Solid Particle Receiver Experiments: Radiant Heat Test," Sandia National Laboratories, SAND84-8251 (1984).

5. Crowe, C.T., M.P. Sharma, and D.E. Stock, "The Particle-Source-in Cell(PSI Cell) Model for Gas-Droplet Flows," Journal of Fluids Engineering, 325, June (1977).

6. Gosman, A.D., and W.M. Pun," Calculation of Recirculating Flow," Lecture Notes, Imperial College of Science and Technology, London, England (1973).

7. Launder, B.E., and D.B. Spalding, Mathematical Models of Turbulence, Academic Press (1972).

8. Abdelmeguid, A.M., and D.B. Spalding, "Turbulent flow and heat transfer in pipes with buoyancy effects," J. Fluid

Mech., **94**, 2, 383 (1979).

9. Spalding, D.B., "A Novel Finite-Difference Formulation for Differential Expressions Involving Both First and Second Derivatives," Int. J. Num. Methods Eng., (1972).

10. Patankar, S.V., Numerical Heat Transfer and Fluid Flow, McGraw-Hill (1980).

11. Wallis, G.B., One-Dimensional Two-Phase Flow, McGraw-Hill (1969).

12. Bird, R.B., W.E. Stewart, and E.N. Lightfoot, Transport Phenomena, Wiley (1960).

13. Siegel, R. and J.R. Howell, Thermal Radiation Heat Transfer, 2nd ed., McGraw-Hill (1981).

14. Houf, W.G. and R. Grief, "Radiation Transfer in a Solar Absorbing Particle Laden Flow," paper presented at the 23rd ASME/AIChE Natl. Heat Transfer Conf., Denver, Co., Aug. 4-7 (1985).

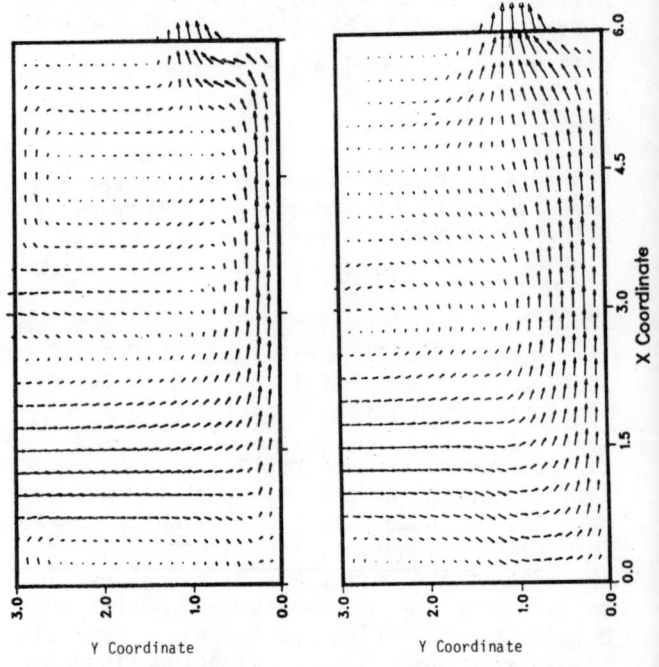

Figure 2. Air flow field without (left) and with (right) particles. Air velocities out top opening are approx. 1 m/s without particles and 4 m/s with particles.

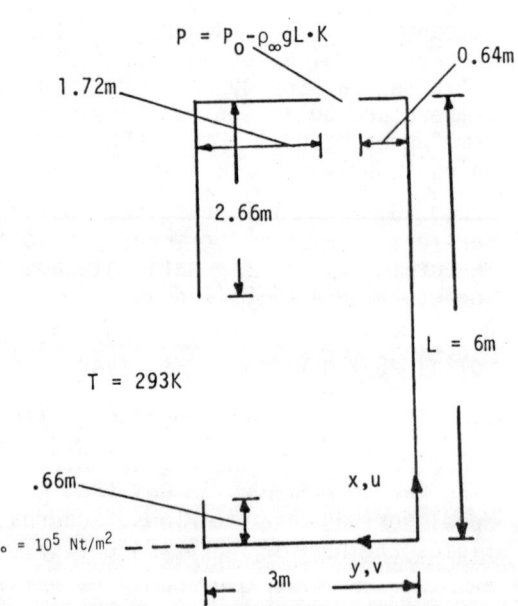

Figure 1. Cavity geometry and boundary conditions for fluid flow.

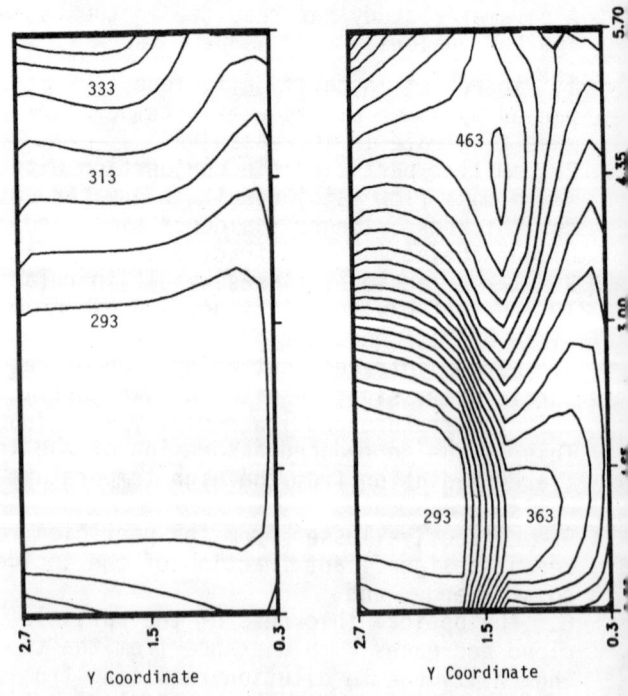

Figure 3. Air isotherms without (left) and with (right) particles (10 K temperature increments).

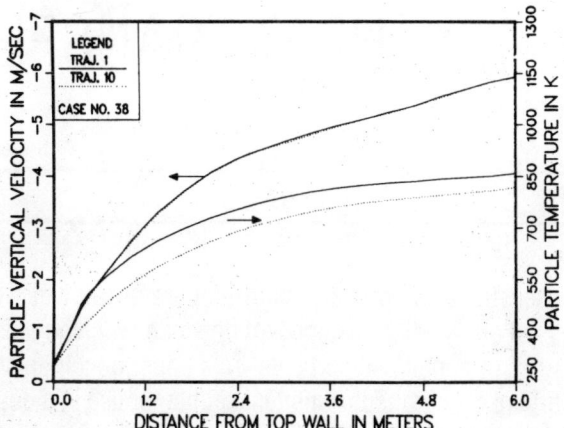

Figure 4. Particle temperature and vertical velocity for the front (——) and back (....) particle trajectories as a function of vertical position in the cavity.

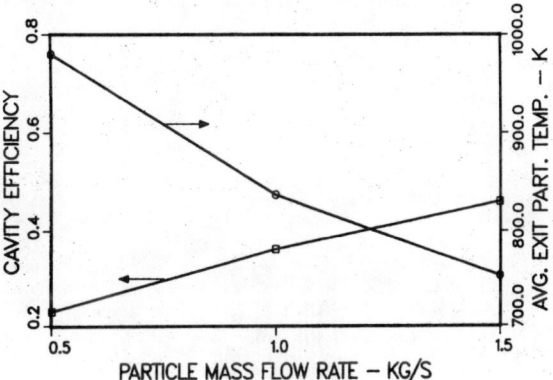

Figure 5. Variation of cavity efficiency and average exit temperature of particles as a function of particle mass flow rate for 650 micron diameter particles and 0.25 mw/m**2 solar flux (symbols indicate values of mass flow rate used in this study).

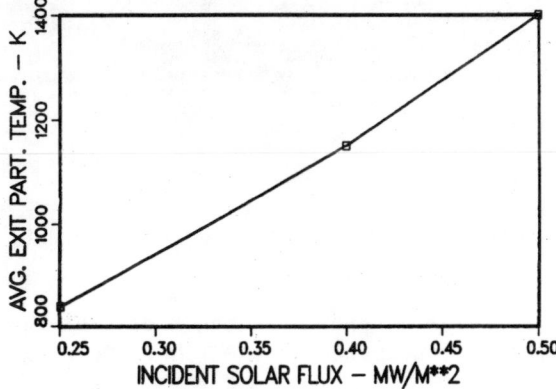

Figure 6. Variation of average exit particle temperature as a function of incident solar flux for 650 micron diameter particles and 1.0 kg/s particle mass flow rate (symbols indicate values of particle size used in this study).

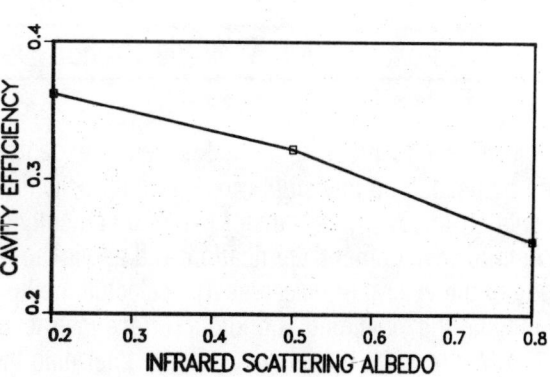

Figure 7. Variation of particle convective loss fraction as a function of particle size for 1.0 kg/s particle mass flow rate and 0.25 mw/m**2 incident solar flux (symbols indicate values of particle size used in this study).

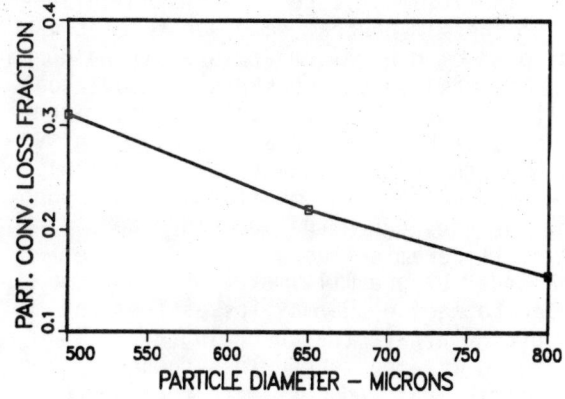

Figure 8. Variation of cavity efficiency as a function of particle infrared scattering albedo for 650 micron diameter particles, 1.0 kg/s particle mass flow rate, and 0.25 mw/m**2 incident solar flux (symbols indicate values of infrared scattering albedo used in this study).

# VOLUMETRIC RECEIVER RADIATION HEAT TRANSFER

M. Kevin Drost and J. F. Welty ■ Oregon State University, Corvallis, Oregon 97330

The volumetric receiver is an advanced solar central receiver concept used for producing high temperature air which can then be used in an industrial process heat application. The successful analysis of this concept depends on being able to calculate radiation heat transfer in a geometrically complicated array of nondiffuse surfaces. This paper documents the selection, development, verification, and application of a Monte Carlo photon transport model suitable for use in a design study of the volumetric receiver. The selected method uses computational cells to reduce the complexity association with identifying the location of a photon/surface interaction.

The Monte Carlo model was used to determine the performance of several types of volumetric receiver designs. The results indicated that the best design consists of a fiber absorbing array with augmented convective heat transfer surrounded by a shroud-type geometric loss reducer.

## INTRODUCTION

The volumetric receiver is an advanced solar thermal central receiver concept used for producing high temperature air which can then be used in an industrial process heat application. The volumetric receiver concept consists of an array of absorbing surfaces arranged in concentric cylindrical rows around an inlet manifold. The absorbing surfaces can either be fin-shaped pins or small ceramic fibers with a diameter on the order of one millimeter or less. In order to reduce thermal losses from the absorbing array, the absorbing array is enclosed in a geometric loss reducer. Two geometric loss reducers were considered; a shroud and wedge-shaped specular reflecting pins. The shroud acts like a cavity-type enclosure while the wedge-shaped reflecting pins tend to reflect incoming radiation into the receiver while inhibiting thermal losses from the absorbing media. A sketch of a volumetric receiver design with fin-shaped absorbing surfaces and wedge-shaped reflecting zones is shown in Figure 1.

The successful analysis of a volumetric concept depends on being able to predict radiation heat transfer in a geometrically complicated array of nondiffuse absorbing surfaces. This paper summarizes the method used to analyze the radiation heat transfer problem and discusses the results with particular attention on the impact of radiation heat transfer on the final design.

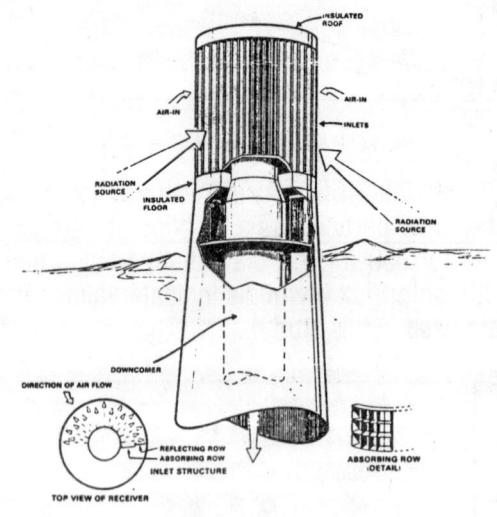

Figure 1. Volumetric Receiver (50 MWt).

## VOLUMETRIC RECEIVER THERMAL PROCESS

Insolation from the collector field is concentrated on the volumetric receiver. The insolation passes the geometric loss reducer with a small amount of the energy being either absorbed or reflected back to ambient. The insolation then strikes the absorbing array where most is absorbed but a small amount is reflected back to the geometric loss reducer. As the temperature of the absorbing array increases, the thermal energy is transferred by convection to air which is being drawn through the absorbing array into the inlet manifold by an induced draft fan.

In the analysis of a volumetric receiver there are two major radiation problems. First, where is the insolation absorbed in the receiver? Second, what is the magnitude of internal radiation heat transfer both between various receiver surfaces and between a receiver surfaces and the surrounding? Both problems involve radiation between geometrically complicated surfaces where directional and spectral variations in surface properties often cannot be ignored.

SELECTION OF METHOD OF ANALYSIS

The key problem in modeling radiation heat transfer between discrete surfaces is in predicting the configuration factor or exchange factor. Emery, et al., (1) identified seven methods for predicting view factors. He concluded that for complicated configurations only numerical methods are appropriate. Of the possible numerical approaches, Emery, et al., concludes that a Monte Carlo approach is best suited to the determination of configuration factors between a single small area and surrounding areas. This is normally the situation encountered in the volumetric receiver. Monte Carlo modeling has the added advantage of allowing the inclusion of non-diffuse, nongray surface effects with relative ease (2).

A variety of generalized Monte Carlo photon and neutron transport codes (3,4) and Monte Carlo radiation heat transfer codes exist (5). In all cases the general applicability of these codes was purchased at the price of substantial complexity. Therefore, it was decided to develop a specialized Monte Carlo computer code which was designed to take advantage of the regular geometry of the volumetric receiver.

MONTE CARLO MODEL

The Monte Carlo approach to radiation heat transfer problems has been widely used and the method is well documented (2,6,7). In this section the Monte Carlo method will be briefly described.

The Monte Carlo approach is a statistical method of solving a physical problem which can be modeled as a series of probabilistic and deterministic events. Energy emitted from a surface is simulated by a large number of energy bundles. The emitted bundles are followed as they proceed from one event to another with the results of each event being recorded until the energy bundles either leave the receiver or are absorbed on a surface. A large number of bundles are simulated with the results of all events being totaled. A sufficiently large number of bundles must be considered to insure that variations in the results due to random events are small. The results can then be used to determine the fraction of the emitted energy which has been absorbed on each surface or has left the receiver.

In this study, the Monte Carlo technique was used to calculate both insolation distribution and exchange factors. The major problems with the Monte Carlo calculation of exchange factors are those associated with geometrical considerations (which surface is struck by a bundle) and surface considerations (what happens when the surface is struck). Methods of modeling the interactions of a bundle with a surface are described by several authors (2,8) and will not be discussed here. Absorbing surface property variations with incident angle and wave length were taken from Modest (9).

The problem associated with receiver geometry involves determining which surface is struck by a photon bundle once the bundle has been emitted. When complex geometries are considered, (5,9) the method consists of describing each surface mathematically and determining which surface intercepts the vector describing the path of the energy bundle. The distance between the emission point and each intercepted surface is calculated and the surface with the shortest distance is identified as the surface impacted by the energy bundle. When a large number of surfaces are included, the computational time associated with determining impact location becomes substantial.

The volumetric receiver includes an arrangement of pins in concentric cylindrical rows. The regular spacing of the pins and the arrangement into rows suggests that a more efficient method of determining impact location can be used. This approach consists of dividing the receiver into computational cells where the cells are arranged so that absorbing surfaces are located on cell boundaries. This simplifies the identification of the impact location because one of the four surfaces in a cell is the emitting surface and only the three remaining cell boundaries can be struck by the emitted photon bundle.

The Monte Carlo model was used as the basis for two computer codes. The VORRUM computer code calculated absorbed insolation distribution while the VORVFM code calculated exchange factors between various receiver surfaces. A detailed description of the codes is included in Drost (10). The computer models were verified by comparison with analytical results for black and diffuse gray enclosures. In addition, the results predicted by the computer model were compared with the experimentally determined transmissivity of two rows of wedge-shaped relecting pins. The comparison with analytical calculation was excellent with the Monte Carlo results being within one standard deviation of the predicted result. The comparison with experimental results indicated that the VORRUM code tended to underestimate the transmissivity of the two rows of wedge-shaped reflecting pins but that given the uncertainties in the experimental investigation, the VORRUM code could be successfully correlated with experimental results.

The Monte Carlo computer codes do not model the impact of the shroud on either insolation distribution or internal exchange factor calculations. The shrouded receiver was modeled as a cylindrical shroud surrounding a cylindrical absorbing core. The Monte Carlo model was used to determine the radiation heat transfer in the core, and these results were combined with configuration factor algebra in order to determine the radiation heat transfer including the shroud.

## RESULTS

The Monte Carlo model was used to analyze both types of absorbing surfaces (fin and fiber) and both types of geometric loss reducers (reflecting zone and shroud). We will now summarize the results of each analysis.

## FIN-SHAPED PIN DESIGNS

The primary radiation heat transfer phenomena associated with fin designs, was the penetration of insolation into the receiver. A typical fin design is shown in Figure 2.

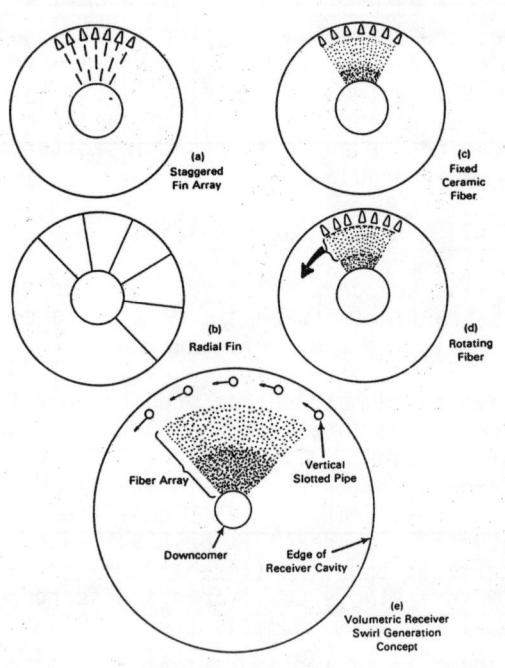

Figure 2. Volumetric Receiver Designs Considered.

A variety of fin spacings were considered and the results are summarized in Table 1. Table 1 presents the fraction of insolation absorbed on various regions, or zones, of a fin-type absorbing surface. These results indicate that no more then 72 fins should be used; otherwise, too small a fraction of the insolation penetrates deeply into the receiver. Additional analysis indicated the wide spacing of the receiver fins did not provide sufficient convective heat transfer area resulting in excessive surface temperature and thermal losses. These results are true for either a shroud or reflecting zone geometric loss reducer.

The analysis on internal radiation heat trasfer showed that internal radiation heat transfer was the major loss mechanism for the hot internal surfaces. The hot surfaces radiated thermal energy to cooler surfaces which ultimately radiated energy to the surrounding.

### FIBER DESIGN

The fiber design consists of a series of concentric cylindrical zones, each of which contains ceramic fibers with a variable packing density. The exterior zones are less density packed with the density increasing towards the interior. An exhaustive optimization of the fiber density distribution was not attempted, but a reasonable design which gave acceptable insolation distribution was identified. When coupled with heat transfer augmentation, such as rotating the fiber zones on inducing an air swirl, the fiber design would appear to have acceptable material temperatures. Characteristics of a typical fiber design with a shroud geometric loss reducer are shown on Table 2. For computational purposes, the continuous distribution of fibers have been combined into 9 zones. The first receiver zone consists of the shroud with the interior zone (zone 11) being the terminal absorber. As with the fin designs, internal radiation heat transfer had a significant impact on receiver losses.

### WEDGE-SHAPED REFLECTING PINS

The fiber absorbing core was combined with both wedge-shaped reflecting pins and shroud. The wedge-shaped reflecting pin concept was the subject of an extensive parametric investigation. The results of the radiation heat transfer analysis indicated the relecting pins tend to absorb too much insolation and reradiated thermal energy from the absorbing core, particularly when the poor convective heat transfer between the pins and the air is considered. This results in excessive reflecting pin temperature. A second major problem with the reflecting pin concept is that these surfaces must maintain a high reflectivity at high surface temperature and no attractive material was identified. Based on these results, the wedge-shaped reflecting pin concept was rejected.

### SHROUD DESIGN

The shroud proved to be a more attractive geometric loss reducer. A typical shroud design is shown in Figure 3.

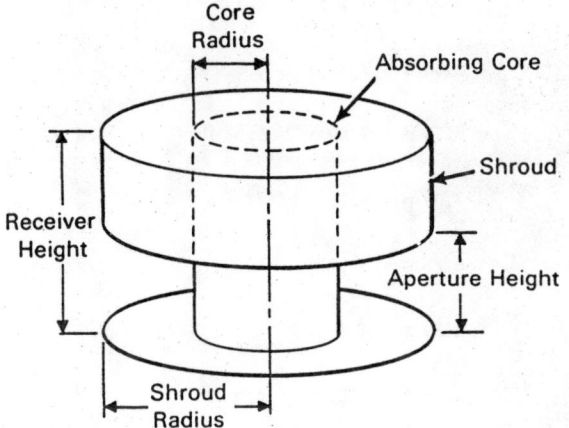

Figure 3. Schematic of Shroud.

The radiation heat transfer analysis showed that the shroud substantially reduced reradiation and reflection losses from the absorbing core. The shroud can have a negative impact by increasing spillage shroud aperture height on thermal efficiency (not including spillage). The calculation of spillage losses was beyond the scope of this study, therefore, the thermal efficiency must be combined with spillage losses in order to determine the optimum shroud aperture height. Flux ratio is the ratio of the maximum allowable energy flux on a surface and the predicted energy flux.

### CONCLUSIONS

The Monte Carlo simulation of radiation heat transfer phenomena used in these tests proved to be an attractive and efficient design tool. The use of a cell to cell method of tracking a photon's path both simplified code development and reduced running time. The use of this approach in similar geometric situations should be considered by other researchers. The results of the analysis as they pertain to the volumetric receiver include:

| Zone | 12 fins (30° between) | 24 fins (15° between) | 36 fins (10° between) | 72 fins (5° between) | 120 fins (3° between) | 360 fins (1° between) |
|---|---|---|---|---|---|---|
| 1<br>10.0 m – 9.75 m | 4.2 | 5.6 | 6.8 | 9.4 | 13.1 | 35.1 |
| 2<br>9.25 m – 8.5 m | 9.1 | 6.6 | 6.9 | 10.4 | 15.6 | 24.4 |
| 3<br>8.5 m – 7.75 m | 5.5 | 6.6 | 7.8 | 11.0 | 15.6 | 14.3 |
| 4<br>7.75 m – 7.00 m | 6.8 | 7.8 | 7.7 | 11.8 | 11.4 | 7.2 |
| 5<br>7.00 m – 6.25 m | 7.4 | 7.0 | 7.2 | 11.0 | 9.5 | 4.2 |
| 6<br>6.25 m – 5.50 m | 6.2 | 7.9 | 9.1 | 8.1 | 8.0 | 3.4 |
| 7<br>5.50 m – 4.75 m | 7.4 | 7.8 | 7.6 | 5.9 | 2.4 | |
| 8<br>4.75 m – 4.00 m | 6.8 | 7.0 | 8.0 | 6.5 | 5.0 | 1.8 |
| 9<br>4.00 m – 3.25 m | 6.6 | 8.3 | 7.7 | 5.4 | 3.6 | 1.5 |
| 10<br>3.25 m – 2.5 m | 7.1 | 7.1 | 6.8 | 5.0 | 3.2 | 0.9 |
| 11 | 25.5 | 22.2 | 19.7 | 10.5 | 6.3 | 2.5 |
| Reflection Loss | 7.4 | 6.1 | 4.8 | 3.4 | 2.9 | 2.2 |

Table 1. Energy Absorbed in Zone (%) Assuming 1 m Image Size.

| Zone Characteristics<br>Zone No. | Type | Height (m) | Depth (m) | Fin Spacing (m) | Convection Area ($m^2$) | Radiation Area ($m^2$) | Emissivity |
|---|---|---|---|---|---|---|---|
| 1 | Shroud | 10.0 | 6.0 | N/A | 879 | 879 | 1.0 |
| 2 | 0.0006 m dia fiber | 10.0 | 0.3 | 0.016 | 34 | 34 | 0.9 |
| 3 | 0.0006 m dia fiber | 10.0 | 0.3 | 0.012 | 43 | 43 | 0.9 |
| 4 | 0.0006 m dia fiber | 10.0 | 0.3 | 0.009 | 52 | 52 | 0.9 |
| 5 | 0.0006 m dia fiber | 10.0 | 0.3 | 0.0055 | 76 | 76 | 0.9 |
| 6 | 0.0006 m dia fiber | 10.0 | 0.3 | 0.0039 | 96 | 96 | 0.9 |
| 7 | 0.0006 m dia fiber | 10.0 | 0.3 | 0.0024 | 132 | 132 | 0.9 |
| 8 | 0.0006 m dia fiber | 10.0 | 0.3 | 0.0011 | 212 | 212 | 0.9 |
| 9 | 0.0006 m dia fiber | 10.0 | 0.3 | 0.0006 | 271 | 271 | 0.9 |
| 10 | 0.0006 m dia fiber | 10.0 | 0.3 | 0.0001 | 415 | 415 | 0.9 |
| 11 | Terminal Absorber | 10.0 | N/A | N/A | 578 | 144 | 0.9 |

Receiver Height = 10 m  
Receiver Radius = 11 m  
Air Flow Rate = 42.8 Kg/s  
Product Design Temperature = 1367K  

Aperture Height = 6.0 m  
Blocking Factor = 0.05  
Shroud Fill = No  

Table 2. Shrouded Fiber Design Characteristics.

- Fin-shaped absorbing surfaces and wedge-shaped pins for geometric loss reduces are not attractive from either a radiation heat transfer or performance point of view.

- The most attractive volumetric receiver design consists of a fiber absorbing core surrounded by a shroud-type geometric loss reducer.

LITERATURE CITED

1. Emery, A.F., et al., "Computation of Radiative View Factors for Surfaces with Obstructed View of Each Other," Paper No. 81-HT-57, ASME, New York, NY, (1981).

2. Seigel, R., and J.R. Howell, "Thermal Radiation Heat Transfer", McGraw-Hill, New York, NY, (1972).

3. Halbleib, "ACCEPT: A Three Dimensional Electron/Photon Monte Carlo Transport Code Using Combinational Geometry," SAND 79-0415, Sandia Laboratories, Albuquerque, NM, (1978).

4. Los Alamos Scientific Laboratories, "MCNP; A General Monte Carlo Code for Neutron and Photon Transport," Los Alamos Scientific Lab, NM (USA) July 1978, (1978).

5. Corlett, R.C., "Direct Monte Carlo Calculations of Radiative Heat Transfer in Vacuum," Journal of Heat Transfer, Vol. 88, No. 4, pp. 376-382, (1966).

6. Tour, J.S. and R. Viskanta, "A Numerical Experiment of Radiation Heat Interaction by the Monte Carlo Method," Journal of Heat and Mass Transfer, Vol. 11, pp. 883-897, (1968).

7. Weiner, M.M., et al., Radiative Interchange Factors by Monte Carlo, Paper No. 65-US/HT-51, ASME, New York, NY, (1965).

8. Yang, R.S., "Heat Transfer Through a Randomly Packed Bed of Spheres by the Monte Carlo Method," Ph.D. thesis, University of Texas, (1981).

9. Modest, M.F., "Three Dimensional Radiative Exchange Factor for Non-Gray, Non-Diffuse Surfaces," Numerical Heat Transfer, Vol. 1, pp. 403-416, (1978).

10. Drost, M.K., "Volumetric Receiver Development," Ph.D. thesis, Oregon State University, (1985).

# A STABLE NUMERICAL METHOD FOR ONE-DIMENSIONAL, TWO-PHASE FLOW

S.T. Free and A.L. Schor ■ Department of Nuclear Engineering
Massachusetts Institute of Technology, Cambridge, Massachusetts 02139

We have developed a numerical method for two-phase flow in which mass and energy convection are treated implicitly. Theoretical analysis and computational testing have suggested that the method is unconditionally stable for subsonic flows. The method has been implemented in an experimental, one-dimensional version of the sodium boiling code THERMIT-4E. A slightly larger amount of computation per time step is required due to the evaluation of a more complex Jacobian matrix, but larger (and hence fewer) time steps are possible, giving a substantial net reduction in computational time.

## INTRODUCTION

The study of two-phase flow using computer simulation plays a major role in the analysis of transients and postulated accidents in nuclear reactor systems. Indeed, proper modeling of the often complex thermal-hydraulic phenomena involved is crucial to a reliable assessment of the sequence, rate of progression, and consequences of events known or postulated to occur. Over the last ten to fifteen years, continuing efforts devoted to the mathematical, physical, and numerical modeling of two-phase flow have produced remarkable progress in developing computer codes for realistic simulation of nuclear reactors. Although two-phase flow calculations have become widely used, they should not yet be considered "routine", in that unusually severe conditions may lead to long and difficult calculations and in some cases algorithm failure. Difficulties are considerably amplified under low pressure conditions, where the intrinsic nonlinearities of the governing equations become extremely severe.

Many computer codes for nuclear reactor analysis approximate the partial differential equations describing two-phase flow by a set of finite difference equations on a staggered mesh. The method used to difference the equations strongly affects the numerical stability of the computational algorithm. In particular, the time levels chosen for the set of dependent variables govern the way that any numerical disturbance (e.g., round-off error) propagates in space and time within this numerical system. Explicit numerical schemes suffer from a time step restriction imposed by sonic propagation. The very high sound speed in pure liquid or nonequilibrium, two-phase mixtures limits the maximum time step allowed for numerical stability to the rate at which a pressure pulse can transit a grid cell (typically $10^{-6}$ to $10^{-5}$ sec). In using these schemes, the large number of time steps required for a calculation offsets the benefits of the relatively small amount of computation per step -- especially for machines without array processing capability. At the other extreme, implicit schemes are generally unconditionally stable. Hence, the time step need only be controlled by the accuracy desired for the calculation. Unfortunately, implicit schemes require a relatively large amount of computation per time step, greatly reducing the savings of using larger steps.

A compromise between these extremes has led to the use of semi-implicit schemes in which only the terms governing phenomena with short transient times (e.g., sonic propagation) are treated implicitly. We note here that this may produce convection terms in the mass and energy equations of mixed-implicitness, where the velocity is taken at the new-time level and the convected quantity is taken at the old-time level. Momentum convection is treated explicitly, but the pressure gradient term in the momentum equations is treated implicitly. Numerical analysis and experience have shown that semi-implicit schemes are stable only for time steps limited by the convective Courant criterion (1,2,3):

$$C_i \leq 1 \text{ (for all i mesh cells, i.e., } 1 \leq i \leq N) \quad (1)$$

where for a one-dimensional, separated flow model:

$$C_i = \text{Max } \{U_{v,i}, U_{\ell,i}\} \frac{\Delta t}{\Delta x_i} \quad (2)$$

Physically, this limits the time step to the shortest convective transport time across any mesh cell in the problem domain. The time steps used in analyzing two-phase flows with high velocities (e.g., sodium vapor under boiling conditions) are particularly sensitive to this restriction. Time steps limited to a few milliseconds can result from the small mesh spacings typically used to model the core region.

Several advanced reactor analysis codes use a semi-implicit difference scheme (1,2,3,4,5). An effective means of solving these equations is the modified ICE method (6,7). The explicit treatment of momentum convection allows the use of the momentum equations to eliminate the velocities in the linearized mass and energy equations in favor of the new-time pressures. This gives a set of equations that are coupled spatially only in pressure. Consequently, the mass and energy equations can be combined to give a system of N equations (where N is the number of mesh cells) to be solved simultaneously for the new-time pressure field. Nonlinearities are handled either by repeating the above process iteratively or by an appropriate time step control strategy which maintains a single linearization about the old-time values adequate within some criterion.

Although semi-implicit numerical schemes have been popular, the convective Courant condition leads to time steps that are frequently less than those required to preserve the accuracy of the calculation. In this paper we will present a new method that is unconditionally stable for subsonic, two-phase flow and yet does not require substantially greater computational time per time step than semi-implicit methods.

## THE SOURCE OF THE CONVECTIVE COURANT CRITERION

A standard method for investigating the stability of numerical schemes is to perform a von Neumann linear stability analysis. Using this method, Fourier components of dependent variables are considered individually with the requirement that any numerical disturbance remain bounded for propagation at any wavelength. Unfortunately, the algebraic complexity for more than two equations becomes severe and computer analysis is necessary. However, some insight into the stability of semi-implicit numerical schemes may be gained from the analysis of a special case. Consider the case of one-dimensional, single-phase flow. If the flow is assumed to be isentropic, the equations describing this flow are:

$$\frac{\partial \rho}{\partial t} + \frac{\partial}{\partial x}(\rho U) = 0 \quad (3)$$

$$\frac{\partial U}{\partial t} + U \frac{\partial U}{\partial x} + \frac{1}{\rho}\frac{\partial p}{\partial x} = 0 \quad (4)$$

$$\rho = \rho(p) \quad (5)$$

Applying a semi-implicit difference scheme to this system, we have:

$$\frac{\rho_i^{n+1} - \rho_i^n}{\Delta t} + \frac{1}{\Delta x}\left[(\rho^n U^{n+1})_{i+1/2} - (\rho^n U^{n+1})_{i-1/2}\right] = 0 \quad (6)$$

$$\frac{U_{i+1/2}^{n+1} - U_{i+1/2}^n}{\Delta t} + U_{i+1/2}^n \frac{U_{i+1}^n - U_i^n}{\Delta x} + \frac{1}{\rho_{i+1/2}^n} \frac{p_{i+1}^{n+1} - p_i^{n+1}}{\Delta x} = 0 \quad (7)$$

Note that the semi-implicit numerical scheme treats mass convection with mixed implicitness while momentum convection is completely explicit. This particular formulation removes any stability restriction on the time step due to sonic effects (1). A convenient choice of dependent variables is the pressure p and the velocity U, with the density given by the equation of state, Equation (5).

To prepare these equations for the stability analysis we linearize Equation (5) about some point j. The density at some point k sufficiently close to point j is given by the Taylor series expansion of Equation (5) about point j:

$$\rho_k = \rho_j + a_j^{-2}(p_k - p_j) \tag{8}$$

where:

$$a_j^2 = \left.\frac{\partial p_j}{\partial \rho_j}\right|_{s_j} = \text{sonic velocity} \tag{9}$$

Since the propagation of any numerical disturbance is governed primarily by finite differences in the dependent variables p and U (which result from derivatives in the differential equations), we can eliminate the density differences in the mass equation, Equation (6), in favor of pressure differences using Equation (8). Before doing this, it is helpful to isolate the density difference in the convection term by adding and subtracting $(\rho_{i-1/2}^n U_{i+1/2}^{n+1})$, which gives:

$$\frac{\rho_i^{n+1} - \rho_i^n}{\Delta t} + \frac{1}{\Delta x}[(\rho_{i+1/2}^n - \rho_{i-1/2}^n)U_{i+1/2}^{n+1}$$

$$+ \rho_{i-1/2}^n(U_{i+1/2}^{n+1} - U_{i-1/2}^{n+1})] = 0 \tag{10}$$

Applying Equation (8) to the temporal and spatial density differences in Equation (10) gives the following equation in p and U:

$$(a_i^n)^{-2}\frac{p_i^{n+1} - p_i^n}{\Delta t}$$

$$+ (a_{i-1/2}^n)^{-2} U_{i+1/2}^{n+1} \frac{p_{i+1/2}^n - p_{i-1/2}^n}{\Delta x}$$

$$+ \rho_{i-1/2}^n \frac{U_{i+1/2}^{n+1} - U_{i-1/2}^{n+1}}{\Delta x} = 0 \tag{11}$$

Finally, we treat the coefficients of difference terms in Equations (7) and (11) as constant and take the spatial finite differences due to convection in an "upwind" sense to obtain the following linear system:

$$\frac{p_i^{n+1} - p_i^n}{\Delta t} + \rho a^2 \frac{U_{i+1/2}^{n+1} - U_{i-1/2}^{n+1}}{\Delta x}$$

$$+ U \frac{p_i^n - p_{i-1}^n}{\Delta x} = 0 \tag{12}$$

$$\frac{U_{i+1/2}^{n+1} - U_{i+1/2}^n}{\Delta t} + U \frac{U_{i+1/2}^n - U_{i-1/2}^n}{\Delta x}$$

$$+ \frac{1}{\rho}\frac{p_{i+1}^{n+1} - p_i^{n+1}}{\Delta x} = 0 \tag{13}$$

Fourier components of p and U are given by:

$$p_i^n = \hat{p}^n e^{ji\theta} \tag{14}$$

$$U_{i+1/2}^n = \hat{U}_{i+1/2}^n e^{ji\theta} \tag{15}$$

where $\theta = k\Delta x$ (k is the wave number) and $j = \sqrt{-1}$.
Substituting for p and U in Equations (12) and (13) gives a system of equations of the form:

$$\bar{V}^{n+1} = \bar{\bar{G}} \bar{V}^n \tag{16}$$

where $\bar{V} = (\hat{p}, \hat{U})^T$ and $\bar{\bar{G}}$ is the amplification matrix of $\bar{V}^n$. For this system to be stable ($\bar{V}^{n+1}$ to remain bounded), every eigenvalue $\lambda_i$ of $\bar{\bar{G}}$ must satisfy the condition:

$$|\lambda_i| \leq 1 \tag{17}$$

Performing the necessary algebra gives the expression for the eigenvalues:

$$\lambda = \frac{1 - rU(1 - e^{-j\theta})}{1 + 4r^2 a^2 \sin^2(\frac{\theta}{2})}\left(1 \pm 2jra \sin(\frac{\theta}{2})\right) \tag{18}$$

where $r = \frac{\Delta t}{\Delta x}$

The stability criterion for this system is:

$$M^2 - C(M^2 - 1) > 0 \tag{19}$$

where $M = \frac{U}{a}$ (Mach Number) and $C = \frac{\Delta t}{\Delta x} U$ (Courant Number)

Clearly, for $M < 1$ (subsonic or sonic flow) this criterion is satisfied for all $C > 0$, and the scheme is unconditionally stable. If $M > 1$ (supersonic flow), the limiting Courant number is always greater than or equal to one. Values of the limiting Courant number are given for various Mach numbers in Table 1. We conclude that the stability criterion for a semi-implicit numerical scheme describing this special case of isentropic flow is significantly less restrictive than the Courant criterion in Equation (1).

This result for isentropic flow has led us to believe that the convective Courant criterion, which is required for the stability of semi-implicit schemes describing the general case of nonisentropic flow, is due to the semi-implicit differencing of the energy convection term. We have postulated that the additional implicit treatment of the energy convection term may provide a scheme that is unconditionally stable (at least for subsonic flow), while causing only a modest increase in the amount of computation per time step.

Hughes and Katsma [8] in their work with the RETRAN code have also suggested the convective Courant criterion is due to the semi-implicit differencing of the energy convection term. Padilla, et al. [9] have developed a numerical scheme for the CAPRICORN code which treats the mass, momentum, and energy convection terms implicitly. The implicit treatment of the momentum convection term required an outer iteration around their extension of the modified ICE solution method to converge the momentum equations. This method allowed time steps that were not limited by the convective Courant criterion, but the outer iteration added a substantial amount of computation per time step. The results of our stability analysis for the case of isentropic flow and all numerical testing to date for the general case of nonisentropic flow have indicated that for numerical stability only energy convection must be treated implicitly. This approach requires a substantially smaller amount of computation per time step than treating all three convection terms implicitly.

## APPLICATION TO TWO-PHASE SODIUM FLOW

We have developed a new numerical scheme with implicit differencing of the energy convection term for use in the sodium boiling code THERMIT-4E [1]. The code uses a two-phase flow model consisting of a mixture mass equation, a mixture internal energy equation, and two phasic momentum equations. Thermal equilibrium on the saturation line between coexisting phases is assumed.

The differential equations describing such flow are:

Mixture Mass Equation

$$\frac{\partial}{\partial t} \rho_m + \nabla \cdot [\alpha \rho_v \vec{U}_v + (1 - \alpha)\rho_\ell \vec{U}_\ell] = 0 \quad (20)$$

Vapor Momentum Equation

$$\frac{\partial}{\partial t}(\alpha \rho_v \vec{U}_v) + \nabla \cdot (\alpha \rho_v \vec{U}_v \vec{U}_v) + \alpha \nabla p = -\vec{F}_{wv} - \vec{F}_i - \alpha \rho_v \vec{g} \quad (21)$$

Liquid Momentum Equation

$$\frac{\partial}{\partial t}[(1-\alpha)\rho_\ell \vec{U}_\ell] + \nabla \cdot [(1-\alpha)\rho_\ell \vec{U}_\ell \vec{U}_\ell]$$
$$+ (1-\alpha)\nabla p = -\vec{F}_{w\ell} + \vec{F}_i - (1-\alpha)\rho_\ell \vec{g} \quad (22)$$

Mixture Internal Energy Equation

$$\frac{\partial}{\partial t}(\rho_m e_m) + \nabla \cdot [\alpha \rho_v e_v \vec{U}_v + (1-\alpha)\rho_\ell e_\ell \vec{U}_\ell]$$
$$+ p\nabla \cdot [\alpha \vec{U}_v + (1-\alpha)\vec{U}_\ell] = Q_w + Q_k \quad (23)$$

where:
$$\rho_m = \alpha \rho_v + (1-\alpha)\rho_\ell$$
$$e_m = [\alpha \rho_v e_v + (1-\alpha)\rho_\ell e_\ell]/\rho_m$$

The interested reader should read Reference 1 for a complete discussion of the flow model and the formulation of the constitutive relations for wall friction ($F_{wv}$, $F_{w\ell}$), interfacial drag ($F_i$), and wall and conduction heat sources ($Q_w$, $Q_k$). In addition, an equation of state is used to relate thermodynamic state variables and provide closure of the set of equations. The code solves these equations for the variables $p$, $e_m$, and the fluid velocities.

To test the stability of our new numerical scheme we have applied it to a special version of THERMIT-4E that considers only one-dimensional flow. This has allowed us to demonstrate the numerical stability of the scheme and assess its performance before extending it to the more complex case of multidimensional flow. The finite difference equations for one-dimensional flow with implicit energy convection (and mass convection -- see comments below) are (with $\alpha_v = \alpha$ and $\alpha_\ell = 1-\alpha$ for notational facility):

Mass:

$$\frac{V_i}{\Delta t}(\rho_m^{n+1} - \rho_m^n)_i + A_{i+1/2}(\alpha_v \rho_v U_v + \alpha_\ell \rho_\ell U_\ell)_{i+1/2}^{n+1}$$
$$- A_{i-1/2}(\alpha_v \rho_v U_v + \alpha_\ell \rho_\ell U_\ell)_{i-1/2}^{n+1} = 0 \quad (24)$$

Vapor Momentum:

$$(\alpha_v \rho_v)_{i+1/2}^n \left[\frac{(U_v^{n+1} - U_v^n)_{i+1/2}}{\Delta t} + (U_v \frac{\Delta U_v}{\Delta x})_{i+1/2}^n\right]$$
$$+ (\alpha_v)_{i+1/2}^n \frac{(p_{i+1} - p_i)^{n+1}}{\Delta x_{i+1/2}}$$
$$= - (F_{wv} + F_{iv})_{i+1/2}^{n+1/2} \quad (25)$$

Liquid Momentum:

$$(\alpha_\ell \rho_\ell)_{i+1/2}^n \left[\frac{(U_\ell^{n+1} - U_\ell^n)_{i+1/2}}{\Delta t} + (U_\ell \frac{\Delta U_\ell}{\Delta x})_{i+1/2}^n\right]$$
$$+ (\alpha_\ell)_{i+1/2}^n \frac{(p_{i+1} - p_i)^{n+1}}{\Delta x_{i+1/2}}$$
$$= - (F_{w\ell} + F_{i\ell})_{i+1/2}^{n+1/2} \quad (26)$$

Energy:

$$\frac{V_i}{\Delta t}[(\rho_m e_m)^{n+1} - (\rho_m e_m)^n]_i + A_{i+1/2}\{(\alpha_v U_v)_{i+1/2}^{n+1}$$
$$\times [p_i^{n+1} + (\rho_v e_v)_{i+1/2}^{n+1}] + (\alpha_\ell U_\ell)_{i+1/2}^{n+1}$$
$$\times [p_i^{n+1} + (\rho_\ell e_\ell)_{i+1/2}^{n+1}]\}$$
$$- A_{i-1/2}\{(\alpha_v U_v)_{i-1/2}^{n+1}[p_i^{n+1} + (\rho_v e_v)_{i-1/2}^{n+1}]$$
$$+ (\alpha_\ell U_\ell)_{i-1/2}^{n+1}[p_i^{n+1} + (\rho_\ell e_\ell)_{i-1/2}^{n+1}]\}$$
$$= (Q_w + Q_k)_i^{n+1/2} \quad (27)$$

This system of finite difference equations can be reduced to a set of 2N equations that are spatially coupled through pressure as well as mixture internal energy. This is accomplished by using the momentum equations to eliminate the new-time velocities in the mass and energy equations in favor of the new-time pressures. Unlike the modified ICE solution method, it is not possible to reduce this system to a pressure field solution because of the additional spatial energy coupling. Since the implicit treatment of energy convection is alone sufficient to spatially couple the system in energy, the additonal implicit treatment of mass convection does not make the solution any more difficult, and has the important advantage of rendering the mass and energy fluxes consistent with respect to time level. The coupled pressure-energy system is solved iteratively using Newton's method. The resulting Jacobian, shown in Figure 1, is a seven-band, nonsymmetrix matrix. It is inverted directly using the IMSL nonsymmetric, banded matrix inversion subroutine, LEQT1B (10).

An analysis of the EB19GR experiment (11), performed in the CF Na loop at Grenoble, France, was used to evaluate the new method. Design data for the experiment is given in Table 2. The inlet flow rate in the experiment was gradually reduced from an initial value of 2.25 Kg/s to a final value of about 0.26 Kg/s, where the sodium temperature reached the saturation line at the end of the heated section.

A quarter of the test section was modeled as a single flow channel of uniform cross section. Mass flow rate and pressure boundary conditions were imposed at the test section inlet and outlet, respectively. The inlet flow rate was decreased through a series of step changes from the initial to the final experimental values, and the calculation was allowed to reach a defined steady-state at each inlet flow rate. The problem was run using both the semi-implicit method and the new implicit method with identical input parameters and boundary conditions. One Newton iteration was performed per time step with time step control based on accuracy and the rate of change of major variables (and the Courant stability criterion in the case of the semi-implicit method). Figure 2 shows the calculated sodium temperature distribution in the test section. Both numerical methods gave essentially the same results and predicted the location where boiling was observed to begin in the channel.

Table 3 summarizes the numerical performance of the two methods. The implicit method proved to be numerically stable for the entire calculation. Because of the extra amount of computation required to evaluate the more complex Jacobian matrix, the implicit method required 12% more CPU time per time step. However, it was able to take considerably larger time steps than the stability-limited, semi-implicit method. Therefore, fewer steps were needed to complete the calculation, and an overall reduction in CPU time of 77% was achieved.

The new method is more robust than the semi-implicit method in the sense that it is less sensitive to large step changes in the boundary conditions and transition from single-phase to two-phase flow. Both of these situations result in hydrodynamic and thermodynamic discontinuities which can require substantially smaller time steps to preserve the accuracy of the solution. With the new method we were able to change the inlet flow rate from the initial to the final value in the first time step. Steady-state was obtained in 223 time steps with the calculation taking 51.55s of CPU time. Step changes of this magnitude in the inlet flow rate caused failure (pressure or energy out of the acceptable range) when using the semi-implicit method. Therefore, the new method will allow the simulation of more severe transients.

## SUMMARY

We have developed a numerical method for one-dimensional, two-phase flow treating mass and energy convection implicitly. Theoretical analysis and computational testing have suggested that the method is unconditionally stable for subsonic flow. The method has been implemented in an experimental version of the sodium boiling code THERMIT-4E and used to simulate an experimental test series. The new method required a slightly greater amount of computation per time step than the semi-implicit method due to the evaluation of a more complex Jacobian matrix, However, larger (and hence fewer) time steps are possible with the new method, giving a substantial reduction in overall computational time. The new method has also proven to be more robust than the semi-implicit method, improving the capability to calculate severe transients in two-phase flow.

## REFERENCES

1. A. L. Schor and N. E. Todreas, "A Four-Equation Two-Phase Model for Sodium Boiling Simulation of LMFBR Fuel Assemblies," MIT Energy Laboratory, MIT-EL 82-039 (1982). See also the paper of the same title given at the Tenth Meeting of the Liquid Metal Boiling Working Group, Karlsruhe, Germany (1982).

2. J. Loomis, W. H. Reed, A. Schor, H. B. Stewart, L. Wolf, "THERMIT: A Computer Program for Three-Dimensional Thermal-Hydraulic Analysis of Light Water Reactor Cores," EPRI NP-2032 (1981).

3. V. H. Ransom et al., RELAP5/MOD1 Code Manual, EGG-2070 (1980).

4. R. G. Zielinski and M. S. Kazimi, "Development of Models for the Two-Dimensional, Two-Fluid Code for Sodium Boiling NATOF-2D," MIT Energy Laboratory Report No. MIT-EL 81-030 (1981).

5. "TRAC-P1: An Advanced Best Estimate Computer Program for PWR LOCA Analysis," LA-7279-MS (1978).

6. F. H. Harlow and A. A. Amsden, "A Numerical Fluid Dynamics Calculation Method for All Flow Speeds," J. Comp. Phys. 8 (1971).

7. D. Liles and W. Reed, "A Semi-Implicit Method for Two-Phase Fluid Dynamics," J. Comp. Phys. 26 (1978).

8. E. D. Hughes and K. R. Katsma, "Numerical Solution Method Improvements for RETRAN," Nuclear Technology, 61 (1983).

9. A. Padilla, Jr., et al., "CAPRICORN Subchannel Code for Sodium Boiling in LMFBR Fuel Bundles," First Proceedings of Nuclear Thermal Hydraulics, 1983

Winter Meeting, Thermal Hydraulics Division, American Nuclear Society, 1983.

10. "The IMSL Library," IMSL, Houston, Texas (1982).

11. G. Basque et al., "Theoretical Analysis and Experimental Evidence of Three Types of Thermohydraulic Incoherency Cluster in Undisturbed Geometry," IAEA Specialists' Meeting in Thermodynamics of Fast Breeder Reactor Fuel Sub-Assemblies under Normal and Non-nominal Operating Conditions," Karlsruhe, Germany (1979).

Table 1: Stability Limitation of a Semi-Implicit Numerical Scheme for Supersonic, Adiabatic Flow

$$C_{max} \leq \frac{M^2}{M^2 - 1}$$

| Mach Number M | 1 | 1.1 | 1.5 | 2.0 | 5.0 | $\infty$ |
|---|---|---|---|---|---|---|
| Maximum Courant Number $C_{max}$ | $\infty$ | 5.76 | 1.80 | 1.33 | 1.04 | 1 |

Table 2: Design Data for the GR19 Experiment

| | |
|---|---|
| Number of Pins | 19 |
| Clad OD (m) | $8.65 \times 10^{-3}$ |
| Heated Length (m) | 0.6 |
| Downstream Unheated Length (m) | 0.494 |
| Upstream Unheated Length (m) | 0.12 |
| Wire Wrap OD (m) | $1.25 \times 10^{-3}$ |
| Wire Wrap Lead (m) | 0.18 |
| Flat to Flat (m) | $4.58 \times 10^{-2}$ |
| Inlet Temperature (°C) | 400 |
| Downstream Pressure (Pa) | $1.4 \times 10^5$ |
| Power (kw) | 170 |
| | (axially uniform) |

Table 3: Results of the EB19GR Test Problem Using the Semi-Implicit and New Implicit Methods

Sequence of Mass Flow Rates Imposed at the Inlet Boundary

| | |
|---|---|
| 561.5 g/s | 87.5 g/s |
| 425.0 g/s | 65.0 g/s |
| 300.0 g/s | 62.5 g/s |
| 151.5 g/s | |

The calculation was run to steady-state for each flow rate.

Definition of Steady-State: The maximum relative changes in pressure, mixture density and mixture internal energy over a time step are all less than $10^{-8}$ and decreasing.

Performance Statistics

| | Total Number of Time Steps | CPU Time Per Time Step (s) | Total CPU Time (s) |
|---|---|---|---|
| Original Method | 1961 | 0.196 | 416 |
| New Method | 346 | 0.220 | 94 |

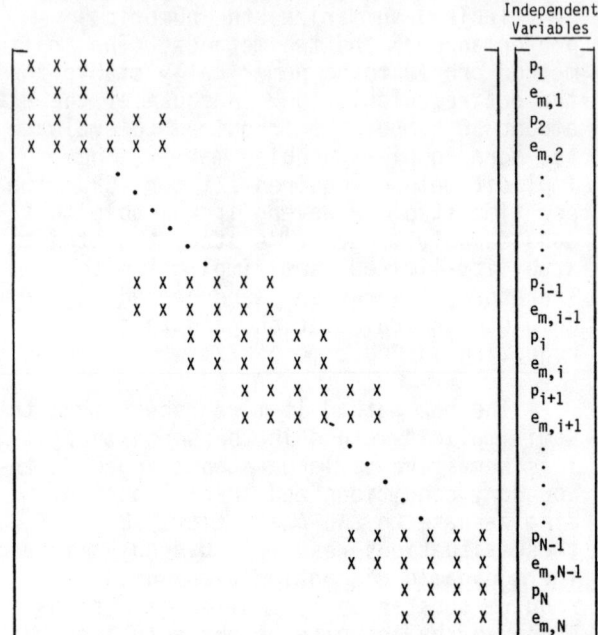

Figure 1: Structure of the Jacobian Matrix

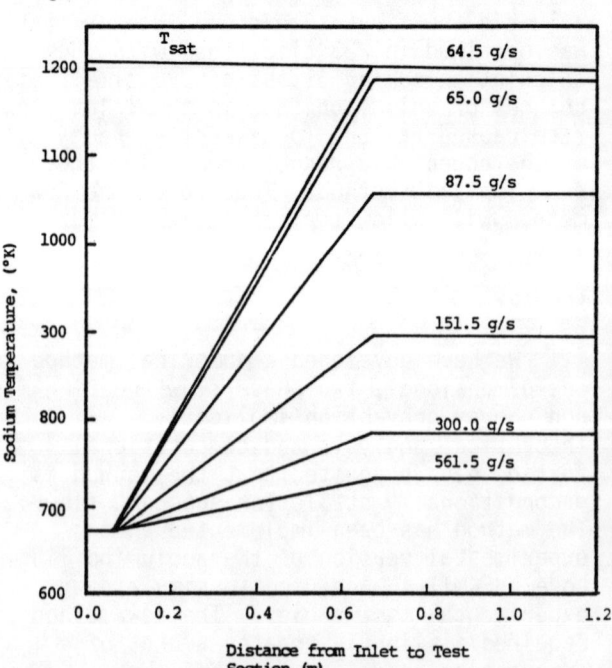

Figure 2: Axial Sodium Temperature Distribution

# BYPASS APPROACH OF FREE SURFACE MODELING

C.C. Miao ■ AT&T Bell Laboratories, Dept. 55224, Naperville, IL 60566
F.F. Chen, W.T. Sha and R.W. Lyczkowski ■ Argonne National Laboratory
9700 South Cass Avenue, Argonne, IL 60439

By introducing a device called the Flexible Momentum Control Volume (FMCV) approach, the continuity-momentum-pressure cycle in the modeling of free surface can be greatly simplified and reduced to single-phase computation. Two sample problems are selected for the test of the proposed scheme and the results are in good agreement with the analytic solutions.

The major achievement of this proposed scheme is that the ill-posed nature of free surface formulation is fully addressed. Instability due to sharp density ratio in the interface area between liquid and gas are bypassed. When flow with free surface is modeled, it provides the same level of stability as single-phase flow. Therefore, to a great extent, the scheme is not problem-dependent.

# VARIATION IN THERMAL HISTORY DURING FREEZING DUE TO THE PATTERN OF LATENT HEAT EVOLUTION

Kenneth R. Diller and Michael E. Crawford ■ Department of Mechanical Engineering
Linda J. Hayes ■ Department of Aerospace Engineering and Engineering Mechanics
The University of Texas at Austin, Austin, TX 78712, U.S.A.

During the freezing of chemically impure noneutectic liquid solutions, latent heat is released nonuniformly over a finite range of temperatures. Accurate thermal modeling of transient heat transfer during the freezing of multicomponent systems requires knowledge of how the latent and sensible heats are liberated as a function of the local thermodynamic state. Here we present an analysis of the variation in the pattern of latent heat release with temperature, demonstrating the dependence of the variation upon (1) the shape of the liquidus curve in the phase diagram and (2) the initial solution concentration. We also illustrate how variations in these patterns may quantitatively affect transient temperature distributions during solution solidification.

## INTRODUCTION

The fraction of a defined liquid melt which has been solidified at a given temperature is determined from the phase diagram by application of the lever rule. In general, as the initial solute concentration of the system is increased, two relevant phenomena are observed: the onset of phase change shifts to a lower temperature and the distribution of the latent heat release becomes biased toward lower temperatures, approaching the eutectic value. The pattern of latent heat loss may have a significant effect on the local thermal history. Therefore, in using analytical methods to model solidification processes, it is important to be able to accurately quantify the most sensitive parameters. In the present study we have analyzed the influence of variations in the pattern of latent heat release on the thermal histories which result from a standard boundary cooling protocol to determine the necessity for accurately incorporating this latent heat pattern into thermal models.

## ANALYSIS OF BINARY PHASE DIAGRAM

Latent heat is released progressively from the liquid solution as freezing occurs. The total latent heat which must be removed to effect solidification may be determined from an equation of state for the system as the difference in the enthalpies for saturated solid and liquid states. Thus,

$$L = n (h_l - h_s) \qquad (1)$$

where n is the number of moles and h is the enthalpy of the system undergoing the phase change. During a typical freezing protocol the latent heat is removed only from liquid water until the eutectic temperature is reached, and at that point latent heat is removed from the remaining impure liquid solution. The liquid and solid phase enthalpies in Equation (1) may be estimated for an impure solution from the enthalpies of the pure components using Euler's theorem ([1]).

$$h = \Sigma_i x_i h_i \qquad (2)$$

where $x_i$ and $h_i$ are the mole fraction and enthalpy respectively for component i.

The progressive release of latent heat as a function of temperature during the freezing of a liquid mixture may be determined by application of the lever rule to the phase diagram for the specific chemical system involved, since latent heat release occurs in direct proportion to the fraction of mass in the solid phase which is formed. It will be assumed for purposes of the present analysis that during the solidification process the system passes through a continuous sequence of quasistationary states as defined by the equilibrium phase diagram. In many practical applications this condition is not satisfied since a system may supercool prior to nucleation of solid and/or the cooling rate is rapid enough to establish significant spatial compositional gradients within the system. However, this simplifying assumption permits phenomena associated with the thermal pattern of latent heat release to be more easily isolated and identified for analysis and discussion.

The latent heat of fusion may be calculated for aqueous sodium chloride mixtures from the Euler theorem, Equation (2), using values for the solid and liquid phase saturation enthalpies of water and salt which are available in the literature. Accordingly, for water (3)

$$\Delta h_{s-\ell} = L_w = 335 \frac{kJ}{kg}$$

and for sodium chloride (5)

$$\Delta h_{s-\ell} = L_{sc} = 491 \frac{kJ}{kg}$$

The latent heat evolved at any supraeutectic temperature, T, is then given in terms of the notation of Figure 1 as

$$L = L_w (1 - x_o) \frac{x_\ell - x_o}{x_\ell - 0} \qquad (3)$$

where $x_o$ and $x_\ell$ are the initial and instantaneous liquid phase salt mole fractions in the solution. At the eutectic state, the latent heat released is

$$L = L_w (1 - x_o) \frac{x_o - 0}{x_e - 0} + L_{sc} x_o \qquad (4)$$

From Equations (3) and (4) it is clear that the larger the initial salt concentration, the greater is the portion of the total latent heat which is released isothermally at the eutectic temperature.

The freezing process will be investigated for the phase diagram of a water-sodium chloride binary system which is of particular interest for numerous applications. The phase diagram is shown in Figure 1, as plotted from standard freezing point depression data (2). Equilibrium freezing processes and corresponding latent heat release patterns will be illustrated for initial solute concentrations which are isotonic ($x_{sc}$ = 0.00278), 5.0 times isotonic ($x_{sc}$ = 0.01380) and 0.50 times eutectic ($x_{sc}^{sc}$ = 0.0442). By applying the lever rule to the phase diagram in Figure 1, the latent heat release was determined as a function of temperature for each of these initial compositions, as plotted in Figure 2. It has been a common practice to assume a linear latent heat release pattern during solidification. A linear approximation for the pattern of latent heat release is shown using broken lines for each of the three cases for comparison. A large variation in the distribution of latent heat release across the range of total possible solidification temperatures is apparent for the thermodynamically derived data, and the actual latent heat release patterns are quite different from their linear approximations.

In the solidification processes for impure systems, the release of latent heat is a specific function of temperature. An objective of the present study is to assess the sensitivity of transient temperature fields to the specific latent heat curves shown in Figure 2

## MODEL FORMULATION AND SOLUTION

The diffusion of heat during freezing in a region $\Omega$ is described by the transient partial differential equation

$$\frac{\partial}{\partial t} (\rho c T + \rho L) = \nabla \cdot k \nabla T \text{ in } \Omega \qquad (5)$$

where initially

$$T (\underline{r},0) = T_o (\underline{r}) \text{ in } \Omega \qquad (6)$$

and on the boundary

$$k \frac{\partial T}{\partial \eta} = h(T_\infty - T) \text{ in } d \Omega \qquad (6)$$

In the present study, only one spatial coordinate is required to illustrate the effect which phase change diagrams have upon thermal histories during solidification; therefore, for the sake of simplicity and brevity only a single coordinate variable, r, will be used to define a cylindrical coordinate system. The thermal conductivity and sensible heat capacitance were different in the liquid and solid phases.

Equation (5) was solved using the apparent heat capacity technique in which both sensible and latent effects are lumped together into a single variable capacitance term (4). Thus, transient changes of the energy stored in the system, which constitute the left-hand side of Equation (5), are written as

$$\frac{\partial}{\partial t}(\rho cT = \rho L) = \rho c \frac{\partial T}{\partial t} + \rho \frac{\partial L}{\partial T} \frac{\partial T}{\partial t}$$

$$= (c + \frac{\partial L}{\partial T}) \rho \frac{\partial T}{\partial t} \quad (8)$$

where $c + \frac{\partial L}{\partial T}$ is defined as the apparent heat capacity, $\bar{c}$. The value $\bar{c}$ is a function of temperature. It is composed of the sensible capacities and of the latent capacity, which varies with the rate of solidification as measured by the lever rule from the phase diagram. The curves in Figure 2 provide a direct evaluation of $\frac{\partial L}{\partial T}$. A revised format for Equation (5) may now be written as

$$\rho \bar{c} \frac{\partial T}{\partial t} = \nabla \cdot k \nabla T \text{ in } \Omega \quad (9)$$

The boundary value problem defined by Equations (6), (7) and (9) was solved for a specific system consisting of a long cylindrical container filled with an aqueous NaCl solution at the three initial compositions defined in Figure 1. The cylinder diameter was 4.33 cm with a wall thickness of 0.1 cm. Initially, the system temperature was at a uniform value of 10.5°C; and the environment was cooled at a rate of -5°C/min until a minimum value of -80°C was reached. This outside temperature was then maintained for the duration of the simulation. A convective film coefficient of 167 W/m$^2$C was assumed between the container surface and the environmental fluid. The thermal properties which were used in this model are given in Table 1. The thermal properties of the NaCl solution were different in the two phases. A linear change in thermal properties was assumed in the temperature range over which latent heat is released. For the system properties presented in Table 1, values of the Biot and Stefan numbers describing the simulation conditions were calculated as

$$Bi_D = \frac{hD}{4k_s} = 6.0, \quad Ste = \frac{c_s(T_{SAT} - T_f)}{L}$$

$$= 0.45$$

where $T_{SAT}$ is the equilibrium phase change temperature for the initial system composition and $T_f$ is the final system temperature.

## RESULTS

The boundary value problem (9) was solved using a one-dimensional, transient finite element technique (5). The finite element grid consisted of 28 quadratic elements defined by 57 nodal points positioned at equal radial increments from r = 0 to r = 4.33 cm. The finite element grid is shown in Figure 3. The co-ordinates of selected points are given in Table 2. The nodes were numbered starting with 1 at the center and proceeding outward. Three different initial solutions were considered: isotonic, 5* isotonic, .5* eutectic. The linear release pattern for the isotonic case is included for comparison purposes. Transient temperature histories were determined at each nodal position, and the results for representative positions are plotted in Figures 4 to 7.

A comparison of temperature histories for an initially isotonic liquid solution implemented for the linear and the nonlinear thermodynamic patterns of latent heat release is illustrated in Figures 4 and 5. Several distinguishing differences can be observed between the two sets of curves. At high sub-zero temperatures, the temperature drop is more rapid in the linear case, since the latent heat release is biased toward lower temperatures, and a greater portion of the initial energy loss is therefore sensible. However, after only a small temperature reduction, the cooling rate becomes noticeable faster for the thermodynamic simulation because the rapid initial phase change results in the presence of more ice, which has a much higher thermal conductivity and diffusivity than water. This effect becomes more pronounced at positions approaching the center line (increasing nodal indices) as a greater percent of the external medium becomes solidified. As a result, the cooling down period for the thermodynamic based model is about 15% faster than for the linear model.

Figure 4 may also be compared with Figures 6 and 7, which simulates higher initial salt concentrations for the thermodynamic derived pattern of latent heat release. Due to the higher salt concentration, the onset of solidification is prolonged to a lower temperature, resulting in greater sensible energy removal by heat diffusion in the water phase. As expected, due to a delayed phase change, the temperature does not drop as rapidly for higher salt concentrations. In the .5* eutectic case (Figure 7) a greater portion of the latent heat release is focused at the eutectic state and a second plateau in the temperature curve is observed associated with the initial and final bursts of rapid solidification.

The analysis presented in this work demonstrates that in solution solidification the pattern of latent heat release implemented in the model may exhibit a strong influence on the calculated thermal histories. Accurate modeling techniques are therefore very important in the simulation of process kinetics that are sensitive to the magnitude of local system temperature.

ACKNOWLEGEMENT

This work was supported by NSF Grant No. MEA 80-23267.

NOTATION

- c = specific heat, J/k mol.°K
- D = characteristic system dimension, m
- h = enthalpy, kJ/kmol
- k = thermal conductivity, W/m.°K
- L = latent heat of fusion, kJ/kmol
- n = moles, kmol
- Ste = Stefan number
- $\rho$ = density, kg/m$^3$
- $\eta$ = unit outward normal vector
- r = spatial coordinate, m
- t = time, sec
- T = temperature, °K
- x = mole fraction

SUBSCRIPTS

- i = chemical species i
- l = liquid phase
- s = solid phase
- sc = sodium chloride
- w = water

TABLE 1. System properties used in the freezing simulaton.

|  |  | SOLUTION | CONTAINER |
|---|---|---|---|
| $\rho c_p \left(\dfrac{J}{cm^3 \, ^\circ C}\right)$ | LIQUID<br>SOLID | 4.08<br>1.88 | 2.08 |
| $K\left(\dfrac{W}{cm \, ^\circ C}\right)$ | LIQUID<br>SOLID | 0.0059<br>0.0222 | 0.00328 |
| $\rho L\left(\dfrac{J}{cm^3}\right)$ | WATER<br>NaCl | 335<br>491 |  |

LITERATURE CITED

1. Guggenheim, E. A., Thermodynamics: An Advanced Treatment for Chemists and Physicists, 6th ed., p. 20, North-Holland Publishing Co., Amsterdam (1977).

2. CRC Handbook of Chemistry and Physics, 54th ed., R. C. Weast, (ed.), CRC Press Cleveland, (1973).

3. Keenan, J. H., Keyes, F. G., Hill, P. G. and Moore, J. G., Steam Tables, Wiley, New York, (1969).

4. O'Neill, K., "A Brief Survey of Computer Programs for Multi-dimensional Freezing and Thawing," (1983).

5. Hayes, L. J. and Diller, K. R., "Implementation of Phase Change in Numerical Models of Heat Transfer," Trans. ASME, Journal of Energy Resources and Technology, Vol. 105, pp. 431-435 (1983).

TABLE 2. Finite element grid and spatial co-ordinates for selected nodes and points in the finite element grid.

| Node Number | r-co-ordinate (cm) |
|---|---|
| 1 | 0.0 |
| 9 | 0.48 |
| 17 | 1.03 |
| 25 | 1.58 |
| 33 | 2.35 |
| 39 | 2.90 |
| 43 | 3.23 |
| 51 | 4.03 |
| 55 | 4.23 |
| 57 | 4.33 |

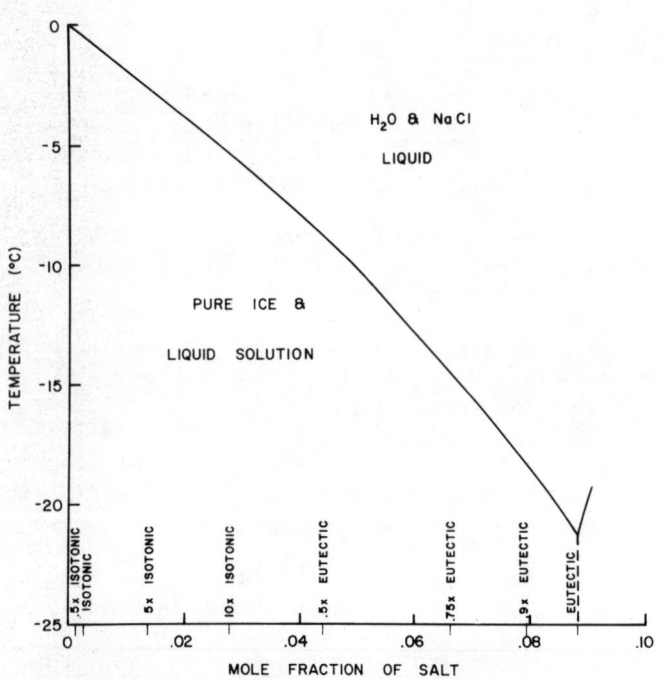

Figure 1. Binary equilibrium phase diagram for water and NaCl. Three initial test solution concentrations are shown in the water rich area of the diagram.

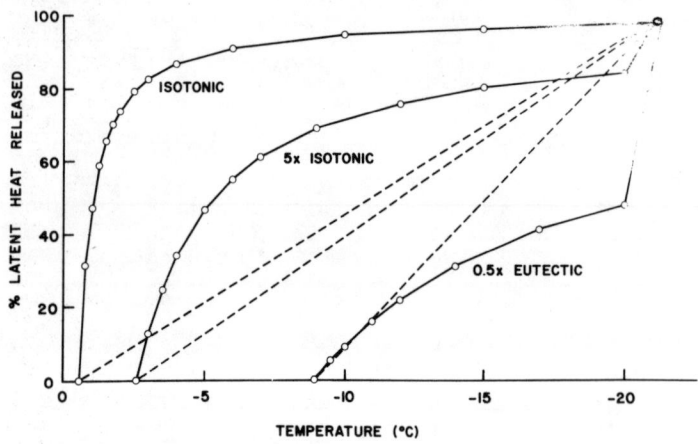

Figure 2. Latent heat release pattern as a function of temperature. Linear functions are shown as broken lines and thermodynamic based functions derived from the phase diagram in Figure 1 as solid lines.

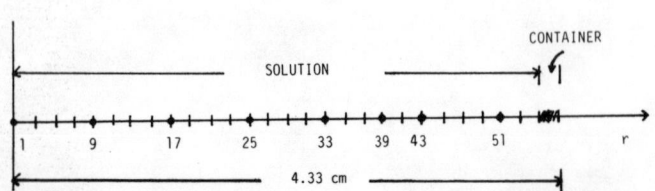

Figure 3. Finite element grid of NaCl solution and container: 28 quadratic elements (57 Nodal Points).

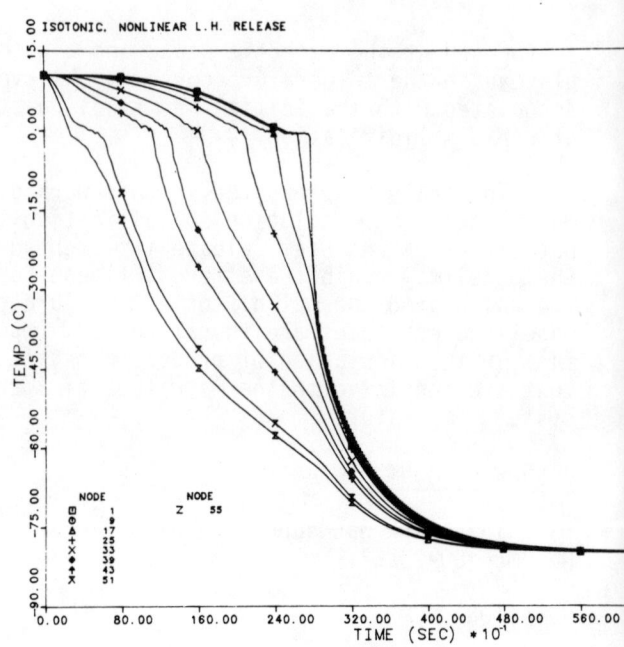

Figure 4. Simulated temperature histories at incrementally spaced radial positions for freezing of initially isotonic concentration with latent heat release pattern derived from the phase diagram.

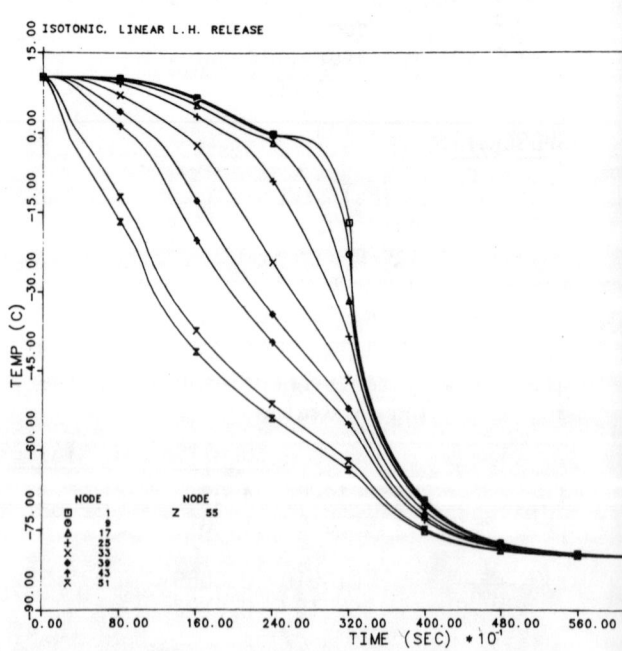

Figure 5. Simulated temperature histories at incrementally spaced radial positions for freezing of initially isotonic concentration with a linear pattern of latent heat release with temperature.

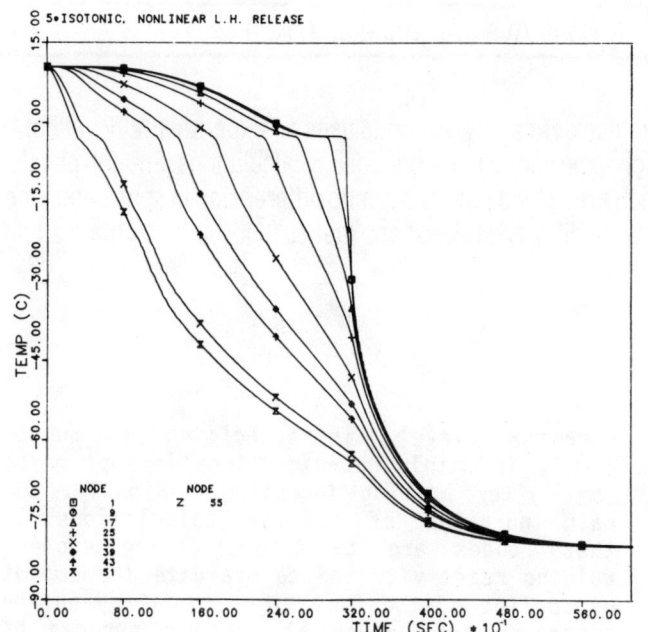

Figure 6. Simulated temperature histories at incrementally spaced radial positions for freezing of initially five times isotonic concentration with latent heat release pattern derived from the phase diagram.

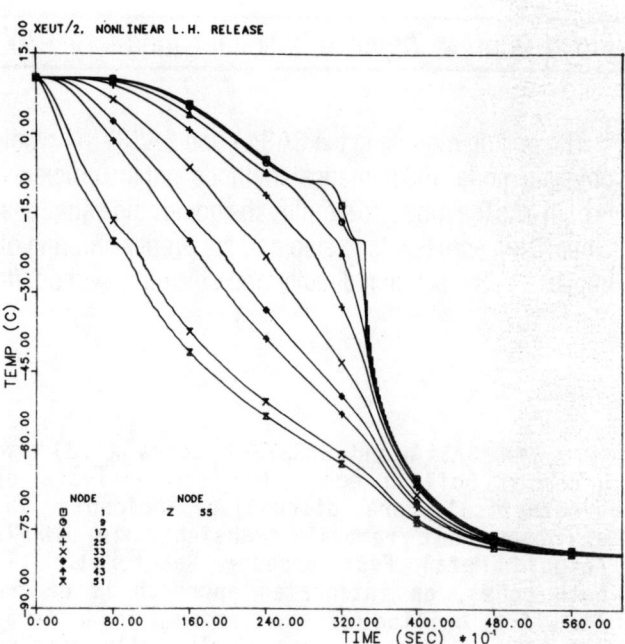

Figure 7. Simulated temperature histories at incrementally spaced radial positions for freezing of intially one-half eutectic concentration with latent heat release pattern derived from the phase diagram.

# THE SAS4A/SASSYS-1 SODIUM BOILING MODEL FOR LMFBR WHOLE CORE ANALYSIS

Floyd Dunn ■ Argonne National Laboratory, 9700 South Cass Ave., Argonne, IL 60439

The boiling module in the SAS4A and SASSYS-1 whole core accident analysis codes uses an experimentally validated physical model and numerical methods that are efficient enough to permit whole-core boiling calculations. Enough physics is built into the model to simplify the governing equations, which are solved mainly by finite differencing in space and time. Linearizing across a time step and the proper ordering of the finite difference equations makes it possible to obtain a fully implicit boiling solution directly and efficiently without iteration.

The SAS4A and SASSYS-1 codes [1,2] use a common boiling model for the analysis of hypothetical core disruptive accidents and shutdown heat removal transients in LMFBRs (Liquid Metal Fast Breeder Reactors). In both codes, an integrated approach is needed in which boiling of the sodium in a large fraction of the core is coupled with single-phase thermal hydraulics for the rest of the primary heat transport system, as well as with fuel pin heat transfer. Providing a boiling model that can be validated reasonably well with experiments and that can also run in a reasonable amount of computer time for a whole-core analysis requires the use of physical models that contain the essential features of the boiling process without being excessively detailed, as well as the use of efficient numerical methods for solving the resulting equations.

Even though whole-core boiling analysis requires treating a very complex three-dimensional system, there are a number of aspects that greatly simplify the task and make whole core calculations feasible in a reasonable amount of computer time. First, unlike the situation in water reactors, boiling in an LMFBR is restricted to the core subassemblies and maybe the radial blanket subassemblies. Second, the subassemblies have solid duct walls, so the only interactions between subassemblies are hydraulic coupling through the inlet plenum, limited heat transfer through the duct walls, and the effects of reactivity feedbacks on the whole-core power level. Third, boiling in a subassembly is mainly one-dimensional except maybe just after boiling inception. Finally, the main purposes of boiling calculations in these codes are to determine the coolant voiding reactivity and to evaluate the amount of cooling of the fuel pins, so it may not be necessary to include all of the aspects of boiling in great detail.

There are some aspects of sodium boiling that can cause problems with many typical numerical methods. First, flow in both the normal direction and reverse flow must be considered. Second, boiling in sodium can lead to sharp interfaces between large vapor bubbles and liquid sodium. A typical numerical scheme with a fixed Eulerian axial mesh may have trouble resolving a sharp interface unless a very large number of closely spaced nodes are used. Third, a small volume of liquid can produce a very large volume of vapor. This can cause problems if iterations are used to obtain consistency between solutions of mass conservation equations and vapor pressure equations. The numerical methods used in SAS4A and SASSYS-1 avoid these problems.

SAS4A and SASSYS-1 use a multi-channel treatment for the reactor core. Each channel represents a subassembly or a group of similar subassemblies. The channel model contains a fuel pin, its associated coolant, and a structure that represents a fraction of the duct wall. Usually, the average fuel pin in the subassembly is used to represent the whole subassembly. By using a number of channels to represent different groups of subassemblies, it is possible to represent a three-dimensional core with models that are mainly one-dimensional.

## BOILING MODEL AND NUMERICAL METHODS

The boiling model approach used in SAS4A and SASSYS-1 is to build much of the physics into a one-dimensional model with a moderate number (up to 48) of axial nodes. This model is a multiple-bubble slug ejection model, as illustrated in Figure 1. A number of vapor bubbles, separated by liquid slugs, can be treated. In addition to a fixed Eulerian mesh of axial nodes, the model uses a Lagrangean node moving with each liquid-vapor interface to resolve the sharp interfaces. Bubbles contain only vapor, except for a liquid film left on the cladding and on the structure or duct wall. The vapor pressures at the liquid-vapor interfaces drive the flow in the liquid slugs, which are treated with a one-dimensional, incompressible flow treatment. Saturation conditions are assumed in a vapor bubble. Evaporation from the liquid-vapor interfaces and the liquid films on the cladding and structure supply the vapor for the bubbles. The boiling model is coupled to heat transfer equations for the fuel pin and for the structure. The cladding surface and structure surface heat fluxes determine the evaporation or condensation rates of the films. The evaporation rate from a liquid-vapor interface is obtained from a semi-analytic approach that gives the heat flux through the boundary of a semi-infinite body, given its initial temperature and the time variation of the interface temperature.

A new bubble is formed at any fixed Eulerian node in the liquid or at a point near an existing liquid-vapor interface whenever the liquid temperature exceeds the local saturation temperature by a specified amount of superheat.

### Small Bubbles

Two models are used for vapor bubbles: one for small bubbles and the other for larger bubbles. For small bubbles, the vapor pressure is uniform within a bubble. Vapor volume changes are determined by the motion of the liquid-vapor interfaces. The pressure in a bubble is determined by the relationship between the rate of change of the volume and the rate of vapor production due to heat flow through the cladding surface and the liquid-vapor interfaces. The transition between the small bubble model and the large bubble model occurs when the bubble length reaches a user-specified value, typically 0.05 m (0.16 ft).

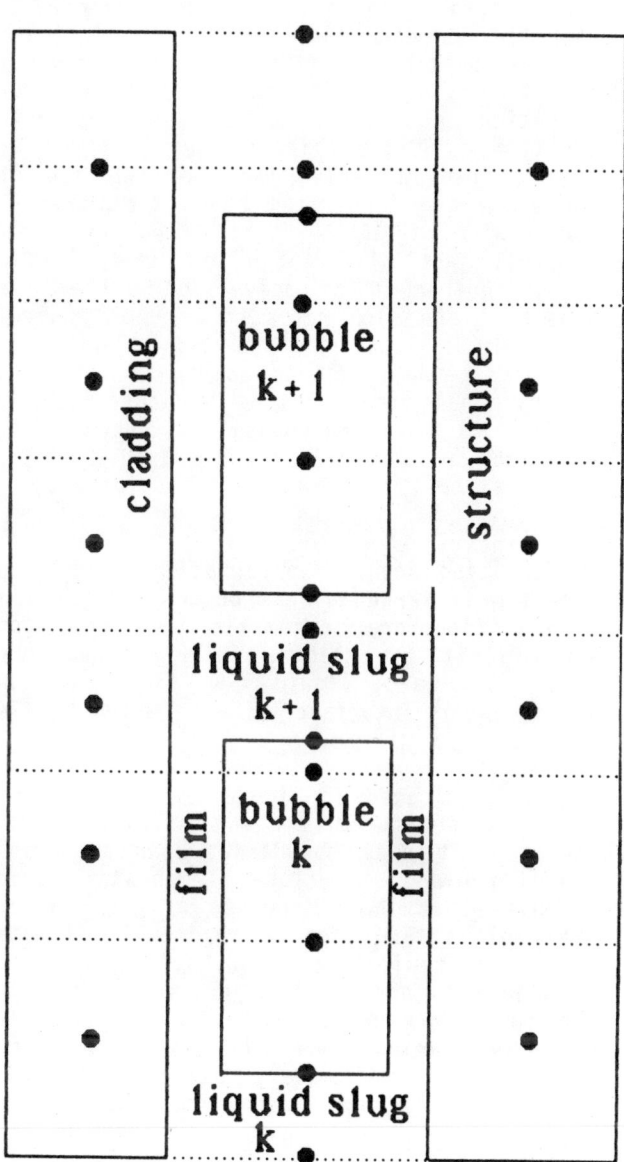

Figure 1. Multiple Bubble Slug Ejection Boiling Model and Axial Node Structure

### Large Bubbles

A more detailed treatment is used for larger bubbles. For these bubbles, the heat flow through the liquid-vapor interfaces is ignored, but vapor velocities and axial pressure gradients within the bubble are computed. After boiling develops, pressure gradients within the vapor bubbles become a dominating factor; and these pressure gradients are mainly determined by vapor flow rates. At each fixed node boundary within a bubble and at each liquid-vapor interface, a vapor pressure and a vapor mass flow rate are computed for each time step. In addition to the saturation condition, the equations solved for each node are a vapor mass conservation equation and a vapor momentum equation. Finite differencing in space and time is used to solve these equations. The equations are linearized about the values at the beginning of the time step, and fully implicit time differencing is used, leading to a requirement to solve for all of the pressures and flow rates in a bubble, simultaneously. Since the non-zero elements in the matrix that must be solved are all near the diagonal, the solution by Gaussian elimination is not time-consuming. Because the equations are linearized and solved directly, no iteration is involved in the solution.

Linearization could lead to numerical inaccuracies and non-conservatism of mass and momentum if large changes occur during a time step. Therefore, automatic time step controls limit the changes in vapor pressure, vapor flow rate, liquid slug flow rate, and liquid-vapor interface motion during a time step.

As indicated in Figure 1, coolant pressures, temperatures, and mass flow rates are computed at node boundaries, whereas fuel, cladding, and structure temperatures are calculated at the mid-points of axial nodes. The coolant flow area and hydraulic diameter are constant within a node, but they can vary from node to node. The vapor mass flow rate per pin (kg/s or lbm/hr) is continuous at node boundaries even if the flow area changes. In this model, the vapor pressure is also continuous across node boundaries. any pressure change due to a sudden area change is accounted for by an orifice pressure drop spread over the nearest node. The vapor mass conservation equation used in this model is

$$A_c \frac{\partial \rho_v}{\partial t} + \frac{\partial w_v}{\partial z} = \frac{QA_c}{\lambda}, \quad (1)$$

and the momentum equation for the vapor is

$$\frac{\partial w_v}{\partial t} + A_c \frac{\partial p}{\partial z} + \frac{1}{A_c}\frac{\partial}{\partial z}\left(\frac{w_v^2}{\rho_v}\right)$$
$$= -\frac{fw_v|w_v|}{2\rho_v A_c D_h} - \frac{w_v|w_v|}{2\rho_v A_c}\frac{\partial K_{or}}{\partial z} + F_c. \quad (2)$$

At a liquid-vapor interface, the vapor velocity is set equal to the interface velocity, so

$$w_v(z_i) = \rho_i A_i v_i \quad (3)$$

The symbols used in these equations are defined in the NOTATION section below. The heat flux contains contributions from both the cladding and the structure. At a given axial node, the heat flux from the cladding to the coolant is based on the difference between the cladding temperature at the middle of the axial node and the average of the coolant temperatures at the top and bottom of the node. The heat flux at the end of a time step is based on cladding and structure temperatures extrapolated to the end of the step and on implicitly calculated coolant temperatures at the end of the step.

In addition to the above equations, the saturation conditions determine the relationships between the vapor temperature, pressure, and density. Also, a momentum equation for the liquid slugs determines the relationship between the interface pressure and the rate of change of the interface velocity.

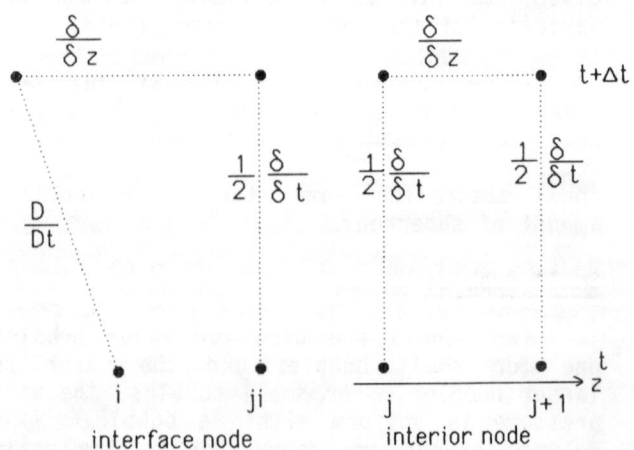

Figure 2. Finite Difference Approximations Used for Large Bubbles

Figure 2 illustrates the manner in which finite differencing of Equations (1) and (2) is carried out. A fully implicit scheme is used, so spatial derivatives are evaluated at the end of the time step. For any quantity x, the finite difference approximation used is

$$\frac{\partial x}{\partial z} = \frac{x_{j+1}(t+\Delta t) - x_j(t+\Delta t)}{z_{j+1} - z_j} \quad (4)$$

For an interior node, the time derivative is evaluated as

$$\frac{\partial x}{\partial t} = [x_j(t+\Delta t) - x_j(t) + x_{j+1}(t+\Delta t) - x_{j+1}(t)]/2\Delta t \quad (5)$$

For the interface node, the Lagrangean total time derivative must be used, since the node is moving. The time derivative at an interface becomes

$$\frac{\partial x}{\partial t} = \frac{1}{2\Delta t} \{x_{ji}(t+\Delta t) - x_{ji}(t) + x_i(t+\Delta t) - x_i(t)\}$$

$$- \frac{v_i[x_{ji}(t+\Delta t) - x_i(t+\Delta t)]}{2(z_{ji} - z_i)} \quad (6)$$

where $ji$ is the fixed node nearest the interface but within the bubble, and $x_i$ is the value of x at the interface.

The vapor mass conservation and momentum equations are non-linear equations, but for each time step they are linearized about the values at the beginning of the time step. For instance, the friction term in the momentum equation for node j becomes

$$\frac{f(t+\Delta t) w_v(t+\Delta t)|w_v(t+\Delta t)|}{2\rho_v A_c D_h}$$

$$= \frac{1}{4A_{cj} D_{hj}} \left\{ \frac{f_j(t) w_{vj}(t)|w_{vj}(t)|}{\rho_{vj}(t)} \right.$$

$$\left[1 + \Delta w_{vj} \left(\frac{2}{w_{vj}(t)} + \frac{1}{f_j(t)} \frac{\delta f_j}{\partial w_j}\right)\right.$$

$$\left. - \frac{\Delta p_j}{\rho_{vj}(t)} \frac{\partial \rho_{vj}}{\partial p_j} \right]$$

$$+ \frac{f_{j+1}(t) w_{vj+1}(t)|w_{vj+1}(t)|}{\rho_{vj+1}(t)}$$

$$\left[1 + \Delta w_{vj+1} \left(\frac{2}{w_{vj+1}(t)} + \frac{1}{f_{j+1}(t)} \frac{\partial f_{j+1}}{\partial w_{j+1}}\right)\right.$$

$$\left. \left. - \frac{\Delta p_{j+1}}{\rho_{vj+1}(t)} \frac{\partial \rho_{vj+1}}{\partial p_{j+1}} \right]\right\} \quad (7)$$

After linearizing and finite differencing the equations solved for a vapor bubble for a time step can be put in matrix form as

$$Ax = B \quad (8)$$

where

$$x = \begin{bmatrix} \Delta p_{j1} \\ \Delta w_{j1} \\ \Delta p_{j1+1} \\ \Delta w_{j1+1} \\ \cdot \\ \cdot \\ \cdot \\ \Delta p_{j2-1} \\ \Delta w_{j2-1} \\ \Delta p_{j2} \\ \Delta w_{j2} \end{bmatrix} \quad (9)$$

$$B = \begin{bmatrix} d_{j1} \\ g_{j1} \\ h_{j1} \\ g_{j1+1} \\ h_{j1+1} \\ \cdot \\ \cdot \\ \cdot \\ g_{j2-1} \\ h_{j2-1} \\ d_{j2} \end{bmatrix} \quad (10)$$

$$A = \begin{bmatrix} a_{1,j1} & a_{2,j1} & 0 & 0 & \cdots & & & & \\ b_{1,j1} & b_{2,j1} & b_{3,j1} & b_{4,j1} & 0 & 0 & \cdots & & \\ c_{1,j1} & c_{2,j1} & c_{3,j1} & c_{4,j1} & 0 & 0 & \cdots & & \\ & & b_{1,j1+1} & b_{2,j1+1} & b_{3,j1+1} & b_{4,j1+1} & 0 & \cdots & 0 \\ & & c_{1,j1+1} & c_{2,j1+1} & c_{3,j1+1} & c_{4,j1+1} & 0 & \cdots & 0 \\ & & & \cdots & & & & & \\ 0 & \cdots & & 0 & b_{1,j2-1} & b_{2,j2-1} & b_{3,j2-1} & b_{4,j2-1} \\ 0 & \cdots & & 0 & c_{1,j2-1} & c_{2,j2-1} & c_{3,j2-1} & c_{4,j2-1} \\ 0 & \cdots & & & & 0 & a_{1,j2} & a_{2,j2} \end{bmatrix} \quad (11)$$

where j1 is the node number of the lower interface, and j2 is the node number of the upper interface. In these equations the coefficients $a_{i,j}$ and $d_j$ come from combining Equation 3 with a liquid slug momentum equation, the coefficients $b_{i,j}$ and $g_j$ come from the finite difference approximation to Equation (1) for node j, and the coefficients $c_{i,j}$ and $h_j$ come from the finite difference approximation to Equation (2) for node j. This matrix equation is solved by Gaussian elimination to obtain all of the vapor pressures and mass flow rates in a bubble simultaneously.

## Liquid Films

The liquid films on the cladding and structure are treated separately, with the film thickness at each node decreasing or increasing due to evaporation or condensation. After cladding film dryout at a node, no more evaporation, and no more heat flow, from the cladding to the coolant occurs at that node until it is rewet by a re-entering liquid slug. Currently, a static film treatment is used in SAS4A and SASSYS-1, but soon a film motion model similar to that (3) used in SAS3A and SAS3D will be added. This model will account for film motion under the influence of gravity, the shear forces from the vapor streaming by, the drag on the cladding or structure, and the coolant pressure gradients.

## Liquid-Vapor Interface Discontinuities

The boiling model handles large discontinuities at liquid-vapor interfaces by tracking the interfaces explicitly and by using Lagrangean nodes that move with the interfaces. The vapor density is about three orders of magnitude lower than the liquid density, and the liquid temperature a short distance from an interface can be appreciably different from the vapor temperature at the interface. In this model, the vapor temperature and density on the vapor side of an interface is calculated separately from the liquid temperature and density on the liquid side of an interface.

## SAMPLE CASE

Figure 3 shows the voiding results from a boiling calculation for the R5 test (4) that was run in the TREAT reactor. This test was a simulated unprotected loss of coolant flow accident with no reactor scram until after cladding melting occurred. The shaded area in this figure represents the voided region. The heavily shaded region represents the region in which the liquid film on the cladding has dried out. The voiding profile was not measured directly in this experiment, but a reconstruction of the voiding profile based on measured inlet and outlet flow rates and on thermocouple data agrees quite well with the calculated voiding profile. Also, the calculated inlet and outlet flow rates show oscillations of about the same amplitude and frequency as those observed in the experiment. This calculation used about one minute of CPU time on an IBM 3033 computer, and the average coolant time step size during boiling was about two milliseconds.

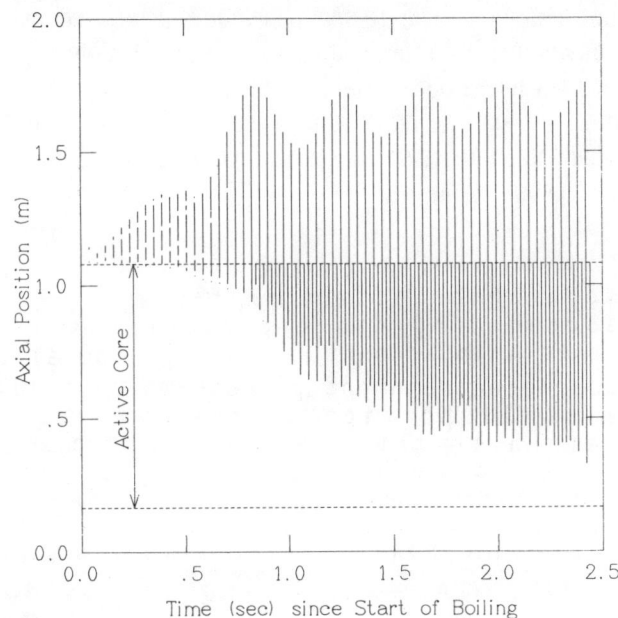

Figure 3. Voiding profile for the R5 experiment.

## NOTATION

$A_c$ = coolant flow area per pin,
$A_i$ = coolant flow area at the interface
$D_h$ = hydraulic diameter,
$f$ = friction factor,
$F_c$ = condensation momentum loss term,
$K_{or}$ = orifice coefficient,
$p$ = vapor pressure,
$Q$ = heat flux, per unit coolant volume, from the clad and structure.

## NOTATION (Contd.)

$t$ = time,
$v_i$ = interface velocity,
$w_v$ = vapor mass flow rate per pin,
$z$ = elevation,
$z_i$ = interface elevation,
$z_j$ = the elevation at the bottom of node j.
$\Delta t$ = time step size
$\Delta P_j$ = change in pressure at the bottom of node j during the time step.
$\Delta w_{vj}$ = change in $W_v$ at the bottom of node j during the time step,
$\lambda$ = heat of vaporization,
$\rho_i$ = vapor density at the interface,
$\rho_v$ = vapor density,

## Subscripts

$i$ = interface value,
$j$ = node number.

## LITERATURE CITED

1. Cahalan, J. E. et al., "The Status and Experimental Basis of the SAS4A Accident Analysis Code System," Proc. International Meeting on Fast Reactor Safety Technology, Seattle, Washington, pp. 603-614 (1979).

2. F. E. Dunn and F. G. Prohammer, "The SASSYS LMFBR Systems Analysis Code," Mathematics and Computation in Simulation, XXVI, 23 (1984).

3. G. Hoeppner, F. E. Dunn, and T. J. Heames, "The SAS3A Sodium Boiling Model and its Experimental Basis," Trans. Am. Nucl. Soc., 20, 519 (1975).

4. Grolmes, M. A. et al., "R-Series Loss-of-Flow Safety Experiments in TREAT," Fast Reactor Safety, USAEC-CONF 740701, Beverly Hills, California pp. 279-302 (April 1974).

# NUMERICAL MODELING OF THE PHASE SEPARATION PROCESSES IN BWR AND PWR STEAM SEPARATORS

Sang S. Wang ■ Wang Software Service, Fremont, California
Govinda S. Srikantiah ■ Electric Power Research Institute, Palo Alto, California

Three dimensional, two phase flow separation in a BWR steam separator is simulated using EPRI's ATHOS code. With the implementation of an energy source term in the model equations, the previous partial flow separation results (1) are improved significantly. Complete flow separation is achieved in the separator barrel.

To model the steam transport phenomena in the separator pool, a steam bubble tracing technique is developed, using which individual bubble motion can be followed. The statistical behavior of a large number of bubbles is used to determine a carryunder factor, which relates the quality in the reactor downcomer to the outlet quality of the separator. The overall system carryunder including the phase separation effects in both the separator barrel and the surrounding pool can be obtained. The calculated values of carryunder at various inlet qualities compare satisfactorily against the experimental results. The method is applicable to studying PWR steam separation systems as well.

## INTRODUCTION

Numerical modeling of steam separators can help analyze the separator performance and improve the separator design. It aims at optimizing the steam separation efficiency in light water reactor systems. Recent work (2-5) in this field indicates an increasing concern and interest to better understand the phase separation processes in such systems. However, these investigations used simplified models which are generally less accurate in describing the separation processes.

In the present study, the flow separation is simulated by the three-dimensional thermal-hydraulic code ATHOS. A two-phase steady-state mixture flow with both axial and radial velocity drifts is calculated in a BWR separator configuration. Reasonable flow evolution with a partial phase separation was calculated previously (1). By including an energy source term contributed from the radial slip gradient in the mixture energy equation, the simulation becomes significantly more realistic. Complete phase separation takes place in and above the axial vanes region. The steam qualities calculated at the downcomer exit and the separator outlet are at acceptable levels.

To calculate the carryunder, the separation effect in the water pool surrounding the separators is also taken into account. That is, the steam bubble entrainment process which enables a small portion of the bubbles to return to the reactor downcomer is modeled numerically. This model establishes a connection between the calculated quality at the separator downcomer exit and the system carryunder.

Since the microscopic mechanism involved in this steam transport process is the bubble movement in the separation pool, a bubble tracing technique is developed to account for the bubble dynamics. The collective bubble behavior will determine the transport result. The influences of buoyancy, drag and flow turbulence are included in the bubble equation of motion. To reduce the magnitude of the effort of tracing an enormous number of steam bubbles actually in the action in the separation pool, only a representative set of bubbles are simulated. The total contribution is calculated from the integrations over the bubble velocity distribution and the bubble size distribution. Only the ranges of size and velocity distribution in which the bubbles are entrained will affect the carryunder.

In summary, the method employed in modeling the phase separation processes in this work is intended to simulate the fundamental phenomena as closely as possible. This is accomplished by a detailed three-dimensional, two-phase flow calculation in a true separator configuration. It is accompanied by a an integrated steam bubble transport simulation in the separation pool to establish the carryunder estimation. The results indicate that the method has captured the underlying physical separation processes.

## THE DERIVATION OF THE RADIAL SLIP SOURCE TERMS

The original ATHOS approach (6) consists of solving three-dimensional mass, momentum and energy conservation equations with a drift-flux model applied in the axial direction. A radial source term was added to the momentum equation later to account for the radial phase relative motion (1). However, only a partial phase separation was obtained in that simulation result. When the radial-slip energy source term is added to the energy equation, a significant phase separation effect begins to take place. The derivation of these two source terms follows. The radial source terms are derived from the general drift-flux equation. First, the mixture momentum equation is given by

$$\frac{\partial}{\partial t}(\rho \bar{V}) + \nabla \cdot (\rho \bar{V}\, \bar{V})$$
$$= -\nabla P + \nabla \cdot (\overleftrightarrow{\tau} + \overleftrightarrow{\tau}^T - \sum_k \alpha_k \rho_k \bar{V}_{km} \bar{V}_{km}) + \rho \bar{g}, \quad (1)$$

where $\bar{V}$ is the mixture velocity vector with components u, v, and w in the circumferential, radial and axial directions, $\overleftrightarrow{\tau}$ and $\overleftrightarrow{\tau}^T$ are the time or ensemble averaged local viscous stress tensor and turbulent diffussion flux of mixture momentum, $\bar{V}_{km}$ is the drift velocity of the k$^{th}$ phase with respect to the mixture velocity and k sums over the gas and the liquid phases. Neglect the viscous stress and the turbulent diffusion terms, the right hand side of Equation (1) can be simplified with the relationships

$$\bar{V}_{lm} = -\frac{\alpha \rho_g}{(1-\alpha)\rho_l} \bar{V}_{gm} \quad (2)$$

and

$$\bar{V}_{gm} = \frac{\rho_l}{\rho} \bar{V}_{gj} \quad (3)$$

where $\bar{V}_{\ell m}$ and $\bar{V}_{gm}$ are the drift velocities of the liquid phase and the gas phase with respect to the mixture velocity $\bar{V}$, and $\bar{V}_{gj}$ is the drift velocity of the the gas phase w.r.t. the volume center of the mixture, that is, $\bar{V}_{gj} = \bar{V}_g - \bar{j}$. The divergence term on the right hand side becomes

$$-\nabla \cdot \{ \frac{\rho \alpha \rho_g}{(1-\alpha)\rho_l} \bar{V}_{gm} \bar{V}_{gm} \} . \quad (4)$$

Substituting (4) into (1), Ishii's (7) three-dimensional drift-flux momentum equation is obtained. Furthermore, by applying the relationship between $\bar{V}_{gm}$ and the liquid versus vapor relative velocity $\bar{V}_{\ell g}$,

$$\bar{V}_{gm} = -\frac{(1-\alpha)\rho_l}{\rho} \bar{V}_{lg} , \quad (5)$$

the right hand side of Equation (4) is reduced to

$$-\nabla \cdot \{ \frac{\alpha(1-\alpha)\rho_g \rho_l}{\rho} \bar{V}_{lg} \bar{V}_{lg} \} .$$

Upon expanding the above equation, and assuming that the gradient of the radial relative velocity component is more dominant than those of the other components, a radial momentum source term results. It is given by:

$$\frac{1}{r}\frac{\partial}{\partial r} \{ r \frac{\rho_l \rho_g \alpha (1-\alpha) v_{lg}^2}{\rho} \} . \quad (6)$$

The new energy source term can be derived in a similar fashion. The general mixture energy equation is :

$$\nabla \cdot (\rho h \bar{V}) = -\nabla \cdot \{ \bar{q} + \bar{q}^T + \sum_k \alpha_k \rho_k h_k \bar{V}_{km} \}$$
$$+ \{ \bar{V} + \frac{\alpha(\rho_l - \rho_g)}{\rho} \bar{V}_{gm} \} \cdot \nabla P + \Phi^\mu \quad (7)$$

where $\bar{q}$ and $\bar{q}^T$ are the time or ensemble averaged conduction and turbulent diffusion flux of energy, respectively, and $\Phi^*$ is the energy dissipation term. It is to be noted that $\bar{V}_{gm}$ and $\bar{V}_{\ell m}$ represent relative velocity vectors and have components $(u_{gm}, v_{gm}, w_{gm})$ and $(u_{\ell m}, v_{\ell m}, w_{\ell m})$ respectively in the three coordinate directions. In the present work, the u-component is neglected. The components $w_{gm}$, $v_{\ell g}$ are computed from drift flux correlation (6) and on the basis of centrifugal force exerted by the bubbles based on the Harmathy drag coefficiency (1). By neglecting the conduction and the diffusion energy fluxes, the dissipation term and the mechanical work term, the right hand side of (7) at steady state becomes

$$-\nabla \cdot \{\alpha \rho_g h_{\ell g} \bar{V}_{gm}\}$$

where $h_{\ell g}$ is the latent heat of vaporization. At thermodynamic equilibrium, $h_{\ell g} = h_g - h_\ell$. Apply (5) to the radial component and neglect the circumferential term, then the energy source term becomes,

$$\frac{1}{r}\frac{\partial}{\partial r}\{r\frac{\alpha(1-\alpha)\rho_\ell \rho_g v_{\ell g} h_{\ell g}}{\rho}\} \quad . \quad (8.a)$$

The contribution from the axial component can be written.

$$-\frac{\partial}{\partial z}\{\rho_g \alpha h_{\ell g} w_{gm}\} \quad . \quad (8.b)$$

The complete energy equation used in our model is thus

$$\nabla \cdot (\rho \bar{V} h) = -\frac{\partial}{\partial z}\{\rho_g \alpha h_{\ell g} w_{gm}\}$$
$$+ \frac{1}{r}\frac{\partial}{\partial r}\{r\frac{\alpha(1-\alpha)\rho_\ell \rho_g v_{\ell g} h_{\ell g}}{\rho}\} \quad (9)$$

The complete momentum equation is solved in the three coordinate directions:

$$\nabla \cdot (\rho \bar{V} ur) = -\frac{\partial P}{\partial \theta} \quad , \quad (10)$$

$$\nabla \cdot (\rho \bar{V} v) = -\frac{\partial P}{\partial r}$$
$$- \rho \frac{(ur)^2}{r^3} + \frac{1}{r}\frac{\partial}{\partial r}\{r\frac{\rho_\ell \rho_g \alpha(1-\alpha)v_{\ell g}^2}{\rho}\} \quad (11)$$

and

$$\nabla \cdot (\rho \bar{V} w) = -\frac{\partial P}{\partial z} - \rho g - \frac{\partial}{\partial z}\{\frac{\rho \rho_g \alpha w_{gm}^2}{\rho_\ell (1-\alpha)}\} \quad (12)$$

The importance of the new energy source term becomes obvious by inspecting Equation (9). It relates the local enthalpy divergence to the radial relative velocity gradient. And, the local void fraction is proportional to the local enthalpy as given by,

$$\alpha = \frac{\rho h - \rho_\ell h_\ell}{\rho_g h_g - \rho_\ell h_\ell} \quad (13)$$

## FLOW SEPARATION SIMULATION RESULT

A non-uniform 8 x 8 x 22 computational grid is used to calculate the two-phase flow evolution in a GE second-generation separator. Barrel, skimmers and downcomer walls are represented by solid baffles in the computation. The boundary conditions consist of the outlet pressure and the inlet flow conditions. At the separator outlet, the dome pressure plus the pressure drop across the dryer are applied. Also, at the downcomer exits, the dome pressure plus the respective hydrostatic pressure heads are imposed. At the separator inlet pipe, the inlet quality X, the mixture velocity W and the mixture enthalpy h are specified. Since the liquid and the vapor densities are determined uniquely by the dome pressure at thermal equilibrium, the mixture velocity can be calculated from the total flow rate G by

$$w_i = \frac{G[\rho_g + X_i(\rho_\ell - \rho_g)]}{A_i \rho_\ell \rho_g} \quad (14)$$

where $A_i$ is the cross-section area of the inlet pipe. The actual operating conditions used in the simulation are :

Dome pressure      $P_d = 6.9 \times 10$ N/m$^2$

Inlet quality      $X_i = 0.095$

Total flow rate    $G = 40.56$ Kg/sec

Water pool level   $z_w = 1.85$ m

Inlet enthalpy     $h_i = 1.41 \times 10$ J/Kg

Liquid density  $\rho_\ell = 7.42 \times 10^2$ Kg/m$^3$
Vapor density   $\rho_g = 3.6 \times 10^2$ Kg/m$^3$.

In addition to the realistic flow field and pressure drop as already reported previously (1), the phase separation effect is clearly manifested in the present simulation result. To quantify these results, the void fration distributions at selected locations are plotted. Figure 1 (d) shows the void distribution just below the axial vanes unit. A uniform distribution with a value of 0.68 is

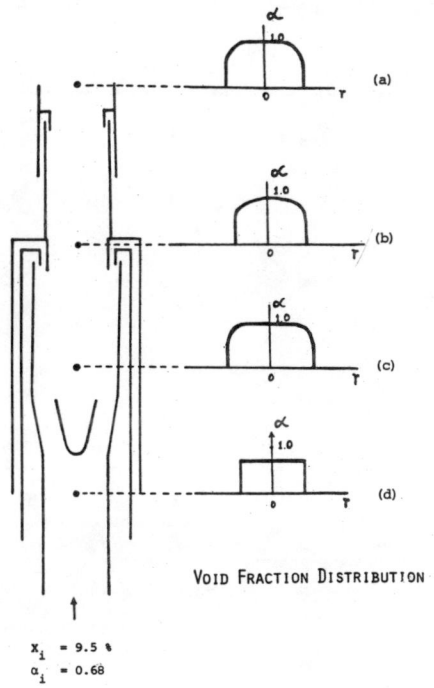

Figure 1. The calculated void fraction distribution at the indicated axial locations in the separator.

used for a typical separator inlet steam flow. This uniformity changes considerably through the vanes region because of the centrifugal separation as shown in Figure 1 (c). The void fraction is nearly unity at the core region indicating that pure steam exists there. It remains close to unity until it approaches the barrel wall where the void fraction drops rapidly to about 0.4. It indicates that a liquid film is formed locally in the cells near the barrel wall. Similar distributions are seen at the second skimmer and the separator exit locations as illustrated in Figures 1 (b) and (a). It shows that the flow is basically remained separated within the barrel, which is indeed what takes place in a separator. The simulation results reveal that a uniform steam-water mixture at the separator inlet is transformed into a steam core and a liquid film by the axial vanes. And, separated flow situation is maintained in the barrel without any notable flow remixing. The physical separation process is thus captured by our numerical simulation.

## BUBBLE TRACING METHOD

The BWR carryunder measures the steam quality in the flow returning to the core via the reactor downcomer. We have calculated the typical steam quality value, which is 0.02, at the separator downcomer exits. A carry-under coefficience is needed to link this value to the quality at the core downcomer inlet. Our approach to determine this factor is explained below.

As viewed from the separator pool, steam bubbles with different sizes and speeds are released in downward two-phase mixture jets from the separator downcomer exits. A turbulent flow is created in the pool by the downward moving jets. Because of the buoyancy force acting on them, most of the steam bubbles will escape to the free surface. Only a small fraction of them will be entrained into the reactor downcomer by the pool flow. This very complex process involves an extremely large number of bubbles in dynamic motion within a turbulent flow field.

To model such a process, we begin by solving the dynamic bubble motion governed by a general equation

$$(\rho_g + \mu \rho_1) V_b \frac{d\bar{u}_b}{dt} = -(\rho_1 - \rho_g) V_b \bar{g} - 1/2 \rho_1 (\bar{u}_b - \bar{u}_1)|\bar{u}_b - \bar{u}_1| C_D \pi r_b^2 \quad (15)$$

where $\mu$ is the added mass coefficient, $V_b$ is the bubble volume, $\bar{g}$ is gravitational acceleration, $C_D$ the drag coefficinet, $\bar{u}_b$ is the bubble velocity, $\bar{u}_\ell$ the local liquid velocity and $r_b$ is the bubble radius. At the steady-state, the buoyancy is balanced by the drag force

$$1/2\, \rho_l\, U_\infty^2\, C_D \pi\, r_b^2 = 4/3\, (\rho_l - \rho_g)\, g\, \pi\, r_b^3, \quad (16)$$

which relates the drag coefficient to the terminal velocity $U_\infty$. Correlations for the terminal velocity of a free rising bubble were presented by Harmathy (8). The drag coefficient $C_D$ can then be uniquely calculated. With a typical value of 0.5 and the calculated local pool flow field, Equation (15) is solved for the individual bubble motion in the separator pool. The pool flow simulation result is described in the next section.

The bubble injection velocity has a gaussian-like probability distribution function. The distribution is centered at a mean injection velocity, which corresponds to the separator downcomer exit mixture velocity. In our model, 10 velocity bins are set up for the bubble velocity distribution function. A random walk procedure is used to assign the bubble velocity starting from the center bin. A gaussian-like velocity distribution is indeed generated as the number of bubbles exceeds 200. The bubble size distribution is formed in the same manner. The mean bubble diameter used in the simualtion is 1 cm.

## SEPARATOR POOL FLOW FIELD CALCULATION

To find the local liquid velocity $\bar{u}_\ell$ for solving Equation (15), the separator pool flow field needs to be calculated. Our method uses the ATHOS code to compute the pool flow field. A 4 x 5 x 10 non-uniform grid is employed to resolve the region above the core shroud and the reactor downcomer inlet. The boundary conditions used in the calculation are the liquid velocity specified at the upper boundary and the core downcomer pressure at the lower boundary.

The submerged structures and the flow constraints implemented in the calculation include : the separator barrels and the separator standpipes represented by flow area reductions in three dimensions and the control cell volume reductions. The cross flow resistance is modeled by the porosity reductions in the circumferential and radial directions. Furthermore, the pool surface outside of the separator region is modeled by a zero inflow velocity at the upper boundary of the outermost cells at the top. A uniform liquid inflow velocity of -0.5 m/sec applied on the top boundary in the separator region. The result of pool flow field calculation is shown in Figure 2, where a large recirculation flow pattern is observed above the core shroud. The flow penetration and the expansion also look realistic. This is pool field used in the bubble tracing calculation.

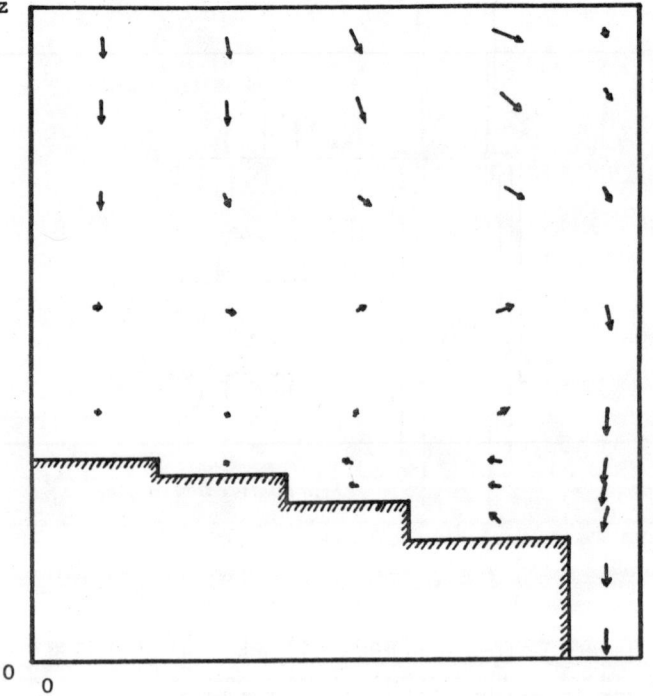

Figure 2. ATHOS calculated pool flow field with internal blockages and a uniform injection velocity of –0.5 m/sec is imposed at the upper boundary.

## SIMULATION RESULTS

### Bubble Trajectory

Figure 3 depicts a typical result of the calculated bubble trajectories. For a better visibility, most of the trajectories are truncated after making their upward turns toward the free surface. Many trajectories end on the

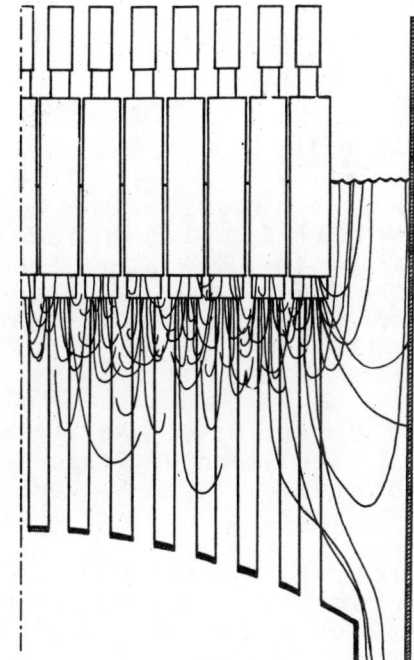

Figure 3. Typical steam bubble trajectories calculated by the bubble tracing code. The carryunder coefficient obtained by following some 2000 particles is 0.067.

reactor barrel wall or the standpipe walls. Few make to the core downcomer. The rest are terminated at the pool surface. The wiggling bubble motion due to the turbulent flow effect is not plotted for clarity. Those trajectories that connect between the separator exits and the reactor downcomer signify the entrained bubbles which contribute to the system carryunder. Figure 3 reveals that these bubblese mostly originate from those separators situated near the rim of the assembly. Those separators located more internally have very little probability to transport steam bubbles all the way to the downcomer inlet as is verified in the simulation. Upon simulating more than 2000 bubble trajectories, the integrated result of the amount of steam transported to the downcomer versus that entering the separator pool converges to 0.067. This is the carryunder coefficient to be used in the following carryunder analysis.

## Carryunder

In a series of flow separation calculations, the inlet steam quality is varied while all other operating conditions are held constant. The range of this quality variation is between 6 % to 14 %. The separator dowmcomer outlet qualities are then multiplied by the carryunder coefficience of 0.067. The result of the system carryunder as a function of the separator inlet quality is shown as the dashed line in Figure 4. It is compared against the experimental result obtained by Wolf and Moen (9) shown as the solid line. Each illustrates a valley of the

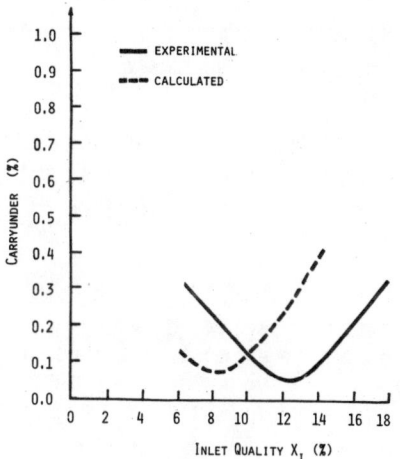

Figure 4. A comparison of the calculated and the measured carryunder variations as the inlet quality is varied.

same magnititude. However, the occurance of the minimum is at an inlet quality of 8 % as predicted by the simulation, compared to 12.5 % as determined from the test results. Aside from this discrepancy, the simulation results look quite encouraging.

## Separator Outlet Quality

The separator outlet quality is also calculated as a function of the inlet quality. Figure 5 shows this result in a dashed line. It decreases as the inlet quality is increased. A similar trend is also observed in the measured carryover result (9) as represented by the solid line in Figure 5. This comparison is not a direct one

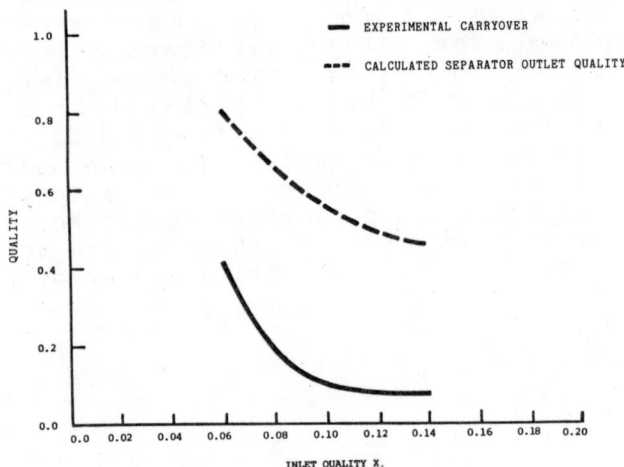

Figure 5. A comparison of the calculated moisture entrainment from the separator outlet and the measured carryover as functions of inlet quality.

because the measured carryover includes the effect of the dryer. To achieve a better agreement between the two resutls, a dryer model must be incorporated in our numerical simulations to further remove the moisture from the separator outlet.

## CONCLUSION

This paper describes the progress that has been made in the flow separation simulation and the carryunder and carryover calculations in a BWR steam separator. The successful flow separation simualtion is accomplished by incorporating an energy source term in the ATHOS code energy equation. This source term is based on the radial velocity slip gradient derived from a general drift-flux equation.

A new bubble tracing technique is developed to calculate the system carryunder. By following the trajectories of a large number of steam bubbles in a calculated pool flow with a modeled turbulence effect, a statistical carryunder coefficience is reduced. The calculated carryunder, based on the separator downcomer exit quality, shows very good agreement with the experimental results.

A qualitative agreement is also achieved in a comparison of the measured carryover and the calculated separator outlet quality.

## LITERATURE CITED

1. Wang, S. S. and Srikantiah, G. S. "Numerical Simulation of Phase Separation in a BWR Steam Separator", Multiphase Flow and Heat Transfer III, Part A, Fundamentals, Ed. T. D. Veziroglu & A. E. Bergles, Elsevier 1984.

2. Vander Vorst, M. J. and Steininger, D. A. "Calculation of Phase Separation by Centrifugal Force", ASME Paper N 82-FE-5, 1982.

3. Bennett, R. R. and Kondic, N. N. "Momentum Flux Model for Liquid Vapor Separation", National Heat Transfer Conference, 1977.

4. Cheung, Y. K. and Anderson, J. G. M., "A Mechanistic Model for Internal Steam Separators in Boiling Water Reactors", ASME Paper 84-WA/HT-83.

5. Kalra, S. P., Yao, L. S. and Davies, E. R. "Flow Behavior in a Stationary Vane Centrifugal Separator: Simulation Experiments and Analysis", Preceedings on Nuclear Reactor Thermal-Hydraulics Vol. 2, P. 979. Paper presented at the Second International Topical Meeting Nuclear Reactor Thermal-Hydraulics Santa Barbara, 1983.

6. Singhal, A. K., Keeton, L. W., Spalding, D. B. and Srikantiah G. "ATHOS - A Computer Program for Thermal-Hydraulic Analysis of Steam Generators", Vols. 1-3, EPRI report NP-2698-CCM, October, 1982.

7. Ishii, M. "One-Dimensional Drift-Flux Model and Constitutive Equations for Relative Motion Between Phases in Various Two-Phase Flow Regimes", ANL 77-47, October, 1977.

8. Harmathy, T. Z. "Velocity of Large Drops and Bubbles in Media of Infinit or Restrict Extent", A. I. Ch. E. Journal, Vol. 6, pp. 281-288, 1960.

9. Wolf, S. and Moen, R. H. "Advances in Steam-Water Separators for Boiling Water Reactors", ASME Paper No. 73-WA/PWR-4, 1973.

# MULTIDIMENSIONAL TWO-PHASE MODELING WITH THE COMMIX-2 COMPUTER PROGRAM

M. Bottoni ∎ Kernforschungszentrum Karlsruhe, Postfach 3640
D-7500 Karlsruhe-1, Federal Republic of Germany
R.W. Lyczkowski, H.N. Chi, T.H. Chien and H.M. Domanus ∎ Argonne National Laboratory
9700 South Cass Avenue, Argonne, IL 60439

This paper summarizes the present state of development of the three-dimensional transient and steady-state COMMIX-2 computer program. The basis of the two two-phase models incorporated into the computer program are given. These two models are 1) a three-equation Slip Model (SM) and 2) a five equation Separated Phases Model (SPM). Validation of the SM has been performed by analyzing two experiments representing quite different conditions. Agreement with the data is good.

## INTRODUCTION

THE COMMIX series of computer programs describe steady-state and transient three-dimensional fluid flows with heat transfer in reactors and in multicomponent systems. The most advanced single-phase version of the code, COMMIX-1A, is described in Ref. (1). The method of modeling complex geometrical configurations relies, in part, upon the concepts of volume porosity, surface permeabilities, distributed resistances, and distributed heat sources (2). COMMIX-1A has provided the basic structure for the development of two two-phase flow models which are incorporated into COMMIX-2.

The first of these is a three-equation Slip Model (SM) which provides for either a constant slip ratio, or a relative velocity normalized by the mixture velocity in each of the coordinate directions. The limiting case of equal phase velocities, known as the Homogeneous Equilibrium Mixture (HEM) model, is a subcase of the Slip Model. In the HEM model, the conservation equations for momentum and specific enthalpy of the two-phase mixture are identical with those of the single-phase flow. In the SM model there are included additional terms which take into account the so-called momentum-slip and enthalpy-slip between the phases. The momentum-slip term is treated implicitly, while the enthalpy-slip term is treated semi-implicitly.

The second model is a five equation Unequal Phase Velocity, Equal Phase Temperature (UVET) seriated two-fluid continuum (3,4), which we will refer to as the Separated Phases Model (SPM). This paper aims at outlining the theoretical foundations of the two approaches followed and at presenting the state of the development work.

The next two sections present the governing equations used in the Slip Model (SM) for the conservation of mass, energy, and momentum of the coolant, and outline briefly their numerical treatment which consists in deriving two Poisson-like equations describing the pressure and enthalpy distributions in the coolant, respectively. The verification of this slip model is discussed next. Finally, the Separated Phases Model (SPM) which is currrently under development is summarized.

## THE GOVERNING EQUATIONS FOR THE SLIP MODEL (SM)

The governing equations for the conservation of mass, momentum, and energy for the Slip Model (SM) as used in COMMIX-2 are given by:

(i) <u>Continuity Equation</u>

$$\frac{\partial \rho_m}{\partial t} + \nabla \cdot (\rho_m \vec{v}_m) = 0 \qquad (1)$$

(ii) **Energy Equation**

$$\underbrace{\frac{\partial}{\partial t}(\rho_m h_m) + \nabla \cdot (\rho_m \vec{v}_m h_m)}_{\text{Term (1)}} +$$

$$\nabla \cdot (x(1-x) \rho_m (h_g - h_\ell) \vec{V}_{SL})$$

$$= \frac{\partial p}{\partial t} + \vec{v}_m \cdot \nabla p - \nabla \cdot \vec{q} + Q$$

$$+ \underbrace{(\frac{1}{\rho_m} \alpha(1-\alpha)(\rho_\ell - \rho_g) \vec{V}_{SL}) \cdot \nabla p}_{\text{Term (2)}} \quad (2)$$

(iii) **Momentum Equations**

$$\underbrace{\frac{\partial}{\partial t}(\rho_m \vec{v}_m) + \nabla \cdot (\rho_m \vec{v}_m \vec{v}_m)}_{} +$$

$$\underbrace{\nabla \cdot (x(1-x) \rho_m \vec{V}_{SL} \vec{V}_{SL}) + \nabla p}_{\text{Term (3)}}$$

$$= \nabla \cdot (\mu \nabla \vec{v}_m) + \rho_m \vec{g} + \vec{F} \quad (3)$$

The symbols are defined in the Notation.

The continuity equation (1) is formally identical with the continuity equation for the single phase flow (1). The energy and momentum equations (2) and (3) are formally identical with those for the single phase flow (1), but contain the additional terms referred to as Term (1) and Term (2) in the energy equation and Term (3) in the momentum equation which depend on the slip velocity. Term (2) in the energy equation is usually small with respect to Term (1) and has so far been neglected. When the slip velocity equals zero the Homogeneous Equilibrium Mixture (HEM) model results. These forms of the energy and momentum equations were used because they are the forms used in COMMIX-1A from which COMMIX-2 was developed.

## NUMERICAL TREATMENT OF THE GOVERNING EQUATIONS

The Implicit Continuous-fluid Eulerian (ICE) technique (5) is used to combine the continuity and momentum equations (1) and (3) and derive a Poisson equation describing the pressure distribution in the coolant. This equation has been derived in terms of pressure increments and can be written in the discrete form as

$$a_o p_o^{n+1} - \sum_{\beta=1}^{6} a_\beta p_\beta^{n+1} = b_o. \quad (4)$$

Index o refers to the center of a cell, while index $\beta$ runs over the indices of the six neighboring cells in the three coordinate directions. All details concerning the derivation of Equation (4), as well as the analytical expressions of the coefficients, are given in Ref. (6). The coolant energy equation (2) is treated implicitly and its discrete form is reduced to an algebraic equation formally identical with Equation (4) which is solved either with the method of Successive Over-Relaxation (SOR) or by means of a matrix inversion technique.

Of paramount importance is the treatment of the compressibility term $\partial \rho_m / \partial t$ contained in the coefficient $a_o$ of Equation (4). It has been discretized according to either of the following formulas:

(i) $\rho_m^{k+1} - \rho_m^k = \left(\frac{\partial \rho_m}{\partial p}\right)_{h_m}^{n} (p^{k+1} - p^k) \quad (5)$

or

(ii) $\rho_m^{k+1} - \rho_m^n = \left(\frac{\partial \rho_m}{\partial p}\right)_{h_m}^{n} (p^{k+1} - p^n)$

$$+ \left(\frac{\partial \rho_m}{\partial h_m}\right)_p^{n} (h_m^{k+1} - h_m^n) \quad (6)$$

where k refers to the iteration index at time step level n+1. Equation (5) linearizes the density over two iterates and Equation (6) linearizes the density over a time step. They were both programmed and may be selected independently by an input flag. They both gave essentially the same results. Theoretically, Equation (5) is more consistent.

The expressions for the derivatives of the coolant mixture density with respect to pressure and enthalpy are lengthy and may be found in Bridgman (7).

In COMMIX-2, two options are available to prescribe either the slip ratio or the slip velocity normalized by the mixture velocity in the Slip Model (SM) as constants in each of the three directions.

The Homogeneous Equilibrium Mixture (HEM) model is obtained as a subcase of the Slip Model (SM) by setting the slip ratio equal to 1 or by setting the normalized slip velocity equal to 0.

VERIFICATION OF THE SLIP MODEL

Two totally different experiments were analyzed in order to demonstrate 1) the flexibility of the COMMIX-2 interchangeable physical properties for water and sodium, 2) the robustness of its numerical solution procedure and 3) the ability to perform steady-state as well as transient two-phase computations in one to three dimensions. A description of these two analyses and the comparison with the data is given in this section.

Thom et al. Steady-State Steam-Water Void Fraction Experiments

As part of an investigation of subcooled boiling, Thom et al. (8) measured the void fraction in a steam-water mixture flowing in an electrically heated tube. Single phase water was introduced at the inlet of the test section, and bulk or saturated boiling was present near the test section outlet.

The test section tube had a heated length of 1.524 m (5 feet) with an inside diameter of 9.754 mm (0.384 inches). Density measurements were made with a Thulium-170 gamma source, and these measurements were used to compute local void fractions. Density measurements were made every 4.45 cm (1.75 inches) between 0.273 m (10.75 inches) from the inlet and 0.139 m (5.5 inches) from the outlet.

One experiment (Test No. 12) was simulated with the COMMIX-2 code. The pressure for this run was $5.2724 \times 10^6$ Pa (764.7 psia) the uniform heat flux was $9.053 \times 10^5$ W/m$^2$ ($2.87 \times 10^5$ Btu/ft$^2$-hr), the mass flux was $9.53 \times 10^2$ kg/(m$^2$s) ($7.03 \times 10^5$ lb/ft$^2$-hr) and the inlet temperature was 225.56°C (438°F). This inlet temperature corresponds to an inlet subcooling of 41.7°C (75°F).

The 1.524 m (5 ft) heated test section was modeled in one dimension using 12 equally sized 0.127 m nodes. The analysis of this problem was performed with the steady-state option in COMMIX-2. With this technique, a very large time step is used, typically $\Delta t$ = 100s which is about 1000 times larger than the transport time through a computational mesh. This demonstrates the robustness of the present numerical solution procedure. Steady state was achieved in a minimum of only 11 time step integrations (for no slip) using one iteration per time step. This required only about 1s CPU on an IBM 3033. The linearization for the density used both Equations (5) and (6). They yielded essentially identical results. The water properties package is based on Agee's functional fits for density, enthalpy and temperature (9).

Several runs were performed to assess the slip model and the two-phase flow friction multiplier which uses the Lockhart-Martinelli correlation (10). This approach has been found suitable for not only water but also sodium two-phase pressure drop (11). The experimental void fraction data and the void fraction computed by the COMMIX-2 code are shown in Figure 1 for three cases: slip ratios equal 1 and 2 and normalized slip velocity equal 2. Clearly the case of a slip ratio equal 1 agrees with most of the data. The case of slip ratio equal 2 agrees with some of the data. The effect of the slip is quite pronounced causing the void fraction to be lowered as the slip is increased. A normalized slip velocity equal 2 produces slip ratios between

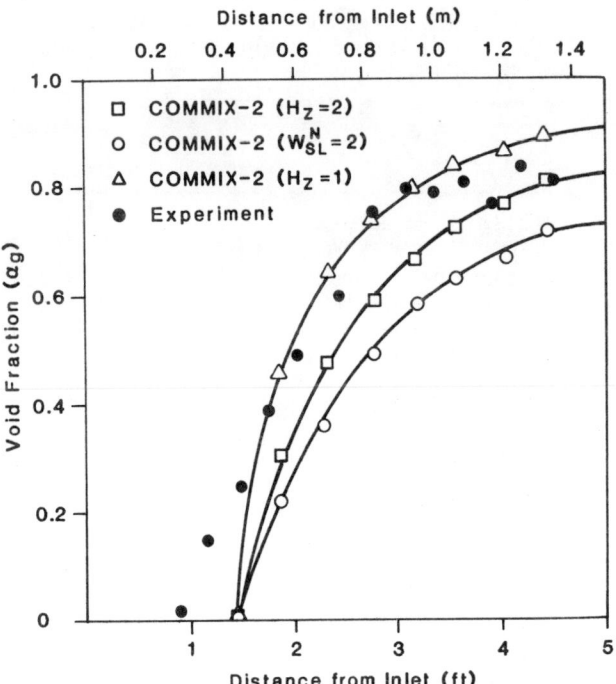

Figure 1 Comparison of Experimental and Computed Steady-State Void Fraction for Test No. 12, Thom et al. (8)

3 and 4. A more mechanistic slip ratio model may yield better results. The effect of the two-phase flow friction multiplier on the void profile for fixed slip ratio was negligible but the pressure drop increase was considerable. All cases begin boiling later than the data (about 0.4 m vs. about 0.2 m) because thermodynamic equilibrium between the phases is assumed. Therefore, subcooled boiling is not predicted. Clearly the data show the subcooling persists only for a short distance (about 0.2m) and then saturated boiling begins.

### KFK 7-Pin Pump Run - Down Sodium Boiling Experiments

During the development of the SM model, the performances of the COMMIX-2 code have been continuously tested against data from the pump run-down sodium boiling experiment 7-2/16 of the NSK (Natrium-Sieden-Kreislauf) series, run at the Kernforschungszentrum Karlsruhe (KfK) (12). This experiment was simulated in one- and three-dimensions. The three-dimensional simulation has been made for one-quarter of the pin bundle using a partition of a cross-section in 3x3 meshes in a rectangular coordinate system. The Sodium properties package is based on Golden and Tokar (13).

A comparison between experimental and simulated steady state conditions is presented in the upper part of Table 1. Further comparison is given in the same table for conditions at boiling inception and during the two-phase flow calculation, which was terminated after power switch-off and recondensation of the sodium vapor. Boiling inception, occurring in the innermost meshes, is predicted accurately by the 3-D simulation. In the 1-D case, where the coolant temperature is averaged over the full cross-section, a delay of 0.5 s is observed. With respect to the time point of the first flow reversal and of dry-out onset, the 3-D simulation is also more accurate than the 1-D. This is explained by the expansion of the two-phase flow in the outward direction of the bundle, which cannot be simulated in the 1-D case.

The calculated time variation of the coolant inlet velocity is compared with the experimental data in Figure 2 for the time interval from boiling inception to re-establishment of single-phase flow conditions. Peak coolant temperatures are compared in Figure 3. The good agreement obtained not only for the single phase flow (0 - 9.15 s) but also during the two-phase flow (where the coolant temperature is the saturation temperature corresponding to the calculated pressure) proves that the calculation of the two-phase pressure drops which relies upon the Lockhart-Martinelli concept of two-phase multiplier - is substantially correct.

The largest discrepancy between experiment and calculations concerns the development in axial direction and the duration of the two-phase region, as shown in Figure 4. In both 1-D and 3-D simulations the calculated voided regions are correct and in fairly good agreement with the experimental data, up to power switch-off (11.0 s). Beyond this point, the 1-D simulation predicts a voiding region spreading upward to the uppermost mesh. The 3-D simulation which allows for the vapor radial expansion overestimates only slightly the vapor axial expansion, but considerably (by about 30%) the duration of the two-phase flow. The main reason for these discrepancies is thought to be the total amount of energy stored in the electrical heaters which are not simulated entirely accurately at present. These

Table 1 Characterization of the Experiment 7-2/16 and of the Simulations at Steady State, at Boiling Inception, and during the Two-Phase Flow

|  | Experiment | Simulation 1-D | Simulation 3-D |
|---|---|---|---|
| **Steady State Coolant Conditions** | | | |
| Peak heat flux density (W/cm2) | 150.3 | 150.3 | 145.5 |
| Total power (kW) | 118.7 | 118.7 | 114.0 |
| Inlet/outlet pressure (bar) | 4.18/1.52 | 4.18/1.57 | 4.18/1.72 |
| Inlet/outlet temperature (°C) | 562/735 | 562/739 | 562/737 |
| Inlet coolant velocity (m/s) | 2.97 | 2.97 | 2.97 |
| Radial heat losses (percent normalized to total power) | <0.01 | 0 | 0 |
| **Experimental Conditions at Boiling Inception** | | | |
| Time of boiling inception (s) | 9.15 | 9.65 | 9.20 |
| Superheat (°C) | 0 | 0 | 0 |
| Coolant velocity (m/s) | 1.13 | 0.96 | 1.10 |
| Peak coolant temperature rising rate (°C/s) | 42 | 48.5 | 44.5 |
| **Two-Phase Flow Conditions** | | | |
| Power off after boiling inception | 1.91 | 1.41 | 1.86 |
| Flow reversal (s) | 10.85 | 10.25 | 10.70 |
| Onset of dry-out (s) | 11.00 | 10.50 | 10.90 |
| Duration of two-phase flow (s) | 4.12 | 5.90 | 5.80 |

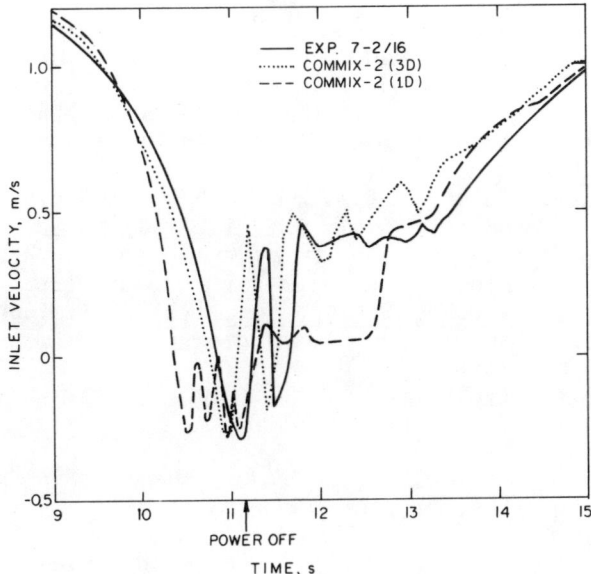

Figure 2  Coolant Inlet Velocity Verus Time

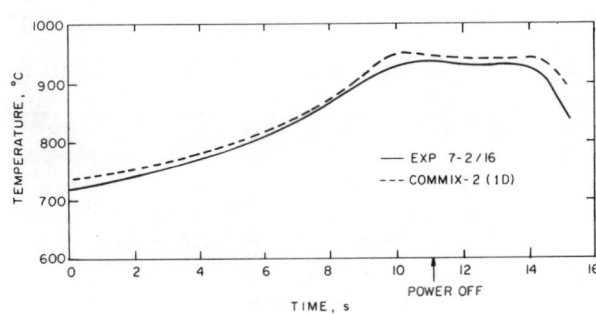

Figure 3  Experiment 7-2/16. Experimental and Calculated Coolant Temperatures Versus Time

calculations have been made using a slip velocity normalized by the mixture velocity equal to 2.

## FURTHER PROGRAM DEVELOPMENTS

The present version of the COMMIX-2 SM(HEM) model is numerically fairly stable. It relies, however, upon heavy numerical under-relaxation during the two-phase flow calculation. From the viewpoint of physical modelling, the strongest limitation of the Slip Model consists in imposing a slip ratio between the phase velocities which is kept constant during the calculation. In reality, the slip ratio depends on the flow regimes and, in general, increases with increasing void fraction. This shortcoming of the SM model is obviously overcome in the more refined Separated Phases Model, in which the momentum and continuity equations are treated separately for the vapor and liquid coolant. The governing equations for the

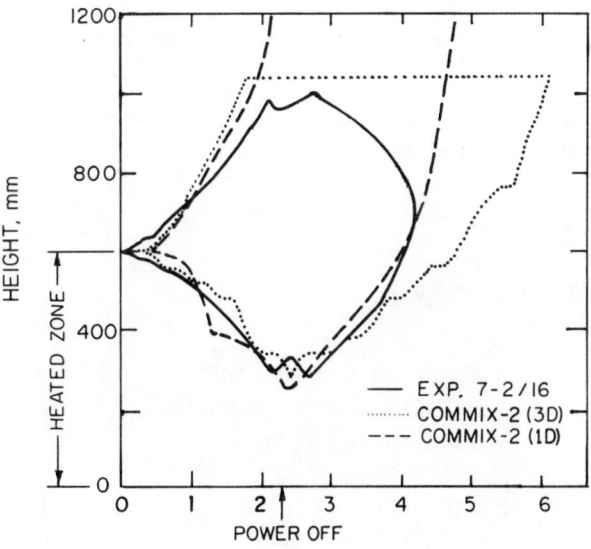

Figure 4  Experiment 7-2/16. Experimental and Calculated Voided Regions Versus Time (since boiling inception)

Separated Phases Model (SPM) are given in Reference (6).

Using the ICE technique, two discrete Poisson-like equations are obtained for describing the pressure distribution in the separated phases (6).

They can be summed to give a combined Poisson equation

$$(a_0^\ell + a_0^g) P_0^{n+1} - \sum_{i=1}^{6} (a_i^\ell + a_i^g) P_i^{n+1} = b_0^\ell + b_0^g . \quad (7)$$

Equation (7) can be solved numerically with the usual Poisson solvers, thus yielding the coolant pressure distribution. Thus, the basic COMMIX-2 modular code structure has been retained for the SPM and the solution flow anticipates a six-equation thermal non-equilibrium model.

## CONCLUSIONS

The three equation Slip Model (SM) and five equation Separated Phases Model (SPM) have been incorporated into the COMMIX-2 computer program retaining the basic modular COMMIX-1A code structure and solution procedure. The systematic investigation and shakedown of these two models has been performed by analyzing two quite different experiments involving steady-state steam-water and transient sodium boiling experiments. Agreement with the data is good. The flexibility of the code's two physical properties packages for water and sodium are demonstrated as is the robust nature of the numerical solution procedure.

NOTATION

Roman Letters

$\vec{F}$ = friction forces [kg/(m²·s²)]
$\vec{g}$ = gravity acceleration [m/s²]
$H$ = Slip ratio, $\vec{v}_g/\vec{v}_\ell$
$h$ = enthalpy [J/kg]
$p$ = pressure [Pa]
$p^{n+1}$ = pressure increment = $p^{n+1} - p^n$ [Pa]
$\vec{q}$ = heat flux [W/m²]
$Q$ = specific power generation in the coolant [W/m³]
$\Delta t$ = time step size [s]
$t$ = time [s]
$u,v,w$ = components of velocity vector [m/s]
$\vec{v}$ = velocity vector [m/s]
$\vec{V}_{SL}$ = $(\vec{v}_g - \vec{v}_\ell)$, slip velocity vector [m/s]
$\vec{V}_{SL}^N$ = $\vec{V}_{SL}/\vec{v}_m$, normalized slip velocity
$x$ = $[(h_m - h_\ell)/(h_g - h_\ell)]$ thermodynamic quality

Greek Letters

$\alpha$ = void fraction
$\mu$ = viscosity [Pa·s]
$\rho_m$ = coolant mixture density = $\alpha\rho_g + (1-\alpha)\rho_\ell$ [kg/m³]

Subscripts and Superscripts

$g,\ell,m$ = vapor, liquid, and coolant mixture, respectively
$k,k+1$ = Iteration index at time step level n+1
$n,n+1$ = time discretation indices at times t and t+$\Delta t$

LITERATURE CITED

1. Domanus, H.M., R.C. Schmitt, W.T. Sha, V.L. Shah, "COMMIX-1A: A Three-Dimensional Transient Single-Phase Computer Code for Thermal-Hydraulic Analysis of Single and Multicomponent Systems, Volume I: User's Manual," NUREG/CR-2896, ANL-82-25 (Dec. 1983).

2. Sha, W.T., B.T. Chao, S.L. Soo, "Local Volume-Averaged Transport Equations for Multiphase Flow in Regions Containing Distributed Solid Structures," NUREG/CR-2354, ANL-81-69 (Dec. 1981).

3. Solbrig, C.W., E.D. Hughes, "Governing Equations for a Seriated Continuum: An Unequal Velocity Model for Two-Phase Flow," in Two-Phase Transport and Reactor Safety, Vol. I, 307, Hemisphere Publ. Corp., Washington (1978).

4. Lyczkowski, R.W., C.W. Solbrig, AIChE J., 26(1), 89 (1980).

5. Harlow, F.H., A.A. Amsden, J. Comp. Phy., 17, 19 (1975).

6. Bottoni, M., H.N. Chi, T.H. Chien, H.M. Domanus, R.W. Lyczkowski, W.T. Sha, and V.L. Shah, ANL unpublished information (1985).

7. Bridgman, P.W., The Thermodynamics of Electrical Phenomena in Metals and a Condensed Collection of Thermodynamic Formulas, Dover Press, Inc., New York (1961).

8. Thom, J.R., W.M. Walker, T.A. Fallon, G.F.S. Reising, Proc. Instn. Mech. Engrs., 180 Pt 2C, 226, (1965-1966).

9. Agee, L.J., Nuclear Eng. and Design, 42, 195 (1977).

10. Lockhart, R.W. and R.C. Martinelli, Chemical Eng. Progress, 45 (1949).

11. Savatteri, C. and H.M. Kottowski, "Two-Phase Flow Metal Boiling Characteristic," pp. 259-286 in Heat Transfer in Nuclear Reactor Safety, S.G. Bankoff and N.H. Afgan Eds., Hemisphere Pub. Corp., Washington (1982).

12. Aberle, J., A.J. Brook, W. Peppler, H. Rohrbacker, K. Schleisiek, "Sodium Boiling Experiments in a 7-Pin Bundle under Flow Rundown Conditions," KFK 2378 (Nov. 1976).

13. Golden, G.H. and J.V. Tokar, "Thermophysical Properties of Sodium," ANL 7323 (Aug. 1967).

# TRIGGERING AND ESCALATION BEHAVIOR OF THERMAL DETONATIONS

C. Carachalios, M. Bürger and H. Unger ■ Abt. Reaktorsicherheit und Umwelt
Institut für Kernenergetik und Energiesysteme (IKE)
University of Stuttgart, Pfaffenwaldring 31
7000 Stuttgart 80, Federal Republic of Germany

For the description of large scale vapor explosions a transient model based on the thermal detonation theory was used. In the multiphase description of the thermal detonation wave a detailed model of the hydrodynamic fragmentation due to stripping of shear flow induced waves is included which assures the best estimate of the energy source term. For a given corium-water mixture and a solid wall as boundary condition, the escalating behavior of the detonation waves is examined under variation of the ignition energy. Thus, escalation length and time needed to reach a steady state behavior can be obtained. Additionally, a comparison with the results of a steady state model, including the same fragmentation description, helps to prove the consistency of both models, especially concerning the stability criterion used by the steady state model.

## INTRODUCTION

An early radioactivity release after a hypothetical core melt down accident seems to be possible only if a large scale vapor explosion could occur which might lead to the failure of the reactor containment (1). For the description of the various phenomena involved in such a Fuel Coolant Interaction (FCI) the thermal detonation theory is currently widely accepted. According to this, a shock wave is propagating in a coarse mixture of fuel and coolant causing vapor collapse and establishing a flow field behind the shock front. The rapid energy release within the detonation wave is due to fine fragmentation of the fuel droplets caused by hydrodynamic or thermal mechanisms.

A transient 1-D multiphase model which describes the processes behind the shock front as well as its propagation in the coarse mixture during a thermal detonation was already presented in (2). Because the most important features of the model and the hydrodynamic code developed especially for this purpose have already been discussed in (2), only the most recent development will be described here. The code enables to calculate the escalation and asymptotic behavior of arbitrary triggers for various boundary conditions. Thus, besides important data about ignition energies, dominance of different fragmentation mechanisms during the escalation, escalation lengths and times, energy release and pressure build-up, a transient approach to the questions of existence and stability of steady state thermal detonation waves can also be performed. The main questions concern the stability of steady supercritical thermal detonations and the existence of stable cases for nonequilibrium conditions (relative velocity ≠ 0), as e.g. discussed in (3) and (4). The formulation of a general stability criterion, especially of the included definition of the speed of sound in the multi-phase mixture at the end of the wave, is also not clear yet. Since in the transient model no additional conditions are used to obtain the steady state cases, which result only from the description of the dynamics of the physical system, a proof of the current stability theories can be undertaken.

## MODEL DEVELOPMENT

To avoid crucial numerical instabilities generated by the shock fitting method, which was used in (2) to describe the propagation of the shock front, as well as by the coupling of the shock front to the flow field behind it, numerical improvements have been introduced, e.g. concerning the discretization and the coupling with the calculation scheme used to solve the system of conservation equations which describes the flow behind the shock front. This improved numerical technique together with the use of a greater computing facility (CRAY 1 M 1200) enabled the extension of the investigated space interval from 0.5 to 5 m with approximately the same computation time.

The fragmentation process, which is the

key part of the modeling, had been described in (2) only by means of a semi-empirical correlation involving a free parameter. This deficiency has now been removed by using a correlation which was adapted to the detailed hydrodynamic fragmentation model of stripping of shear flow induced waves, described e.g. in (5). The fragment size is not important as long as (for reasons of conservativity) instantaneous heat transfer from the fragments to the coolant is assumed. Further, a model describing thermal fragmentation based on a modification of the theory of local pressurization (6) is included in the transient code. The main result of these fragmentation models is the fragmentation speed under actual conditions inside the wave. The thermal fragmentation model gives a decrease of the fragmentation speed with increasing pressures and is finally restricted to subcritical wave pressures. Thus, its importance must decrease during the escalation process.

Another improvement of the physical description has been performed by allowing for the generation of vapor in the relaxation zone which follows after the reaction zone, assuming thermal equilibrium at saturation conditions between the liquid and the vapor phase. By appearance of vapor in some location, the hydrodynamic fragmentation is assumed to cease up to this place even though the velocity equilibration is not completed yet (in the reaction zone no vapor exists). By means of this description the development of an initial trigger pulse, starting near a solid wall, to a steady state case with subcritical end pressure and saturation conditions at the end of the wave is possible. Thus, detailed investigations on the establishment of steady state waves can now be performed. This gives a good means for examining stability assumptions in steady state codes (e.g. (7), (8)).

The detailed modeling without adjusted parameters justifies the use of the transient model for simulation of various physical events depending e.g. on the boundary conditions (triggering at an open end, a solid wall, etc.), the ignition energy, the composition of the coarse mixture, the mode of propagation, etc. With respect to triggering, the inclusion of the thermal fragmentation model also allows to consider spontaneous triggering events. Finally, the performed extension of the propagation description to cylindrical and spherical waves as well as the possibility to include inhomogeneities of the coarse premixture allows to take into account further realistic restrictions to the energy yield of a thermal detonation. Although the included model of hydrodynamic fragmentation has already been experimentally validated for the special case of single drops, assuring thus the best estimate of the energy release term, the global behavior of the model remains to be investigated and tested yet.

A first comparison to the experiment has already been presented in (2) and further comparisons are scheduled with the experiments of SANDIA (9). In this work, besides the examination of the triggering and escalation behavior of thermal detonation waves, a comparison between the results of the transient model and a steady state model (7), which is based on the same theory and contains the same fragmentation models, is performed.

## PERFORMANCE OF THE CALCULATIONS, ASSUMPTIONS AND RESULTS

For a corium-water mixture (for data see Figure 1) and a solid wall as boundary condition behind the wave the ignition energy is parametrized. Since with the present numerical scheme, a shock front must be assumed to exist from the beginning, a starting procedure must be established, including also a velocity distribution from the beginning. For this, a small pressurized layer near the wall is assumed in which the values of the phase velocities increase from zero at the wall up to the jump values directly behind the shock front of the same pressure. Because the relative velocity is under these conditions too small yet to support the wave by hydrodynamic fragmentation, an additional energy source is required during the initial phase of the detonation. This can be supplied by an external trigger or by spontaneous triggering due to thermal fragmentation, e.g. according to the thermal fragmentation model already mentioned above.

For the following calculations the additionally needed energy support at the beginning of the interaction is supposed to be delivered by an external trigger and is therefore parametrized. The interaction starts for all these cases from a region of 7.5 cm length near the wall initially compressed to 2.5 MPa. The energy needed for this compression could be supplied by the fragmentation of 7.16 % of the melt mass contained in this layer, before the propagation begins, and corresponds to 6.25 $MJ/m^2$ cross section available.

Figure 1 shows the history of the shock front pressure for different ignition powers E, expressed in the part of the melt contained in the layer of 7.5 cm thickness near the wall which fragments because of the external trigger in the first 200 µs.

In Table 1 the most important steady state data for the cases of Figure 1 are shown. For E < 1.7 % a steady state propagation with a shock front pressure of 5.6 MPa and a constant reaction length was established. The evaporation under saturation conditions occured long before the equilibration of the velocities and limited the reaction zone to the length of 1.9 cm. Since the shock front velocity was equal to 165 m/s, the time of fragmentation of 9 % of the melt mass results to 115 μs. The possibility for the existence of such steady state detonation waves without velocity equilibration at the end of the wave was discussed e.g. by Sharon and Bankoff in (3) on the basis of stability considerations related to the method of characteristics. Another steady state case of the same type, but with higher energy release and shock front pressure, was reached for 2.7 % < E < 3.1 % (see Case 3,4).

For 3.9 % < E < 4.5 % a last case with greater energy release (but still without velocity equilibration) for saturation conditions at the end of the reaction zone was reached. Although its shock front pressure differed by a factor of 2.4 from the value of the former case, the fragmented mass increased only by 1 %. This is due to the description of the physical counteracting effects between fragmentation and velocity equilibration in the present model. The greater initial fragmentation in the case with higher shock front pressure leads to a steeper pressure gradient within the wave which causes a faster velocity equilibration, limiting thus the effect of the higher shock front pressure. The small difference between the end pressures however corresponds to the small difference of the fragmented mass, as could be expected from a Hugoniot consideration between the state ahead the shock front and the state at the end of the wave.

In the interval 3.1 % < E < 3.9 % one of the two stable cases with a shock front pressure of 12.1 or 29.2 MPa can be reached apparently due to arbitrary disturbances. In this region also cases with larger oscillations of the pressure occured. The frequency of these pressure oscillations is low compared to the frequency of usual numerical disturbances caused by the shock fitting, which in addition creates only oscillations with negligible amplitudes. It is not clear yet whether the existence of the observed strong instabilities is of numerical or physical nature.

The steady state case as indicated by Case 7 in Table 1 with a shock front pressure of 1045 MPa was finally reached for E > 4.8 % after an escalation time of about 1.5 ms and an escalation length of 2 m. A much higher grade of fragmentation of 76 % resulted within 107 μs and 21 cm behind the shock front, where velocity equilibration occured. The pressure level between the shock front and the wall remained very high at supercritical values. For a region behind the shock front, which approximately corresponded to the length of the reaction zone, the spatial pressure profile remained unaltered in time. This behavior can be seen in Figure 2, which shows the pressure development of the supercritical case. After a length of approximately 22 cm behind the shock front (length of the reaction zone: 21 cm) where the pressure distribution shows a characteristic shape, an unsteady expansion zone begins. Thus, the choking seems to occur at the plane of velocity equilibration or at least very near to it. In Figure 2 and also in Table 1 the corresponding values from the model (7) are given, which calculates the steady state case by means of a stability criterion (indeterminacy of the pressure gradient, in this case identical with the simultaneous occurence of sonic condition and velocity equilibration at the end of the wave). As can be seen from Table 1, there is a discrepancy between the results of the steady state and the transient model. This may be caused by deficiencies of the stability criterion or also by the uncertainties in the values of the speed of sound in the water under extreme conditions as well as in the definition of the speed of sound in mixtures which is included in the stability criterion. Figure 3 shows the time development of the velocity field, the pressure and the fragmentation at a place reached by the steady state wave calculated with the transient model. For comparison, the corresponding curves calculated by the steady state model are also plotted.

Case 7 seems to be the only detonation case with supercritical end pressures. Indeed, a simple stability consideration shows, that in cases, for which the sonic condition is reached before velocity equilibration, an arbitrary displacement of the sonic condition plane away from the shock front will lead to further escalation due to additional fragmentation and heat transfer. Thus, the plane of velocity equilibration is displaced towards the plane of velocity equilibrium. The instability of such cases with nonequilibrium end states under supercritical conditions was already predicted by Sharon and Bankoff (3). In addition, they doubted the existence of stable supercritical cases at all, which however contradicts to the present results.

The steady state case with velocity equilibration and saturation conditions at the end of the wave predicted by the steady state model (see Table 1) could not be detected by means of the transient model. For the possible interval 4.5 % < E < 4.8 % oscillating cases as well as slowly escalating cases resulted.

The latter did not remain at the expected steady state situation but continued to escalate for the time of investigation. It is not clear yet whether this behavior is caused by physical or numerical instabilities.

For the same mixture but with a halved rate of fragmentation, as it results from the hydrodynamic fragmentation model, a similar investigation has also been performed in (10). Again more than one stable case resulted with saturation conditions at the end of the wave and only a single steady state case with supercritical end pressure occured. Especially, the case with maximum energy release (i.e. velocity equilibrium) and saturation conditions at the end of the wave was also detected in this investigation series with the halved fragmentation rate. A comparison with the steady state model showed an excellent consistency of all results (pressures, fragmentation grade, length of the reaction zone, etc.). This assures the numerical schemes as well as the stability criterion used in the steady state model.

As a further step of work the influence of the boundary condition behind the wave on the stability behavior should be investigated systematically. This could be done by assuming e.g. an open end instead of a solid wall behind the relaxation zone and comparing the patterns of the resulting steady state cases. A first comparison for the halved fragmentation rate showed that with a permeable plane as boundary condition the same steady state cases result as with a solid wall but more ignition energy is needed for the corresponding cases with the permeable plane.

For the three cases in Table 1 with saturation and non-equilibrium conditions at the end of the wave, calculations have also been performed by means of the model (7). They delivered exactly the same data (e.g. fragmentation grade, reaction length, maximum pressures, etc.) by use of the shock front pressure from the transient model. The stability analysis has however still to be done. While Sharon and Bankoff excluded in (3) stable cases with non-equilibrium conditions and supercritical end pressures because of an apparent lack of restoring forces, they considered the non-equilibrium cases with subcritical pressures to be stable because of restoring forces due to the influence of the vapor blanket around the drops (and fragments) on the heat transfer. A small decrease in vapor blanket thickness, due to a small increase in pressure, would increase the heat transfer and thus tend to restore the vapor film. Although this special phenomenon is not described in the present model, the used procedure leads to similar effects. E.g. local overproduction of vapor, caused by additional heat release, e.g. due to an arbitrary increase of the length of the reaction zone, leads to a faster propagation of the vapor front towards the shock front with a subsequent decrease of the reaction zone. This behavior produces for the most investigated lower values of E automatically such steady state cases, shown in Figure 1, which also remain very stable for the entire investigation time. However, this fluctuation behavior may in other cases lead to larger oscillations as observed for some intervals of the values of E.

## CONCLUSIONS

For a given corium-water mixture and a fragmentation behavior delivered by a hydrodynamic fragmentation model escalation to a pattern of steady state cases was obtained, depending on the trigger strength. While several cases were found with saturation conditions at the end of the wave, only a single one resulted with supercritical end states.

Although for most of the values of the ignition power there existed a clear tendency concerning the escalation and steady state behavior, for some intervals of E no systematic results could be obtained. Especially for the lower values of E different steady state cases are reached in an alternative way, apparently influenced by arbitrary disturbances. It is not clear yet whether this behavior was primarily caused by physical or numerical instabilities. Once however the steady state case with velocity equilibrium and saturation conditions at the end of the wave was obtained, the consistency with the corresponding case of the steady state model (7) was excellent (see e.g. results in (10)).

Above a certain value of E only a single steady state with velocity equilibrium and a significantly higher fragmentation and pressure build-up exists. In this case a pressure at the end of the wave occured, which remained supercritical throughout the pressurized region onto the wall. The corresponding steady state case as calculated from the steady state model lay in the same range. The total ignition energy needed for the escalation to the supercritical steady state was 10.55 MJ/m$^2$ which could be delivered by the fragmentation of 12.1 % of the melt mass contained in the layer of 7.5 cm length near the wall. If a relevant cross section under reactor conditions of e.g. 1 m$^2$ for larger interactions is taken, then 7.5 kg melt mass should fragment in the initial phase of the interaction due to thermal mechanisms. Since during the coarse mixing local events such as bubble collapse, entrapment, etc. are likely to occur, the question is whether these events could lead to the relatively coherent reaction of such an amount of the melt mass required to ignite the detonation or not. Another severe restriction for the occurence of such cases in a reactor pressure vessel after a hypothetical core melt down accident seems to be the needed escalation length of about 2 meters. Additional restrictions should result from the inhomogeneities of the mixture which are expected to weaken the wave strongly and also from a more probable spherical or cylindrical than the investigated plane propagation behavior. In addition, conservative approaches are still included, e.g. an instantaneous heat release from the fragments is assumed, the existence of inert gas in the mixture is not considered, etc.

Since the option of considering a finite heat transfer is already included in the code, this conservativity can however be released in further investigations. The effect of inert gas on the pressure development within the wave could also be taken into account by some extension of the multi-phase description.

Although the case with ignition at a solid wall seems to deliver the lowest threshold, further investigations with other boundary conditions should be carried out in order to get a more profound understanding of the effects. Although the consistency of the steady state and the transient model was proved in (10) to be excellent for the lower pressure cases with equilibration and saturation conditions at the end of the wave, further improvement is necessary concerning the stability criterion in the steady state model for the supercritical cases and nonequilibrium subcritical cases. Finally, instead of the trigger variation, the thermal fragmentation model (6) can be applied alternatively in order to investigate the initial phase of the escalation.

Since the transient model delivers stable steady state cases without any additional condition, but only on the basis of the physical dynamics included in the multiphase description, a useful tool is established which can be applied to verify the current stability considerations and possibly answer the open questions posed e.g. in (3), (4) and (8). The present transient calculations prooved the existence of stable non-equilibrium detonation cases under subcritical conditions, as predicted in (3). Fluctuations and further escalation however hindered the establishment of the corresponding equilibrium case calculated by the steady state model. This indicates possibly a limited stability of this case. On the other hand a stable supercritical case was obtained in contrast to predictions of Bankoff and Sharon (3), (8), while their prediction of the non-existence of supercritical non-equilibrium cases could be affirmed.

Stating that the models based on the theory of thermal detonation have reached an advanced degree of maturity, a validation of the theory should be carried out by means of comparisons with experiments of sufficient large scale. Corresponding requirements can be derived from calculations with the transient detonation code. An appropriate instrumentation of the experiments is of course indispensable, especially the registration of the pressure history by a series of pressure transducers in the propagation direction of the detonation wave, in order to enable a detailed comparison between theoretical and experimental results. In case of a satisfactory prediction of the experimental results by means of the models, the risk probability due to a vapor explosion could be calculated without the current uncertainty in the value of the energy release.

ACKNOWLEDGEMENTS

This report is based on a project, which is sponsored by the Bundesministerium für Forschung und Technologie (Project No. BMFT 1500 639 0). The authors however are responsible for its scientific content.

The part of this work concerning the fragmentation modeling was sponsored by the Joint Research Center, Ispra, Italy.

LITERATURE CITED

1. Unger, H., et al.,"The Role of Steam Vapor Explosions During Core Meltdown of LWR's", Int. Mtg. on Thermal Nuclear Safety, NUREG/CP-0027, Vol. 2, 1357 (1982).

2. Carachalios, C., M. Bürger and H. Unger, "A Transient Two-Phase Model to Describe Thermal Detonations Based on Hydrodynamic Fragmentation", Int. Mtg. on LWR Sev. Acc. Eval., Cambridge, Mass., USA (1983).

3. Sharon, A. and S.G. Bankoff, Int. J. Heat Mass Transfer, 24 (10), 1561 (1981).

4. Condiff, D.W., Int. J. Heat Mass Transfer, 25 (1), 87 (1982).

5. Bürger, M. et al., "Two-Phase Description of Hydrodynamic Fragmentation Processes within Thermal Detonation Waves", 21st Nat. Heat Transf. Conf., HTD-Vol. 23, 45 (1981).

6. Corradini, M.L.,"Analysis and Modelling of Steam Explosion Experiments", SAND 80-2131, Sandia Laboratories, Albuquerque, N.M. 87185, USA (April 1981).

7. Schwalbe, W., et al., "Ein stationäres thermisches Detonationsmodell zur Beschreibung von Dampfexplosionen mit großen beteiligten Schmelze- und Kühlmittelmengen", 1. Techn. Fachbericht BMFT 150 371, IKE, Univ. Stuttgart (Sept. 1980).

8. Bankoff, S.G. and A. Sharon, "Modelling of Steady State Plane Thermal Detonation", ANS/ENS Int. Meeting on Fast Reactor Safety Technology, Seattle, Wash. (1979).

9. Mitchell, D.E., M.L. Corradini and W.W. Tarbell, "Intermediate Scale Steam Explosion Phenomena: Experiments and Analysis", NUREG/CR-2145 (1981).

10. Carachalios, C., M. Bürger and H. Unger, "Untersuchung des Trigger- und Eskalationsverhaltens von thermischen Detonationswellen mit Hilfe eines transienten Modells", Jahrestagung Kerntechnik, München, FRG (21-23 May, 1985).

TABLES

Table 1: Data for the various steady state cases calculated by the transient model (+ results of the steady state model (7) for comparison).

| Cases | Shock front pressure/MPa | Reaction length/cm | Shock front velocity/(m/s) | Reaction time/$\mu$s | End pressure /MPa | Fragmentation grade % | Maximum pressure/MPa |
|---|---|---|---|---|---|---|---|
| 1,2 | 5.6 | 1.9 | 165. | 115. | 4. | 9. | 5.8 |
| 3,4 | 12.1 | 2.6 | 245. | 106. | 6.6 | 10.7 | 13.3 |
| 5,6 | 29.2 | 3.5 | 380. | 92. | 9.2 | 12.3 | 30.3 |
| + | 55. | 4.4 | 513. | 86. | 10.2 | 13.2 | 56.8 |
| 7 | 1045. | 21. | 1963. | 107. | 1380. | 77. | 1750. |
| + | 910. | 16. | 1833. | 87. | 1185. | 68. | 1681. |

FIGURES

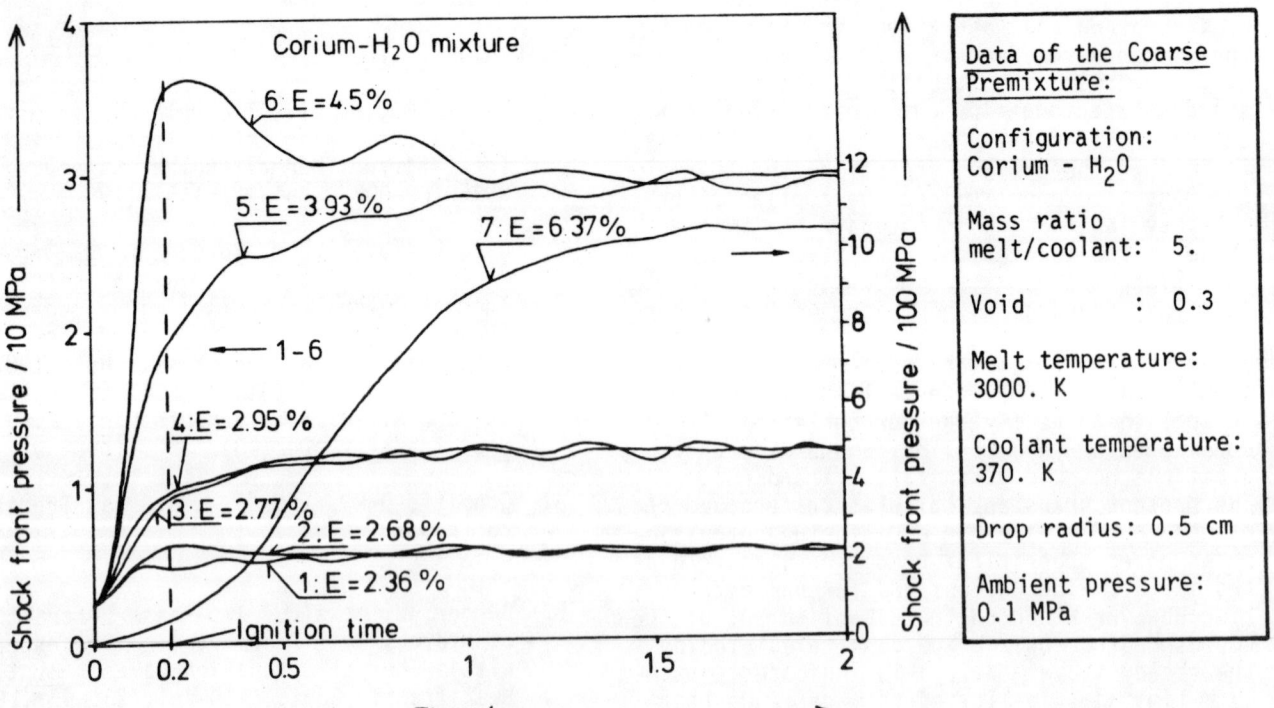

Figure 1: Histories of the shock front pressure calculated by the transient model for different values of E (expressed in percent of the initially fragmented melt mass).

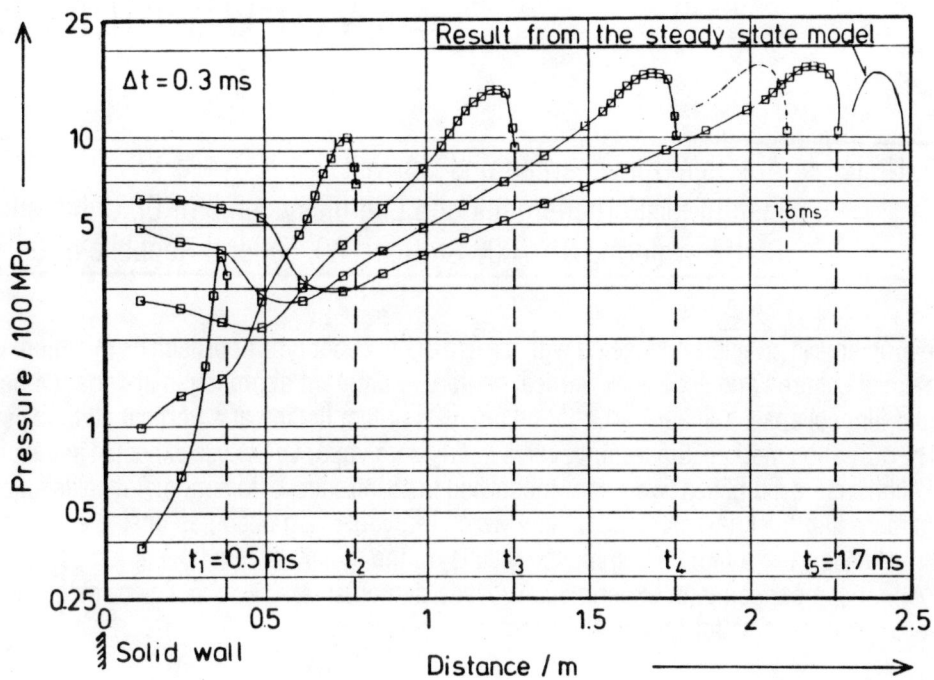

Figure 2: Pressure development for an example case with E = 6.37, as calculated by the transient model; comparison with the pressure field calculated by the steady state model (7).

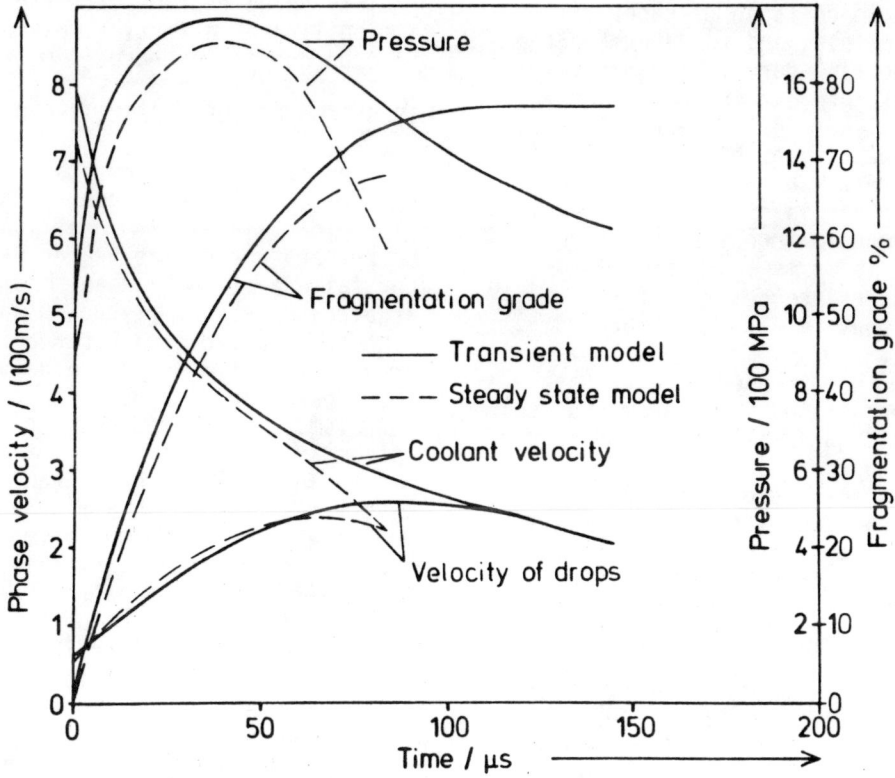

Figure 3: Histories of the pressure, the fragmentation grade and the phase velocities calculated by the transient as well as the steady state model at a fixed place in the labor system (time = 0 corresponds to the arrival of the shock front).

# ON THE PREDICTION OF MINIMUM VAPOR FILM BOILING

M. Bürger and H. Unger ■ Abt. Reaktorsicherheit und Umwelt
Institut für Kernenergetik und Energiesysteme (IKE), University of Stuttgart
Pfaffenwaldring 31, 7000 Stuttgart 80, Federal Republic of Germany

Because of some problems in principle connected with the use of the Zuber type approach on minimum vapor film boiling for geometries like small spheres and especially vertical heaters, a different approach is undertaken here. In analogy to analyses of the liquid film collapse, a linear perturbation analysis of film boiling at a vertical plate is performed, using a steady-state small perturbation method. Since, however, a nearly zero vapor velocity is expected at the phase boundary in contrast to the liquid film case, a simplified two-phase model had to be introduced. Minimum Reynolds numbers of the vapor flow as well as minimum film thicknesses result, below which the vapor film becomes unstable. By comparison with a characteristic film thickness from a film boiling model, results on the minimum film boiling temperature are derived for spherical heaters. A first comparison with experimental results demonstrates some deficiencies of the present state of modeling but also promising aspects for further development.

## INTRODUCTION

Various approaches have been tried in the past to analyse the collapse of vapor film boiling on hot surfaces and to determine the minimum of the boiling curve. This problem attracted special interest in the frame of post-accident analyses for nuclear reactors, especially in rewetting problems and in the vapor explosion research, however is also practically important in other fields, e.g. in the frame of cryogenic technics and in metallurgy. Moreover, it forms a key problem in understanding the existence of different boiling regimes and thus is also of interest in principle.

In spite of the extended work on this problem the uncertainties in experimental and theoretical results are still large. Most of the theoretical approaches are essentially based on the classical theory of Zuber [1]. This theory, which assumes hydrodynamic effects as decisive for vapor film collapse, was originally developed for the case of vapor film boiling on a horizontal plate and then extended to horizontal cylinders [2] and to spheres [3]. Correlations based on the concept of Zuber are often also used for other geometrical configurations, such as vertical plates or cylinders. However, since in this concept the existence of a Taylor unstable liquid-vapor configuration plays a dominant role, its application to small spheres becomes already problematic and must finally be dropped totally for vertical heaters. Thus, at least for these geometrical situations different explanations must be searched. This should also help to get more insight into the general problem. As an extreme case the stability of film boiling at a vertical plate shall therefore be considered here.

## DISCUSSION OF THE STABILITY PROBLEM

The theory of Zuber states that vapor film boiling will collapse catastrophically below a certain minimum heat flux resp. minimum temperature of the heater due to an inherent failure of the mechanism of film boiling. Large perturbations producing contacts with the heater surface are not required for destroying the film boiling mode at this point. This idea is pursued here, in that instability mechanisms without recourse to large perturbations are searched.

As in the Zuber model the instability mechanism must establish a minimum vapor production rate resp. a minimum film thickness, below which the film becomes unstable. Correspondingly a minimum velocity of the vapor flow must exist. Similar problems constitute the transition between slug flow and annular flow in vertical pipes and the transition between isolated bubbles and continuous vapor columns in nucleate boiling. While a minimum rate of vapor production and bubble formation is necessary to establish continuous vapor columns by coalescence of subsequent bubbles resp. vapor slugs, collapse of the columns should be expected below a certain Reynolds number of the vapor flow. Surface tension has a collapsing effect for these fluid columns as can be seen from simple surface energy considerations. The vapor flow should then have a stabilizing influence. However, usual theoretical analyses of such jet or parallel flows only show destabilizing effects of velocity on the fluid interface, e.g. by shear flow effects or Kelvin-Helmholtz instabilities. From the above considerations a minimum as well as a maximum velocity of interface stability should however establish.

The same is true for liquid films flowing down an inclined or vertical plate. In this case, which corresponds most closely to the film boiling problem, a collapsing effect of surface tension is however not obvious. In works on the breakdown of such liquid films below a minimum flow rate, the stability of assumed dry patches is analysed by means of force balance or energy considerations (see e.g. (4), Part II). In principle, surface tension leads to breakdown, while the flow stabilizes the film. However, these analyses can say nothing about the mechanism of appearance of dry patches. Thus, the question is whether the film breakdown can be obtained within a linear stability theory or whether it is only caused by nonlinear effects.

The classical form of perturbation used in linear stability analyses of parallel flows is a time-dependent, spatially periodic one in the directions in which the fluid extends to infinity (x-direction in Figure 1 for the present case). In contrast to this time-dependent instability, also a spatial instability analysis is however used in the problems of breakup of liquid jets and films (see e.g. (4),(5)), based on the observation of waves growing spatially and not in time. In this case, the wave number is taken to be complex, thus allowing spatial growth of the wave amplitude while the wave speed is considered to be real. Both types of perturbations give however only interface instabilities above a certain Reynolds number of the flow.

Therefore, Anshus and Ruckenstein (6) introduced another, purely spatial type of perturbation:

$$f(x,y,z) = f^{\dagger}(y) e^{ik_z z} e^{\varkappa x} \qquad (1)$$

which in fact yielded an instability of the liquid film below a certain Reynolds number. In spite of a severe discussion on the validity of this type of perturbation (see (7),(8)), it shall be applied here for an approach to analyse vapor film collapse at a vertical plate.

## LINEAR STABILITY ANALYSIS OF VAPOR FILM BOILING AT A VERTICAL PLATE

A first approach to the present problem has already been performed in (9). However, a deficiency in principle must be noted. Like in the liquid film case, a slip condition had been used for the vapor velocity at the phase boundary. Actually however, in the considered case of water as coolant, it should be expected that the vapor flow will adopt nearly zero velocity at the interface (nonslip condition). This boundary condition leads however to a failure of the perturbation analysis, since the coupling between the perturbation velocity perpendicular to the interface and the deflection of the interface is dropped hereby. A finite - although arbitrarily small - velocity at the interface must therefore be considered. For an exact solution of the problem the perturbation analysis has therefore to be applied to the complete two-phase description of vapor film boiling. A simplified two-phase model of the flow is in a first step chosen as follows.

### Simplified Two-Phase Model

Considering the undisturbed, steady state of film boiling at a vertical plate, it is assumed for simplification that the velocity component of the vapor flow in y-direction may be neglected compared to the one in x- direction (see Figure 1). Correspondingly, only small changes of the film thickness $\delta$ with x are admitted. Assuming incompressible fluid flow, fixed walls with nonslip conditions in y = $\delta$ and y = $-\ell$ according to Figure 1b as well as equal velocities and shear stresses in the interface y = 0, the velocity profiles in the vapor and the liquid flow result in dimensionless form as follows:

$$\widetilde{U}_d = -\widetilde{\omega}\widetilde{y}^2 + \widetilde{\omega}\chi\widetilde{y} + \widetilde{\omega}(1-\chi), \qquad (2)$$

$$\widetilde{U}_f = \gamma\widetilde{\omega}\chi\widetilde{y} + \widetilde{\omega}(1-\chi), \qquad (3)$$

where

$$\widetilde{\omega} = (\tfrac{2}{3} - \tfrac{\chi}{2})^{-1}, \quad \chi = (1+\gamma\widetilde{\ell})^{-1},$$

$$\gamma = \frac{\mu_d}{\mu_f}, \quad \widetilde{\ell} = \frac{\ell}{\delta}. \qquad (4)$$

The nonslip, resp. slip case as boundary conditions at the interface result for $\gamma = 0$, resp. $\gamma \to \infty$. Within an exact description, the boundary condition in y = $-\ell$ must be replaced by $U_f \to 0$ for $y \to -\infty$. This is however not possible under the above simplifying assumptions leading to a linear velocity profile in the liquid. In the present approach, the distance $\ell$, left undetermined, defines a solid boundary to the liquid flow. The film thickness $\delta(x)$ derives finally from an energy balance at the interface $y = 0$, where convective heat transfer to a subcooled liquid is taken into account.

## Linear Perturbation Analysis

By means of including the perturbed quantities $\{u,v,w,p\} = \{U+u_s, v_s, w_s, P+p_s\}$ - according to the spatial perturbation approach - in the steady-state Navier-Stokes and continuity equations and linearizing, the equation set for the perturbation quantities $\{u_s, v_s, w_s, p_s\}$ results. Since the instability is assumed to be essentially caused by hydrodynamic effects, temperature fluctuations are not considered. For the perturbation quantities time-independent expressions of the type given in Equation (1), which can grow or decay in downstream-(x)-direction, are now introduced. The wave part $\exp(iK_z z)$ in z-direction describing the formation of vertical ripples on the interface is however neglected here, since it showed only small effects on the stability results in (9).

After inclusion in the equation set and elimination of $p^+$ and $u^+$, perturbation equations of Orr-Sommerfeld type in the perturbation amplitude $v^+$ of the velocity component perpendicular to the plate result for the vapor as well as the liquid region:

$$(\tilde{D}^2 + \tilde{\varkappa}^2)^2 \tilde{v}^+ = \gamma^* \tilde{\varkappa} \mathrm{Re}\,[\tilde{U}(\tilde{D}^2 + \tilde{\varkappa}^2)\tilde{v}^+ - (\tilde{D}^2 \tilde{U})\tilde{v}^+]. \tag{5}$$

In case of vapor $\gamma_d^* = 1$, in case of liquid $\gamma_f^* = \nu_d/\nu_f$ have to be chosen together with the corresponding expressions for the basic flow, Equations (2), (3).

The boundary conditions for the perturbation equations, obtained from the same procedure, are finally given in non-dimensional form by

$$\underline{\tilde{y} = 1}: \quad \tilde{v}_d^+ = \tilde{D}\tilde{v}_d^+ = 0 ; \tag{6}$$

$$\underline{\tilde{y} = -\tilde{\ell}}: \quad \tilde{v}_f^+ = \tilde{D}\tilde{v}_f^+ = 0 ; \tag{7}$$

$$\underline{\tilde{y} = 0}: \quad \tilde{v}_f^+ = \tilde{v}_d^+ , \tag{8}$$

$$\tilde{D}\tilde{v}_f^+ - \tilde{D}\tilde{v}_d^+ - (\gamma-1)\frac{\tilde{D}\tilde{U}_d}{\tilde{U}_d}\tilde{v}_d^+ = 0 , \tag{9}$$

$$\tilde{D}^2\tilde{v}_f^+ - \gamma \tilde{D}^2\tilde{v}_d^+ + [\tilde{\varkappa}^2(\gamma-1) + \gamma\frac{\tilde{D}^2\tilde{U}_d}{\tilde{U}_d}]\tilde{v}_d^+ = 0 , \tag{10}$$

$$\tilde{D}^3\tilde{v}_d^+ - \frac{1}{\gamma}\tilde{D}^3\tilde{v}_f^+ + [3\tilde{\varkappa}^2(1-\frac{1}{\gamma}) - (1-\frac{\gamma_f^*}{\gamma})\tilde{\varkappa}\,\mathrm{Re}\,\tilde{U}_d]\tilde{D}\tilde{v}_d^+$$
$$+ [(1-\frac{\gamma_f^*}{\gamma})\tilde{\varkappa}\,\mathrm{Re}\,\tilde{D}\tilde{U}_d + \frac{\xi\tilde{\varkappa}^3}{\tilde{U}_d} - 3(1-\frac{1}{\gamma})\frac{\tilde{D}\tilde{U}_d}{\tilde{U}_d}\tilde{\varkappa}^2]\tilde{v}_d^+ = 0. \tag{11}$$

Equations (6) and (7) design the vanishing of the velocity perturbations at the plate as well as at the right-hand boundary in the liquid. The kinematic conditions at the interface are given by Equations (8), (9), where Equation (8) expresses the equality of normal fluid velocities and Equation (9) the connection between interface deflection and fluid motion. Here, as in the other boundary conditions, the amplitude of surface deflection $\eta^+$ has been replaced due to the kinematic condition $v^+ = \varkappa U_d \eta^+$. The collapse of the procedure for nonslip conditions, that is for $U_d = 0$ in $y = 0$, can be seen from this equation, connecting the normal perturbation velocity to the surface deflection. Equations (10) and (11) finally are derived from the balance of tangential and normal stresses. As an additional dimensionless quantity

$$\xi = \frac{\sigma}{\mu_d \tilde{U}_d} = \mathrm{Re}^{\frac{2}{3}} \zeta \tag{12}$$

with the parameter $\zeta$ of surface tension:

$$\zeta = \sigma / [\rho_d^2 (\rho_f - \rho_d) \frac{1}{3} v_d^4 g (1 - \frac{3}{4} \cdot \frac{1}{1+\gamma\tilde{\ell}})]^{\frac{1}{3}} \tag{13}$$

was introduced.

In the sense of the arguments of Ruckenstein and Anshus (8), solutions of the eigenvalue-problem with eigenvalues $\tilde{\varkappa}_R > 0$, indicating instability of the film, have now to be searched in dependence of the parameters of the problem. These are especially the Reynolds number Re, the parameter $\zeta$ of surface tension and the viscosity ratios $\gamma$ and $\gamma_f^*$. As follows, an approximate analytical solution of the problem for small Re will be performed since film collapse is expected to occur for small Re. The solution procedure follows the method of Anshus and Ruckenstein (6).

## Approximate Analytical Solution

Replacing $\tilde{U}$ in the first right-hand term of Equation (5) by a constant value, the differential equation has only constant coefficients and thus can be solved analytically. This procedure can be based on the assumption that surface processes are governing the instability. However, in the present case this would mean to take practically a zero velocity, which seems not to be adequate to the problem if considering the above discusssion on the stabilizing effect of the flow. Indeed, as will be seen, instability would result in this case for arbitrary film thicknesses.

Thus, the choice of a fixed value for $\tilde{U}$ is here not as straight-forward as in the case of liquid films, for which the slip condition is appropriate, delivering a maximum value of $\tilde{U}$ at the interface. Nevertheless, a constant value is also chosen here, e.g. given by the maximum or mean value of the vapor velocity. For the liquid however the interface value is taken.

The general solution can be composed of the fundamental system of four solutions. The four free integration constants have to be determined from the boundary conditions (6) to (11). The homogeneous boundary condition problem however only leads to a non-trivial solution for the integration constants, if the determinant of their coefficients vanishes. This constitutes an eigenvalue problem for the amplification factor $\tilde{\varkappa}$ in dependence of the system parameters.

Since breakdown of the film is expected for small Reynolds numbers, an approximate solution procedure is performed according to the procedure in (6). After the transformation

$$\tilde{y}^* = \xi \tilde{y} , \quad \tilde{\varkappa}^* = \frac{\tilde{\varkappa}}{\xi} , \quad (14)$$

different terms in the perturbation equation (5) as well as in the boundary conditions (6) to (11) can be neglected by scaling with the large quantity $\xi$ (for small Re). As approximate solution of the eigenvalue problem results:

$$\tilde{\varkappa} = \frac{\gamma^2}{(\gamma-1)^2} \cdot \frac{\phi_d}{\tilde{\omega}(1-\chi)} \zeta Re^{\frac{2}{3}} \frac{Re}{8} \cdot \frac{\sin(0.5\phi_d Re)}{1-\cos(0.5\phi_d Re)} \cdot (15)$$

While this expression becomes zero in the nonslip case ($\gamma = 0$), an exponential increase of the perturbation occurs for an arbitrarily small, but finite value of $\gamma$. This happens, if the Reynolds number becomes smaller than the critical value $2\pi/\phi_d$. In this case, the sign of $\tilde{\varkappa}$ changes from − to +, which means a transition from stability to instability. For the slip case ($\gamma \to \infty$), the analogous result to that of Anshus and Ruckenstein (6) for breakdown of liquid films derives from Equation (15), choosing $\phi_d = \tilde{U}|_0$.

Finally, with the expressions for the Reynolds number and the mean vapor velocity, a minimum film thickness can be derived, below which instability occurs:

$$\delta_{min} = (\frac{6\pi}{\phi_d})^{\frac{1}{3}} [\rho_d(\rho_f-\rho_d)g(1-\frac{3}{4}\chi)]^{-\frac{1}{3}} \mu_d^{\frac{2}{3}} . \quad (16)$$

Choosing $\phi_d = \tilde{U}|_0$ with $\chi \to 1$ resp. $\gamma \to 0$ (nonslip), the critical film thickness goes to infinity: $\delta_{min} \to \infty$ (and correspondingly also the Reynolds number: $Re_{min} \to \infty$), i.e. the film becomes unstable for all film thicknesses. While the indifferent stability behavior for choosing exactly $\gamma = 0$ can be explained by the decoupling between the surface deflection and the velocity perturbation, the behavior for $\chi \to 1$, resp. $\gamma \to 0$ is explainable from the supposed stabilizing effect of the vapor flow discussed above. This effect must vanish as $\phi_d \to 0$. However, following the above arguments, choosing a maximum or a mean value of the vapor velocity for $\phi_d$ should be more appropriate in this case. With the mean value the following expression for the minimum vapor film thickness results:

$$\delta_{min} = (C\pi)^{\frac{1}{3}} (\frac{\mu_d^2}{g\rho_d(\rho_f-\rho_d)})^{\frac{1}{3}} \quad (17)$$

with C = 24 in the nonslip case (to be understood here as $\gamma \approx 0$, $\chi \approx 1$) and C = 6 in the slip case.

Before in the next chapter first results will be discussed, a curious point must still be mentioned. Although the final reason of the film collapse is thought to be the effect of surface tension, the critical Reynolds number as well as the critical film thickness do not depend on surface tension. On the other hand, the growth factor is proportional to the surface tension and thus indifferent stability would exist for zero surface tension. Further analyses, especially a more complete solution by numerical methods, must therefore be performed. It may however also be concluded that a certain growth strength should be taken as a criterion to determine the transition point. Such a procedure was chosen by Chung and Bankoff ((4), Part I) in the liquid film problem.

## RESULTS

The results on a minimum stable film thickness derived here can be used for determining a critical or minimum point of vapor film boiling. However, in the present simplified approach, using the assumption of parallel flow (V≪U), a region with small film thicknesses at the lower region of the heater must be excluded. Thus, heuristic approaches are necessary for determining the region which decides on film collapse.

Such an approach was tried for spherical heaters. The location determining the stability was assumed to be the equator of the sphere ($\gtrsim 90°$ from the lowest point). At this location, the situation of the vertical plate model is given approximately, thus justifying the use of the present results. On the other hand, a lower or upper region may of course actually decide on stability. However, stabilizing effects of evaporation and gravity should dominate at least near the lowest point. At this point the vapor flow parallel to the surface becomes zero, and only the radial evaporation flow exists. Thus, the present model is not applicable at this point.

For a sphere diameter of 1 cm, results of the present approach are shown in Figure 2, depending on the subcooling of the water (the film boiling model used is similar to that of (10); see (11)). For comparison, experimental correlations of different authors are included, showing a large scatter. Nevertheless, the temperatures from the more realistic nonslip case are obviously too high. However, from the theory an increase of $T_{w,min}$ with decreasing radius results which must be taken into account in a future, more detailed comparison. Besides the strong scattering in experiment, one has also to take into account that the theoretical results are rather sensitive with respect to the film thickness. Thus, a film boiling model based on an exact two-phase description should be used. The tendencies however agree with the experimental findings for the dependence on subcooling as well as - with less assurance - for the radius of the sphere.

Finally, the question arises whether the instability point derived from the linear perturbation theory must really coincide with the minimum of vapor film boiling. A short discussion on this point will follow in the next chapter, which allows to classify the present type of modeling of vapor film collapse.

## VAPOR FILM INSTABILITY AND THE PROBLEM OF MINIMUM FILM BOILING

The present approach deals with the instability of a vapor film, with respect to arbitrarily small perturbations. However, nonlinear effects may finally stabilize it again, e.g. by increased heat transfer as the interface approaches the heater surface, perhaps in connection with rapid vapor production at short-time contacts. Under these conditions the vapor film, although not collapsing, must appear unstable and wavy since the stability conditions with respect to small perturbations are not fulfilled. This agrees with experimental observations under steady-state conditions, from which e.g. Stephan (15) concludes the existence of stable - although wavy - film boiling below the minimum surface temperature. Corresponding ideas were also put forward by Ruckenstein (16), assuming the onset of film boiling due to a bubble coalescence criterion, even though a stability condition of the film may not be fulfilled. Higher values of the critical temperature of instability than experimental minimum temperatures may be explained hereby. This may be more valid for saturation conditions than for subcooled coolant. In the latter case, smooth film boiling without waviness is usally observed before catastrophic collapses occur in coincidence with first contacts with the surface. This may indicate that in contrast to the saturation case a linearly unstable film cannot be stabilized due to coalescence conditions.

On the other hand, this behavior also indicates that the film may remain stable in some range due to linear stability, although the coalescence condition is not fulfilled. Under such conditions the film must collapse catastrophically at the first contact which may occur due to large perturbations. Film boiling would in this case be metastable and could only be maintained by decreasing the temperature under avoiding disturbances.

The hysteresis behavior may be described in the general Landau picture of phase-transition like phenomena (see e.g. (17)), as can be seen from the scheme in Figure 3. As order parameter $\varepsilon$, a measure of the deviation from film boiling may be chosen, such as an amplitude of interface deflections and finally the time-average contact surface. Then the first scheme (a) shows the stable film boiling state for $\varepsilon = 0$ at the minimum of the potential function. Decreasing the temperature, as the external parameter of the system, a second minimum may develop. This corresponds to the behavior of a first-order phase transition, producing the possibility of metastable states. The corresponding situation can be seen in the third scheme (c), where the second minimum representing nucleate boiling has already developed and possesses a lower potential, thus becoming the more stable case. Film boiling is then metastable; larger fluctuations may be necessary to destroy it. This situation corresponds to the case in which a coalescence condition is no longer fulfilled. In Scheme d the limit of metastability is reached. The present approach of linear stability analysis corresponds to this situation, as does in principle also the Zuber type of analysis. Other types of analysis, introducing stability considerations of the collapsed state, have already been mentioned for the liquid film case. While the force balance analysis would correspond to the situation of Scheme b in Figure 3, Scheme c illustrates the energy considerations in rivulet theories.

## CONCLUSIONS

Various situations exist for which the classical minimum film boiling theory of Zuber becomes problematic respectively is not applicable in principle. Therefore a new approach to the problem is tried for the most critical case of a vertical plate, using methods known from the problem of breakup of falling liquid films. A simplified two-phase model had to be introduced for a realistic treatment of the conditions at the phase boundary. The approximate analytical solution gives a minimum Reynolds number of vapor flow, resp. a minimum film thickness. By comparison with a characteristic film thickness from the film boiling model, a critical film boiling temperature can be derived. However, heuristic approaches for the characteristic film thickness had to be made due to some rough approximations at the present state of modeling. A more complete solution, using a full two-phase description, must therefore be performed in the future. Because of these deficiencies of the present modeling and also possibly because of rough approximations in the analytical solution procedure, discrepancies in the comparison with experimental results can be understood. However, also promising results must be noted, such as the size range of the critical film thicknesses as well as reasonable tendencies with different system parameters.

## ACKNOWLEDGEMENT

This report is based on a project, which was sponsored by the Bundesministerium für Forschung und Technologie (Project No. BMFT 150 371). The authors however are responsible for its scientific content.

## NOTATION

| | |
|---|---|
| c | phase velocity of the perturbation wave |
| $\tilde{D}$ | dimensionless differential, $\tilde{D} = d/d\tilde{y}$ |
| g | constant of gravity |
| k | wave number |
| $\ell$ | length (see Fig. 1) |
| P | pressure in basic flow |
| p | pressure |
| R | radius |
| Re | Reynolds number, $Re = \bar{U}_d \delta / \nu_d$ |
| T | temperature |
| U,V | velocity components of basic flow (Fig.1) |
| u,v,w | velocity components in x,y,z-direction |
| x,y,z | coordinates of space (Fig. 1) |
| $\gamma$ | ratio of dynamic viscosities, $\gamma = \mu_d/\mu_f$ |
| $\gamma^*$ | factor ($\gamma^* = \nu_d/\nu_f$) |
| $\delta$ | film thickness |
| $\zeta$ | dimensionless quantity, Eq. (13) |
| $\eta$ | deflection of interface |
| $\varkappa$ | growth factor |
| $\mu$ | dynamic viscosity |
| $\nu$ | kinematic viscosity |
| $\xi$ | dimensionless quantity, $\xi = Re^{-2/3} \cdot \zeta$ |
| $\rho$ | density |
| $\sigma$ | surface tension |
| $\phi$ | constant, replacing $\tilde{U}$ in Eq. (5) |
| $\chi$ | dimensionless quantity, $\chi = (1+\gamma\ell)^{-1}$ |
| $\omega$ | factor, $\omega = g(\rho_f - \rho_d)/2\mu_d$ |

SUBSCRIPTS

| | |
|---|---|
| d | vapor |
| f | liquid |
| min | minimum, critical value |
| R | real part |
| s | perturbation quantity |
| W | wall |

SUPERSCRIPTS

| | |
|---|---|
| ~ | dimensionless quantity, lengths related to $\delta$, velocities related to $\bar{U}_d$ |
| — | mean value over cross-section |
| + | amplitude of perturbation quantities |

## LITERATURE CITED

1. Zuber, N., "Hydrodynamic Aspects of Boiling Heat Transfer", AEC Report No. AECU-4439, Physics and Mathematics (1959).

2. Lienhard, J.H. and P.T.Y. Wong, Heat Transfer, Trans. ASME, Ser. C, 86 (2), 220 (1964).

3. Gunnerson, F.S. and A.W. Cronenberg, "A Prediction of the Minimum Film Boiling Conditions for Spherical and Horizontal Flat Plate Heaters", ASME 79-HT-45 (1980).

4. Chung, J.C. and S.G. Bankoff, Chem. Eng. Commun., 4, Part I: 433, Part II: 455 (1980).

5. Bogy, D.B., Phys. Fluids, 21 (2), 190 (1978).

6. Anshus, B.E. and E. Ruckenstein, Colloid and Interface Science, 51 (1), 12 (1975).

7. Wilson, S.D.R., . Colloid and Interface Science, 59, 188 (1977).

8. Ruckenstein, E. and B.E. Anshus, J. Colloid and Interface Science, 59, 191 (1977).

9. Sharon, A., M. Bürger and W. Schwalbe, Israel J. of Technology, 19, 147 (1981).

10. Frederking, T.H.K. and J.A. Clark, Adv.

Cryog.Eng., 8, 501 (1963).

11. Kammerer, E., M. Bürger and W. Schwalbe, "Wärmeübergang beim Dampffilmsieden an Kugeln", IKE-Bericht Nr. 2-54, IKE, Univ. Stuttgart (1981).

12. Dhir, V.K. and G.P. Purohit, "Subcooled Film-Boiling Heat Transfer from Spheres", ASME 19-HT-78 (1978).

13. Bradfield, W.S., Heat Transfer, 89, 269 (1967).

14. Benz, R., Bürger, M. and O. Zach, "Vergleich experimenteller Daten zur minimalen Filmsiedetemperatur für Kugeln", IKE-Bericht Nr. 2-41, IKE, Univ. Stuttgart (1977).

15. Stephan, K., "Bubble Formation and Heat Transfer in Natural Convection Boiling", in Heat Transfer in Boiling, Hahne, Grigull (Ed.), Academic Press, New York (1977).

16. Ruckenstein, E., Int. J. Heat Mass Transfer, 10, 911 (1967).

17. Haken, H., "Synergetics - An Introduction", Springer-Verlag, New York (1978).

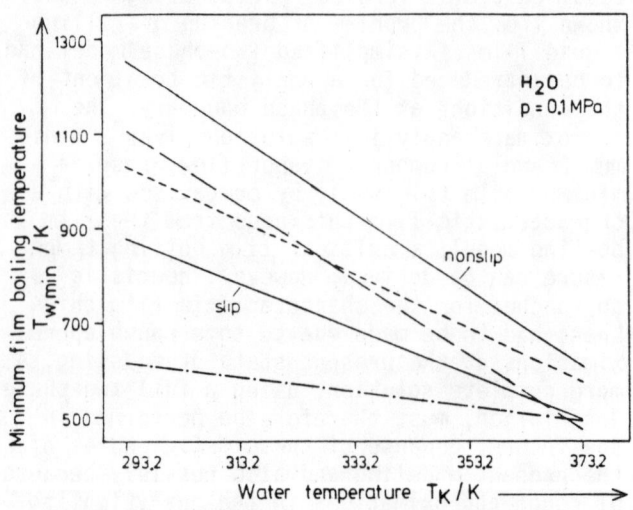

Figure 2: Effect of water temperature on the minimum film boiling temperature of hot spheres in water:
—— Present theory (R = 5 mm).
Experimental correlations:
– – Dhir and Purohit (12) (R = 9.5 and 12.7 mm);
- - - Bradfield (13) (R = 3 mm);
-·- Benz, Bürger and Zach (14) (R = 10 mm).

FIGURES

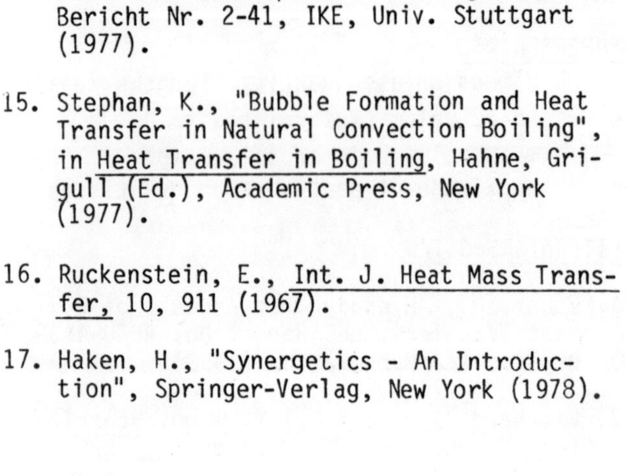

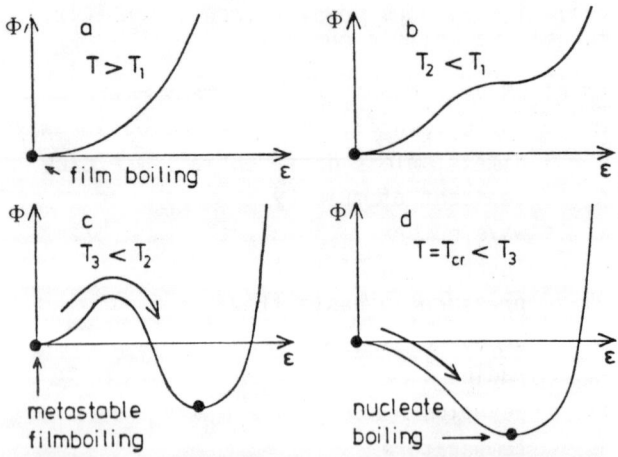

Figure 1: Vapor film boiling at a vertical plate:
a) Scheme of undisturbed basic flow.
b) Scheme of simplified two-phase model.

Figure 3: Scheme on a phase-transition analogy (Landau picture) of the transition between film and nucleate boiling ($\varepsilon$ = order parameter, $\Phi$ = potential, $T_{cr}$ = critical temperature of linear instability).

# ANALYSIS OF INFLUENCE OF STEAM SUPERHEATING ON PACKED BED QUENCH PHENOMENA

T. Ginsberg ■ Brookhaven National Laboratory, Department of Nuclear Energy, Upton, New York 11973

Experimental data suggest that steam produced within a particle bed, which is being quenched by flow from an overlying pool of water, has a strong potential for being superheated. This is a result of heat transfer from the unquenched particulate in the dry channels of the bed. A model is presented for the debris bed quench process, which considers the effects of steam superheat on both the bed heat flux and on the quench front propagation characteristics. Calculations are presented which demonstrate the effects of steam superheat. A preliminary comparison of the model with experimental data from packed bed quench experiments is also presented.

## INTRODUCTION

Light water reactor core meltdown accident sequence studies consider the coolability of superheated debris beds which may exist on the concrete floor beneath the reactor vessel [1]. In some accident sequences water would be present in the region beneath the reactor vessel and the superheated debris would be cooled by an overlying pool of water. The process of debris quenching (reduction of the debris temperature to the water saturation temperature) would be accompanied by steam pressurization of the containment building. Debris bed quenching experiments [2,3] have led to development of analytical models of the quench process which include descriptions of both the steam generation rate and the particle bed temperature distribution [4,5,6]. A model has been implemented in the MEDICI ex-vessel interactions model of the CONTAIN computer code [7]. The analytical models of Refs. 4-7 are based upon the assumption that the steam generated within the bed does not remove heat from the hot debris and exits the bed at its saturation temperature. This paper discusses a model of debris bed quench which considers heat transfer between the steam and the hot debris and allows the steam to be superheated.

BNL debris bed quenching experiments [2] suggest that a superheated debris bed which is cooled by liquid supplied from an overlying pool of water is cooled in a two-stage quench front propagation process, represented schematically in Figure 1. Coolant is postulated to initially penetrate the bed, leaving dry regions of particles. If the bed were internally heated then the particles would continue to heat. Upon arrival of the downward front to the base of the bed, a final upward-directed front propagates up the bed, removing the remaining stored energy. The experiments further indicated that the steam generation rate, proportional to the bed cooling rate, was reasonably constant during the entire period of the quench process. The bed cooling rate was found to be reasonably well predicted by the steady-state Lipinski debris bed model [8]. The reasonable agreement between the bed cooling data and the Lipinski model prediction was taken to imply that the heat removal process is controlled by a countercurrent two-phase flow mechanism, that the limiting flow condition occurs at the top of the bed and that the steam is at its saturation temperature at this location.

The bed quench data [2] indicate that in the unquenched regions of the bed the particles may remain at their initial temperature for extended periods of time. During the extended dry periods, the dry channels are postulated to serve as flow channels through which steam, produced at the quench front, flows upward along a path which leads to the overlying

---

Work performed under the auspices of the U.S. Nuclear Regulatory Commission.

pool of water. Since the steam is generated at the water saturation temperature and the particles would be at some elevated temperature, perhaps close to their initial temperature, the potential exists for heat transfer between the steam and the particles. Steam-particle heat transfer calculations were performed using single-phase particle bed heat transfer data (9). These calculations indicate that, for particles in the millimeter diameter range, the steam would heat to nearly the particle temperature as it flows a distance of only millimeters or centimeters. These calculations strongly suggest that steam produced within a particle bed which is being quenched has a strong potential for being superheated as a result of heat transfer from the unquenched particulate in the dry channels or pockets of the bed.

If the steam is indeed superheated as it leaves the particle bed, then the countercurrent flow mechanism which controls the global cooling rate of the bed and, also, the quench propagation rate, could be affected by the elevated temperature of the steam through the influence of vapor density and viscosity. In addition, the particles in the dry regions of the bed would transfer energy to the steam. Both of these effects were modeled in a modification of the basic theory which was presented earlier (4). The modifications to the prior analysis are presented below ("ANALYSIS" section) along with the results of calculations to study the effects of steam superheating on the debris bed heat removal processes ("MODEL CALCULATIONS..." section). Conclusions based upon the analysis are discussed in the final section of this paper.

ANALYSIS

Effect of Steam Superheating on Debris Bed Heat Flux

Consider the debris bed shown in Figure 1. The heat flux at the quench front is given by

$$q''_{QF} = (\rho_g j_g)_{QF} h_{fg} \quad (1)$$

where $\rho_g$ is the steam density, $h_{fg}$ is the heat of vaporization, both evaluated at the water saturation temperature, and $j_g$ is the steam superficial velocity at the quench front. At the top of the bed it is assumed that the steam exits at temperature $T_g$. The heat flux at the top of the bed is, therefore,

$$q''_{TB} = (\rho_g j_g)_{TB} h_{fg} \left[1 + \frac{c_g(T_g - T_{SAT})}{h_{fg}}\right] \quad (2)$$

where $T_{SAT}$ is the saturation temperature and $c_g$ is the steam specific heat. The quantities $\rho_g$ and $j_g$ are evaluated at the top of the bed. The mass fluxes at the quench front and at the top of the bed are equal, and, therefore,

$$(\rho_g j_g)_{QF} = (\rho_g j_g)_{TB} \quad . \quad (3)$$

In terms of the mass flux at the top of the bed, the heat fluxes at the quench front and at the top of the bed are

$$q''_{QF} = (\rho_g j_g)_{TB} h_{fg} \quad (4)$$

$$q''_{TB} = (\rho_g j_g)_{TB} h_{fg} \left[1 + \frac{c_g(T_g - T_{SAT})}{h_{fg}}\right] \quad . \quad (5)$$

Both $\rho_g$ and $j_g$ are evaluated at temperature of the steam leaving the bed, $T_g$.

The two-phase countercurrent flow conditions at the top of the bed are assumed to limit the bed heat removal rate, since that is the location of maximum volumetric flux of both liquid and vapor. The vapor-liquid, separated-flow formulation proposed by Lipinski was used to characterize the bed heat removal process. It was assumed that all steam properties at the top of the bed are evaluated at the specified steam temperature. Calculation results using this model are presented in the "MODEL CALCULATIONS"... section. The two-phase relative permeabilities used in these calculations were those proposed by Lipinski as a best fit for prediction of the world steady-state debris bed heat flux data. Upon computing the vapor flux at the top of the bed, $(j_g)_{TB}$, the heat flux at the top of the bed is computed using Equation (2). The heat flux at the quench front is computed using Equation (4).

Model for Debris Bed Quench Phenomena Including the Effects of Steam Superheating and Steam-Debris Heat Transfer

As discussed in the Introduction, the data indicate that packed beds of superheated particles are quenched in a two-stage cooling process. Figure 1 is a schematic view of a superheated bed under quench conditions. Water initially penetrates the bed during the initial downward frontal progression period. The process is irregular and leaves channels or pockets of dry particles. It is estimated that approximately 30 to 40% of the initial stored energy is transferred to the water during this time period. During the initial frontal period the bed is assumed to consist of three regions. The uppermost partially quenched region, shown schematically in Figure 1, consists of wetted channels of particles which are quenched to the saturation temperature and channels of unquenched particles close to their initial temperature. A two-phase region follows below. The particles in the wetted channels are not yet quenched to the saturation temperature and the surrounding fluid is composed of steam and water. The bottom-most region is completely dry. No liquid has yet penetrated to the particles in this region. The speed of the downward-progressing front is $v_d$. A final upward-progressing front, moving at speed $v_u$, begins its traverse subsequent to completion of the downward process. During this final upward frontal progression, the remaining stored energy is removed from the particles and the bed is completely quenched and filled with water.

The above description characterizes the observed quench behavior of superheated beds of particles 1- to 6-mm in diameter. Experiments with beds of 12-mm diameter particles suggest a single-stage quench process. The model presented below applies to the two-stage process characteristic of the smaller particle sizes. The experimental data, it is noted, were not adequate to identify the geometry of the dry channels or pockets of particles.

A model based upon the above observations is presented in Ref. 4. The model characterizes the quench of a superheated packed bed of uniform-size particles and uniform porosity which produce heat with internal heat generation $Q'''$. The bed cooling rate was modeled using the Lipinski formulation (8), modified to account for transient liquid storage within the bed which takes place during the quench process. The quench fronts were assumed to propagate at a rate limited by the rate of liquid supply to the front. Steam cooling of the debris and steam superheating were neglected. The model employs an energy equation for the particles that assumes, irrespective of shape of the dry channels or pockets, that 30 to 40% of the particles are quenched as the quench front propagates a unit distance downward. The liquid momentum and continuity equations, on the other hand, allows the liquid and vapor volume fractions at the top of the bed to be computed independently of the 30 to 40% quench fractions.

The basic model presented in Ref. 4 was modified to allow steam superheating and steam cooling of the debris. This was accomplished by writing an energy equation for the unquenched debris above the quench front which accounts for heat transfer between the steam and the debris. The formulation, described in the Appendix, is a lumped-parameter energy equation for the entire volume of unquenched debris above the quench front under the additional assumption that the heat transfer coupling between steam and particles is so efficient that the steam exits the bed at the particle temperature. The complete formulation of the model includes: (i) a coolant mass conservation equation, (ii) a pair of momentum equations, one for liquid and the other for vapor, (iii) quench front propagation equations, (iv) an energy equation for the unquenched particles beneath the downward-progressing quench front and (v) a lumped-parameter energy equation for the particles in the vapor channels above the quench front. The mathematical formulation is summarized in the Appendix.

## MODEL CALCULATIONS AND COMPARISON WITH EXPERIMENTAL DATA

### Demonstration of Effect of Steam Superheat on Debris Bed Cooling Rate

The quench front and debris bed heat fluxes are shown in Figure 2 as a function of steam exit temperature for quench cooling of a superheated bed of porosity 0.4 and height of 1 m. Note that the heat flux at zero steam superheat is approximately that at the intercept. As indicated above, the quench front heat flux is smaller than the overall bed heat flux as a result of the assumption that the steam is superheated as it flows up the channels of unquenched particles. The advance of the quench front would be determined by the quench front heat flux. Note that for small particles the effect of superheat is significant, both for the quench front and the overanl bed heat fluxes. For the large particles the effect of steam superheat is less pronounced.

The effect of steam superheating on the quench front and debris bed heat fluxes can be understood with the aid of Equations (4) and (5) and Figure 3, which shows the behavior of the steam volume flux computed using the Lipinski model with steam superheat. The parameters affecting the heat fluxes are the steam density, the volume flux of steam at the top of the bed and the steam superheat. The steam density is a decreasing function of steam temperature. The dependence of volume flux of steam on temperature comes about through the effects of density and steam viscosity. For beds of particles in the range of millimeter in diameter the volume flux of vapor is a relatively weak function of steam temperature, as shown in Figure 3. For small particles (~1-mm diameter) the vapor flux from the bed decreases with temperature as a result of the dependence on the viscosity, whereas for the large particles (~12-mm diameter) the vapor flux increases with temperature due to the effect of decreasing vapor density. Since both the steam density and volume flux decrease with temperature for small diameter particles, the quench front heat flux also decreases relatively strongly with temperature. For the large particles the steam flux increases with temperature and the combined effect with vapor density is a weaker dependence of quench front heat flux on steam temperature. The heat flux at the top of the bed is influenced by the superheat multiplier given in Equation (2). This leads to the weaker dependence on temperature than computed for the quench front heat flux.

The above calculations indicate that the effect of steam superheat on heat flux during bed quench is a significant one for small particles under conditions of large steam superheat.

## Preliminary Comparison of Bed Quench Model with Experimental Data

The bed quench model described above, which includes the effects of steam superheating on the quench front and debris bed heat fluxes, is compared with experimental observations of bed heat flux and quench front propagation.

The apparatus used in the bed quench experiments has been described elsewhere (2,10). Experiments were performed using packed beds of stainless steel spheres. No internal heating was employed in the experiments. The test vessel, which contained the beds was a Schedule 10 stainless-steel pipe, 1.219-m long, 108.2-mm i.d. with a 3.05-mm wall thickness. The test section was instrumented with thermocouples that penetrated through the wall into the test container. Thermocouples were also mounted on the outer wall of the pipe. A turbine flowmeter was used to monitor the flow of steam during the particle quench process and provided the measurement of bed heat flux. A typical experiment was performed by first heating the particles to the desired temperature in an oven and the water to the saturation temperature. The water was then released from the holding vessel onto the particle bed, thus initiating the quench process. The steam flow rate provided a measure of the bed heat flux and the thermocouple traces were interpreted to provide plots of the frontal cooling behavior (2).

Figures 4 and 5 present a set of data from a set of quench experiments using a bed of 3.18-mm spheres packed to a height of 300 mm. Figure 4 presents the time-averaged bed heat flux from several experiments as a function of initial particle superheat. Figure 5 shows the frontal propagation plot obtained from a single experiment. Note the observation of downward frontal propagation followed by the final upward frontal progression.

Figures 4 and 5 also contain predictions based on the model described above. A central feature of the model is the use of the Lipinski separated flow formulation for the debris bed hydrodynamics and heat flux. This formulation requires specification of the two-phase relative permeability parameters for the porous medium under consideration. The formulation requires that, as indicated in the Appendix, the dependence of each of four parameters on the local vapor fraction within the packed bed be specified. Lipinski has proposed a set of parameters based upon the available steady-state dryout heat flux data. These are called "Lipinski" parameters in the Appendix. A set of parameters can be specified based upon the assumption that the two-phase flow pattern within the debris bed is composed of separate channels (11). This is referred to as the "linear" parameters. Calculations were performed using both of these sets of parameters. It was assumed in these calculations that 40% of the bed stored energy was removed during the downward frontal propagation period.

The comparison of data and predictions shown in Figure 4 indicates that the bed heat flux data lies between predictions based upon the "Lipinski" and "separated flow" parameters. Preliminary analysis of additional data from experiments with beds of 1-mm, 6-mm and 12-mm particles, not shown here, indicate a similar trend. The predictions of bed heat flux shown in Figure 4 indicate a weak dependence on particle superheat within the range of superheat shown in the figure. Within the scatter of the data no significant dependence of bed heat flux on superheat is observed. From the standpoint of dependence of heat flux on particle superheat, therefore, the comparison between data and model is inconclusive.

Experiments at elevated initial particle temperatures would be necessary for an evaluation of the model based on the effect of temperature. Figure 5 compares the frontal propagation data with the model predictions. As in the case of the bed heat flux, the data lie between predictions based upon the "Lipinski" and "linear" parameters. Work is continuing in the direction of obtaining a set of parameters which will provide a "best fit" to all of the bed quench experiments.

## SUMMARY AND CONCLUSIONS

A model is presented which characterizes the quench behavior of superheated packed beds of particles. A model presented in earlier work was modified to account for heat transfer from the hot debris to the steam and for superheating of the steam as it flows through the unquenched channels within the bed above the quench front. This model predicts the particle bed cooling rate, the heat flux at the quench front, the frontal propagation characteristics and the solid temperature within the bed. Calculations indicate that the effect of steam superheat is a significant one for small particles and for large superheats.

## NOTATION

| | |
|---|---|
| $A$ | bed cross-sectional area |
| $c$ | specific heat at constant pressure |
| $d$ | particle diameter |
| $e$ | specific internal energy (per unit mass) |
| $f_d$ | fraction of bed stored energy removed during downward frontal period |
| $F_{gs}$ | vapor-solid drag force per unit volume bed |
| $F_{\ell s}$ | liquid-solid drag force per unit volume bed |
| $g$ | acceleration of gravity |
| $h_{fg}$ | latent heat of vaporization |
| $H$ | bed height |
| $j$ | volume flux; superficial velocity |
| $j_{go}$ | steam volume flux at the top of bed |
| $j_{\ell o}$ | liquid volume flux at the top of bed |
| $j_\ell$ | liquid volume flux at quench front |
| $m_p$ | mass of particles in dry region above quench front |
| $\dot{m}_p$ | rate of change of mass of particles in dry region above quench front |
| $\dot{m}_s$ | mass flow rate steam leaving bed |
| $p$ | pressure |
| $q''$ | heat flux |
| $Q'''$ | internal heat generation per unit volume bed |
| $t$ | time |
| $T_{IN}$ | temperature of steam entering dry region above quench front |
| $T_{OUT}$ | temperature of steam leaving bed |
| $T_p$ | particle temperature |
| $T_{SAT}$ | saturation temperature of water |
| $v$ | speed of quench front |
| $V$ | volume of bed above quench front |
| $z$ | axial position of bed |
| $z_d^*$ | position of downward quench front |
| $z_u^*$ | position of upward quench front |
| $\alpha$ | steam volume fraction in bed |
| $\epsilon$ | bed porosity |
| $\kappa$ | bed permeability |
| $\kappa_g$ | relative permeability to vapor phase |
| $\kappa_\ell$ | relative permeability to liquid phase |
| $\mu$ | viscosity |
| $\eta$ | bed passability |
| $\eta_g$ | relative passability to vapor phase |
| $\eta_\ell$ | relative passability to liquid phase |
| $\rho$ | density |
| $\rho_{go}$ | vapor density at top of bed |
| $\sigma$ | surface tension |

### Subscripts

| | |
|---|---|
| d | downward front |
| g | vapor |
| $\ell$ | liquid |
| p | particle |
| QF | quench front |
| s | steam |
| TB | top of bed |
| u | upward front |

## LITERATURE CITED

1. Pratt, W.T. and Bari, R.A., "Containment Response During Degraded Core Accidents Initiated by Transients and Small Break LOCA in the Zion/Indian Point Reactor Plants," Brookhaven National Laboratory, BNL-NUREG-51415 (July 1981).

2. Ginsberg, T., et al., "LWR Steam Spike Phenomenology: Debris Bed Quenching Experiments," Brookhaven National Laboratory, BNL-NUREG-51571, NUREG/CR-2857 (June 1982).

3. Cho, D.H., et al., "Debris Bed Quenching Experiments," Proceedings of International Meeting on Thermal Nuclear Reactor Safety, Chicago, IL, NUREG/CR-0027, Vol. 2, p. 987 (August 1982).

4. Ginsberg, T., et al., "Transient Core Debris Heat Removal Experiments and Analysis," Proceedings of International

Meeting on Thermal Nuclear Reactor Safety, Chicago, IL, NUREG/CR-0027, Vol. 3, p. 996 (August 1982).

5. Ginsberg, T. and Chen, J.C., "A Model for Superheated Debris Bed Quench for Severe Accident Containment Calculations," First Proceedings of Nuclear Thermal Hydraulics, American Nuclear Society Annual Winter Meeting, October 31-November 3, 1983, p. 227 (1983).

6. Gorham-Bergeron, E., "An Analytical Model for Predicting Dryout and Quench Behavior in a Volumetrically Heated Particle Bed," Proceedings of International Meeting on Light Water Reactor Severe Accident Evaluation, Cambridge, MA, TS-15.4 (August 1983).

7. Bergeron, K.D. and Trebilock, W., "The MEDICI Reactor Cavity Model," Proceedings of the International Meeting on Light Water Reactor Severe Accident Evaluation, Cambridge, MA, TS-5.6 (1983).

8. Lipinski, R.J., "A Coolability Model for Postaccident Nuclear Reactor Debris," Nuclear Technology, 65, p. 53 (April 1984).

9. Kunii, D. and Levenspiel, I., "Fluidization Engineering," John Wiley & Sons (1977).

## APPENDIX: FORMULATION OF BASIC EQUATIONS

Consider a superheated debris bed of uniform size particles and uniform porosity with decay heat generation $Q'''$. The bed assumed to be dry initially and at temperature $T_0$. Particles in the unquenched region of the bed beneath the quench front are assumed to heat adiabatically due to $Q'''$. Particles in the dry channels transfer energy to the steam. The liquid supply to the bed and the bed heat flux is determined by the hydrodynamics of the countercurrent two-phase flow at the top of the bed and is characterized by the Lipinski separated flow formulation.

### Mass Balances

A mass balance on the liquid phase yields

$$\rho_\ell \varepsilon \frac{dj_\ell}{dz} = - \frac{Q''' f_d}{h_{fg}}, \quad (A.1)$$

which when integrated from $z^*$ to $H$ yields

$$j_\ell^* = j_{\ell o} - \frac{Q'''(H-z_d^*)f_d}{\rho_\ell h_{fg} \varepsilon} \quad (A.2)$$

The overall coolant conservation equations are

$$j_{\ell o}\rho_\ell + j_{go}\rho_g = (1-\alpha)v_d(\rho_\ell-\rho_g) \quad \text{down-front}$$

$$= (\alpha)v_u(\rho_\ell-\rho_g) \quad \text{up-front} \quad (A.3)$$

where $j_{\ell o}$ and $j_{go}$ are the liquid and vapor superficial velocities at the top of the bed. The right-hand side of Equations (A.3) accounts for the increase in coolant mass within the bed with time due to fillup with coolant of the newly quenched region of particles.

### Frontal Balance Equation

It is assumed that the quench front is propagated at a speed limited by the rate of liquid flow to the quench front, and that the thickness of the two-phase front is negligible. The front is assumed located at $z=z^*(t)$, as shown in Figure 1. During the downward front propagation period

liquid available    liquid flow    liquid flow to
     for         =     to       - fill voids for
 vaporization       front       quenched region

An energy balance across the quench front then yields

$$(1-\varepsilon)f_d\rho_p c_p(T_p-T_{SAT}) \frac{dz_d^*}{dt}$$

$$= [\varepsilon j_\ell^* \rho_\ell - \varepsilon(1-\alpha)v_d\rho_\ell]h_{fg}. \quad (A.4)$$

Combining Equations (A.2) and (A.4), with $v_d = dz_d^*/dt$ gives

$$\frac{dz_d^*}{dt} = \frac{(1-\varepsilon)\rho_\ell h_{fg}}{(1-\varepsilon)f_d\rho_p c_p \Delta T + \varepsilon(1-\alpha)\rho_\ell h_{fg}}$$

$$\times [j_{\ell o} - \frac{Q'''(H-z_d^*)f_d}{h_{fg}\rho_\ell \varepsilon}], \quad (A.5)$$

A similar set of equations can be written for the time period of the upward-progressing front. During this time period, liquid supplied from the pool is used to remove decay heat from the debris which is already quenched, including those for which $z < z_u^*$. The resulting frontal propagation equation is

$$\frac{dz_u^*}{dt} = \frac{\epsilon \rho j_\ell^* h_{fg} - Q''' z_u^*}{(1-\epsilon)(1-f_d)\rho_p c_p \Delta T + \epsilon \alpha \rho_\ell h_{fg}}. \quad (A.6)$$

### Particle Energy Equation

The energy equations for the particles in the uncooled region beneath the quench front is

$$\rho_p c_p (1-\epsilon) \frac{dT_{po}}{dt} = Q''' \quad (A.7)$$

for the quenched regions of the bed $T_p = T_{SAT}$.

Consider the dry channels of the bed above the quench front shown schematically in Figure 1. A lumped-parameter energy balance approach is applied to the particles in this region. The stored energy of these particles increases due to internal heating and decreases due to heat transfer to the steam. During the downward frontal progression period particles are added to this region at temperature $T_{po}$ due to frontal propagation. This mechanism serves to increase the stored energy within the dry channel control volume.

A lumped-parameter energy balance on the dry channel particles during the downward frontal period is given by

$$\frac{d}{dt} m_p c_p T_p = Q''' V - \dot{m}_s c_s (T_{OUT} - T_{IN})$$

$$+ \frac{d}{dt} m_p c_p T_{po} . \quad (A.8)$$

Assuming that the steam enters the dry zone at $T_{SAT}$ and leaves the region in thermal equilibrium with the particles, Equation (A.8) can be written

$$m_p c_p \frac{dT_p}{dt} = Q''' V - \dot{m}_s c_s (T_p - T_{SAT})$$

$$+ \dot{m}_p c_p (T_{po} - T_p) \quad (A.9)$$

where

$$m_p = A \alpha (1-\epsilon) \rho_p z_d^* \quad (A.10)$$

$$V = \alpha A z_d^* \quad (A.11)$$

$$\dot{m}_p = A \alpha (1-\epsilon) \rho_p v_d \quad (A.12)$$

and $\dot{m}_s$ is the steam flow rate at the top of the bed. The steam flow rate is the sum of contributions from quenching and from removal of internal heat in the quenched portion of the bed, i.e.,

$$\dot{m}_s = \dot{m}_q + \dot{m}_Q \quad (A.13)$$

where

$$\dot{m}_q = v_d \rho_p c_p (1-\epsilon) f_d (T_{po} - T_{SAT}) A / h_{fg} \quad (A.14)$$

$$\dot{m}_Q = Q''' f_d A z_d^* / h_{fg} . \quad (A.15)$$

A similar particle energy equation can be written for the upward-propagation period. The result is

$$m_p c_p \frac{dT_p}{dt} = Q''' V - \dot{m}_s c_s (T_p - T_{SAT}) \quad (A.16)$$

where

$$m_p = A \alpha (1-\epsilon) \rho_p z_u^* \quad (A.17)$$

$$V = \alpha A z_u^* \quad (A.18)$$

$$\dot{m}_s = \dot{m}_q + \dot{m}_Q \quad (A.19)$$

$$\dot{m}_q = v_u \rho_p c_p (1-\epsilon)(1-f_d)(T_p - T_{SAT}) A / h_{fg} \quad (A.20)$$

$$\dot{m}_Q = \frac{Q''' A}{h_{fg}} [f_d z_u^* + (H - z_u^*)] . \quad (A.21)$$

In this case steam is generated by the quenched particles both below and above the upward-moving quench front.

### Lipinski Separated Flow Momentum Equations

The steady-state vapor momentum balance equation is given by (8)

$$-\frac{dp_g}{dz} = \rho_g g + F_{gs} \quad (A.22)$$

and the liquid momentum equation is

$$-\frac{dp_\ell}{dz} = \rho_\ell g - F_{\ell s} \quad (A.23)$$

where

$$F_{gs} = \frac{\rho_g j_g^2}{\eta \eta_g} + \frac{\mu_g j_g}{\kappa \kappa_g} \quad (A.24)$$

$$F_{\ell s} = \frac{\rho_\ell j_\ell^2}{\eta \eta_\ell} - \frac{\mu_\ell j_\ell}{\kappa \kappa_\ell} \quad (A.25)$$

$$\Delta P_g - \Delta P_\ell = \sigma \left(\frac{\varepsilon}{5\kappa}\right)^{1/2} . \quad (A.26)$$

The bed permeabilities are given by

$$\eta = \frac{d \, \varepsilon^3}{1.75(1-\varepsilon)} \quad (A.27)$$

$$\kappa = \frac{d^2 \, \varepsilon^3}{180(1-\varepsilon)^2} . \quad (A.28)$$

Lipinski's latest formulation (8) for the relative permeabilities are

$$\left. \begin{array}{ll} \kappa_g = \alpha^3 & \eta_g = \alpha^5 \\ \kappa_\ell = (1-\alpha)^3 & \eta_\ell = (1-\alpha)^5 \end{array} \right\} \quad (A.29)$$

These are referred to in this paper as the "Lipinski parameters." They were chosen by Lipinski to provide agreement with steady-state debris bed dryout experimental data.

An earlier formulation for the relative permeabilities (11) is equivalent to a "separated flow" model. The permeabilities are given by

$$\left. \begin{array}{ll} \kappa_g = \alpha & \eta_g = \alpha^2 \\ \kappa_\ell = 1-\alpha & \eta_\ell = (1-\alpha)^2 \end{array} \right\} \quad (A.30)$$

Both of the above formulations are considered in the paper.

### Heat Flux Relationships

The heat flux at the top of the bed is given by

$$q''_{TB} = \rho_{go} j_{go} h_{fg} \quad (A.31)$$

where $\rho_{go}$ and $j_{go}$ are evaluated at conditions at the top of the bed where the steam is superheated.

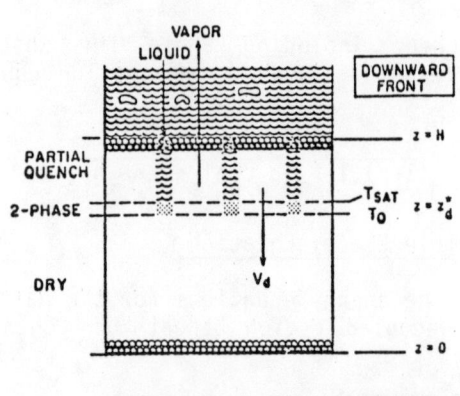
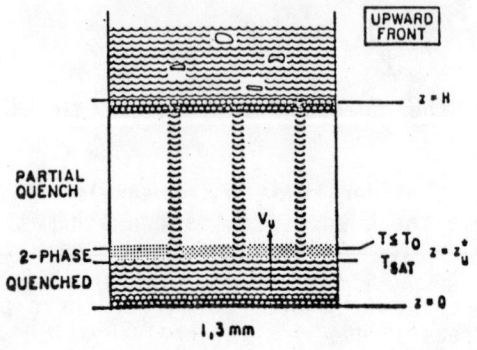

Figure 1. Schematic of superheated packed bed quench process.

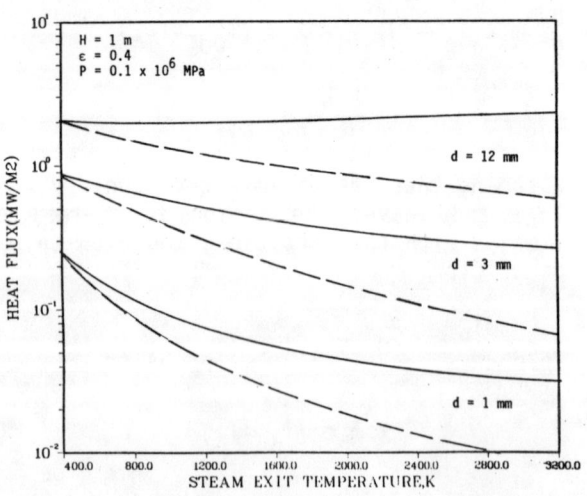

Figure 2. Variation of bed heat flux with steam temperature during quench process: ——— heat flux at top of bed; ——— heat flux at quench front.

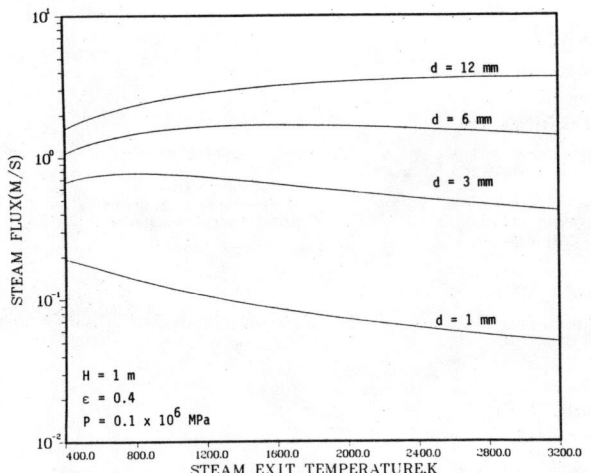

Figure 3. Effect of steam superheat on steam volume flux.

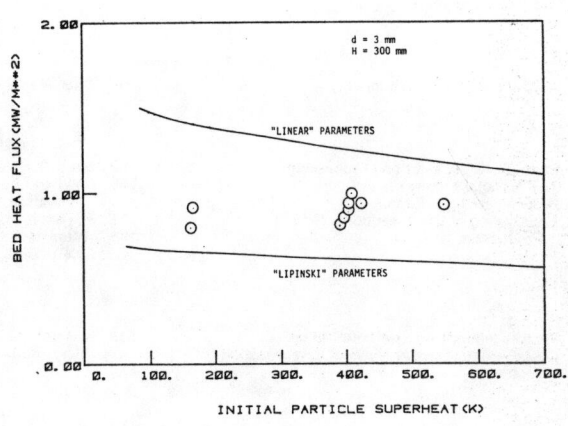

Figure 4. Comparison of bed heat flux data with model predictions including steam superheating.

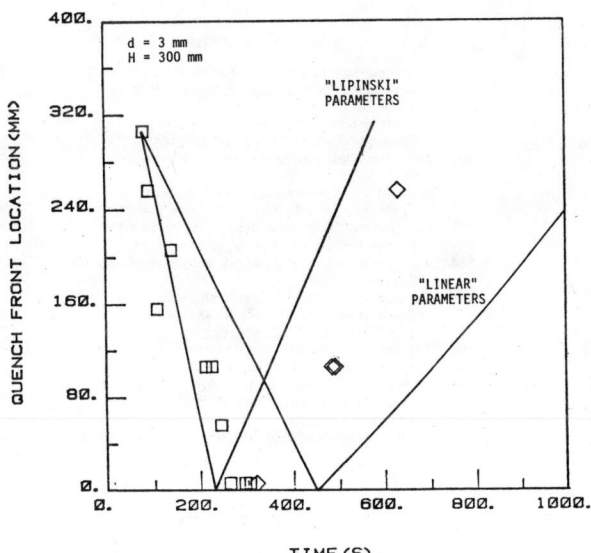

Figure 5. Comparison of bed quench propagation data with model incorporating steam superheat.

# SYMPOSIUM SERIES

## ADSORPTION

- 96 Developments in Physical Adsorption
- 117 Adsorption Technology
- 219 Recent Advances in Adsorption and Ion Exchange
- 230 Adsorption and Ion Exchange—'83
- 233 Adsorption and Ion Exchange— Progress and Future Prospects
- 242 Adsorption and Ion Exchange: Recent Developments

## AEROSPACE

- 33 Rocket and Missile Technology
- 52 Chemical Engineering Techniques in Aerospace

## BIOENGINEERING

- 69 Bioengineering and Food Processing
- 84 The Artificial Kidney
- 86 Bioengineering ... Food
- 93 Engineering of Unconventional Protein Production
- 99 Mass Transfer in Biological Systems
- 108 Food and Bioengineering—Fundamental and Industrial Aspects
- 114 Advances in Bioengineering
- 163 Water Removal Processes: Drying and Concentration of Foods and Other Materials
- 172 Food, pharmaceutical and bioengineering— 1976/77
- 181 Biochemical Engineering Renewable Sources of Energy and Chemical Feedstocks

## CRYSTALLIZATION

- 110 Factors Influencing Size Distribution
- 193 Design Control and Analysis of Crystallization Processes
- 215 Nucleation, Growth and Impurity Effects in Crystallization Process Engineering
- 240 Advances in Crystallization From Solutions

## DRAG REDUCTION

- 111 Drag Reduction
- 130 Drag Reduction in Polymer Solutions

## ENERGY

### Conversion and Transfer

- 5 Heat Transfer, Atlantic City
- 57 Heat Transfer, Boston
- 59 Heat Transfer, Cleveland
- 75 Energy Conversion Systems
- 79 Heat Transfer with Phase Change
- 87 Advances in Cryogenic Heat Transfer
- 113 Convective and Interfacial Heat Transfer
- 118 Heat Transfer—Tulsa
- 119 Commercial Power Generation
- 138 Heat Transfer—Research and Design
- 162 Energy and Resource Recovery from Industrial and Municipal Solid Wastes
- 174 Heat Transfer: Research and Application
- 189 Heat Transfer—San Diego 1979
- 202 Transport with Chemical Reactions
- 208 Heat Transfer—Milwaukee 1981
- 216 Processing of Energy and Metallic Minerals
- 225 Heat Transfer—Seattle 1983
- 236 Heat Transfer—Niagara Falls 1984
- 245 Heat Transfer—Denver 1985

### Nuclear Engineering

- 53 Part XIII
- 56 Part XIV
- 94 Part XX
- 104 Part XXI
- 106 Part XXII
- 119 Commercial Power Generation
- 168 Heat Transfer in Thermonuclear Power Systems
- 169 Developments in Uranium Enrichment
- 191 Nuclear Engineering Questions Power Reprocessing, Waste, Decontamination Fusion
- 221 Recent Developments in Uranium Enrichment

## ENVIRONMENT

- 78 Water Reuse
- 97 Water—1969
- 115 Important Chemical Reactions in Air Pollution Control
- 122 Chemical Engineering Applications of Solid Waste Treatment
- 124 Water—1971
- 126 Air Pollution and its Control
- 133 Forest Products and the Environment
- 137 Recent Advances in Air Pollution Control
- 139 Advances In Processing and Utilization of Forest Products
- 144 Water—1974: I. Industrial Wastewater Treatment
- 145 Water—1974: II. Municipal Wastewater Treatment
- 146 Forest Product Residuals
- 147 Air: I. Pollution Control and Clean Energy
- 148 Air: II. Control of $NO_{xx}$ and $SO_x$ Emissions
- 149 Trace Contaminants in the Environment
- 151 Water—1975
- 156 Air Pollution Control and Clean Energy
- 157 New Horizons for the Chemical Engineer in Pulp and Paper Technology
- 165 Dispersion and Control of Atmospheric Emissions, New-Energy-Source Pollution Potential
- 170 Intermaterials Competition in the Management of Shrinking Resources
- 171 What the Filterman Needs to Know About Filtration
- 175 Control and Dispersion of Air Pollutants: Emphasis on $NO_X$ and Particulate Emissions
- 177 Energy and Environmental Concerns in the Forest Products Industry
- 184 Advances in the Utilization and Processing of Forest Products
- 188 Control of Emissions from Stationary Combustion Sources Pollutant Detection and Behavior in the Atmosphere
- 195 The Role of Chemical Engineering in Utilizing the Nation's Forest Resources
- 196 Implications of the Clean Air Amendments of 1977 and of Energy Considerations for Air Pollution Control
- 198 Fundamentals and Applications of Solar Energy
- 200 New Process Alternatives in the Forest Products Industries
- 201 Emission Control from Stationary Power Sources: Technical, Economic and Environmental Assessments
- 207 The Use and Processing of Renewable Resources—Chemical Engineering Challenge of the Future
- 209 Water—1980
- 210 Fundamentals and Applications of Solar Energy II
- 211 Research Trends in Air Pollution Control: Scrubbing, Hot Gas Clean-up, Sampling and Analysis
- 213 Three Mile Island Cleanup
- 223 Advances in Production of Forest Products
- 232 Applications of Chemical Engineering in the Forest Products Industry
- 239 The Impact of Energy and Environmental Concerns on Chemical Engineering in the Forest Products Industry
- 243 Separation of Heavy Metals and Other Trace Contaminants

## FLUIDIZATION

- 101 Fundamental Processes in Fluidized Beds
- 105 Fluidization Fundamentals and Application
- 116 Fluidization: Fundamental Studies Solid-Fluid Reactions, and Applications
- 176 Fluidization Application to Coal Conversion Processes
- 205 Recent Advances in Fluidization and Fluid-Particle Systems
- 234 Fluidization and Fluid Particle Systems: Theories and Applications
- 241 Fluidization and Fluid Particle Systems: Recent Advances

## HISTORY OF CHEMICAL ENGINEERING

100   The History of Penicillin Production
235   Diamond Jubilee Historical/Review Volume

## ION EXCHANGE

179   Adsorption and Ion Exchange Separations
219   Recent Advances in Adsorption and Ion Exchange
230   Adsorption and Ion Exchange—'83
233   Adsorption and Ion Exchange—Progress and Future Prospects

## KINETICS

25    Reaction Kinetics and Unit Operations
73    Kinetics and Catalysis

## MATHEMATICS

## MINERALS

15    Mineral Engineering Techniques
85    Fossil Hydrocarbon and Mineral Processing
173   Fundamental Aspects of Hydrometallurgical Processes
180   Spinning Wire from Molten Metals
216   Processing of Energy and Metallic Minerals

## PETROCHEMICALS

49    Polymer Processing
127   Declining Domestic Reserves—Effect on Petroleum and Petrochemical Industry
135   The Petroleum/Petrochemical Industry and the Ecological Challenge
142   Optimum Use of World Petroleum
212   Interfacial Phenomena in Enhanced Oil Recovery

## PETROLEUM PROCESSING

103   $C_4$ Hydrocarbon Production and Distribution
127   Declining Domestic Reserves—Effect on Petroleum and Petrochemical Industry
135   The Petroleum/Petrochemical Industry and the Ecological Challenge
142   Optimum Use of World Petroleum
155   Oil Shale and Tar Sands
226   Underground Coal Gasification: The State of the Art

## PHASE EQUILIBRIA

2     Pittsburgh and Houston
6     Collected Research Papers
88    Phase Equilibria and Gas Mixtures Properties

## PROCESS DYNAMICS

36    Process Dynamics and Control
46    Process Systems Engineering
55    Process Control and Applied Mathematics
214   Selected Topics on Computer-Aided Process Design and Analysis

## SEPARATION

120   Recent Advances in Separation Techniques
192   Recent Advances in Separation Techniques—II

## SONICS

109   Sonochemical Engineering

## THERMODYNAMICS

## TRANSPORT PROPERTIES

56    Selected Topics in Transport Phenomena

## MISCELLANEOUS

48    Chemical Engineering Reviews
70    Small-Scale Equipment for Chemical Engineering Laboratories
112   Engineering, Chemistry, and Use of Plasma Reactors
125   Vacuum Technology at Low Temperatures
143   Standardization of Catalyst Test Methods
182   Biorheology
183   The Modern Undergradate Laboratory Innovative Techniques
185   Electro Organic Synthesis Technology
186   Plasma Chemical Processing
187   Chronic Replacement of Kidney Function
194   Hazardous Chemical—Spills and Waterborne Transportation
203   A Review of AIChE's Design Institute for Physical Property Data (DIPPR) and Worldwide Affiliated Activities
204   Tutorial Lectures in Electrochemical Engineering and Technology
206   Controlled Release Systems
217   New Composite Materials and Technology
220   Uncertainty Analysis for Engineers
228   Problem Solving
229   Tutorial Lectures in Electrochemical Engineering and Technology—II
231   Data Base Implementation and Application
237   Awareness of Information Sources
238   New Developments in Liquid-Liquid Extractors: Selected Papers From ISEC '83
244   Experimental Results from the Design Institute for Physical Property Data. I: Phase Equilibria

## MONOGRAPH SERIES

3     The Manufacture of Nitric Acid by the Oxidation of Ammonia—The DuPont Pressure Process by Thomas H. Chilton
4     Experiences and Experiments with Process Dynamics by Joel O. Hougen
5     Present Past, and Future Property Estimation Techniques by Robert C. Reid
6     Catalysts and Reactors by James Wei
7     The 'Calculated' Loss-of-Coolant Accident by L.J. Ybarrondo, C.W. Solbrig, H.S. Isbin
8     Understanding and Conceiving Chemical Process by C. Judson King
9     Ecosystem Technology: Theory and Practice by Aaron J. Teller
10    Fundamentals of Fire and Explosion by Daniel R. Stull
11    Lumps, Models and Kinetics in Practice by Vern W. Weekman, Jr.
12    Lectures in Atmospheric Chemistry by John H. Seinfeld
13    Advanced Process Engineering by James R. Fair
14    Synfuels from Coal by Bernard S. Lee